*The Protein Folding Problem
and Tertiary Structure Prediction*

The Protein Folding Problem
and Tertiary Structure Prediction

Kenneth M. Merz, Jr.
Scott M. Le Grand
Editors

Birkhäuser
Boston • Basel • Berlin

Kenneth M. Merz, Jr.
Department of Chemistry
The Pennsylvania State University
152 Davey Laboratory
University Park, PA 16802-6300

Scott M. Le Grand
Department of Chemistry
The Pennsylvania State University
152 Davey Laboratory
University Park, PA 16802-6300

Library of Congress Cataloging-in-Publication Data

The Protein folding problem and tertiary structure prediction /
    Kenneth M. Merz, Jr., Scott M. Le Grand, editors.
        p.   cm.
    Includes bibliographical references and index.
    ISBN 0-8176-3693-5 (acid-free)
    1. Protein folding.   I. Merz, Kenneth M. 1959-   . II. Le Grand,
Scott M.
QP551.P6958216 1994   94-41522
547.7'5--dc20   CIP

Birkhäuser ®

QP
551
.P6958216
1994

Printed on acid-free paper
© Birkhäuser Boston 1994

ISBN 0-8176-3693-5
ISBN 3-7643-3693-5

Typeset by TEXniques, Inc.
Printed and bound by Braun-Brumfield, Ann Arbor, MI.
Printed in the U.S.A.

9 8 7 6 5 4 3 2 1

Contents

Preface

A solution to the protein folding problem has eluded researchers for more than 30 years. The stakes are high. Such a solution will make 40,000 more tertiary structures available for immediate study by translating the DNA sequence information in the sequence databases into three-dimensional protein structures. This translation will be indispensable for the analysis of results from the Human Genome Project, *de novo* protein design, and many other areas of biotechnological research. Finally, an in-depth study of the rules of protein folding should provide vital clues to the protein folding process. The search for these rules is therefore an important objective for theoretical molecular biology. Both experimental and theoretical approaches have been used in the search for a solution, with many promising results but no general solution. In recent years, there has been an exponential increase in the power of computers. This has triggered an incredible outburst of theoretical approaches to solving the protein folding problem ranging from molecular dynamics-based studies of proteins in solution to the actual prediction of protein structures from first principles. This volume attempts to present a concise overview of these advances.

Adrian Roitberg and Ron Elber describe the locally enhanced sampling/simulated annealing conformational search algorithm (Chapter 1), which is potentially useful for the rapid conformational search of larger molecular systems. Stephen Wilson and Weili Cui described their work with simulated annealing (Chapter 2) to perform conformational searches on a number of model systems and their improvements to the basic algorithm. Trevor Hart and Randy Read describe their multiple-start Monte Carlo docking algorithm (Chapter 3) for the docking of flexible ligands to proteins. Scott Le Grand and Ken Merz provide a review of current research focused on applying the genetic algorithm (Chapter 4) to protein tertiary structure prediction. Robert Bruccoleri (Chapter 5) describes the application of his program CONGEN to the study of antigen-combining sites of antibodies as well as to peptide conformation and side-chain placement in proteins. William Taylor and András Aszódi describe their work adapting distance geometry methods (Chapter 6) into a general approach for structure prediction and modeling. Amedeo Caflisch and Martin Karplus describe recent molecular dynamics-based studies (Chapter 7) of protein folding as well as their recent work on the unfolding of barnase using high-temperature molecular dynamics. Gordon Crippen and Vladimir Maiorov describe a contact potential function (Chapter 8) that can identify the native fold of a protein within a large pool of alternative conformers. Cathy Wu describes her work with feed-forward neural networks (Chapter 9) for the purpose of identifying protein structural motifs. Johan Desmet, Marc De Maeyer, and Ignace Lasters describe the "Dead-End Elimination" theorem (Chapter 10), which can significantly reduce the computational complexity of the side-chain packing problem by eliminating a large pool of conformations that can be proven not to be in the lowest energy conformation. Ron Unger describes the identification of short structural motifs in proteins (Chapter 11) and how they may be used in protein tertiary structure prediction algorithms. Manfred Sippl, Sabine Weitckus, and Hannes Flöckner describe the use of knowledge-based potential functions (Chapter 12) to identify the native structure of a protein when presented with a large number of alternative conformations. Sandor Vajda and Charles Delisi describe an adaptation of dynamic programming using branch and bound techniques (Chapter 13) to perform reliable conformational searches of molecules. Thomas Ngo, Joe Marks, and Martin Karplus provide a rigorous discussion of the computational complexity of the protein structure prediction problem (Chapter 14) and its relation to the Levinthal Paradox. Teresa Head-Gordon describes the Antlion conformational search algorithm (Chapter 15) and its application to the conformational search of several proteins. Finally, James Hurley provides a discussion of the role of interior side-chain packing (Chapter 16) in protein folding and stability.

The editors would like to thank Andrea Merz for preliminary editing and formatting, the staff at Birkhäuser-Boston, and all of the contributors for helping us produce an outstanding volume on recent advances toward understanding how the primary sequence of a protein determines its three-dimensional fold.

Scott M. Le Grand
Kenneth M. Merz, Jr.
Spring 1994

Contributors

András Aszódi, Laboratory of Mathematical Biology, National Institute for Medical Research, The Ridgeway, Mill Hill, London NW7 1AA, United Kingdom

Robert Bruccoleri, Bristol-Myers Squibb Pharmaceutical Research Institute, P.O. Box 4000, Princeton, NJ 08543-4000, USA

Amedeo Caflisch, Department of Chemistry, Harvard University, 12 Oxford Street, Cambridge, MA 02138, USA

Gordon M. Crippen, College of Pharmacy, University of Michigan, Ann Arbor, MI 48109-1065, USA

Weili Cui, Department of Medicinal Chemistry, Synaptic Pharmaceuticals Corporation, W. College Road, Paramus, NJ 07625-1043, USA

Charles Delisi, Department of Biomedical Engineering, Boston University, 44 Cummington Street, Boston, MA 02215, USA

Marc De Maeyer, CORVAS International NV, Jozef Plateaustraat 22, B-9000 Gent, Belgium

Johan Desmet, K.U. Leuven Campus Kortrijk, Interdisciplinary Research Center, B-8500 Kortrijk, Belgium

Ron Elber, Department of Chemistry M/C 111, University of Illinois at Chicago, P.O. Box 4348, Chicago, IL 60680 and Department of Physical Chemistry, Fritz Haber Research Center and Institute of Life Sciences, The Hebrew University, Jerusalem 91904, Israel

Hannes Flöckner, Center of Applied Molecular Engineering, University of Salzburg, Salzburg, Austria

Trevor N. Hart, Department of Medical Microbiology and Infectious Diseases, University of Alberta, Edmonton, Alberta, Canada T6G 2H7

Teresa Head-Gordon, Lawrence Berkeley Laboratories, 1 Cyclotron Road 1-314, Berkeley, CA 94720, USA

James H. Hurley, Laboratory of Molecular Biology, National Institute of Diabetes, Digestive and Kidney Diseases, National Institutes of Health, 9000 Rockville Park, Building 5 Room 324, Bethesda, MD 20892 USA

Martin Karplus, Department of Chemistry, Harvard University, 12 Oxford Street, Cambridge, MA 02138, USA

Ignace Lasters, CORVAS International NV, Jozef Plateaustraat 22, B-9000 Gent, Belgium

Scott M. Le Grand, Department of Chemistry, The Pennsylvania State University, 152 Davey Laboratory, University Park, PA 16802-6300, USA

Vladimir N. Maiorov, College of Pharmacy, University of Michigan, Ann Arbor, MI 48109-1065, USA

Joe Marks, Cambridge Research Laboratory, Digital Equipment Corporation, 1 Kendall Square, Building 700, Cambridge, MA 02139, USA

Kenneth M. Merz, Jr., Department of Chemistry, The Pennsylvania State University, 152 Davey Laboratory, University Park, PA 16802-6300, USA

J. Thomas Ngo, Interval Research Company, 1801-C Page Mill Road, Palo Alto, CA 94304, USA

Randy J. Read, Department of Medical Microbiology and Infectious Diseases, University of Alberta, Edmonton, Alberta, Canada T6G 2H7

Adrian Roitberg, Department of Chemistry, Northwestern University, 2145 Sheridan Road, Evanston, IL 60626-3113, USA

Manfred J. Sippl, Center of Applied Molecular Engineering, University of Salzburg, Salzburg, Austria

William R. Taylor, Laboratory of Mathematical Biology, National Institute for Medical Research, The Ridgeway, Mill Hill, London NW7 1AA, United Kingdom

Ron Unger, Center for Advanced Research in Biotechnology, Maryland Biotechnology Institute, University of Maryland, 9600 Gudelsky Drive, Rockville, MD 20850 and Institute for Advanced Computer Studies, University of Maryland, College Park, MD 20742, USA

Sandor Vajda, Department of Biomathematical Sciences, Mount Sinai School of Medicine, 1 Gustave L. Levy Place, New York, NY 10029 USA

Sabine Weitckus, Center of Applied Molecular Engineering, University of Salzburg, Salzburg, Austria

Stephen R. Wilson, Department of Chemistry, New York University, Washington Square, New York, NY 10029, USA

Cathy H. Wu, Department of Epidemiology/Biomathematics, The University of Texas Health Center at Tyler, Tyler, TX 75710, USA

1

Modeling Side Chains in Peptides and Proteins with the Locally Enhanced Sampling/Simulated Annealing Method

Adrian Roitberg and Ron Elber

1. Introduction

In this chapter, we present the formal basis of a new optimization method that we call LES (locally enhanced sampling) together with applications to side-chain modeling in peptides and proteins. We examine the relationship between an energy function that is derived from data on small model systems and the correct structure of the protein. The question is: Given a functional form for the potential energy of a macromolecule, are the side-chain coordinates at the global energy minimum similar to the x-ray coordinates? This comparison enables us to detect inaccuracies in the force fields that we used (Brooks et al., 1983; Jorgensen and Tirado-Rives, 1988) and possibly to improve them.

In the present approach we therefore insist on using an atomic detail off-lattice model. This is in conjunction with a global search method that employs a complete force field (Brooks et al., 1983; Jorgensen and Tirado-Rives, 1988). We first demonstrate that our global search method is capable of locating reasonable energy minima for a number of examples and establish that the potential energy and the optimization algorithm are sound for proteins. We then continue to derive some thermodynamic properties of the folding process, assuming a specific thermodynamic path suggested originally by Shakhnovich and Finkelstein (1982). We study a path similar but not identical to it, as explained later. Two thermodynamic steps are assumed: In the first step (which we do not model) the backbone folds to

The Protein Folding Problem and Tertiary Structure Prediction
K. Merz, Jr. and S. Le Grand, Editors
© Birkhäuser Boston 1994

the correct structure, and in the second step the side chains freeze to their final configuration. Detailed modeling of the second step is carried out.

We emphasize that the present optimization protocol and the calculation of approximate thermodynamic properties employ full atomic-level microscopic force field. In this way the present approach is different from other protocols modeling side chains (see below). Obviously, a method that uses a full potential, provides approximate thermodynamic properties, and can also be employed in other structural problems (Simmerling and Elber, 1994) is not necessarily the most efficient in doing parts of the job. For the present application, the LES approach is computationally less efficient in determining side-chain structure compared to approaches that were specifically designed to do only that.

Side-chain modeling can be useful in a number of contexts, some of which are

(1) refinement of structures from low-resolution x-ray crystallographic data;

(2) refinement of structures from incomplete NMR data; and

(3) structural predictions using homology modeling.

In the past, many independent researchers have applied a variety of methodologies to the problem of side-chain modeling (Zimmerman et al., 1977; Richards, 1977; Janin et al., 1978; Warme and Morgan, 1978; Greer, 1981; Blow, 1983; James and Sielecki, 1983; Narayana and Argos, 1984; Lesk and Chothia, 1986; Blundell et al., 1987; Bruccoleri and Karplus, 1987; Ponder and Richards, 1987; Summers et al., 1987; Reid and Thornton, 1989; Summers and Karplus, 1989; Singh and Thornton, 1990; Lee and Levitt, 1991; Desmet et al., 1992; Holm and Sander, 1992). These methods can be grouped into two main categories: statistical and energy-based methods.

For methods using the first strategy, the goal is to greatly reduce the number of conformers available for each side chain to only a few basic rotamers instead of the three possible conformers per χ_1 angle. This method was first described by Janin et al. (1978), and is based on the sharp peaks in the statistical distribution of χ_1 extracted from the database of known protein structures.

Since then, many researchers have been developing efficient techniques using the statistical approach. Among the recent ones, one may mention the work done by Tuffery et al. (1991). They made use of a rotamer library similar in spirit and accuracy to that compiled by Ponder and Richards (1987), but instead of trying to optimize the structure by exploring the full combinatorial space, they resorted to the use of a genetic algorithm technique, which they called a selection-mutation-focusing (SMF) algorithm.

Also recently, an important work by Holm and Sander (1992) has been published. They used a Monte Carlo procedure coupled with a simulated annealing-type method to optimize side-chain positions in proteins. The conformational space explored was restricted to those side-chain rotamers found in the database of Tuffery et al. (1991), while the potential-energy form used was a 6–9 potential with high-energy cutoffs.

The database approach is based on the assumption that only a handful of conformations are available for the side chains. This assumption restricts the search space, not allowing for conformations out of the ordinary that might be present in a particular protein. Also, the idea of a search in discrete conformational space tends to overweight the center of the side-chain statistical distribution while disregarding the usual deviations. Local energy minimization after the side-chain positioning might overcome this problem. In any event, the approach described above relies on the accuracy and availability of statistical data and cannot provide first-principle understanding of the physical rules governing the packing.

An alternative approach is based (at least partially) on energy functions. However, the protocols below cannot address thermodynamic questions and are also restricted to side-chain modeling (Gelin and Karplus, 1975; Zimmermann et al., 1977; Gelin and Karplus, 1979; Bruccoleri and Karplus, 1987; Summers and Karplus, 1989; Lee and Subbiah, 1991). Excluding the study of Lee and Subbiah, which uses Monte Carlo and simplified energy functions, the other investigations solve the side-chain positions one by one. This approach solves the problem of the large number of permutations by ignoring them. This advantage is also its main weakness: since the side-chain conformations in the protein core are usually coupled, combinations of side-chain packing should be studied. For instance, in a side-chain prediction for the protein flavodoxin (Reid and Thornton, 1989), an initial misplacement of a phenylalanine residue led to a variety of errors, causing the internal core to be incorrectly packed. A similar effect will be demonstrated in this work.

Quoting Lee and Subbiah (1991): "In the general case, the best conformation for all side-chains is not that deducted from the best conformation of individual side-chains, but the other way around."

There are additional problems inherent to the approaches mentioned above. First, the simulations should be repeated a number of times, in order to establish the convergence of the results. The predictions from a single run can be different from those obtained with another set of initial conditions. Second, the methods are unable to deal, in a single run, with the problem of structural heterogeneity. Some of these problems are addressed here.

Several side chains are observed experimentally in more than one conformation. Smith et al. (1986), in a study on structural heterogeneity in protein crystals, found that extensive conformational heterogeneity exists in highly refined crystallographic models for the proteins crambin, erabutoxin, myohemerythrin, and lamprey hemoglobin. From 6 to 13% of the amino-acid side chains of these proteins are seen in multiple, discrete conformations. These authors suggest that, since the electron density is missing or very weak for only a few side chains in these proteins crystals, there is a strong preference for discrete over continuous conformational perturbations.

A method that is consistent, fairly independent of the initial conditions chosen for the modeling, and relying on optimization of energy functions and not on statistical data, is complementary to existing approaches and is desirable. Such a method should be based on full atomic level force field.

The basis and some tests for such a method were previously introduced (Roitberg and Elber, 1991). The method is an implementation of the LES approximation (Elber and Karplus, 1990) that was used in a number of different contexts: to search for diffusion pathways of a ligand through proteins (Elber and Karplus, 1990; Czerminski and Elber, 1991), studies of ligand recombination dynamics (Gibson et al., 1992), and free-energy calculations (Verkhivker et al., 1992). The corresponding energy of the system was optimized by a simulated annealing (SA) method in the usual molecular dynamics form.

The method is based on a mean-field theory, where the part of the protein that requires modeling (the side chains in the present case) is oversampled. In the present implementation, we attach to a given C_α a number of copies of the same side chain. If modeling of several side chains is needed, each of the side chains is "multiplied." The system is set in such a way that the copies of the same side chain do not interact with each other, while the interaction with the other side-chain bundles is defined in an average way, i.e., the force acting on a single copy is the sum of all forces of the copies of the other residues divided by the number of the other copies. This provides a new potential energy surface for the system. As we demonstrated (Roitberg and Elber, 1991), the global energy minimum for the new mean-field system and the original one is the same. This means that the global optimization of the LES energy will find (if properly executed) a global energy minimum that is identical to that of the real system. Moreover, the barriers separating the minima in the mean-field calculation are lower than in the original system. Finally, the use of this protocol increases the sampling significantly, since more combinations of the different copies are taken into account. This is done at a computational cost much lower than the one needed to sample the system similarly using the single copy approach.

In the present chapter, we shall introduce the basis for the LES method, along with the exact derivation of the equations of motion for the mean-field system. We further show that the barriers separating the local minima are lower in the mean-field system, thus helping the optimization. Results will be presented for conformational searches in model tetrapeptides.

Also, an application of the LES/SA method to the modeling of side chains in a significantly larger system, the core of carboxymyoglobin, is presented (Roitberg, 1992). Forty-three side chains were selected according to their surface accessible areas determined by Lee and Richards' method (1971). A core residue is defined as an amino acid with exposed surface area less than $4\,\text{\AA}^2$. The following points will be addressed in this part of the work:

(a) A comparison of the computed structure against known, predetermined x-ray coordinates of carboxymyoglobin.

(b) Different force fields are compared (CHARMM and MOIL): Even if the global energy minimum of the system is found, it is possible that the potential energy surface used is not sufficiently accurate to make precise structural predictions. To test the sensitivity of the results to the choice of the potential energy surface, two different force fields were used for predicting the side-chain positions. The differences and the similarities are discussed.

(c) Influence of the backbone flexibility: The influence of increased mobility in the backbone of the protein on the optimized structure is tested. Another simulation is presented in which the backbone associated with the residues being modeled is allowed to move. From this simulation, we found a problem with the force fields associated with the aromatic rings. The exclusion of explicit hydrogen atoms in aromatic rings favors a relative conformation between pairs of rings that is different from the one found in crystallographic data.

(d) Thermodynamic considerations: Sharp transitions to a single energy minimum were observed for most of the side chains studied as a function of temperature, suggesting a freezing-like transition during the annealing. A comparison with the theory of Shakhnovich and Finkelstein (1982) for protein folding was attempted, but no evidence of a pseudo phase transition was found. It should be noted that the thermodynamic variable we considered (temperature) is different from that of Shakhnovich and Finkelstein (volume).

This chapter is organized as follows. In Section 2 we present formal derivation of the equations of motion for the mean-field system, as well as some properties of the approximation. Section 3 deals with the computational implementation in a regular molecular dynamics program. Section 4 is devoted to some results obtained on side chains in a tetrapeptide. In Section 5 the results of three simulations for the modeling of 43 side chains in the core of carboxymyoglobin are presented. These simulations differ in the force field used, as well as in the amount of mobility permitted to the protein backbone. Section 5 also deals with the analysis of a freezing-like transition in the system during the annealing. Conclusions and final remarks are given in Sections 6 and 7.

2. The LES Approximation

The optimization method described in this chapter is based on a mean-field approximation for portions of the macromolecule that require modeling. The first application of a mean-field classical approximation in protein simulations was pursued by Elber and Karplus (1990). In that study, approximate mean-field treatment of protein plus ligand dynamics enabled detailed study of the diffusion pathways of carbon monoxide in myoglobin. In that approximation multiple ligands are allowed to share a single protein matrix. In a later work, Czerminski and Elber (1991) investigated the diffusion pathways in another protein—leghemoglobin. They demonstrated that the diffusion rate in leghemoglobin is faster than in myoglobin in accord with experiment. The approximation is called locally enhanced sampling (LES). Formal properties of the approximation were investigated by Straub and Karplus (1991) and by Ulitsky and Elber (1993). LES is based on the "trajectory bundles" method of Gerber et al. (1982), which was modified to include enhanced sampling of only a part of the system, an implementation that is appropriate for large molecules. Recently, an application of the LES protocol to the calculation of accurate free-energy differences has been reported (Verkhivker et al., 1992). The applications of LES to search for diffusion pathways employ approximate dynamics (Elber and Karplus, 1990; Czerminski and Elber, 1991). Here a variant of LES is used for efficient calculation of the minima of the *exact* potential energy surface using an effective (and larger) system.

The effective system is used only as a computational tool. For the optimization problem it is not essential to understand how and why the artificial system was constructed. It is of interest, however, to see the connection to the commonly used time-dependent Hartree (TDH) approximation (Gerber et al., 1982) and to the LES equations of motion (Elber and Karplus, 1990; Ulitsky and Elber, 1993). This connection puts the present investigation in a somewhat broader perspective. It also provides a simple physical realization to the annealing of the artificial system. We therefore pursue the analogy below.

We define the coordinate vector of all the atoms in the molecule as $\mathbf{X}$ and the corresponding velocity vector as $\mathbf{V}$. The probability of finding the molecular coordinates between $\mathbf{X}$ and $\mathbf{X} + d\mathbf{X}$ and the velocities between $\mathbf{V}$ and $\mathbf{V} + d\mathbf{V}$ at time t is denoted by $\rho(\mathbf{X}, \mathbf{V}, t)\, d\mathbf{X}\, d\mathbf{V}$, where ρ is the (time-dependent) probability density of coordinates and velocities; ρ is normalized as usual: $\int \rho\, d\mathbf{X}\, d\mathbf{V} = 1$. Like any other function, ρ can be expanded into a complete set of delta functions. We use

$$\rho(\mathbf{X}, \mathbf{V}, t) = \sum_{k=1}^{K} w_k\, \delta\big(\mathbf{X} - \mathbf{X}_{0k}(t)\big)\, \delta\big(\mathbf{V} - \mathbf{V}_{0k}(t)\big), \qquad (1)$$

where w_k's are constant coefficients that are determined by the initial distribution (at $t = 0$). K is the number of delta functions used in the expansion. For practical purposes K is (of course) finite. The vectors of parameters $\mathbf{X}_{0k}(t)$ and $\mathbf{V}_{0k}(t)$ depend on time. If the above expression for ρ is substituted in the Liouville equation, it yields, after some manipulation, the usual Newtonian equations of motion for K trajectories (Elber and Karplus, 1990). It is also possible to use the classical variation principle for the K trajectories to extract the equations of motion (Ulitsky and Elber, 1993). Here we use the last approach for reasons that will become clear later. We consider the classical action for a collection of trajectories

$$S = \text{stationary point} = \int \rho(\mathbf{X}, \mathbf{V}, t)\, L\, d\mathbf{X}\, d\mathbf{V}\, dt \qquad (2)$$

$$L = \tfrac{1}{2} \sum_{i=1}^{N} m_i v_i^2 - U(\mathbf{X}).$$

L is the Lagrangian expressed as a function of the spatial variables; m_i and v_i are the mass and the velocity of the ith atom. Integration over $\mathbf{X}$ and $\mathbf{V}$ yields

$$S = \int \left(\tfrac{1}{2} \sum_{k=1}^{K} \sum_{i=1}^{N} w_k m_i v_{0ij}^2 - \sum_k w_k U(\mathbf{X}_{0k}) \right) dt; \qquad \delta S = 0. \qquad (3)$$

The requirement of S being stationary gives the *exact* equations of motion, for the K independent trajectories. Equation (3) is exact and equivalent to Newton's equations of motion; however there is little (more precisely, no) computational advantage to equation (3) as compared to the usual molecular dynamics.

Nevertheless, equation (3) is a convenient starting point to introduce the main idea of the LES protocol: the construction of the artificial system. In the mean-field approximation we employ the following *ansatz*:

$$\rho(\mathbf{X}, \mathbf{V}, t) = \prod_{j=1}^{J} \rho_j(\mathbf{X}_j, \mathbf{V}_j, t), \tag{4}$$

where $\mathbf{X}_j$ and $\mathbf{V}_j$ are the coordinates and the velocities of the jth subspace of the system. Thus, we approximate the total probability density by a product of probability densities of different subspaces. An example of a possible partition to subspaces is of a ligand and a protein (Elber and Karplus, 1990; Czerminski and Elber, 1991; Gibson et al., 1992). An example more relevant to the present study is the partition of the protein into backbone, and into side chains. Similarly to equation (1), each of the ρ_js is expanded in delta functions

$$\rho_j(\mathbf{X}_j, \mathbf{V}_j, t) = \sum_{k_j=1}^{K_j} w_{k_j} \delta\big(\mathbf{X}_j - \mathbf{X}_{0k_j}(t)\big) \delta\big(\mathbf{V}_j - \mathbf{V}_{0k_j}(t)\big), \tag{5}$$

where the index k_j runs over all the copies of the jth subspace. For example, if the jth subspace corresponds to a side chain, there are multiple side-chain copies connected to the same C_α—the backbone atom (Figure 1-1). Substituting equations (4) and (5) into (2), and integrating over the spatial variables, gives

$$S = \int dt \, (T_{\text{eff}} - U_{\text{eff}}),$$

$$T_{\text{eff}} = \frac{1}{2} \sum_{\substack{k_j=1 \\ j=1,\dots,J}}^{K_j} \sum_{i=1}^{N_i} w_{k_j} m_{ij} v_{i0k_j}^2 \tag{6}$$

$$U_{\text{eff}} = \sum_{\substack{k_j=1 \\ j=1,\dots,J}}^{K_j} \left(\prod_{l=1}^{J} w_{k_l} \right) U(\mathbf{X}_{0k_1}, \mathbf{X}_{0k_2}, \dots, \mathbf{X}_{0k_J}).$$

The "effective" kinetic energy—T_{eff}—can be viewed as the kinetic energy of $\sum_j N_j K_j$ particles (instead of $\sum_j N_j$ in the original system), each with a modified mass $w_{k_j} m_{ij}$. The "effective potential" includes a sum of "real" potential energies calculated at different points and multiplied by constant normalization factors. The different points at which the potential energies are calculated are obtained from the delta-function expansion of

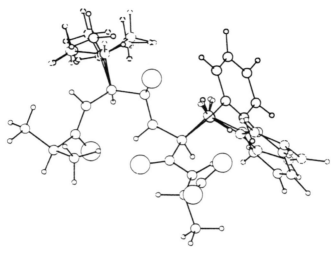

Figure 1-1. A snapshot in time of the mean-field structure of the tetrapeptide AVFA. Sticks and balls model with the side chains of the valine and the phenylalanine "multiplied" is shown. For clarity only 3 of the 10 side-chain copies that were used in the simulation are shown.

the probability density. They correspond to all possible combinations of the coordinates with the different k_js. There are $K_j \times K_{j'}$ alternative configurations of side-chain j and side-chain j' that can be examined using a *single configuration* of the effective system. This should be contrasted with a *single combination* that is generated at each time step if the exact Hamiltonian is used. Hence it is possible to optimize (by any optimization algorithm) the effective system and to obtain significantly more sampling statistics than one obtains by optimizing the exact potential.

By pursuing the Euler–Lagrange procedure to find the stationary point for the functional S, we obtain the Newton-like equations of motion for the vector of parameters—$\mathbf{x}_{0k_j}(t)$:

$$
(w_{k_{j'}} m_{j'}) \frac{d^2 \mathbf{x}_{0k_{j'}}}{dt^2} = -\frac{\partial U_{\text{eff}}}{\partial \mathbf{x}_{0k_{j'}}}
\tag{7}
$$

$$
= -\sum_{\substack{k_j \\ j=1,\dots,J}}^{K_j} \left(\prod_{l=1}^{J} w_{k_l} \right) \frac{\partial}{\partial \mathbf{x}_{0k_{j'}}} U(\mathbf{X}_{0k_1}, \dots, \mathbf{X}_{0k_j}, \dots, \mathbf{X}_{0k_J})
$$

or, in a more compact form,

$$
m_{j'} \frac{d^2 \mathbf{x}_{0k_{j'}}}{dt^2} = -\left\langle \frac{\partial U}{\partial \mathbf{x}_{0k_{j'}}} \right\rangle_{j \neq j'},
\tag{8}
$$

where the average $\langle \ldots \rangle$ denotes an average over all subspaces j, excluding j'. Equation (8) resembles an equation of motion of one particle moving in the average force of the multiple copies of the other subspaces. Equation (8) was introduced by Gerber et al. (1982) using the equations of the quantum mechanical analog as a starting point. In our specific implementation it corresponds to the motion of one of the copies of the j'th side chain interacting with all the other side-chain copies (Figure 1-1). Note also that in equation (8) the "real" mass of the particle was recovered by canceling from both sides the coefficient $w_{k_{j'}}$. In the calculations, the scaled mass, equation (7), is used to ensure that the forces obtained are symmetric and therefore derivable from a potential. This is done for computational convenience and cannot (of course) affect the results.

The equations of motion obtained from the *ansatz* of equation (4) are clearly approximate. The particles are moving on a smeared potential surface rather than on the exact one. Exact description of the system dynamics would be obtained only in a special case—when all the copies occupy the same point in phase space. This is, however, uninteresting, since we would lose all the advantages of LES.

The effective potential, equation (6), is worth further examination. We first note that the global energy minimum of the effective potential is the same as that of the exact potential. This can be shown in a number of ways. Consider first an exact representation of the system. Only one copy of the backbone and one copy of each of the side chains is employed. In other words, all the K_j's are set to 1. The system is placed in the global energy minimum (we assume that there is only one). We now add one more copy of one of the side chains and we check the value of the effective energy. Possible differences between the energy of the system with one copy and the system with two copies may come from different internal side-chain interactions and different interactions of the side-chain copies with the rest of the protein. The two copies do not "see" each other. If the total energy decreases, a better global energy minimum for the *real* system can be constructed by removing the old copy. This will preserve only the lower, more favorable interactions of the new copy. Such "improvement" of the energy is, however, impossible since it contradicts the initial setup in which the "real" system was in the global energy minimum. Hence, the effective energy must be greater than or equal to the energy of the global minimum. The effective energy is equal to the energy of the global minimum if the two copies occupy identical positions. Thus, the global minimum of the system with two copies is exactly the same as that with one copy. The addition of side-chain copies (and the test for the global minimum by eliminating side-chain copies) is continued using the structure of the global energy minimum of the effective system with two copies. Following a similar argument as above, after each addition the global minimum of the effective system is found to be identical to that of the "real" system.

Another (shorter) way to prove that the global energy minimum in the real and the effective system is the same is by examining the effective energy. The effective energy is a weighted sum of many evaluations of the real energy. If each of the terms in the sum is in its global minimum, so is the sum.

Based on the last argument, it is evident that any local minimum of the real system is also a local minimum of the mean-field system. It is also clear that the LES energy may have a larger number of minima if the backbone and the side chains do not interact strongly. In this case, a combination of different positions of the same side-chain copies will be a new minimum of the mean-field system. We shall show that our method allows a study of a number of alternate conformers in one structure.

Another property of the effective potential that is of interest is the reduction of barrier heights as compared to the real system. In a search for alternative minima on the potential energy surface, it is helpful if the barriers separating the different minima are as low as possible. To overcome barriers it is necessary to heat up the molecule. However, heating up the system may distort the molecule to undesirable regions of configuration space. Careful heating and cooling procedures must be designed to ensure convergence to the global minimum. The lower the barriers are, the less heating is needed, and the required "baking" of the molecule is less extensive.

The barrier height is estimated as the energy value of the lowest-energy saddle point that connects the two corresponding minima. Let U_B be the value of a saddle point in the real potential. We take the energy of the two minima connected by the saddle point to be U_0 and U_1. U_1 is larger than U_0 and both are smaller than the energy of the saddle point—U_B. The energy barrier for moving from state 0 to state 1 in the real system is simply $U_B - U_0$. The value of the barrier in the effective system would be the same as in the real system, if all the copies were simultaneously placed on the relevant transition state. This is, however, not a saddle point of the effective system, since the matrix of the second derivatives of the potential has more than one negative eigenvalue at that point. Hence, minimization in all directions except one must yield a lower energy barrier. As a result, the barrier in the effective system is always lower than in the real system. In the effective system, crossing the barrier can be converted to a sequence of transitions in which one of the copies is passing at a time. We do not know how to derive an exact relation for the barriers in the real and in the

LES–mean-field systems. An approximate expression for the saddle points in the effective system is given below. We shall show in the computational examples that the estimate works quite well for side-chain transitions in the protein BPTI. The estimate is as follows. Let the number of copies be N and for simplicity let us assume that the weight—w_{k_l}—is identical for all the copies and is equal to $\frac{1}{N}$. We now assume that the copies at the minimum-energy positions do not perturb strongly the structure of the rest of the protein (this will be demonstrated in the results section). When the first copy is passing from state 0 ,to 1, the barrier for its passage is $\frac{1}{N}U_B - \frac{1}{N}U_0$. This is significantly lower than the barrier in the real system. The highest barrier in the LES system (taking the reference point at which all the copies are at U_0) is when the last copy is passing, which is equal to $\frac{1}{N}U_B + \frac{N-1}{N}U_1 - U_0$. Since the energies of the minima are always smaller than the energy of the barriers, the barrier height in the LES system is significantly reduced by approximately a factor of N, where N is the number of copies.

Another favorable feature of the LES protocol is the large number of configurations that are explored at each time step. We consider a typical application of LES in which only a small fraction of the complete system (protein) requires modeling and enhancement (e.g., several side chains). We denote the number of atoms in the rest of the system (which is not enhanced) by N_b and the enhanced subsets by N_j, where $j = 1, \ldots, J$. Let the number of copies for each subset j be K_j. Then the number of configurations sampled at each step is $\prod_j K_j$, while the size of the system is $N_b + \sum_j N_j \times K_j$. The computational effort associated with the energy evaluation is proportional to the first, or at most second, power of the system size. In our protocol, the size of the system increases only linearly with N_b and $\sum_j N_j \times K_j$, while the sampling statistics obtained increases geometrically.

3. Computational Implementation of LES

As we have shown in Section 2, a Lagrangian for the LES approximation can be explicitly written—equation (6). The steps that need to be taken for the implementation of the LES in a regular molecular dynamics program will be explained next.

During the present study, we have introduced these changes in our local version of the program CHARMM (Brooks et al., 1983), and into a general molecular dynamics program called MOIL, developed in our research group (Elber et al., 1993).

The steps that must be taken in order to include the LES Lagrangian in an molecular dynamics program are as follows

(1) The atoms that belong to the part of the system to be oversampled (the side chains in our case) should be labeled accordingly, and their masses scaled down by the weighting factor w_{k_i} (usually $w_{k_i} = \frac{1}{N_k}$, where N_k is the number of copies of subspace k.)

(2) The covalent structure (topology) of the system should be altered in order to exclude interactions between atoms belonging to different copies of the same LES subspace. For example, the angle between C_β's, where the C_β's belong to different copies of the same residue, should not be included.

(3) The generation of the nonbonded list should be modified so as not to include interactions between atoms belonging to different copies of the same LES subspace, while allowing for interactions between atoms belonging to the same copy of the same LES subspace.

(4) The force constants for the potential energy terms and other potential parameters should be scaled according to equation (6).

4. Tests of the Method

In this section we describe a test of the locally enhanced sampling protocol/simulated annealing (LES/SA) method on the problem of conformational search in a tetrapeptide. In particular we will focus on the positioning of the two middle side-chains in the zwiterionic tetrapeptide alanine-valine-phenylalanine-alanine (AVFA). This system was chosen because it is small enough that an exact solution (within the force field) can be found by systematic search, making comparison with an exact solution possible.

We will also demonstrate the lowering of the barriers for conformational transitions predicted in Section 1, as applied to the transition between two different conformations in an aromatic side chain in the protein BPTI.

The exact solution (the location of the global energy minimum for the system) can be obtained by fully exploring the conformational space available for the two middle side chains. In this particular case, the conformational space can be described as a combination of dihedral angles: χ_1 for valine, and χ_1 and χ_2 for phenylalanine.

An adiabatic map was constructed for the χ_1 angles in valine and phenylalanine side chains. The map was constructed by rotating the χ_1 rigidly in steps of 20° and minimizing the energy of the other degrees of freedom.

The CHARMM program (Brooks et al., 1983) was used for this purpose. One thousand steps of the Powell conjugate gradient minimizer were used to optimize a given dihedral angle combination. The final norm of the gradient of the potential energy was set to 10^{-3} kcal/molÅ^{-1}. For both the adiabatic map calculation and the dynamics, the all-hydrogen version of the CHARMM force field was used, with the distance-dependent dielectric option. No cutoff for the electrostatic and nonbonded interactions was used. We can expect to find $9 = 3^2$ minima for this small system. In Figure 1-2 the adiabatic map described here with the minima labeled A–I is shown, and in Table 1-1 we show the values of the dihedral angles and the energy (as a difference from the global energy minimum) corresponding to the minima described in Figure 1-2.

Table 1-1. Dihedral angles corresponding to the minima labeled A–I in Figure 1-2. The energy difference is taken from the global energy minimum in kcal/mol.

Label	χ_1^{val} (degrees)	χ_1^{phe} (degrees)	χ_2^{phe} (degrees)	$\Delta Energy$ (kcal/mol)
A	56.12	-157.91	-100.49	0.2
B	-68.00	-167.14	-163.78	1.7
C	179.69	-167.12	-163.78	0.0
D	55.89	-57.23	170.37	1.6
E	-68.08	-57.94	170.99	1.7
F	179.58	-57.24	170.48	0.1
G	56.14	73.44	62.73	5.7
H	-67.95	73.43	62.69	5.7
I	179.67	73.47	62.69	5.3

By examining Figure 1-2, a high symmetry in the location of the minima is evident, suggesting that the two side chains do not interact strongly with each other. To a reasonable approximation we may then write the total energy as a sum of two independent terms: $U_{total} = U_{val} + U_{phe}$. We have shown in Section 2 that the LES protocol becomes exact for systems with decoupled coordinates. Hence, if the LES/SA protocol can give good results for any system, this molecule should be a good candidate.

The effective LES Lagrangian that we constructed for this system includes 10 copies of valine and 10 copies of phenylalanine side chains. A typical view of the effective system is shown in Figure 1-1. For clarity, only three copies of the ten used in the calculation were plotted.

The position of the side chains is now optimized by means of the simulated annealing method. In order to anneal the system, we must heat it up first, and slowly cool it down. The initial system was heated to a temperature of 1000 K over a period of 20 ps. Then, equilibration was achieved

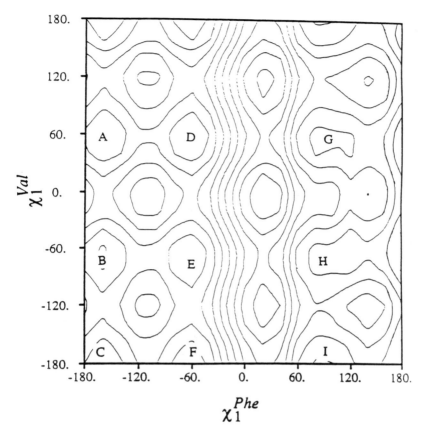

Figure 1-2. Adiabatic energy map for the side-chain dihedral angles (χ_1) of the valine and the phenylalanine. The minima are denoted by A–I. The adiabatic energy was calculated for a fixed backbone. The values of the backbone torsion angles are given in Table 1-1.

at that high temperature by running the simulation for an additional 40 ps. The annealing was pursued by decreasing the temperature by 100 K, and running a trajectory at the new temperature for 20 ps. This procedure was repeated until the final assigned temperature was 0 K.

In order to test the efficiency of the method, the behavior of some relevant parameters should be considered. The first prerequisite for the simulated annealing to run correctly is to start from a temperature high enough that the system will be effectively "melted," visiting all allowed conformations during the annealing period. In Figure 1-3 we show the behavior of the dihedral angle χ_1 for one copy of valine as a function of temperature during the annealing. At some time during the simulation, this copy of the side chain visited all three conformations allowed. This is true for all of the copies of both side chains.

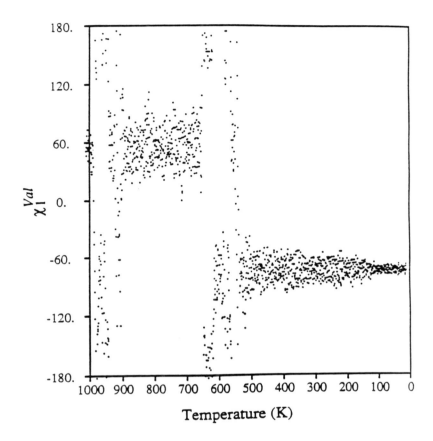

Figure 1-3. The variation in the valine side-chain dihedral angle as a function of the annealing temperature (and time) in AVFA.

Another question remains to be addressed: If we want to draw some conclusions not only about the global energy minimum but also about the dynamics of the system during the annealing, we must find a way to check that the LES potential energy surface is not extremely distorted from that of the physical system. One possible test for this consists of following a trajectory of the LES system and compare the configurations sampled to the allowed regions of the potential energy surface for the original system. In Figure 1-4 we show a single trajectory for one copy of the valine side

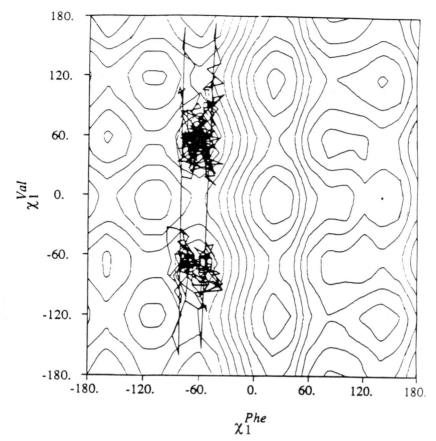

Figure 1-4. A typical trajectory in dihedral angle space (χ_1^{val} vs. χ_1^{phe}) for one copy in the LES Lagrangian. The trajectory was run at 600 K for 20 ps. It is superimposed with the adiabatic energy surface of Figure 1-2.

chain with one copy of the phenylalanine side chain. Since we have 10 copies of each side chain, 100 (i.e., 10^2) such trajectories are obtained. The trajectory corresponds to 20 ps dynamics at 600 K in the LES Lagrangian. Note that the trajectory closely follows the adiabatic potential energy surface, spending long times near two of the minima, and undergoing some conformational transitions during that time. We should expect the rate of conformational transitions to increase in this case with respect to the exact Lagrangian, since the barriers separating the energy minima are reduced (roughly by 10).

The last issue that we should address in order to test the method is related to its predictive power. In Table 1-2 we show the results of the run.

The copy combination (CC) number is defined as the number of trajectories that finally freeze in a given minimum. The sum of all CC numbers is 100. We show, for all energy minima, the number of CC that were annealed to that minimum as well as the energy difference from the global energy minimum. No copies froze in the $\chi_1^{phe} = 60$ minimum (g^+). This is *not* due to inadequate sampling since this conformation was visited during the annealing.

Table 1-2. Minima of the adiabatic energy map for AVFA as labeled in Figure 1-2. The energy (with respect to the global energy minimum) and the CC number (see text for details) are shown.

Minimum	ΔEnergy (kcal/mol)	CC#
A	0.2	12
B	1.7	12
C	0.0	36
D	1.6	8
E	1.7	8
F	0.1	24
G	5.7	0
H	5.7	0
I	5.3	0

The minima with the lowest energies are also the ones with highest population. Furthermore, the occupation number correlates very well with the energy of the corresponding minimum. In this particular case, a distribution, instead of a single conformation, was obtained. The reason is related to the fact that 6 out of the 9 minima differ by only 1.7 kcal/mol from the global energy minimum. We should keep in mind that once the annealing is finished a listing of energy for those minima that were highly populated can give the proper answer for the global energy minimum. Now that we have shown that the method is capable of finding the conformation of the global energy minimum, we shall discuss the computational gain of the protocol.

The number of degrees of freedom for the effective system is larger, by (approximately) a factor of 10 compared to the physical system with only a single side chain. Twenty ps of simulation required 7.5 min on a Stardent 3010 for the "real" system compared to 83 min for the LES system with 10 copies. For 10 independent "real" trajectories 10 different combinations of the side chains are obtained. In LES the number of side-chain combinations is $10^2 = 100$. This corresponds to a reduction of the computational effort by a factor of approximately 10. (Note that the

LES calculation should be somewhat more expensive than 10 independent trajectories since each copy sees all the copies of the other side chain.) Nevertheless, the computer time for the LES simulation with 10 copies only increases by a factor of 11 compared to a single-copy calculation. For $10^2 = 100$ trajectories this is a significant saving in computational effort.

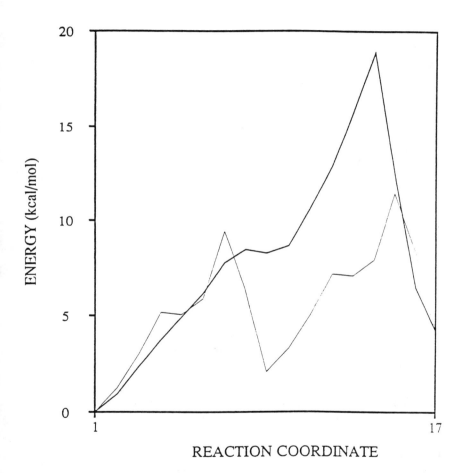

Figure 1-5. A minimum energy path for the transition of a phenylalanine side chain (Phe 33 BPTI) between the two alternative minima found in the simulated annealing. *Thick line*: A minimum energy path for the "real" system (only one copy of the side chain of phenylalanine is employed). *Thin line*: Sequential flips of two copies of the phenylalanine side chain. A mean-field energy for two copies as described in the text was employed. Note the significant reduction in the barrier height by approximately a factor of two when comparing the mean field and the exact energies.

Another aspect of the LES Lagrangian described in Section 2 is the lowering of the energy barriers for conformational transitions by a factor of the order of the number of copies used. During a series of simulations carried out to predict conformations of three aromatic side chains in the hydrophobic core of the protein BPTI, we found two different low-energy configurations for the side chain in the residue Phe 33. In Figure 1-5 we show the minimum energy path between the two alternative minima in this system. One calculation is for the "real system" (thick line), in which only one copy of the side chain exists, and the other is for a calculation with two side-chain copies (thin line). The paths were calculated including all the protein degrees of freedom using the SPW algorithm (Czerminski and Elber, 1990) for 15 grid points. In the notation of Czerminski and Elber, the parameters for the calculation were: $\gamma = 128$ kcal/molÅ^{-2}, $\rho = 200$ kcal/mol, and $\lambda = 2$. The chain was optimized for 10,000 Powell-conjugate gradient steps or until the path-energy gradient was smaller than 10^{-3} kcal/molÅ^{-1}. The figure shows that the large barrier in the case of a single side chain is indeed reduced to two smaller barriers. This is another advantage of LES that was pointed out in the discussion on the method.

5. Myoglobin Modeling

In this section, results and analysis of modeling of 43 side chains in the core of carboxymyoglobin are presented. Some thermodynamic properties of the system are studied, in order to look for a freezing-like transition during the annealing.

Most methods for structure prediction perform poorly when trying to model portions of a system that are exposed to the solvent. This is due to a large number of alternate conformations, non trivial solvent effects and crystal contacts. This is why we attempted to model only residues at the core of the protein, that is, the part that is not accessible to the solvent in the static x-ray structure. The identity of the core residues was determined by the method of Lee and Richards (1971) and was applied to the x-ray structure of carboxymyoglobin as determined by Kuriyan et al. (1986). The amino-acids with accessible surface area lower than 4 Å^2 were defined as core residues. The positions of all protein atoms, including the side chains, were initialized at the x-ray structure. This choice did not bias the final results since complete randomisation was achieved during the initial heating period. In Table 1-3 we list the residues that were modeled in the present study together with their accessible surface area.

Figure 1-6 shows the backbone trace for the protein under study, with the side chains of the residues listed in Table 1-3 highlighted. Also shown

Table 1-3. Residues in carboxymyoglobin to be modeled using the LES/SA approach. Their accessible surface area is shown (in Å^2).

Residue type	Residue #	Acc.Surf.Area
VAL	10	0.0
	13	0.2
	68	0.0
	114	2.9
ILE	28	0.0
	75	0.2
	99	0.0
	107	0.0
	111	0.0
	142	0.0
LEU	2	0.3
	29	0.0
	32	0.1
	40	3.4
	49	0.6
	61	0.0
	69	0.2
	72	0.0
	76	0.0
	86	1.5
	89	1.2
	104	0.0
	115	0.0
	135	0.0
PHE	33	0.0
	43	0.0
	46	0.0
	123	0.0
	138	0.0
MET	55	1.2
	131	0.0
TRP	7	0.5
	14	0.3
SER	108	0.0
THR	39	0.0
TYR	103	3.4
	146	0.8
HIS	24	0.3
	64	0.2
	82	0.3
	97	2.6
	113	2.6
	119	2.9

is the heme group of the protein. Note that the residues to be modeled are distributed over the whole tertiary structure corresponding to the core of the protein.

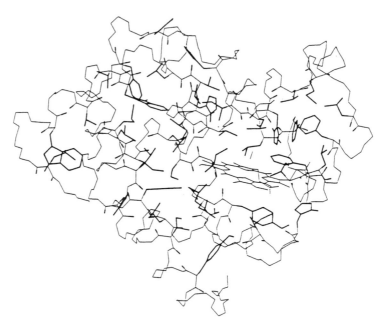

Figure 1-6. Backbone trace for the protein carboxymyoglobin, with the side chains of Table 1-1 highlighted. Also shown are the heme group and the carbon monoxide molecule.

Three simulations were carried out for the protein carboxymyoglobin. The first simulation, hereafter referred to as simulation A, was carried out with a fixed (in the x-ray-determined position) backbone. Four copies of each of the side chains being modeled were used and patched onto a single copy of the C_α backbone corresponding to that side chain. That is, only the atoms belonging to the side chains being modeled were multiplied.

In simulation A, a local modification of the CHARMM20 program (Brooks et al., 1983), MAD, was used. The LES protocol was introduced in the MAD package by making the necessary changes in the definition of the covalent structure and normalization of forces and masses. Simulation A employed the CHARMM force field. A distance-dependent dielectric option was used with a 1–4 scaling parameter of 1. The nonbonded interactions were truncated using a 9 Å cutoff distance. The long-range forces were smoothly truncated with a shift cutoff function. The time step was 1 fs, and the bonds that included an H atom were constrained to a fixed distance using the SHAKE algorithm (Ryckaert et al., 1977). The part of the protein that was not being modeled was kept fixed during the whole simulation. Native carboxymyoglobin contains 1543 atoms, whereas this simulation was done on a system containing 2256 atoms, of which 940 were allowed to move.

The protein was heated from 0 to 8000 K in 20 ps, and maintained at that temperature for 4 ps. At this temperature the side-chain positions become effectively randomized, with an almost free-rotor behavior with regard to the rotational barriers. (Even though the conditions for simulations B and C were slightly different than those of simulation A, in all three cases the initial temperature before the annealing was high enough to ensure effective randomization of the rotational isomer distribution.) The annealing was then started by cooling the system from 4800 K to 400 K over a period of 600 ps. The decrease in temperature was achieved by scaling of the velocities at each time step. The total CPU time required was 40 days on a Stardent 3010 computer.

The simulation described here as B, was carried out with the program MOIL (Elber et al., 1993). The parameters for the force field used by this program are a combination of a covalent part taken primarily from the AMBER force field (Weiner et al., 1984), with the improper torsion parameters taken from CHARMM20 (Brooks et al., 1983), and with the nonbonded parameters of the OPLS force field (Jorgensen and Tirado-Rives, 1988). The LES protocol is naturally included in the MOIL program.

Four copies of the whole residue for which the side chain was modeled were included in the simulation (in contrast to the side chains only in simulation A). Atoms that do not belong to the side chains being modeled were kept fixed during the annealing. The total number of atoms in the system was 2913, of which 940 were allowed to move. A dielectric constant with a value of 1 was used, with a cutoff of 9 Å for the long-range forces. The 1–4 scaling factors used for the van der Waals and electrostatic forces were 8 and 2 respectively. The time step used was 0.5 fs, and bonds that included an H atom were kept rigid using the SHAKE algorithm (Ryckaert et al., 1977).

The protein was annealed from a starting temperature of 4800 K to 0 K over a period of 600 ps. The temperature was scaled down after each dynamic step. The total CPU time required was 24 days on a Hewlett-Packard 9000/720 workstation.

The last simulation C was set up in the same way as simulation B, except that all atoms belonging to the modeled residues were allowed to move, including the backbone atoms.

The number of moving atoms in the system increased to 1824 (out of 2913 total). The time step used in this case was of 2 fs. The annealing was performed from 4000 K to 0 K over a period of 600 ps. The total CPU time was 6.5 days on a Hewlett-Packard 9000/720 workstation. The difference in CPU time between simulations B and C is solely due to the change in the time step. These simulations, while computationally intensive, provide information on approximate thermodynamic properties in addition to structure.

6. Results

A detailed analysis of the results obtained in the simulation labeled A (modeling the side chains with the CHARMM20 force field, with the backbone kept rigid) is presented in this section, along with the structural predictions in simulations B and C. Most of the general behaviors displayed by the first simulation were reproduced in simulations B and C, so comments about the latter will be made only when pertinent.

An important parameter to follow during the simulation is the total potential energy of the side chains during the annealing (Figure 1-7). The decrease with time (or temperature) is linear, and therefore the specific heat of the mobile system is constant during the simulation. The fact that the potential energy, as shown on Figure 1-7, decreases along the annealing, is not surprising. In order to check that the annealing effectively finds minima that are lower in energy as the cooling progresses, we compare the energy obtained from minimization of the individual structures during the annealing process.

Figure 1-8 shows the potential energy for the side chains after minimizaton of individual structures during the annealing process. The minimizations were carried out using a conjugate gradient minimizer, until the norm of the gradient was smaller than 10^{-3} kcal/molÅ^{-1}. Note the change in scale as compared to Figure 1-7. The fact that the minimized potential energy for the sidechains decreases monotonically along the annealing procedure is an indication that the optimization procedure has actually worked towards the improvement of the structure.

Shakhnovich and Finkelstein (1989) suggested that the transition observed during protein folding corresponds to ordering of side chains, from a state in which conformational isomerization is possible (the molten globule state) to an ordered state that they identify as the final fold of the protein. This final transition occurs when the protein backbone is already in the correct fold. Using a model system, these authors showed that a free-energy

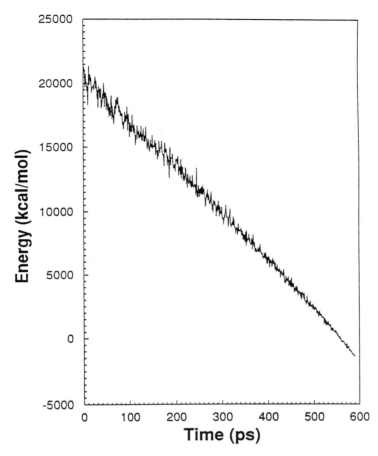

Figure 1-7. The potential energy of the side chains as a function of time. Temperature decreases linearly as a function of time from 4800 K to 400 K.

barrier is obtained for the last step in protein folding (assumed to be packing of the side chains). If this hypothesis is correct for the present atomic detail model, we should be able to find some indication of a transition during the annealing. Figure 1-7 shows that there is no change in the slope of the potential energy as a function of time (or temperature) during the annealing, and therefore the specific heat is constant. This indicates that no first-order phase transition is taking place during the simulation.

In order to check for structural fingerprints of higher-order phase transitions, we followed the values of the χ_1, angles during the annealing for all side chains studied. Figure 1-9 shows the value of χ_1 for all the copies of the residue ILE 28. From this plot it is possible, in principle, to de-

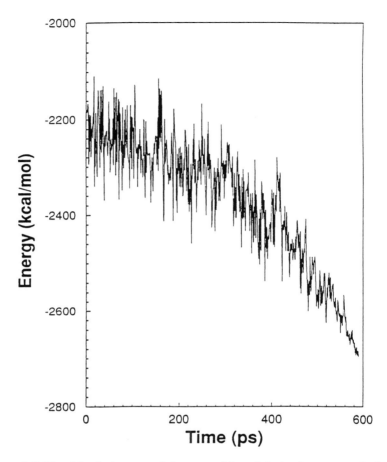

Figure 1-8. The side-chains potential energy of the minimized structures during the annealing. Note the difference in scale as compared to Figure 1-2.

fine a "freezing temperature" for a given residue, that is, a point at which the last transition between minimum-energy wells happened. However, in some cases, such as the ILE 107 (Figure 1-10), the freezing temperature is not clearly defined. In order to define a freezing temperature for a given residue in a consistent way, we followed the values of the χ_1 angles for the minimized structures during the annealing. Figure 1-11 shows the result of the application of this procedure to the system shown in Figure 1-10. Now the freezing temperature (280 ps = 2300 K) is clear, and easily assigned.

The existence of such a definite transition in the side chains studied suggests, contrary to the linear decrease of the potential energy, that "sharp" ordering happens in the system. Individual side-chain transitions may be,

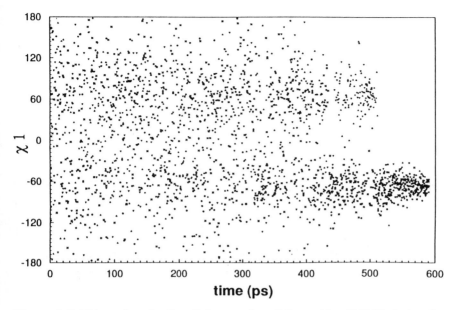

Figure 1-9. Value of χ_1 for for all four copies of the residue ILE 28 during the annealing. The side chains freeze after 510 ps (660 K).

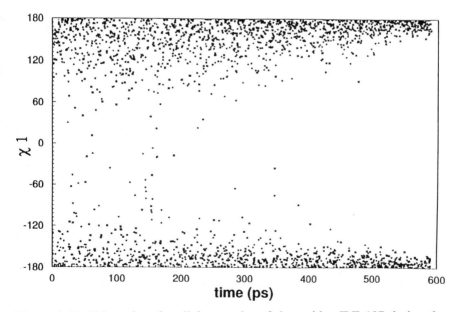

Figure 1-10. Value of χ_1 for all four copies of the residue ILE 107 during the annealing.

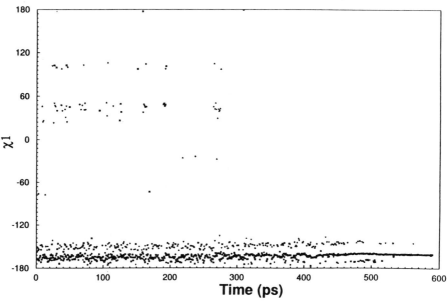

Figure 1-11. Value of χ_1 for all four copies of the residue ILE 107 for the minimized structures during the annealing. The side chains freeze after 280 ps (2300 K).

however, misleading, since we should look for a cooperative transition for the system as a whole.

A histogram of the number of side chains that freeze at a given temperature was constructed (Figure 1-12). An approximate freezing temperature (the average) can be defined and used to study thermal properties of the system if the distribution is narrow.

Since the distribution of freezing temperatures shown in Figure 1-7 is broad, the distribution of energy barriers is also expected to be broad. This is in disagreement with the model of Shakhnovich and Finkelstein (1989) that was discussed above. We comment that the backbone of the protein is kept fixed during the annealing, restricting the conformational space available to the side chains. The fixed backbone structure may make the interaction between the side chains weaker than in reality (since no through-backbone interaction is allowed), so that the total interaction between the side chains may not be strong enough to trigger a cooperative transition. On the other hand, freezing the backbone keeps the highly compact structure of the correctly folded state, which maximizes the side-chain interactions. Hence, the present simulation cannot prove conclusively the existence or nonexistence of the transition in the physical system; however, it is a reasonable approach to try to address this problem.

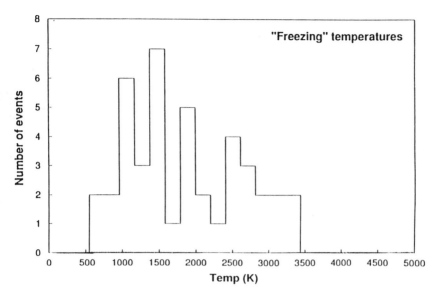

Figure 1-12. Histogram of freezing temperatures for all side chains studied. Note the broad distribution of freezing temperatures.

Another question, having more practical consequences, is related to the behavior of different types of residues during the annealing. Even though the total distribution of freezing temperatures is broad, we might expect some correlation between side-chain sizes and their freezing temperatures. That is, the bulkier side chains (e.g., tryptophan) may freeze at temperatures that are higher than those for the smaller side chains (e.g., leucine). Moreover, if the distribution for individual residue types is narrow, a cooling schedule could be devised according to the type of side chains that are present in the system, which will allow more efficient calculations to be performed. Figure 1-13 shows the average, minimum, and maximum freezing temperatures for the residue types studied. It is obvious that a separation according to residue type cannot be found, and that the distribution for single-residue type is also broad.

We now return to the structural question: the prediction of side-chain positions starting from backbone coordinates. To assess the predictive capabilities of the LES/SA method, its "success rate" is determined as follows. Let us consider the side chain of residue ILE 28. The value of the χ_1 dihedral angle in the x-ray structure was $-71.0°$. As shown in Figure 1-9, the four copies used for the side chain in the present study were found

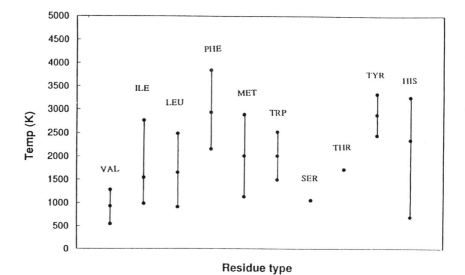

Figure 1-13. Distribution of freezing temperatures as a function of residue type. Presented are the minimum, maximum, and average freezing temperatures for the residue types studied. There was only one THR and one SER residue in the study.

(after the annealing) in the same minimum, with a value for χ_1 of $-70.8°$. This is considered a "hit," since all four copies have the same χ_1 value as the x-ray structure. The maximum deviation from the crystallographic value of χ_1 allowed in a "hit" is $25°$. If three of the copies found the right minimum and only one copy is out of range, we will count that prediction as a "hit" as well. Of course, if none or only one copy finds the proper minimum, we will count this as a "miss." The case in which two copies found the right minimum is slightly more complex. If the two copies that did not find the proper minimum are in different energy wells, then it will be counted as a "hit." If, however, they both are in the same well, we have to use energy criteria to distinguish between the two sets. If the pair with the lowest energy has the correct χ_1 angle, then it is counted as a "hit" too, and a "miss" otherwise.

Using the definitions described above, the simulation labeled A correctly predicted the value of the χ_1 dihedral angles for 38 out of the 43 side chains. This represents a correct prediction for 88% of the modeled residues. Figure 1-14 shows the value of $\Delta\chi_1 = \left|\chi_1^{\text{predicted}} - \chi_1^{\text{x-ray}}\right|$ for the residues modeled. A value between zero and $25°$ for this property was defined as a "correct prediction" of the χ_1 value for that side chain.

The residues whose positions were wrongly predicted in simulation A

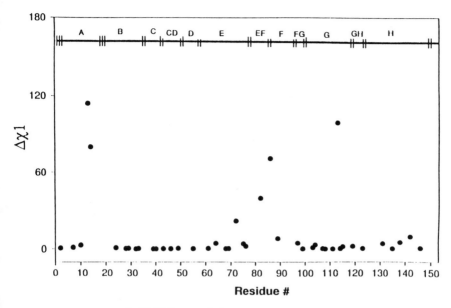

Figure 1-14. $\Delta\chi_1 = \left| \chi_1^{\text{predicted}} - \chi_1^{\text{x-ray}} \right|$ for the residues modeled in simulation A. Also shown is the secondary structure assignment.

are: VAL 13 (A Helix), TRP 14 (A Helix), HIS 82 (EF Loop), LEU 86 (F Helix), and HIS 113 (H Helix). The distance between atoms in VAL 13 and TRP 14 is around 5 Å while the side chains in the pair HIS 82 and LEU 86 are in contact. This means that an error in placing one of the side chains provoked an error in the placement of another. Hence side-chain positioning can be coupled and it is important to take these interactions into account. When the interactions are poorly represented (for example, due to inaccurate force fields), significant errors can be found. This point will be further exemplified for simulation C.

The heme pocket in the globin family plays an important biological role. A closer look at the predictions at that region is then warranted. Figure 1-15 shows the heme pocket. The heme group and the carbon monoxide ligand are included in the picture as a reference, and were not modeled during the simulation. The thick lines correspond to the x-ray structure, whereas the thin lines correspond to the predicted structure for the side chains. The side chains shown are: LEU 29, PHE 43, HIS 64, VAL 68, and ILE 107. The positions of PHE 43, ILE 107, and VAL 68 were predicted correctly for all atoms in the side chains. In the case of LEU 29, the method predicted the χ_1 for three of the copies used within 2° of the crystallographic value,

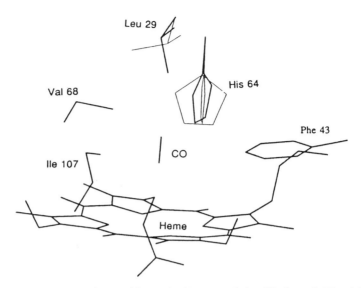

Figure 1-15. The Heme pocket residues, the Heme and the CO ligand. The thick lines correspond to the x-ray structure, whereas the thin lines correspond to the predicted structure for the side chain, which deviate from the experimental data (single thick lines correspond to calculations whose predictions overlap with the x-ray structure). Simulation A.

and the fourth copy to have a χ_1 with a 25° deviation from the crystal structure. This deviation corresponds to a different χ_2 value for that copy, as shown in Figure 1-10. In the case of HIS 64, the value of χ_1 for all four copies of the side chain lies within 7° of the crystallographic value. Three of the copies have a value of χ_2 that deviated 120° from the experimental structure, while the fourth copy's χ_2 deviated 50° from the x-ray structure. HIS is an asymmetric side chain, but since the diffraction pattern for N and C atoms is almost the same, and the H atoms are not detectable in x-ray crystallography, a 180° flip of the χ_2 angle gives the same x-ray structure. Furthermore, HIS 64 is known to be quite mobile even in x-ray crystallography (Kuriyan et al., 1986), and therefore its precise orientation is uncertain.

The adequacy of the χ_2 prediction was also examined. Of the 38 side chains whose χ_1 angles were predicted correctly, 35 also possess a χ_2 angle. Of those 35 side chains, 29 were found to have an optimized χ_2 within 20° of their crystallographic values, that is, 83% "hit" probability.

Using the definitions outlined before, simulation B (similar to simulation A, except that the MOIL program was used) correctly predicted positions of 37 out of 43 side chains. That represents an 86% prediction.

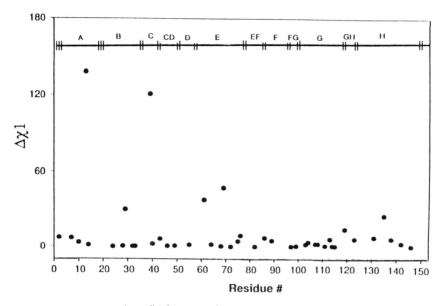

Figure 1-16. $\Delta\chi_1 = \left| \chi_1^{\text{predicted}} - \chi_1^{\text{x-ray}} \right|$ for the residues modeled in simulation B. Also shown is the secondary structure assignment.

This number is close to that obtained in simulation A. Figure 1-16 shows the value of $\Delta\chi_1 = \left| \chi_1^{\text{predicted}} - \chi_1^{\text{x-ray}} \right|$ for the residues modeled.

The χ_1 for 6 residues of the 43 modeled were incorrectly calculated by simulation B. These residues were VAL 13, LEU 29, THR 39, LEU 61, LEU 69, and LEU 135. The distance between the pair LEU 29/LEU 61 was approximately 5 Å, so one can assign the problem to a single coupled error. All four other "miss" events are uncoupled. Except for VAL 13, all the other residues that were wrongly placed are different from those of simulation A. As previously, the predicted conformation around the Heme pocket is examined in more detail (Figure 1-17). The positions of the side chains PHE 43, ILE 107, and VAL 68 were placed appropriately. For HIS 64, the value of χ_1 was predicted within 1° of the crystallographic value, while its χ_2 was correct within 20°. LEU 29 was incorrectly assigned.

Of the 37 side chains whose χ_1 was correctly predicted, 34 also have a χ_2 angle. Of those, 30 were predicted to have a χ_2 value within 15° of their crystallographic position, that is, 86% predictive ability for χ_2 angles.

Our final simulation C was carried out using the MOIL program, and differs from simulations A and B in the mobility of the backbone. More precisely, the backbone of the residues whose side chains were modeled was allowed to move during the annealing.

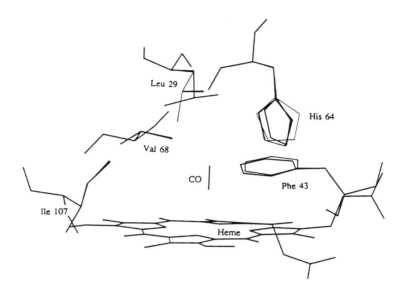

Figure 1-17. The Heme pocket, the Heme group and the CO ligand. The thick lines correspond to the x-ray structure and the thin lines to the structure of the side chains as predicted in simulation B.

Backbone rigidity may have a double effect. First, it reduces the size of the conformational space to the extent that the annealing becomes feasible in a short time. The second and the less desirable effect is of kinetics, in which the rigidity presents formidable barriers for transitions altering the structures that should be explored.

Even though in simulation C the backbone was allowed some freedom, no backbone transitions were observed during annealing. The ϕ and ψ fluctuations had an average amplitude of about 20° during the annealing.

Using the definitions outlined above, this simulation correctly found the value of χ_1 for 32 out of 43 total residues modeled. This represents a correct prediction in 74% of the cases. This number is significantly lower than the one obtained for simulation A or B and the reason for this poorer performance will be discussed below. In Figure 1-18 we show the value of $\Delta\chi_1 = \left| \chi_1^{\text{predicted}} - \chi_1^{\text{x-ray}} \right|$ for the residues modeled during simulation C.

Of the 10 residues whose χ_1 were incorrectly predicted, 7 were found in a 19-residue segment. By visual inspection of that section of the protein (Figure 1-19) one finds an important difference between the predicted and the x-ray structure. The CD loop contains two phenylalanine residues (PHE 43 and PHE 46) that are misplaced. This misplacement may have

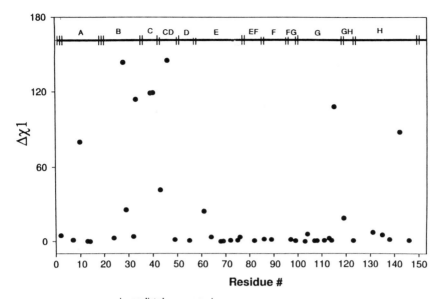

Figure 1-18. $\Delta\chi_1 = \left| \chi_1^{\text{predicted}} - \chi_1^{\text{x-ray}} \right|$ for the residues modeled in simulation C. Also shown is the secondary structure assignment.

induced significant strain in the nearby moving residues. This accounts for the difference in the final number of correctly placed side chains. *This is a cooperative effect in which one or two wrongly placed side chains influence others.* Figure 1-19 shows the x-ray structure of a part of the CD loop, superimposed with the predicted structure. The annealed structure (the thick line) has the two phenylalanines in a face-to-face arrangement. Within the force field employed the only interactions between the two phenylalanine residues are of a van der Waals type. In this potential form, a face-to-face arrangement for the aromatic rings maximizes the attractive contacts.

This is in disagreement with the findings of Burley and Petsko (1986), who showed that the minimum energy conformation for a pair of aromatic side chains strongly depends on their relative orientation, favoring the arrangements of edge-to-face type. The x-ray structure (thin line) in Figure 1-19 is an example of edge-to-face arrangement. The flexibility allowed to the system by the mobility of the backbone was enough to permit the two phenylalanines to explore the low-energy region that includes the minimum with the face-to-face structure. This means that the freezing of the backbone described above provided a kinetic barrier to a lower-energy configuration that is experimentally incorrect. We have to conclude then

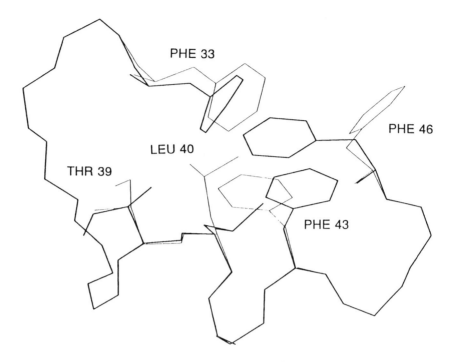

Figure 1-19. Structure of a portion of the CD loop in carboxymyoglobin. The thin line corresponds to the x-ray structure and the thick line to the structure predicted after the annealing. Note that in the x-ray case, the two phenyalanines are in an edge-to-face conformation, while in the predicted structure they are in a face-to-face arrangement.

that the LES/SA method effectively found the minimum energy conformation within the limitation of the force field employed. In fact, the new minimum is significantly lower in energy (by 10 kcal/mol) compared to a minimized x-ray structure. The fact that this configuration does not correspond to the x-ray structure is attributed to a problem in the design of the force field and not to a failure of the optimization algorithm.

A possible solution to this problem was already proposed by (among others) Burley and Petsko (1986). The explicit inclusion of H atoms as partially charged particles in the aromatic rings is crucial for representation of the quadrupole moments that play a significant role in determining the relative orientation of the two aromatic rings. The minimum energy conformation is then obtained with an H atom pointing into the π-cloud of the aromatic ring, an edge-to-face conformation. As an example of the magnitude of this correction, Petterson and Liljefors (1987) solved the same problem for a benzene dimer using the MM2 force field, by adding a charge of -0.15 on the C atom, and a charge of $+0.15$ on the H atom. This gives good agreement with an SCF-CI structure determined by the same authors. The OPLS potential on which MOIL is based was extended to include hydrogens in the aromatic rings. These extensions were included in MOIL.

7. Conclusions

In this chapter we described a new optimization algorithm that is based on a combination of a mean-field approximation and simulated annealing. We emphasize that the novel feature of our algorithm is the use of a new effective energy that is significantly smoother compared to the original potential but has the same global energy minimum. We used simulated annealing to optimize the effective energy surface, since we were interested in obtaining thermodynamic properties in addition to the final structure. Nevertheless, different optimization protocols such as the genetic algorithm (Tuffery et al., 1991), the diffusion equation (Piela et al., 1989), the renormalization group ideas (Shalloway, 1992), or imaginary time Schrödinger equation (Amara et al., 1993) can be also used to optimize the effective energy that we introduced. The advantages of LES are that it gives more than just the structure and it is relatively simple and general to use. The disadvantage is that it is LESs effective in smoothing the energy surface compared to (for example) the diffusion equation. Some combination is therefore desirable if only the final structure is of interest.

We also comment that we employ atomic-level energy surface rather than a drastically simplified Hamiltonian. The detailed potential was derived from data on small model systems. Thus, we are taking the *ab initio* approach to structure determination. This is in the sense that we do not use statistical data on large molecules in order to extract information on an effective energy. We remain with a "physical chemistry" model that enables us to investigate thermodynamic properties and to study the relative contributions of different energies that determine the structure. This is with more control on the approximations that we made compared to models that use a simplified representation of the macromolecule and/or employ reduced energy functions.

We demonstrated in small model systems (tetrapeptides) that the method locates the global energy minimum. Furthermore, in a single run it provides the distribution of the lowest-energy minima. In addition to a test of the optimization algorithm, we pursued a test of the quality of the atomic detail potentials by calculating the positions of the core residues in myoglobin. The method was successful in predicting about 88% of the χ_1 angles of the side chains studied. The success rate was investigated as a function of the force field and the protein flexibility.

For each side chain, a temperature was found at which all transitions between conformational wells freeze sharply. This prompted us to look for a collective, freezing-like transition of the system as a whole. No evidence of such a collective transition was found under the conditions of the simulation. Different side chains freeze at different temperatures.

The distribution of freezing temperatures for individual side chains did not depend on the size or the residue type, ruling out the possibility of designing an annealing schedule that takes advantage of the particular protein residue composition.

The atomic model that we employ does not treat aromatic–aromatic interactions well. As a result, two phenylalanine rings were incorrectly packed and influenced conformations of other side chains in the CD loop of myoglobin. We anticipate an improvement in the prediction ability of the methodology if a more accurate potential for the rings becomes available. It is the new optimization protocol that demonstrated the importance of this interaction when *global* packing of side chains is considered. It is very unlikely that a method that is based on hard-core interactions only and searches for optimal packing will correctly predict the structure of the CD loop. In a preliminary study (Elber, unpublished) the model for the phenylalanin was improved to include hydrogens and quadrupole moment. An LES/SA simulation with the new model correctly predicts the packing of the CD loop, giving a 91% prediction.

The LES/SA method suggested itself as a useful tool for a determination of structural segments as well as for investigation of approximate thermodynamic properties.

Acknowledgments. This research was supported by NIH grant GM41905 to R.E., who is a University of Illinois West Scholar and Allon Fellow in the Hebrew University. The Fritz Haber Research Center is supported by a Minnerva fund.

REFERENCES

Amara P, Hsu D, Straub JE (1993): Global energy minimum searches using an approximate solution of the imaginary time Schrödinger equation. *J Phys Chem* 97:6715

Blow D (1983): Molecular structure. Computer cues to combat hypertension. *Nature* 304:213–214

Blundell TL, Sibanda BL, Sternberg MJE, Thornton JM (1987): Knowledge based prediction of protein structures and the design of novel molecules. *Nature* 326:347–352

Brooks SR, Bruccoleri RE, Olafson SD, States DJ, Swaminathan S, Karplus M (1983): CHARMM: A program for macromolecular energy, minimizalion and dynamic calculations. *J Comput Chem* 4:187–217

Bruccoleri RE, Karplus MK (1987): Predicting the folding of short polypeptide segments by uniform conformational sampling. *Biopolymers* 26:137–168

Burley SK, Pelsko GA (1986): Dimerization energetics of benzene and aromatic amino acid side-chains. *J Am Chem Soc* 108:7995–8001

Czerminski R, Elber R (1990): Self-avoiding walk between two fixed-points as a tool to calculate reaction paths in large molecular-systems. *Int J Quant Chem* 24:167–186

Czerminski R, Elber R (1991): Computational studies of ligand diffusion in globins: 1. Leghemoglobin. *Proteins: Structure, Function and Genetics* 10:70–80

Desmet J, De Maeyer M, Hazes B, Lasters I (1992): The dead-end elimination theorem and its use in protein side-chain positioning. *Nature* 356:539–542

Elber R, Karplus M (1990): Enhanced sampling in molecular dynamics: use of the time dependent Hartree approximation for a simulation of carbon monoxide through myoglobin. *J Am Chem Soc* 112:9161–9175

Elber R, Roitberg A, Simmering C, Goldstein R, Verkhiver G, Li H, Ulitsky A (1993): MOIL: A molecular dynamics program with emphasis in conformational searches and reaction path calculations. To be published in the proceedings of the NATO conference: *Statistical Mechanics, Protein Structure and Protein–Substrate Interactions*, Doniac S, ed., New York: Plenum Press. This program is available via anonymous ftp from 128.248.186.70

Gelin BR, Karplus MK (1975): Side-chain torsional potentials and motion of amino acids in proteins: Bovine pancreatic trypsin inhibitor. *Proc Natl Acad Sci* 72:2002–2006

Gelin BR, Karplus MK (1979): Side-chain torsional potentials: effect of dipeptide, protein and solvent environment. *Biochemistry* 18:1257–1268

Gerber RB, Buch V, Ratner MA (1982): Time dependent self-consistent field approach for intramolecular energy transfer. I. Formulation and application to dissociation of van der Waals molecules. *J Chem Phys* 77:3302–3030

Gibson QH, Regan R, Elber R, Olson JS, Carver TE (1992): Distal pocket residues affect picosecond ligand recombination in myoglobin: An experiment and molecular dynamics study of position 29 mutants. *J Biol Chem* 267:22022–22034

Greer J (1981): Comparative model building of mammalian serine proteases. *J Mol Biol* 153:1027–1042

Holm L, Sander C (1992): Fast and simple Monte Carlo algorithm for side-chain optimization in proteins: Application to model building by homology. *Proteins: Structure, Function and Genetics* 14:213–223

Janin J, Wodak S, Levitt M, Maigret B (1978): Conformation of amino acid side-chains in proteins. *J Mol Biol* 125:357–386

James MNG, Sielecki AR (1983): Structure refinement of penicillopepsin at 1.8 Å resolution. *J Mol Biol* 163:299–361

Jorgensen WL, Tirado-Rives J (1988): The OPLS potential functions for proteins, energy minimizations for crystals of cyclic peptides and crambin. *J Am Chem Soc* 110:1657–1666

Kuriyan J, Wilz S, Karplus M, Petsko GA (1986): X-ray structure and refinement of carbon-monoxy (Fe II)-myoglobin at 1.5 Å resolution. *J Mol Biol* 192:133–154

Lee B, Richards FM (1971): The interpretation of protein structures: estimation of static accessibility. *J Mol Biol* 55:379–400

Lee C, Levitt M (1991): Accurate prediction of the stability effects of site-directed mutagenesis on a protein core. *Nature* 352:448–451

Lee C, Subbiah S (1991): Prediction of side-chain conformation by packing optimization. *J Mol Biol* 217:373–388

Lesk AM, Chothia C (1986): The response of protein structures to amino-acid sequence changes. *Phil Trans Roy Soc A* 317:345–356

Narayana SV, Argos P (1984): Residue contacts in protein structures and implications for protein folding. *Int J Pept Prot Res* 24:25–39

Petterson I, Liljefors T (1987): Benzene–Benzene (Phenyl–Phenyl) interactions in MM2/MMP2 molecular mechanics calculations. *J Comp Chem* 8:1139–1145

Piela L, Kostrowicki J, Scheraga HA (1989): The multiple-minima problem in the conformational analysis of molecules. Determination of the potential energy surface by the diffusion equation. *J Phys Chem* 93:3339–3346

Ponder JW, Richards FM (1987): Tertiary template for proteins. Use of packing criteria in the enumeration of allowed sequences for different structural classes. *J Mol Biol* 193:775–791

Reid L, Thornton JM (1989): Rebuilding flavodoxin from C_α coordinates: A test study. *Proteins: Structure, Function and Genetics* 5:170–182

Richards FM (1977): Volumes, packing and protein structure. *Annu Rev Biophys Bioeng* 6:151–176

Roitberg A (1992): Ph.D. thesis, University of Illinois at Chicago

Roitberg A, Elber R (1991): Modeling side-chains in peptides and proteins: applicaton of the locally enhanced sampling and the simulated annealing methods to find minimum energy conformations. *J Chem Phys* 95:9277–9287

Ryckaert JP, Ciccotti C, Berendsen HJC (1977): Numerical integration of the Cartesian equations of motion of a system with constraints: Molecular dynamics of n-alkanes. *J Comput Phys* 23:327–341

Shakhnovich EI, Finkelstein AV (1989): Theory of cooperative transitions in protein molecules. I. Why denaturation of globular proteins is a first-order phase transition. *Biopolymers* 28:1667

Simmerling C, Elber R (1994): Hydrophobic "collapse" in a cyclic hexapeptide: Computer simulations of CHDLFC and CAAAAC in water. *JACS* 116:253–254

Singh J, Thornton JM (1990): SIRIUS: An automated method for the analysis of the preferred packing arrangements between protein groups. *J Mol Biol* 211:595–615

Smith J, Hendrickson WA, Honzatko R, Sheriff S (1986): Structural heterogeneity in protein crystals. *Biochemistry* 25:5018–5027

Straub J, Karplus M (1991): Energy equipartition in the classical time-dependent Hartree approximation. *J Chem Phys* 94:6737–6739

Shalloway D (1992): In *Recent Advances in Global Optimization*, Floudas A, Pardalos PM, eds. Princeton, NJ: Princeton University Press, pp. 433–477

Summers NL, Carlson WD, Karplus MK (1987): Analysis of side-chain orientations in homologous proteins. *J Mol Biol* 196:175–198

Summers NL, Karplus MK (1989): Construction of side-chains in homology modeling. Application to the C-terminal lobe of rhizopuspepsin. *J Mol Biol* 210:785–811

Tuffery P, Etchebest C, Hazout S, Lavery R (1991): A new approach to the rapid determination of protein side-chain conformations. *J Biomo/Struc Dynamics* 8:1267–1289

Ulitsky A, Elber R (1993): The thermal equilibrium properties of the time dependent Hartree and the locally enhanced sampling approximations. Formal properties, a correction and computational examples of rare gas clusters. *J Chem Phys* 98:3380–3388

Verkhivker G, Elber R, Nowak W (1992): Locally enhanced sampling in free energy calculations: Application of mean field approximation to accurate calculation of free energy differences. *J Chem Phys* 97:7838–7841

Warme PK, Morgan RS (1978): A survey of side-chain interactions in 21 proteins. *J Mol Biol* 118:289–304

Weiner SJ, Kollman PA, Case DA, Chandra Singh U, Ghio C, Alagona G, Profeta S, Weiner P (1984): A new force field for molecular mechanical simulation of nucleic acids and proteins. *J Am Chem Soc* 106:765–784

Zimmerman SS, Pottle MS, Nemethy G, Scheraga HA (1977): Conformational analysis of the 20 naturally occurring amino acid residues using ECEPP. *Macromolecules* 10:1–9

2

Conformation Searching
Using Simulated Annealing

Stephen R. Wilson and Weili Cui

1. Introduction

Protein folding by direct computer-directed conformation searching is not yet possible (Brunger and Karplus, 1991). Two fundamental problems challenge researchers in this area. The first involves the general issue of energy calculations for large molecules, i.e., parameters, solvents, etc. The second is a more fundamental one called the multiple minimum problem. Since the number of possible minima is an exponential function of the number of degrees of freedom (rotatable angles), an *a priori* search for the lowest-energy solution to such a problem is impossible for more than about 15 rotatable angles, even with today's fastest computers. A classic example of this type of problem is known as the "Traveling Salesman Problem" (Kalos and Whitlock, 1986). The problem goes as follows: Try to find the shortest route (i.e., cheapest airfare) for a hypothetical salesman as he/she visits an increasing number of cities.

One of the pioneers of molecular mechanics force fields, Norman Allinger, has said that "since a complete conformation search is impossible, the results are dependent not so much on the force field being used but on the intuition of the person doing the calculations" (Burkert and Allinger, 1982). In his paper on the conformation searching of metenkephalin, Harold Scheraga (Isogai et al., 1977) lists nine strategies for picking starting points for his conformation search. These strategies sound rather arbitrary, but represent the best possible application of expert insight.

The Protein Folding Problem and Tertiary Structure Prediction
K. Merz, Jr. and S. Le Grand, Editors
© Birkhäuser Boston 1994

In a paper published in *Science* in 1983, Kirkpatrick originated the concept of simulated annealing and formulated a practical algorithm based on the concept (Kirkpatrick et al., 1983). He demonstrated the power of this algorithm in several classical examples of multivariable optimization. Due to its appealing concept, simple implementation, and broad applicability, this algorithm has been adapted by researchers in many disciplines and applied to a wide range of optimization problems. Simulated annealing attracted such attention that a monograph was written with over 292 citations only five years later (Johnson, 1988)

In this review, we will concentrate on the application of simulated annealing to conformational search and protein folding. We will begin with a brief introduction of the multiple minimum problem as it exists for conformational search and then discuss the most basic conformational search method—the grid search. We will then describe an alternative approach, a Monte Carlo approach using the Metropolis algorithm. Finally, we show how we have implemented the Monte Carlo simulated annealing algorithm for conformational search in our own program, Anneal-Conformer.

Simulated annealing has been used either in its original form based on the Metropolis algorithm or with molecular dynamics. In the area of optimization of molecular structures, it has been applied to the conformational search of small molecules (Wilson et al., 1988a) and small peptides (Wilson and Cui, 1990), simulation of protein main-chain folding (Wilson and Doniach, 1989), packing of protein side chains (Lee and Subbiah, 1991), molecular docking (Goodsell and Olson, 1990; Yue, 1990; Hart and Read, 1992), simulation of crystal packing (Deem and Newsam, 1992), molecular similarity searches (Barakat and Dean, 1991), and modeling polymer systems (Karasawa and Goddard, 1988).

1.1 Multiple Minimum Problem

For any molecule having torsional freedom, there always exist multiple minima on its potential surface. Take a simple molecule such as n-butane. Its torsional energy curve shows the existence of three minima separated by energy barriers (Figure 2-1). Energy minimization procedures, as implemented in most molecular modeling software packages, are designed for local minimization. That is, from a given starting structure, the energy minimization will move the molecule downhill toward the bottom of the local energy well. Thus, starting from point A on the curve in Figure 2-1, only the local minimum at point B will be found. The global minimum at point C will not be located by an energy minimization procedure starting from point A. This creates a problem if your objective is to locate the global

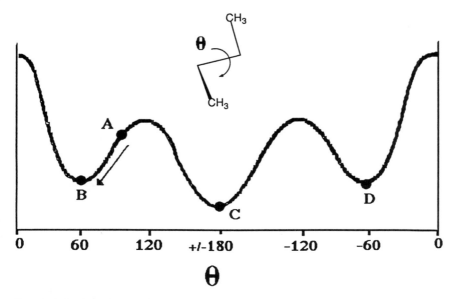

Figure 2-1. Plot of the molecular mechanics energy (*vertical axis*) vs. dihedral angle (Θ) for butane, a molecule with only one rotational degree of freedom. Point C is the global minimum and points B and D are local minima.

minimum. This problem is referred as the multiple minimum problem in the literature of conformational analysis.

1.2 The Systematic Search

A straightforward solution to the multiple minimum problem would be to sample systematically along the torsional coordinate by a defined increment. At a resolution of 30° for n-butane, 12 starting conformations would be generated and then be minimized one after another. Four of them would lead to the global minimum C. Others would lead to the local minimum A and B. This approach is called the systematic search method or grid search (Lipton and Still, 1988). For very small molecules, this approach works very well and is still the method of choice. However, with more than 10 torsional angles to sample, this method is still prohibitively expensive even on modern superworkstations.

1.3 Monte Carlo Sampling—The Metropolis Algorithm

Still has reported a variation of the grid search that randomly samples the energy surface (Chang et al., 1989). One hopes that one is lucky enough to sample the right part of the surface and locate the global minimum.

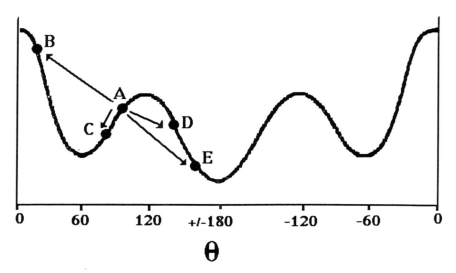

Figure 2-2. Plot of the molecular mechanics energy (*vertical axis*) vs. dihedral angle (Θ) for butane showing Monte Carlo moves from randomly selected point A. Note local and global minima are the same as in Figure 2-1.

Clearly, however, a Monte Carlo sampling without any constraints or direction is very inefficient, and it is rarely used in modern computational studies. Metropolis and co-workers (Metropolis et al., 1953) first proposed an efficient Monte Carlo sampling technique. They realized that for the purpose of sampling a molecular system in thermodynamic equilibration, one could efficiently sample the significant states by directing the sampling process according to a Boltzmann distribution. This algorithm has become one of the most widely used Monte Carlo sampling techniques. To apply to conformational sampling, the implementation of the Metropolis algorithm goes as follows. For a given arbitrary structure R_0 (point A in Figure 2-2), the energy of this structure is calculated as E_0. By randomly changing the dihedral angle, a new conformation R is generated. This new conformation could be located at point B, C, D, or any other point along the curve. The energy of this new conformation is calculated as E_{new}. Whether this new conformation is accepted depends on an energy criterion. The probability of accepting this new conformation is calculated in two steps. First, a probability density function is calculated as a Boltzmann factor:

$$\text{pdf} = e^{-(E-E_0)/RT}. \tag{1}$$

This pdf value is then compared with a random number RM ($0 < \text{RM} < 1$). If pdf $\geq$ RM, this new conformation is accepted; otherwise, this new conformation is rejected. Note that if $E \leq E_0$, then pdf $\geq 1 >$ RM, this new conformation is to be accepted. This means that the downhill move on the energy surface is always accepted; only in the case of the uphill move is the acceptance of this conformation determined randomly. In essence, the Metropolis algorithm is different from the normal random walk in that the Metropolis algorithm directs the walk toward low-energy regions of the potential surface. Yet the uphill moves are allowed randomly to move out of local minima in search of the global minimum. It is in this sense that we can characterize the Metropolis algorithm as an energy-directed Monte Carlo sampling.

1.4 Simulated Annealing

The Metropolis algorithm greatly improves the sampling efficiency of the Monte Carlo sampling process. However, its sampling efficiency largely depends on choosing a suitable temperature factor T (see equation (1)) and the selection of proper T is largely based on trial and error. Once a value of T is chosen, it remains the same for the whole sampling process. If the value of T is too high, there will be too many uphill moves allowed and therefore the low-energy region might not be sampled sufficiently. If T is too low, there will be too few uphill moves allowed and the system could be trapped in a local minimum. Until Kirkpatrick proposed a scheme to gradually decrease the temperature during the Metropolis sampling process and called it "simulated annealing" (Kirkpatrick et al., 1983), the Metropolis algorithm had been used in its original form for almost 30 years.

    Kirkpatrick made an intuitive, yet very consequential analogy between computational optimization of a multivariable function and the experimental optimization of a crystal. Annealing is an experimental technique routinely used for the improvement of the quality of crystals. In a typical annealing process, an imperfect crystal is heated until it melts, then cooled very slowly to form a perfect crystal. The initial heat destroys the imperfect crystal structure and the slow cooling allows the system to sample low-energy packing arrangements in forming a perfect crystal. Applying this heating and slow cooling strategy to the computational optimization of a multivariable function, Kirkpatrick started the Metropolis sampling at a high temperature to allow many uphill moves in order to sample the global space. He then slowly decreased the temperature to anneal the system to its global minimum.

2. Anneal-Conformer

We were attracted to simulated annealing since it involves a global rather

Metropolis Sampling of Conformations

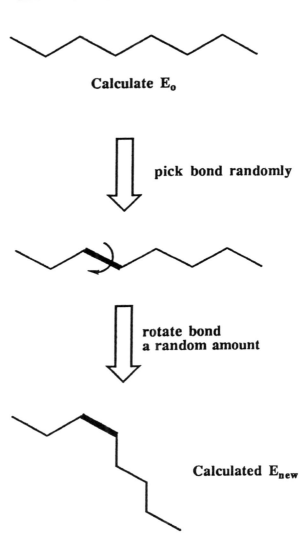

Calculate E_0

pick bond randomly

rotate bond
a random amount

Calculated E_{new}

Figure 2-3. Method for using the Metropolis algorithm to sample conformation space.

than a local optimization. This means that no matter what your starting point, you will always find the global minimum point (best conformation). Extension of the method to molecules was simply accomplished as shown in Figure 2-3. First, a molecule in any starting state (geometry) is read by the program and its starting energy E_0 is computed. Then a bond is picked at random and rotated a random number of degrees. The energy of the new geometry E_{new} is then calculated. If $E_{new} - E_0 < 0$ (i.e., the movement is downhill energetically), then the movement is accepted. If $E_{new} - E_0 > 0$, then the movement is only accepted with the Boltzmann probability $e^{-(E_{new}-E_0)/kT}$ (cf. discussion of pdf above). This means that at high temperature, both uphill and downhill moves are accepted. As the system is cooled, the movement uphill is constrained so that eventually the system freezes into the global minimum.

We were skeptical that this simple idea would actually work. It seemed initially counterintuitive that a totally random process could give an exact solution. We were supposed to be convinced by the following classic example. Imagine that you want to calculate the exact value of π using raindrops. First you need a square 8-in pie pan (no pun intended) and a

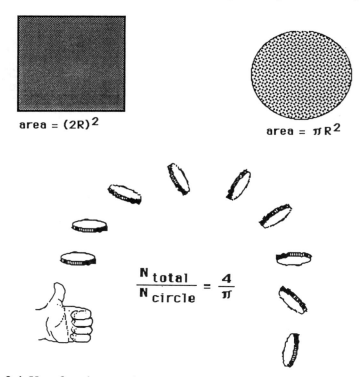

area = $(2R)^2$

area = πR^2

$$\frac{N_{total}}{N_{circle}} = \frac{4}{\pi}$$

Figure 2-4. Use of random numbers to determine the value of π: If the square has side = $2R$ and the circle a radius of R, a series of random numbers between 0 and 1 will converge to the exact value of π (cf. Table 2-1)

round 8-in pie pan. Place the round one inside the square one and put the two outside in a light rain. It is best to try this experiment during a sprinkle, since you must collect a countable number of drops in the pans. After collecting some drops you count the total number of drops collected and also the number inside the round pan. Using the following formula, you can calculated the value of π:

$$N_{\text{total}}/N_{\text{round}} = 4/\pi. \tag{2}$$

We have written a small computer program to do the same thing. A random number generator produces numbers between 0 and 1. Each pair of numbers defines a point x, y that is either in or is not in the circle (cf. Figure 2-3). Using the above formula one can quickly accumulate the data in Table 2-1. Note that after only 10 steps (raindrops!) the value of π is rather inaccurate. On the other hand, you can see that 100 steps (or better 1000 steps) converges on a recognizable value of π.

Table 2-1. Calculation of π with random numbers.

Number of Steps	π-Calculated	% Error
10	2.80000	10.87324
100	3.09091	1.61331
1,000	3.12432	0.54967
10,000	3.13591	0.18077
100,000	3.14442	0.08985
1,000,000	3.14201	0.01321

Application to a small polyalanine model peptide was one of our first test systems for Anneal-Conformer (Wilson et al., 1988a). An arbitrary model peptide Ala_7 was built and used as input to our program. Simulated annealing of Ala_7 using the AMBER force field (Weiner et al., 1984) leads to formation of an α-helix (Figure 2-5). A graph of the total energy vs. random walk is shown. Clearly the energy begins at a high point and at some point rapidly drops. This point is the formation of an α-helix. One can see this in a similar plot of the α-helix 1-13 H-bond distance as a function of the same random walk (Figure 2-6). The critical 1-13 H-bond characteristic of the α-helix was found early in the search, then lost at high temperature, then found again when the helix nucleates at about step 1000.

2.1 Testing on Amino Acids

Anneal-Conformer was then tested on the conformations of N-acetyl-N-methylamide amino acid dipeptide models (Zimmerman et al., 1977). A rigorous conformation search on "dipeptide models" of the 20 natural amino

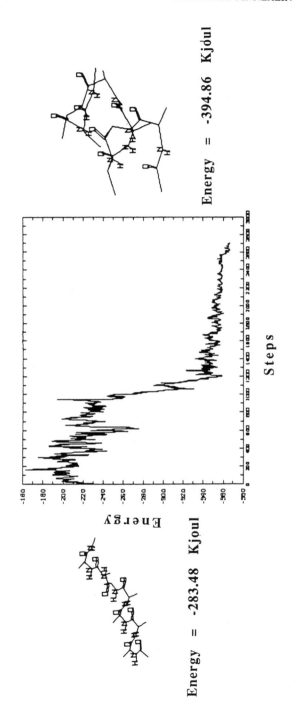

Figure 2-5. Anneal-Conformer (simulated annealing) run on Ala$_7$ polypeptide model. Starting geometry, plot of energy vs. steps of the random walk and final geometry are shown. Cooling during the course of the run causes Ala$_7$ to freeze out in the global minimum (an α-helix).

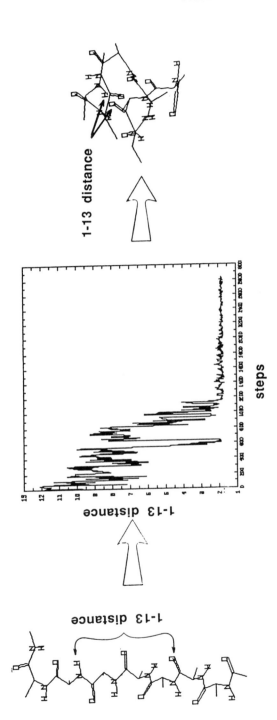

Figure 2-6. Same Anneal-Conformer (simulated annealing) run on Ala$_7$ polypeptide model as shown in Figure 2-5 but monitoring the 1-13 H-bond distance characteristic of an α-helix. Note the H-bond was found at relatively high temperature, broken and found again as the system freezes into the global minimum.

acids using the systematic grid search (Lipton and Still, 1988) option of Macromodel (Mohamadi et al., 1990) and the AMBER force field provides a standard set of molecules for which the global minimum was known. The number of starting geometries, CPU time and global minima found are shown in Table 2-2. (Glu, Lys, and Arg were beyond the reach of the Macromodel systematic grid search.) Using this set of data as the "correct answer," we have carried out simulated annealing with our program, using as starting geometry various high-energy local minima. For the 17 "dipeptides" for which the global minimum could be found, the *same* global minimum was found by Anneal-Conformer and by Macromodel.

Our Anneal-Conformer program shows considerable savings in CPU time for larger structures, such as Asp or Met. Note, however, that the simulated annealing random walk used by Anneal-Conformer means that multiple runs are needed to develop confidence that a global minimum is found. Table 2-2 indicates that out of 10 runs the global minimum is found 10 times out of 10 for the simple residues but less frequently for more complex ones.

2.2 Tracing the Random Walk

As stated earlier, simulated annealing allows a broad search of the whole conformation space and then, as the temperature is lowered, the search is focused only on the low-energy regions of the conformation space. Figure 2-7 shows a display of the trace of a random walk for the N-Ac-Ala-NCH_3 dipeptide model in an Anneal-Conformer run. One can note that the simulated annealing process scans the energy surface widely at high temperature (cf. the Ala Ramachandran plot). When the temperature is lowered, the search becomes concentrated in the low-energy regions and eventually at the global minimum.

2.3 Testing on Polyalanines

The performance of Anneal-Conformer with larger peptides was examined by carrying out conformation searches for a series of polyalanine models Ala_n where $n = 1$–10 (see Table 2-3). For Ala_1, Ala_2, and Ala_3 the global minimum found by Macromodel could be also located with Anneal-Conformer. Ala_3 was the largest poly-Ala that could be systematically searched (Cui, 1988). (Minimization of $\sim$30,000 starting geometries of Ala_3 [6 dihedrals] required two Microvax II *months* of CPU time.) The global minimum found by Anneal-Conformer for Ala_3 to Ala_{10} are all α-helices. Our previous publications also show that Anneal-Conformer was effective in finding α-helices for Ala_{20}, Ala_{40}, and Ala_{80} (Wilson and Cui, 1990).

Table 2-2. Simulated annealing of amino acid dipeptide models. Reprinted with permission of John Wiley & Sons from Wilson and Cui (1990).

DIPEPTIDE MODELS	# DIHEDRALS[a] (# CONFORMERS)	MACROMOD[b] CPUTIME	MACROMOD[c,d] GLOBAL MININA	ANNEALING[d,f] MINIMA (# RUNS)
ALA	2 (26)	0:02:00	−61.09	−61.09(10)
GLY	2 (29)	0:01:23	−61.88	−61.88(10)
PRO	3 (33)	0:01:39	5.72	5.72(10)
VAL	3 (129)	0:13:00	−58.71	−57.64(2)[f], −53.06(8)
PHE	4 (617)	2:40:00	−55.54	−55.54(2), −55.23(5) −52.94(3)
SER	4 (729)	0:49:26	−67.16	−67.16(4), −61.96(5), −61.94(1).
ASN	4 (586)	1:45:00	−117.65	−117.65(8), −104.83, −100.19(1)
THR	4 (619)	0:42:03	−68.08	−68.08(6), −61.44(1). −60.97(1),−46.49(1)
CYS	4 (684)	0:43:44	−64.88	−64.88(3), −64.16(4), −63.08(2), −59.92(1).
LEU	4 (597)	0:58:00	−64.73	−64.73(6), −63.13(4).
ILE	4 (608)	1:04:00	−56.23	−55.43(1)[f], −54.81(4), −48.67(5)
TRP	4 (640)	4:00:00	−67.15	−67.15(2), −65.68(1), −65.67(1),−65.66(1), −65.63(1), −63.07(2), −62.78(1), −61.67(1).
HIS	4(580)	2:00:00	−91.52	−91.52(3),−91.14(6) −90.90(1)
TYR	4 (618)	3:00:00	−51.22,	−51.22(3), −51.02(3) −50.53(1), −48.61(1) −47.84(2)
ASP	5 (3690)	5:33:46	−117.46	−117.46(1), −114.57(1), −105.84(1), −105.01(5), −100.21(2).
MET	5 (4091)	9:30:00	−69.14	−69.14(1), −67.04(3), −65.61(1),−65.17(1), −63.21(2), −62.90(1), −61.73(1).
GLN	5 (3626)	12:12:00	−115.81	−111.11(3)[f], −109.96(2), −104.94(4),−96.65 (1).
GLU	6 (23600)			−108.72(1), −102.96(3), −102.38(4),−101.56(2).
LYS	6 (18169)			−98.13(1), −95.58(1) −88.80(3), −84.78(1) −81.16(2), −80.68(1) −80.47(1)
ARG	7 (87245)			−53.11(3), −49.78(1), −28.02(6).

Notes:

a. The number in parentheses refers to the number of starting conformations generated with Macromodel by rotation at a 60° resolution about the indicated number of bonds (i.e., phi, psi, $omega_1$,...$omega_n$).

b. CPU time for batch minimization of indicated number of conformations on a VAX 8600 using Macromodel.

c. The lowest energy conformation from Macromodel in 11 cases was superimposable with that from Anneal-Conformer. For Val and Ile, Anneal-Conformer converged to the second lowest Macromodel minima. For Gln a conformation not found with Macromodel was found.

d. All amber energies are in Kjoul/mol.

e. CPU time for Anneal-Conformer is 1–3 minutes per run on a VAX 8600. Each amino acid "dipeptide" was run 10 times in order to provide the statistics. The random walk involved 250 steps at each of 50 temperatures.

f. Simulated annealing with full minimization at each step (i.e., 250 iterations of Block-Diagonal Newton–Raphson) led to convergence to the Macromodel global minimum but at the expense of CPU time.

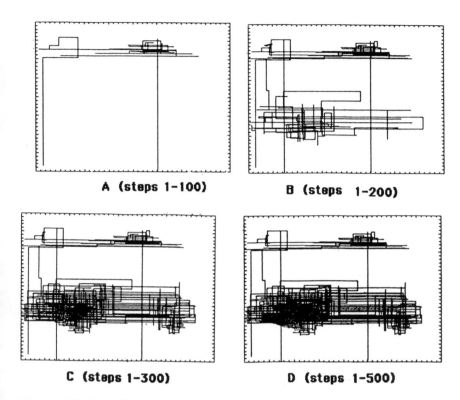

A (steps 1-100)   B (steps 1-200)

C (steps 1-300)   D (steps 1-500)

Figure 2-7. Steps of the random walk for the conformation search of (N-Ac)-Ala-N(CH$_3$) (cf. Table 2-2) plotted as a Ramachandran plot. A: Initial steps 1–100 (high temperature). B: Steps 1–200. C: Steps 1–300. D: Steps 1–500. At long time the system is trapped (freezes).

2.4 Conformation Search for Complex Peptides

We also tested Anneal-Conformer on more complex peptides such as met-enkephalin, bradykinin, mellitin, and a helical antifreeze peptide (Cui, 1988). Using the AMBER force field, polar side chains bury themselves in the backbone, preventing formation of helices. Thus lack of solvation in the force field is a limitation (Cui, 1988).

On the other hand, careful studies of the conformational search for met-enkephalin were quite successful. Metenkephalin is a pentapeptide (Tyr-Gly-Gly-Phe-Met) neurotransmitter whose conformation has beenstudied previously by most known conformational search methods. The global minimum-energy conformation of metenkephalin on the ECEPP energy surface was shown to be a folded conformation with a type II' β-turn at Gly3-Phe4 (Isogai et al., 1977). Because of its biological significance and

Table 2-3. Simulated annealing[a] of polyalanines: AcNH-(Ala)$_n$-CONHCH$_3$. Reprinted with permission of John Wiley & Sons from Wilson and Cui (1990).

n	# DIHEDRALS (10 runs)	CPUTIME	ENERGY(kj/mole)[b] (# of runs)	# of 1-13 H-BONDS[c] (helical residues)
2	4	0:24:12	−103.11(5), −97.31(5)	0
3	6	0:39:16	−151.83(3) −145.65(3) −140.54(2) −137.78(1) −108.70(1)	1
4	8	0:43:28	−210.83(7) −199.03(1) −194.81(1) −186.81(1)	2 2 1
5	10	0:59:57	−269.48(3) −256.42(1) −246.34(1) −236.47(1) −227.35(2) −219.74(1) −216.92(1)	3 1 1
6	12	2:34:04	−332.02(5) −311.85(3) −294.50(1) −289.49(1)	4 3 2 1
7	14	1:39:47	−394.95(5) −374.95(1) −366.90(1) −364.96(2) −336.08(1)	5 4 3 3
8	16	2:03:22	−458.42(4) −437.83(1) −411.40(1) −408.89(1) −409.17(1) −404.39(1) −374.28 (1)	6 5 4 4 3 3 2
9	18	2:20:18	−521.72(3) −500.82(1) −439.54(1) −436.86(1) −433.05(1) −419.47(1) −416.88(1) −395.67(1)	7 6 1 1 1
10	20	2:25:06	−585.63(7) −544.04(1) −533.41(1) −502.33(1)	8 5 4 1

Notes:
a.   The number of steps in the random walk was 250 at each temperature except for (ala)$_6$, where 500 steps were used.
b.   The lowest-energy conformation for cases n=3–20 was 100 % alpha helix confirmed by phi/psi angles the characteristic hydrogen-bonding pattern (1–13).
c.   The C=O of residue i is H-bonded to the N-H of residue i+4. A complete alpha helix shows n–2 hydrogen bonds. All helices are right-handed.

these accumulated theoretical studies, metenkephalin has become a model peptide for testing any new conformational search method.

In our initial efforts to validate Monte Carlo simulated annealing, we carried out a conformational search of metenkephalin on the AMBER potential surface (Wilson et al., 1988a, 1988b). We found the lowest-energy conformation of metenkephalin on the AMBER energy surface (Figure 2-8a) to be very similar to the global energy minimum conformation on the ECEPP energy surface (Figure 2-8b). Using our program Anneal-

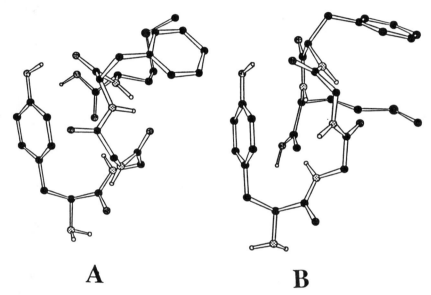

A   **B**

Figure 2-8. Global mimimum found for metenkephalin on the (A) AMBER energy surface and (B) ECEPP energy surface.

Conformer and the AMBER force field, we were able to efficiently locate several new families of metenkephalin conformations (Figure 2-9). Further studies of metenkephalin in our group lead to location of potential active conformations of this peptide (Montcalm et al., 1993).

Kawai et al. (1989) carried out a conformational search of metenkephalin on a simplified ECEPP surface by the simulated annealing method and demonstrated the efficiency of simulated annealing for ECEPP.

The efficiency of simulated annealing for conformational searching was challenged by a study from Scheraga's group (Nayeem et al., 1991). Nayeem did a conformational search of metenkephalin using his technique called "Monte Carlo minimization" (MCM) and compared it to simulated annealing. He concluded that, while simulated annealing converges to low

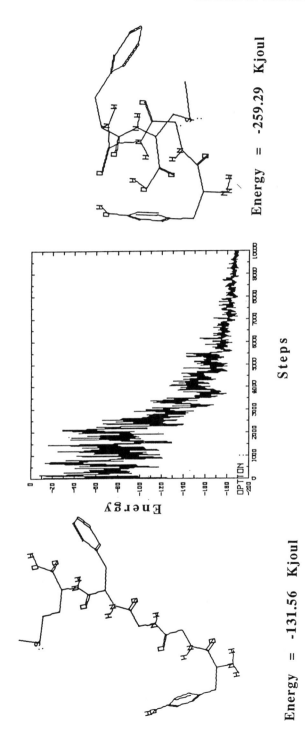

Figure 2-9. Anneal-Conformer (simulated annealing) run on metenkephalin model. Starting geometry, plot of energy vs. steps of the random walk and final geometry are shown. Cooling during the course of the run causes metenkephalin to freeze out in the global minimum shown in Figure 2-8A.

energy conformations significantly faster than MCM, it does not converge to a unique minimum whereas MCM does. This conclusion was refuted by a recent publication from the Kawai group (Okamoto et al., 1992). In this publication, the efficiency of simulated annealing for prediction of low-energy structures of metenkephalin was examined with the complete ECEPP force field. Eleven out of 40 runs of simulated annealing converged to a global minimum-energy conformation cluster. With an excellent convergency rate and considering that each simulated annealing run was 240 times faster than the corresponding MCM run, the authors rejected the previous conclusion from the Scheraga group that simulated annealing is inferior to MCM for the conformational search of peptides.

The efficiency of simulated annealing for conformational searching was further supported by additional studies of metenkephalin by several groups (Deng et al., 1991; Morales et al., 1991; Okamoto et al., 1992).

2.5 Additional Conformation Searches Using Simulated Annealing

Simulated annealing has also been applied to conformational searching of dipeptide models of Gly, Ala, Asp, pentaglycine and Leu-enkephalin (Morales et al., 1991); n-decaoctane and an analog of vasopressin (Deng, et al., 1991); (3S, 4S)-statin (Lelj and Cristinziano, 1991); cycloalkanes (Morely et al., 1992); analogs of thyrotropin releasing hormone (Garduno-Juarez and Perez-Neri, 1991); C-peptide of ribonuclease A (Okamoto et al., 1991); tetrahydroionone and octane (Wilson at al., 1991); and cycloheptadecane (Guarnieri et al.,1991).

3. Simulated Annealing with Quantum Mechanical Potentials

In the studies described in previous sections, simulated annealing was used for the conformation searching on energy surfaces of common molecular mechanics force fields. A number of reports extend the use of simulated annealing to the study of molecular structure with quantum mechanical potentials (Dutta and Bhattacharyya, 1990; Field, 1990; Dutta et al., 1991). The ability of simulated annealing methods to seek out the global energy minimum was demonstrated in the case of formamide and hydroperoxide on the MNDO potential surface (Dutta et al., 1991).

4. Simulated Annealing and Protein Folding

The power of simulated annealing demonstrated in conformational search studies of short peptides encouraged the use of this method for direct simulation of protein folding. Over the last four years, simulated annealing has been applied to the folding simulation of a number of small proteins and has also been applied to the study of helix–helix packing and global equilibrium configuration of supercoiled DNA. These studies can be grouped into two categories. First, the global folding of protein or DNA with a residue based folding potential. Second, optimization of side-chain conformations with an atomic-detailed force field.

The first such study appeared in a 1989 publication (Wilson and Doniach, 1989). In this report, a simplified force field for protein folding was formulated using a single point to present the side chains of amino acids. Simulated annealing was used to search for the global minimum of crambin, a small 32-residue globular protein, on the hypersurface of the simplified force field. The global minimum-energy conformation found had an RMS deviation of 4 Å from the known crystal structure. In addition, many structural features, such as secondary structure, the disulfide bridge and the packing of secondary structural elements were correctly predicted by this simulated folding procedure.

Snow also successfully used simulated annealing with a protein-folding potential on a 21-amino-acid fragment of BPTI (Snow, 1992). The latest publication on this simulated folding potential strategy is a by Garrett (Garrett et al., 1992). This simulated-annealing procedure found the same global minimum of a 22-residue protein previously identified by a previous, more elaborate simulation.

One study has appeared on the application of simulated annealing to the folding of DNA (Hao and Olson, 1989).

In a series of papers on protein packing and protein side-chain conformation optimization, Chou employed the ECEPP/2 force field, which treats the protein structure at atomic level (Chou and Carlacci, 1991; Chou, 1992). Other studies of protein side chains have been reported (Lee and Subbiah, 1991; Roitberg and Elber, 1991).

5. Optimization of Molecular Interactions

Simulated annealing has been quite successfully applied to the optimization of molecular interactions. An early demonstration of the utility of simulated annealing in this regard was Donnelly's study of the interaction of two propane molecules (Donnelly, 1987). Simulated annealing has been used for docking flexible ligands into the binding site on a protein (Goodsell and Olson, 1990) as well as for rigid-body docking with distance constraints (Yue, 1990). Docking of a rigid body without distance constraints has also recently been explored (Hart and Read, 1992). The related problem of crystal structure packing has been studied by Deem (Deem and Newsam, 1992) and Gdanitz (Gdanitz 1992).

6. Simulated Annealing of Rings and Loops

While peptides and proteins are often acyclic (open-chain systems), actual deformations usually take place in loops. Such partial movement of protein loops introduces a unique mathematical constraint on how the computer model may be manipulated. Deformation of a portion of a protein chain while keeping the rest of the model rigid is identical to the deformation of a simple ring. While deforming an *acyclic* flexible structure simply requires rotating a dihedral angle, rotating a dihedral angle in a *cyclic* structure will inevitably distort some of the bond lengths and bond angles of the molecule (Figure 2-10).

Figure 2-10. (a) Rotation of one ring bond distorts bond 1–2 and associated bond angles. Energy minimization is required after every step to fix the structure. (b) In our program Anneal-Ring, three atoms are exactly repositioned to create a new conformation with no bond-length or bond-angle distortion.

We have recently reported an algorithm for the smooth distortion of a ring model with no energy minimization, using a program called Anneal-Ring (Guarnieri, 1991). The mathematical consequences of transforming one good conformation of a flexible ring into another good conformation, i.e., keeping all of the bond lengths and bond angles of the molecule at standard values, are a set of six nonlinear, simultaneous equations in six unknowns (Go and Scheraga, 1970). Solving for the six dependent dihedrals to create a new conformation requires repositioning exactly three atoms of the old conformation. For example, in Figure 2-10 the exact recalculation and repositioning of atoms 2, 3, and 4 creates a new conformation with no bond-length or bond-angle distortion. Hence, unlike other loop deformation methods, no energy minimization during an Anneal-Ring run is required. We have reported simulated annealing studies of rings systems with up to 17 members (Guarnieri and Wilson, 1992). We have also used Anneal-Ring to deform a loop of the protein P-21 (Figure 2-11) (Guarneri, 1991).

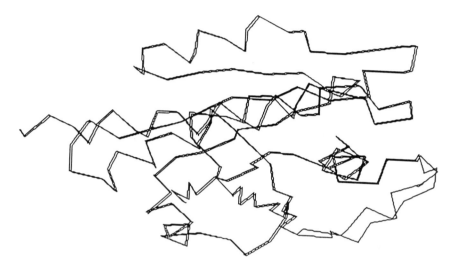

Figure 2-11. An α-carbon plot of the x-ray structure of the protein P-21 superimposed on a loop-deformed structure P-21. Exact deformation of the loop was carried out using Anneal-Ring.

7. New Simulated Annealing Methodology

In most of the studies described above, the Metropolis simulated-annealing procedure as originally proposed by Kirkpatrick was used. Some new methodology introduced over the last few years will now be discussed (Sibani et al., 1989; Somorjai, 1991; Sutter and Kalivas, 1991; Sylvain and Somorjai, 1991; Kalivas, 1992; Marinari and Parisi, 1992).

7.1 Molecular Dynamics Simulated Annealing

The most important feature of simulated annealing is the annealing schedule (cooling rate, number and size of the steps, etc.). In Monte Carlo simulated annealing, annealing is realized by changing the temperature factor in the probability density function (pdf) criterion. In molecular dynamics, temperature is related to the kinetic energy, which is represented by the velocities of the atoms in the dynamics integration procedure. Thus, in molecular dynamics simulated annealing, the annealing schedule may be realized by gradually scaling atomic velocities.

Brunger first implemented molecular dynamics simulated annealing in his molecular dynamics simulation package X-plore (Nilges et al., 1988; Brunger, 1991). The molecular dynamics was started at a very high temperature. At high temperature, atoms can move relatively freely because the large velocity overcomes the constraints imposed by the force field. This allows the conformational space to be explored globally. As the simulation proceeds, the temperature was decreased gradually by scaling (reducing) the atomic velocities. The constraints imposed by the potential force field dominate the force directing the movement of each atom. This results in a refinement of the structures locally. The molecular dynamics simulated annealing implemented in X-plore has since been widely used in x-ray crystal structure refinement and the derivation and refinement of solution conformation of proteins and peptides from the NOE distance constraints determined by NMR experiments.

Molecular dynamics simulated annealing has also been applied to the global conformational search of small molecules. Lelj applied this technique to the conformational study of a molecule of biological interest, (3S,4S)-statine and its diastereomer (Lelj and Cristinziano, 1991). He proposed a new criterion for monitoring conformational transition called "fractional energy fluctuation," as an alternative to the specific heat used in the original algorithm of Kirkpatrick (Kirkpatrick et al., 1983).

7.2 Hybrid Dynamics/Monte Carlo with Simulated Annealing

In a recent publication, a new general simulation algorithm was presented that combines aspects of molecular dynamics, Metropolis Monte Carlo sampling, and simulated annealing (Morely et al., 1992). In this method, trial conformations are generated by short bursts of concerted molecular dynamics, with the kinetic energy concentrated into one randomly selected bond rotation. The dynamics trajectories are allowed to continue as long as the energies of the new structures satisfy the Metropolis test. A simulated annealing protocol can build upon this combined sampling procedure by gradually decreasing the temperature of the molecular dynamics and the temperature factor of the Metropolis test.

7.3 Simulated Shocking

The effective use of temperature as a control factor in the Metropolis Monte Carlo process is the most important factor for the success of the simulated annealing method. This has inspired an attempt to exploit temperature in a new way. Von Freyberg (von Freyberg and Braun, 1991) has proposed a "Simulated Shocking" protocol for the efficient search for conformations of polypeptides. In this protocol, the temperature starts at normal room temperature ($\sim$300 K), then jumps between a very low temperature (5 K) and a very high temperature (2000 K). The high temperature is used when one finds the system is trapped in a local minimum, while the low temperature is used to thoroughly search of the vicinity of each new low-energy state. This protocol was used in the search of all the low-energy conformations of met-enkephalin using ECEPP/2. The author claims that the Metropolis Monte Carlo method with this variable temperature schedule works better than one with constant temperature (MCM) or continuously decreasing temperature (simulated annealing) for the search of all the low-energy conformations for peptides.

7.4 New Work on Anneal-Conformer

Using most conformation search techniques, an unsymmetrical molecule containing 12 flexible dihedrals is presently the limit (Saunders et al., 1990). Even using simulated annealing, only a few additional dihedrals are achievable. We have recently implemented a method that combines simulated annealing and importance sampling (Guarnieri, 1992). An initial simulated annealing run provides a knowledge-base in the form of what we have called a "Flex-Map" (Wilson and Guarnieri, 1991). These plots contain the complete population distribution for each bond as a function of temperature. For example, a run using a Phe dipeptide model shows the Flex-Map plots in Figure 2-12. The Flex-Map rapidly reveals occupied regions of dihedral space and "dead zones," where no conformations were found at any temperature. Since most conformation search methods sample from the whole space uniformly, "dead zone" sampling is obviously a waste of time. The implementation of importance sampling with simulated annealing was carried out by Guarnieri (1992) using a new algorithm that maps the simulated annealing probability density function onto the Flex-Map distributions. The program selects new conformations only from the populated regions, in a percentage related to the population of that region. Using the same simulated annealing control data, the results are shown in Figure 2-13. Faster lowering of the conformer energy and more rapid energy convergence is usually observed.

8. Conclusions

We have attempted in this review to cover all applications of simulated

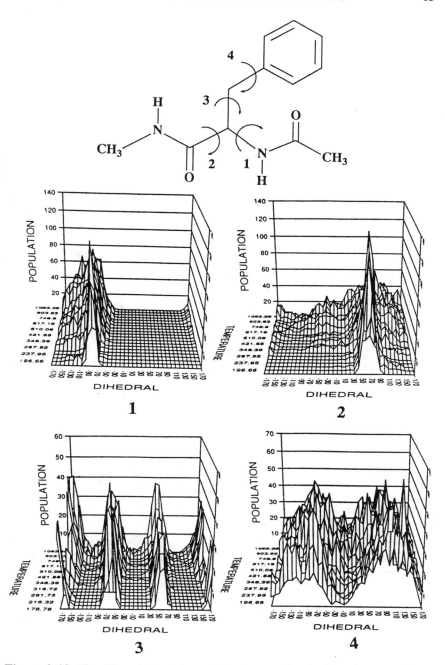

Figure 2-12. Flex-Map plots of the rotational states for the four degrees of free-dom of a Phe model. Plots show rotational state population vs. Θ at decreasing temperature.

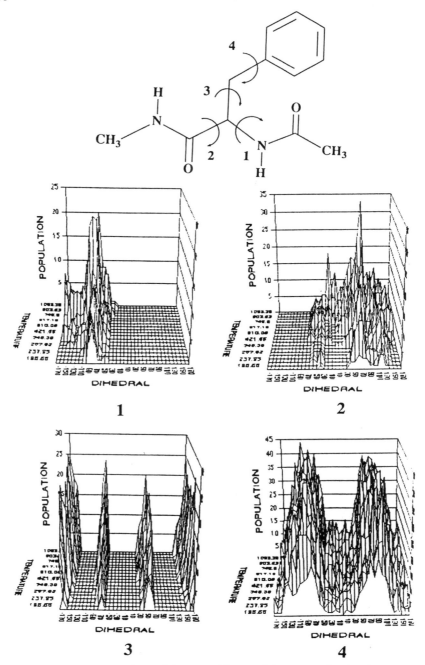

Figure 2-13. Flex-Map plots of the same Phe model generated using importance sampling. Note that Anneal-Conformer randomly picked angles from regions chosen "learned" from Figure 2-12.

annealing to conformation searching, protein folding, and related problems that have appeared in the literature up to early 1993. Our review was limited specifically to work on molecular geometry. We have emphasized our own work. We hope to have established that the technique of simulated annealing is a useful tool for optimization of global conformations, but in addition, it is evolving and being improved. We hope that readers will avail themselves of some of the commercial packages that have included simulated annealing routines, obtain source code from QCPE, from the authors, or write their own program.

REFERENCES

Barakat MT, Dean PM (1991): Molecular structure matching by Simulated Annealing III. The incorporation of null correspondences into the matching problem. *J Comput-Aided Mol Des* 5(2):107–17

Brunger AT and Karplus M (1991): Molecular Dynamics Simulations with Experimental Constraints. *Acc Chem Res* 24:54–61

Brunger AT (1991): Simulated Annealing in Crystallography. *Ann Rev Phys Chem* 42:197–223

Burkert U, Allinger N (1982): *Molecular Mechanics*. Washington, DC: American Chemical Society

Chang G, Guida WC, Still WC (1989): An internal-coordinate Monte Carlo method for searching conformational space. *J Amer Chem Soc* 111:4379

Chou KC, Carlacci L (1991): Simulated Annealing approach to the study of protein structures. *Protein Eng* 4:661–7

Chou KC (1992): Energy-optimized structure of antifreeze protein and its binding mechanism. *J Mol Biol* 223(2):509–17

Cui W (1988): *Computer-assisted Structure-Property Studies of Conformationally Flexible Molecules*. PhD Thesis, New York University

Deem MW, Newsam JM (1992): Framework crystal structure solution by Simulated Annealing: Test application to known zeolite structure. *J Am Chem Soc* 114(18):7189–98

Deng Q, Han Y, Lai L, Xu X, Tang Y, Hao M (1991): Application of Monte-Carlo Simulated Annealing to conformation analysis. *Chin Chem Lett* 2(10):809–12

Donnelly RA (1987): Geometry optimization by Simulated Annealing. *Chem Phys Lett* 136(3–4):274–8

Dutta P, Bhattacharyya SP (1990): A new strategy for the calculation of configuration interaction wave functions: Direct search involving Metropolis Simulated Annealing. *Phys Lett A* 148(6–7):331–7

Dutta P, Majumdar D, Bhattacharyya SP (1991): Global optimization of molecular geometry: A new avenue involving the use of Metropolis Simulated Annealing. *Chem Phys Lett* 181(4):293–7

Field MJ (1990): Simulated Annealing, classical molecular dynamics, and the Hartree–Fock method: The NDDO approximation. *Chem Phys Lett* 172(1):83–8

Garduno-Juarez R, Perez-Neri F (1991): Global minimum energy conformations of thyrotropin releasing hormone analogs by Simulated Annealing-II. *J Biomol Struct Dyn* 8(4):737–58

Garrett DG, Kastella K, Ferguson DM (1992): New results on protein folding from Simulated Annealing. *J Am Chem Soc* 114(16):6555–6

Gdanitz RJ (1992): Prediction of molecular crystal structures by Monte Carlo Simulated Annealing without reference to diffraction data. *Chem Phys Lett* 190(3–4):391–6

Go N, Scheraga H (1970): Ring closure and local conformational deformations of chain molecules. *Macromolecules* 3:178–187

Goodsell DS, Olson AJ (1990): Automated docking of substrates to proteins by Simulated Annealing. *Proteins: Struct, Funct, Genet* 8(3):195–202

Guarnieri F (1991) *Computer-assisted Studies of Flexible Molecules Using the Method of Simulated Annealing*. PhD Thesis, New York University

Guarnieri F, Cui W, Wilson SR (1991): ANNEAL-RING: A new algorithm for optimization of rings using Simulated Annealing. *J Chem Soc, Chem Commun* 1542–3

Guarnieri F, Wilson SR (1992): Simulated Annealing of rings using an exact ring closure algorithm. *Tetrahedron* 48(21):4271–82

Hao MH, Olson WK (1989): The global equilibrium configurations of supercoiled DNA. *Macromolecules* 22(8):3292–303

Hart TN, Read RJ (1992): A multiple-start Monte Carlo docking method for proteins. *Proteins: Struct, Funct, Genet* 13:206–222

Isogai Y, Nemethy G, Scheraga HA (1977): Enkephalin: Conformational analysis by means of empirical energy calculations. *Proc Natl Acad Sci US* 74: 414–418

Johnson MW (1988): *Simulated Annealing and Optimization*. Syracuse, NY: American Sciences Press

Kalivas JH (1992): Optimization using variations of Simulated Annealing. *Chemom Intell Lab Syst* 15(1):1–12

Kalos MH, Whitlock PA (1986): *Monte Carlo Methods, Volume I*. New York: John Wiley and Sons

Karasawa N, Goddard WA III (1988): Phase transitions of polymethylene single chains from Monte Carlo Simulated Annealing. *J Phys Chem* 92(20):5828–32

Kawai H, Kikuchi T, Okamoto Y (1989): A prediction of tertiary structures of peptide by the Monte Carlo Simulated Annealing method. *Protein Eng* 3(2):85–94

Kirkpatrick S, Gelatt Jr CD, Vecchi MP (1983): Optimization by Simulated Annealing. *Science* 220:671

Lee C, Subbiah S (1991): Prediction of protein side-chain conformation by packing optimization. *J Mol Biol* 217(2):373–88

Lelj F, Cristinziano PL (1991): Conformational energy minimization by Simulated Annealing using molecular dynamics: Some improvements to the monitoring procedure. *Biopolymers* 31(6):663–70

Lipton M, Still WC (1988): The multiple minimum problem in molecular modeling. Tree searching internal coordinate space. *J Comp Chem* 9(4):343–55

Marinari E, Parisi G (1992): Simulated tempering: a new Monte-Carlo scheme. *Europhys Lett* 19:451–8

Metropolis N, Rosenbluth AW, Rosenbluth MN, Teller AH (1953): Equation of state calculation by fast computing machines. *J Chem Phys* 21:1087–1092

Mohamadi F, Richards NGJ, Guida WC, Liskamp R, Lipton M, Caulfield C, Chang G, Hendrickson T, Still WC (1990): MacroModel—An integrated software system for modeling organic and bioorganic molecules using molecular mechanics. *J Comp Chem* 11:440–467

Montcalm T, Cui W, Zhao H, Guarnieri F, Wilson SR (1993): The low energy conformations of metenkephalin and their relevance to the membrane-bound solution and solid-state conformations. *Theochem* (in press)

Morales LB, Garduno-Juarez R, Romero D (1991): Applications of Simulated Annealing to the multiple-minima problem in small peptides. *J Biomol Struct Dyn* 8:721–35

Morely SD, Jackson DE, Saunders MR, Vinter JG (1992): DMC: A multifunctional hybrid dynamics/Monte Carlo simulation algorithm for the evaluation of conformational space. *J Comput Chem* 13(6):693–703

Nayeem A, Villa J, Scheraga HA (1991): A comparative study of the Simulated-Annealing and Monte Carlo-with-minimization approaches to the minimum-energy structures of polypeptides: Metenkephalin. *J Comput Chem* 12(5):594–605

Nilges MG, Angela M, Bruenger AT, Clore GM (1988): Determination of three-dimensional structures of proteins by Simulated Annealing with interproton distance restraints. Application to crambin, potato carboxypeptidase inhibitor and barley serine proteinase inhibitor 2. *Protein Eng* 2(1):27–38

Okamoto Y, Fukugita M, Nakataka T, Kawai H (1991): α-Helix folding by Monte Carlo Simulated Annealing in isolated C-peptide of ribonuclease A. *Protein Eng* 4(6):639–47

Okamoto Y, Kikuchi T, Kawai H (1992): Prediction of low-energy structures of Met-enkephalin by Monte Carlo Simulated Annealing. *Chem Lett* 7:1275–8

Roitberg A, Elber R (1991): Modeling side chains in peptides and proteins: application of the locally enhanced sampling and the Simulated Annealing methods to find minimum energy conformations. *J Chem Phys* 95(12):9277–87

Saunders M, Houk KN, Wu Y, Still WC, Lipton M, Chang G, Guida WC (1990): Conformations of cycloheptadecane. A comparison of methods for conformational searching. *J Amer Chem Soc* 112:1419–1427

Sibani P, Pedersen JM, Hoffmann KH, Salamon P (1989): Scaling concepts in Simulated Annealing. *Rep-Univ Copenhagen, Phys Lab* 89–9, 16 pp

Snow ME (1992): Powerful simulated-annealing algorithm locates global minimum of protein-folding potentials from multiple starting conformations. *J Comput Chem* 13(5):579–84

Somorjai RL (1991): Novel approach for computing the global minimum of proteins. 1. General concepts, methods, and approximations. *J Phys Chem* 95(10):4141–6

Sutter JM, Kalivas JH (1991): Convergence of generalized Simulated Annealing with variable step size with application towards parameter estimations of linear and nonlinear models. *Anal Chem* 63(20):2383–6

Sylvain M, Somorjai RL (1991): Novel approach for computing the global minimum of proteins. 2. One-dimensional test cases. *J Phys Chem* 95(10):4147–52

Von Freyberg B, Braun W (1991): Efficient search for all low-energy conformations of polypeptides by Monte Carlo methods. *J Comp Chem* 12(9):1065–1079

Weiner SJ, Kollman PA, Case DA, Singh UC, Ghio C, Alagona G, Profeta S Jr, Weiner P (1984): A new force field for molecular mechanical simulation of nucleic acids and proteins. *J Amer Chem Soc* 106:765–784

Wilson SR, Cui W (1990): Applications of Simulated Annealing to peptides. *Biopolymers* 29:255

Wilson SR, Cui W, Moskowitz J, Schmidt K (1988a): Conformational analysis of flexible molecules: Location of the global minimum energy conformation by the Simulated Annealing method. *Tetrahedron Lett* 4343–6

Wilson SR, Cui W, Moskowitz J, Schmidt K (1988b): The application of Simulated Annealing to problems of molecular mechanics. *Int Jour of Quant Chem* 22:611–617

Wilson SR, Cui W, Moskowitz J, Schmidt K (1991): Applications of Simulated Annealing to the conformational analysis of flexible molecules. *J Comput Chem* 12(3):342–9

Wilson C, Doniach S (1989): A computer model to dynamically simulate protein folding: studies with crambin. *Proteins: Struct, Funct, Genet* 6(2):193–209

Wilson SR, Guarnieri F (1991): Calculation of rotational states of flexible molecules using Simulated Annealing. *Tetrahedron Lett* 3601–4

Yue SY (1990): Distance-constrained molecular docking by Simulated Annealing. *Protein Eng* 4(2):177–84

Zimmerman SS, Pottle MS, Nemethy G, Scheraga HA (1977): Conformational analysis of the 20 naturally occuring amino acid residues using ECEPP. *Macromolecules* 10:1–9

3

Multiple-Start Monte Carlo Docking of Flexible Ligands

Trevor N. Hart and Randy J. Read

1. Introduction

The docking problem has received a great deal of attention over the last few years, with the appearance of a number of automated docking methods. These methods can be divided into two classes: shape-based methods, which use a simplified representation of the molecular surfaces as a means to guide docking, and energy-based methods, which search for good dockings based on favorable interaction energy. Each type of approach has its advantages and disadvantages. We give a detailed presentation of our method, which is essentially a combination of shape-based and energy-based approaches, and uses the method of simulated annealing to optimize dockings. We also present a recent study that examines a new approach to the problem of dealing with flexible ligands.

One of the goals of structural biochemistry is to understand how macromolecules bind with each other and to small molecules of biological interest. In many applications, the structure of the two interacting molecules is known, and we wish to predict whether or not they can form a reasonably stable complex and what is the exact binding mode if they do. Generally speaking, this is the docking problem. It has generated a tremendous amount of activity and interest in recent years and given rise to a wide range of different approaches to the problem. In this review, we will outline some of these approaches and then focus on our own method: energy-driven, multiple-start simulated annealing.

Why is molecular docking important? Basically, there are three main applications for automated docking techniques. First is to understand and

The Protein Folding Problem and Tertiary Structure Prediction
K. Merz, Jr. and S. Le Grand, Editors
© Birkhäuser Boston 1994

predict how a binary complex is formed given structural knowledge of the two molecules forming the complex. In biophysical applications, the two molecules are usually either a protein and a small molecule (most often an inhibitor), or a protein or peptide and another protein. Normally, we have crystal or modeled structures of the two members of the complex, as well as other information about where the binding may occur (e.g., knowledge about residues important for binding from point mutation studies).

A second application is to use docking studies as a tool to gauge our understanding of which factors are important in intermolecular interactions. Since docking involves ranking different modes of interaction, comparison with experiment gives us a means of assessing the validity of a given ranking method. To date, no ranking method used in docking techniques has totally succeeded in predicting the correct binding modes in all cases. Thus, docking experiments can give us insight into which effects are most important for correct prediction of binding modes.

The third application, and one that has attracted considerable attention, is the use of docking in the design of new ligands for a given binding site on a protein. These docking problems fall into two classes: modification of a known ligand in order to achieve better binding affinity, and discovery of an entirely new ligand without reference to any known ligand. In the first class, the modified ligand will be expected to bind in a similar mode to the original ligand; in this case, the new binding mode can be determined by the use of standard molecular mechanics or molecular dynamics methods and does not require automated docking techniques. In the second class, however, where no knowledge of the binding mode is known, automated docking methods are essential.

Thus, the docking problem may be defined as the problem of finding the mode of optimal complementarity for two molecules. Since we generally have limited information about how the molecules might interact (except, possibly, a general binding region on one molecule, typically the protein), we have to be willing to consider many possible binding modes in the search. There are several factors that contribute to the complexity of this problem. First, the size and complexity of the molecules of interest (proteins) make for a large number of possible ways for two molecules to be put together with at least partial complementarity. Second, all current methods for evaluating molecular complementarity are fundamentally local in nature; thus, an initial guess will have to be very close to the optimal binding mode in order to recognize it as such. The third problem is that the size of the problem—essentially the number of possible ways of orienting two molecules—is much greater than current computer technology; "brute force" methods are at best very inefficient and at worst totally impractical.

Another difficulty is finding an appropriate measure of molecular complementarity. Since the true binding mode is determined by the free energy of interaction, the best possible method would be to approximate this quantity in some way. Unfortunately, the factors that contribute to free energy include bulk effects, such as solvent effects, as well as thermodynamic effects, such as entropy, which are difficult to model in a computationally cheap way. An alternative is to evaluate dockings strictly in terms of the interaction energy, usually in terms of a pairwise atom model. This is a common method, although it suffers in particular from an inability to account for effects of a solvent. Another approach is to evaluate dockings in terms of matching of shapes, using a simplified representation of the molecular surfaces. During the search portion of the docking procedure, complementary shapes are matched without particular regard to the chemical properties of the molecules; screening in terms of interaction energy is then done afterwards. In either approach, there is a compromise that will give rise to incorrect results in some circumstances.

Thus, there are two fundamental aspects to the docking problem: the problem of evaluating molecular complementarity, and the problem of searching through the possible binding modes for an optimal one. Given a means to evaluate complementarity, the second is basically a well-defined optimization problem. The nature and scope of this optimization problem depends largely on the method of evaluation. For example, if complementarity is evaluated in terms of surface shapes, it becomes possible to perform a systematic search by restricting which parts of the surface are allowed to interact. On the other hand, if complementarity is evaluated by interaction energy, the optimization problem becomes more complex, and more sophisticated algorithms, such as simulated annealing, are required.

In some docking applications, the ligand has some inherent flexibility in solution, and the binding mode may involve a considerably different conformation from the one we may find from crystal structures or modeling studies. The docking of these flexible ligands poses particular problems. If the two molecules are considered rigid, then the docking problem has six degrees of freedom (three translational and three rotational), of one molecule relative to the other. If the two molecules together have N internal degrees of freedom, then the docking problem will have $6 + N$ degrees of freedom. There are several ways to approach this problem. One is to allow the internal degrees of freedom to vary just as the rotational and translational ones do and perform optimization in the full $6 + N$ dimensions. This is impractical if N is at all large (> 10) but is difficult for even small N. Another approach is to find all the low-energy conformations by independent modeling and then dock each of these conformations as rigid molecules.

This approach is much simpler than the first, although it has the risk of missing intermediate-energy conformations that might bind particularly well. Yet another approach is to divide the molecule up into approximately rigid parts, each of which is to be docked as an independent molecule, and then reassemble the molecule in the bound state from the independent parts, in whichever conformation(s) it is found. The suitability of the conformation would then be assessed after-the-fact. This method relies on each group having a suitable binding affinity and also on a means of assembling the final molecule from the various groups.

The remainder of this review, after this introduction, is divided into three parts. The next section will contain a general review of the docking field, with an eye to understanding the diversity of different approaches that have been used. The third section will be a detailed discussion of our own method. Here we outline the basic steps and decisions involved in performing a simulation. In the fourth section, we will present some recent studies on the docking of flexible ligands, where we attempt to evaluate a particular approach to this problem.

2. A Brief Overview of Docking Methods

In recent years there has been a great deal of activity in developing docking methods, giving rise to a wide variety of approaches to the problem, each with its advantages and disadvantages. However, there are two basic features to any docking method: there must be practical criteria for distinguishing good dockings from poor ones, and there must be some algorithm for finding good dockings, based on those criteria. In this section, we will give an overview of the current state of the docking field. In particular, we will discuss the basic principles of some of the more important methods and classify them according to different approaches to these two aspects of the problem.

Choosing criteria for evaluating molecular configurations in a docking procedure is perhaps the most difficult part of developing a docking method. This is because evaluations usually have to be calculated very many times and therefore need to be computationally cheap. Therefore, the challenge is to put as much chemical information into the evaluation method as possible, while keeping the computation time short. In many cases, several criteria are used, so that one criterion is used in generating collection of candidates, while a second is implemented to select from among them.

Ultimately, the best evaluation method would be one that could be directly correlated with the experimental binding free energy, since this is the physical quantity that determines the effectiveness of binding. Unfortunately, the only computational method of estimating the free energy in solution, thermodynamic integration using molecular dynamics simulations, is far too computationally costly to be used in automated docking methods in the near future. Hence, we must look for some other way of characterizing good molecular interactions. The primary evaluation methods have been based on complementarity of molecular surfaces, use of simple chemical complementarity (defining hydrophobic and hydrogen-bonding regions), and, more recently, pairwise atom and grid empirical energy calculations. Each of the methods has its advantages and disadvantages in terms of computational cost and faithfulness to the actual molecular interaction.

2.1 Shape-based Methods

By far the most common method of evaluating dockings has been using some form of comparing molecular shapes. In these methods, the primary procedure is first to evaluate dockings (whether generated systematically or by some other method) for complementarity of shape and then to perform further evaluations, based on chemical information, on the subset that shows some degree of complementarity. One reason these methods have been so widely used is that only recently have computers become fast enough for docking methods to be based primarily on energy calculations. However, this approach has another key advantage: the number of possible binding modes can be significantly reduced by using a simplified model for the shapes of the receptor and ligand. The challenge of developing such models is to make complementarity simple to evaluate while keeping representation sufficiently faithful so that good dockings might not be inadvertently rejected.

Almost all methods of shape comparison are based, at least in part, on the concept of the solvent-accessible surface of a molecule, introduced first by Lee and Richards (1971) and later modified by Connolly (1983a, 1983b). The solvent-accessible surface, under Connolly's treatment, is a set of discrete points near the van der Waals surface of a molecule where a solvent molecule (usually a spherical atom, with the van der Waals radius of a water molecule) would be able to make contact with the molecule without clashing sterically. The number of surface points used is determined by the sampling density; in principle, the surface may be sampled as coarsely or as finely as we wish. To evaluate the complementarity of an arbitrary ligand with a receptor, we first calculate the solvent-accessible surface for the receptor (this is done only once at the beginning) and then count how many atoms of the ligand make contact with at least one surface point. One problem is that the ligand atoms will generally be of different sizes than the probe molecule used to generate the surface. This is handled by counting atoms as making contact if they are within their van der Waals radius of the surface, plus or minus a small distance tolerance.

Although the above scheme illustrates the principle of how a solvent-accessible surface can be used, it is not very practical. First, it requires that the distance of each ligand atom to each surface point be calculated, involving considerable computational cost. Second, this scheme does not take advantage of the principal virtue of using surfaces: the ability to reduce the number of possible binding modes that need to be considered. How do surface-based methods make this possible? First, consider the problem of systematic search: there are six degrees of freedom (three rotational plus three translational) associated with docking one molecule to another. If a surface representation can perform comparisons requiring the surfaces to be in contact this will reduce the degrees of freedom to five (two for the surface of each molecule plus one for the twist angle). Furthermore, since meaningful dockings will make mutual contact at many surface points, every pair of surface points need not be considered. If we know the receptor binding location fairly well, for example, we can require the ligand to make contact with one particular point; this will further reduce the degrees of freedom to three (two for the ligand surface plus one for twist angle). It is the general method of approach to this problem that distinguishes among shape-based docking methods, which we now discuss in detail.

This scheme has been implemented, originally by Connolly (1986) and later, in a much more sophisticated way, by Bacon and Moult (1992). In Connolly's implementation, he used his representation of the solvent-accessible surface to derive a method for defining and comparing local shape at any point on the surface. A sphere is placed at each surface point and the volume of the sphere inside the surface is calculated. This value, together with a unit vector that extends from the centroid of the enclosed spherical volume to the surface point, is used to characterize the local shape. Complementarity of two surfaces is measured by comparing the local shape functions; there will be perfect (local) complementarity if the unit vectors are oppositely aligned and the volumes sum to $\frac{4}{3} \pi R^3$. To perform dockings, the method only searches for complementary matches at "knobs and depressions"—points of local maximum or minimum volume—which brings the task of systematically comparing surface points down to a manageable size. Since a single match does not uniquely define a rigid body transformation, several pairs of points are compared simultaneously. The best matches are further screened in detail for bad steric contacts and then ranked in terms of the total amount of surface area in contact, which is taken as the final method for ranking the dockings.

The method of Bacon and Moult (1992) is similar to Connolly's, but uses a much more detailed representation of the surfaces and performs much more sophisticated and extensive screening. Their method is based on the concept of a *web*, which uses the Connolly surface representation to define a set of surface points in a local coordinate system. The local coordinate system is built around a given point on the surface and parameterized by radius and angle (r, θ). One web is centered at a particular point on the receptor (typically, the center of the binding site) and a collection of webs are defined on the ligand, centered on a set of points covering the surface of the ligand. A systematic search is then performed by matching the ligand and web centers. This search involves the three degrees of freedom: two for the surface points and one for the twist angle. Once a web center on the ligand and a twist angle is selected, the surface points can be matched up one-for-one in the local coordinate system and are oriented by a least-squares method. Selected dockings are then screened first for steric clash and then ranked according to electrostatic complementarity, using an image-charge model to handle solvent effects.

Another recent method based directly on the solvent-accessible surface is the method of Jiang and Kim (1991). Instead of using the surface to reduce the degrees of freedom, they represent the molecular surfaces on a regular three-dimensional grid, making the translation searches very efficient. The representation starts with the Connolly surface representation and assigns the accessible surface points to cubes in a regular grid. For a given relative orientation of the molecules, the translation search is performed by translating one molecule relative to the other in units commensurate with the grid. A discrete set of relative orientations are systematically generated for the ligand molecule, at which time a new grid representation must be generated. Complementarity is tested by counting the number of surface cubes that overlap, while bad steric contacts are detected by overlap of "volume" cubes, cubes that represent the inside of the molecules. Dockings are ranked by the total surface area in contact. Further ranking can be performed by assigning chemical properties to the each cube (charged, H-bond donor/acceptor, hydrophobic) and scoring dockings according to the favorability or unfavorability of the overlapping cubes on each surface.

The final shape-based method we will discuss is the method developed by Kuntz and co-workers (Kuntz et al., 1982; DesJarlais et al., 1986, 1988; Shoichet and Kuntz, 1991). The method starts with the Connolly surface of the receptor and generates a representation of the receptor site as a collection of spheres that make contact with the surface but lie outside of the receptor molecule. The ligand is represented as a collection of spheres centered on the ligand atoms. The idea is that, having represented the receptor site as a complementary image of the receptor molecule, the problem of docking the ligand becomes the simpler one of finding like, rather than complementary, shapes. Although the concepts behind generating the collections of spheres is simple, the actual process is complex due to the need to generate spheres of optimal number and size. The distances between the sphere centers in both the ligand and receptor collections are saved, and docking is performed by systematically pairing up ligand and receptor spheres and comparing intersphere distances. Highly ranking dockings are then rotated into the receptor site and screened for bad steric contact and favorable chemical interactions. This method has been under continual development over the past 10 years, particularly with respect to the method of screening the dockings. Originally, screening was performed by eliminating bad steric contacts and counting hydrogen-bond pairings. More recent versions include optimizing dockings with molecular mechanics, and testing for charge complementarity using Poisson–Boltzmann methods (Gilson and Honig, 1988).

2.2 Energy-based Methods

In contrast to the shape-based methods are the docking methods based on energy calculations. These methods have only recently become feasible due to improvements in computer technology. Energy-based methods necessarily have to be more brute-force in nature than shape-based methods because we do not have a molecular representation that identifies points of interaction. On the other hand, they arguably represent a more faithful way of evaluating dockings since they are more closely related to the free energy, although many important effects, such as the influence of solvent, are usually neglected. Furthermore, they are in a better position to deal with conformational changes than shape-based methods, which tend to lose internal bond information in the process of molecular shape representation. There are also differences in the typical docking procedure between the two types of methods. While shape-based methods usually involve a multistep screening of dockings, in which several different evaluation techniques are used, energy-based methods generally use a single-step procedure based solely on the interaction energy between the two molecules as the measure of goodness of fit. At the present time, there have been no implementations of systematic search using energy evaluations; it is perhaps only marginally feasible with current computer technology. Instead, current energy-based methods have used the more efficient method of Monte Carlo simulated annealing to search for favorable dockings.

One of the principal problems in energy-based docking is the complexity of the molecular surfaces. This complexity means that there are a great number of local energy minima, most of relatively high energy, in the six-dimensional space of rigid rotations and translations of the ligand. If we start the ligand at an arbitrary state (by state, we mean a given position and orientation) and perform minimization along the energy gradient, we will likely end up at one of these high-energy local minima. Many of these minima involve some degree of complementarity and might be of negative energy, although hardly low enough to overcome the solvation energy and loss of entropy in binding. Hence, the problem is to search for the small number of very low energy states (among them, the global minimum) without becoming trapped in higher-energy local minima.

The method of simulated annealing has been very successful in searching for global minima in optimization problems involving many local minima. The method, as the name implies, simulates the process of slowly cooling a glass in order to produce a more stable, lower-energy state. The implementation of simulated annealing uses the Monte Carlo method of Metropolis to generate a sequence of states that follow a Boltzmann probability distribution $p(x) = \exp(-E(x)/kT)$, where $E(x)$ is the energy of the system in state x, k is Boltzmann's constant, and T is the system temperature. Each state in the sequence is generated by a small random change from the previous state, and is accepted on the basis of the change in energy and the parameter T (the details of the implementation will be discussed in the next section). The process proceeds in cycles, with each cycle involving a sequence of Monte Carlo steps generated at a fixed temperature. The temperature for the initial cycles is normally quite high, allowing the system to move over large energy barriers between local minima. At later cycles, the temperature is systematically lowered to the point where only very small increases in energy are allowed. The procedure of going from high to low temperatures means that a large region of state space can be sampled without becoming trapped in local minima until the low-temperature phase is reached.

Goodsell and Olson (1990) were the first to present a general docking method based on simulated annealing. One of the key features of their method is the use of precalculated energy grids, using the method of Goodford (1985) to perform the energy calculations. During the docking procedure the energy for each ligand atom is calculated by grid interpolation, involving considerable computational savings over pairwise atom calculations. The docking procedure consists of a single, long simulated annealing run from a given starting position outside the receptor. The annealing protocol consists of a sequence of constant-temperature cycles in which the lowest-energy configuration found in each cycle is used as the starting point for the subsequent cycle. Their method allows for some conformational freedom of the ligand, although no internal bonded energy is used. In their tests,they used several different starting positions in order to duplicate the final docking, since simulated annealing, unlike a systematic search, does not guarantee finding the global energy minimum.

Our own method (Hart and Read, 1992) uses a multiple start approach to energy-based simulated annealing. Instead of using a small number of long simulated annealing runs, we use a large number of short runs. The starting state of the ligand for each run is generated completely randomly within a specified cube in space—in essence, our intention is to generate the results of a systematic search using a random search method. To deal with bad initial overlaps with the receptor molecule, the ligand is relaxed by simulated annealing using a score function that evaluates steric contact using a precalculated grid. The energy-based simulated annealing uses an empirical pairwise atom energy calculation with a distance cutoff. Both the ligand and receptor are rigid; however, ligand flexibility can be handled by dividing the ligand into approximately rigid fragments and subsequently assembling them to form molecules (unpublished results). Final dockings are saved if they fall below a specified energy cutoff.

Caflisch et al. (1992) have applied simulated annealing techniques in studying the docking of flexible peptides with proteins. They use a two-step procedure involving long simulated annealing runs from a single starting position. Energy calculations are performed with pairwise calculation including van der Waals, hydrogen bond and electrostatic interactions as well as bonded interactions in the ligand. As with Goodsell and Olson's approach, docking is performed from an arbitrary starting position.

Which method is better, evaluating dockings by shape complementarity or interaction energy? Each method clearly has its advantages; shape methods are able to reduce the complexity of the search problem and allow for systematic search, while energy methods give a more realistic means of evaluating dockings and are better prepared to deal with flexibility. On the other hand some of these problems can be overcome; molecular flexibility has been approached using shape-based methods (DesJarlais et al., 1986), while our energy-based method can duplicate a systematic search. Ultimately, there is room for both approaches.

One consideration in choosing a docking method is the type of docking problem one is interested in solving. There are two principal applications of docking: determining the mode of binding in protein–protein interactions, and determining the binding of small molecules to proteins for the purposes of structure-based drug design. The first problem is the more combinatorially difficult of the two, since the ligand molecule is much larger and the mode of binding often unknown. Additional nonstructural experimental information (such as results from site-directed mutagenesis studies) and intelligent guesswork are often helpful in bringing the problem down to a manageable size. Shape-based methods tend to be better for this problem because of their ability to reduce the size of the combinatorial problem. Energy-based methods are able to handle protein–protein docking problems, however, and have the advantage that empirical energy functions do work well for protein–protein interactions.

The task of docking small molecules to proteins is in some respects easier than protein–protein docking; we generally know where the binding site on the receptor is located, and the smaller size of the ligand means that the overall size of the combinatorial problem is less. On the other hand, small molecules are often quite flexible, which increases the number of degrees of freedom in the search and tends to pose problems for shape-based methods. Both methods suffer from the fact that empirical energy functions are less effective for nonpeptide molecules and that drug–protein interactions can involve considerable hydrophobic interactions, which many empirical energy functions do not handle at all.

What is the current status and future of docking methods? All of the above docking methods are successful, in most cases, at locating the correct binding mode and ranking it among the most favorable dockings found. This means that, basically, the search problem has been solved; these methods can deal with the combinatorial problem of searching through the many possible binding modes for the correct one. On the other hand, the fact that the binding mode of highest ranking is often not correct means that considerable improvement in the evaluation of dockings is required. Since both energy-based and shape-based methods use energy calculations to determine final dockings, it is here that improvement is needed. The main problem is the need to include effects of solvent and hydrophobic interactions in the calculations without introducing great computational cost. Fortunately, recent efforts to develop surface-area dependent terms in energy calculations (Eisenberg and McLachlan, 1986; Still et al., 1990; Wesson and Eisenberg, 1992) give the hope that the problems of dealing with hydrophobic and solvent effects may be to some degree overcome. Other significant effects include accounting for entropy loss due to the freezing of internal degrees of freedom of the ligand and solvent screening of charges.

Another significant problem is that of handling molecular flexibility. There are two aspects to this problem: dealing with small conformational shifts as the two molecules bind together, and large conformational changes in either flexible side chains of the receptor or the ligand itself. The first problem might be handled by using a "soft" van der Waals potential function, which approximate flexibility by using a form of the potential function that increases less rapidly, as the atoms come into close contact, than the standard R^{-12} or R^{-9} terms that are commonly used. This approach suffers from its neglect of stereochemistry, since an atom's ability to change position depends on the relative positions of its internal bonded and nonbonded contacts.

The second problem is a much greater one, however, since large conformational changes involve overall changes in the shape of the molecule, and cannot be handled using soft potentials or other such means. On the other hand, introducing all the conformational degrees of freedom on the ligand and receptor to the docking process is computationally beyond our means. However, it is possible to deal with a limited number of internal degrees of freedom in the ligand, as has been demonstrated by Goodsell and Olson (1990) and particularly Caflisch et al. (1992). Our method of docking the molecule in rigid pieces and assembling them after docking has shown some promise, although the automated assembly process still needs to be perfected. Another alternative is to perform detailed modeling of the ligand to determine its most stable conformations in solution and dock each one independently as a rigid molecule. This approach relies on the ligand to bind in a reasonably stable internal conformation, which is likely true in most cases.

2.3 Automated De Novo Drug Design

Although our primary concern is with docking, it is important to discuss the related problem of automated *de novo* drug design, since it has an important influence on the design of docking methods. One of the main applications of automated docking is the design of new ligands for a specific protein binding site of known structure. The primary method of developing new drugs is based on selectively modifying compounds that display binding activity but for various reasons are not suitable as therapeutic agents. At present, there are fairly well developed methods for systematically modifying ligands, usually based on comparing binding affinities and structural data of related compounds (QSAR). When a crystallographic structure of the receptor is available, modeling is usually performed with standard molecular mechanics and molecular dynamics techniques, since a modified ligand should bind in approximately the same binding mode as the original ligand, which is usually known. In these cases, automated docking methods play a limited role, except perhaps to confirm results that have been found by other methods. Automated docking can play a much greater role when we wish to determine the binding of a ligand that is not directly related to other known ligands. Here we wish to use automated docking to determine the correct binding mode of the ligand and assess its overall effectiveness in binding, compared with other ligands.

Given an effective docking method, the main problem becomes one of designing an appropriate ligand to test. Since there are an enormous number of potential candidates, and since the actual docking process is automated, it makes sense to also utilize the computer in the design process. Basically there have been two approaches to this problem: perform a systematic search through a structural database of small molecules, docking each one, or docking molecular fragments that are subsequently assembled into ligands by some automated process.

The first approach has been implemented in the laboratory of Kuntz (DesJarlais et al., 1988). They use the Cambridge Structural Database (Allen et al., 1983) as a source of chemical structures, and perform the docking using their shape-based docking method with screening performed by energy calculation. The searches cover approximately 2,700 chemical structures from the database. One disadvantage of this method is that the structures are docked as rigid bodies, in the crystallographic conformation found in the database; thus, it does not deal with the problem of ligand flexibility.

The second approach was, in spirit at least, initiated by Goodford (1985). His approach to drug design is to understand the role of functional groups in binding to a receptor site by graphically displaying the potential of interaction of a group as grid contours around the receptor site. The functional groups are represented essentially as spherical probes, and the interaction is calculated using hydrogen-bond potential functions. The design of a ligand proceeds graphically by examining where various functional groups are of optimal energy and building up a ligand by joining the functional groups together.

Moon and Howe (1991) have developed an automated design method based on building peptides into a receptor site from a specified starting point. The method uses an extensive library of amino acids, with each amino acid represented by a collection of low-energy conformations. The program sequentially adds residues from the starting point by systematically evaluating each member of the library in turn. At each stage, only a small number of the highest-scoring peptides are kept. Scoring is done in terms of the internal energy of the peptide and the interaction energy with the receptor, and includes the use of buried surface area terms to approximate solvent effects. The starting position is represented by an acetyl group, which determines the position and direction of the peptide growth. Automated docking is needed only to define this starting position.

Recently, an automated fragment assembly method has been developed by Böhm (1992a, 1992b). A simple, rule-based docking method is used to dock small molecular fragments from a database into a specified region of the receptor. Atoms of the ligand are superimposed on "interaction sites" for hydrophobic and hydrogen bond interactions. Suitable matches of the interaction site with the atoms are then checked for van der Waals contact with the receptor but are otherwise not ranked. The fragments are then assembled into larger molecules using small linker fragments to join them together. The method is able to join fragments to form a new molecule, or to add fragments to an existing molecule that has been docked by other means.

Our approach to fragment assembly is along similar lines. Our methods involve first docking fragments in such a way that all significant energy minima for each fragment are located. The docked fragments are then fed into another program that searches for ways to join them together, given a set of joining points on each fragment as defined in a library. This program then acts recursively to form pairs, triples, etc. of fragments into larger molecules. We originally outlined this approach to fragment assembly in the paper reporting our Monte Carlo docking method (Hart and Read, 1992). Since then, we have developed a program to perform the automatic fragment assembly, although at present it suffers from several technical problems and remains incomplete.

The goal of designing ligands by automated fragment assembly has been a major motivating factor in the development of our docking method. The requirements of a docking method for this task are (1) it must be able to locate *all* significant binding modes for the fragment in the receptor site, (2) it should be able to rank them accurately according to the energy of interaction, and (3) it should be completely free of user bias in the initial positioning of the dockings. In the next section, we will present our solution to this problem that, in addition to fulfilling these requirements, also serves as a general docking method that can handle many different kinds of docking problems.

3. Multiple-Start Monte Carlo Docking

In this section, we give an extensive description of our docking method and present some examples of its use. We first give a detailed overview of our method, including a description of the programs in the current implementation and the preparation necessary to perform a docking simulation. We will then present some examples illustrating the benefits and problems of the method.

The basic technique we use for searching and optimizing the ligand-receptor interaction is Monte Carlo simulated annealing, an optimization method specifically designed for dealing with problems involving many local minima (for example, see Kirkpatrick et al., 1983). To motivate this approach, consider the problem of finding the lowest-energy binding mode for a ligand by positioning it outside the receptor and performing standard energy minimization. Unless the initial positioning is very near to the correct one, the ligand will become trapped in a local energy minimum near the surface of the receptor, because it will simply proceed down along the local gradient until a local minimum is reached, and then stop. Simulated annealing, however, is able to overcome this problem by allowing the ligand to randomly proceed against the local gradient, and thus move from the attracting region of one local minimum to another.

Simulated annealing has its origins in the Monte Carlo methods used in statistical mechanics. The fundamental principle of statistical mechanics, which is concerned with the bulk properties of many-body systems, is that those properties can be expressed as weighted averages over all the microscopic states of the system. The relative weighting of each state depends on the energy of the state E_s and the temperature of the system T, and is given by the familiar Boltzmann factor, $\exp(-E_s/kT)$. In all but the simplest systems, it is impossible to explicitly calculate these averages; Monte Carlo methods overcome this problem by performing a random sampling that favors those states with the highest weight and therefore make the largest contribution to the average. In essence, Monte Carlo sampling simulates a bulk system near equilibrium, where the system spends time proportional to its Boltzmann factor in each state.

The implementation of the Monte Carlo method, due to Metropolis (Metropolis et al., 1953), is amazingly simple. Assume the system is at a temperature T. The system starts in a state s_i and a new state s' is randomly generated by making a small change in s_i. For a molecular system, this might involve a small translation and rotation from the initial position. The difference in energy ΔE of the two states is calculated and the new state for the system s_{i+1} is determined by the following rule: If $\Delta E < 0$, then take s' as the new state, i.e., set $s_{i+1} = s'$; if $\Delta E > 0$, then take s' as the new state according to the probability distribution given by $p = \exp(-\Delta E/kT)$; otherwise use the current state as the new state, i.e., set $s_{i+1} = s_i$. The second case is implemented by generating a random number $0 \leq r < 1$ and accepting the state s' if $r < \exp(-\Delta E/kT)$. Thus, the new state is likely to be accepted if $\Delta E \ll kT$, but unlikely to be accepted if $\Delta E \gg kT$. This procedure generates a sequence of states s_i that can be shown to be statistically distributed according to the relative Boltzmann weighting $\exp(-\Delta E/kT)$. A Monte Carlo calculation is performed by starting the system in some state, performing a number of preliminary runs so the system can achieve equilibrium, and then calculating average quantities by simply summing them over the successive states and dividing by the number of states sampled. No Boltzmann weighting is required in the average since the method samples states according to the relative Boltzmann distribution.

The Monte Carlo method allows us to simulate a thermodynamic system at a given temperature as a markov process. If we want to find the global energy minimum of the system, the ground state, we can use an analogous procedure to that used when forming crystals by slow cooling; we start with a high temperature and perform Monte Carlo runs, successively lowering the temperature until the system becomes "frozen in" at the ground state. This method is effective because the high temperature runs allow the system to avoid becoming trapped in higher-energy local minima by randomly jumping energy barriers between the minima's attracting regions. This is the method of simulated annealing. This procedure can be easily adopted for any optimization problem that involves finding a maximum or minimum of a score function $\xi(s)$, by simply replacing the energy function E by ξ. The temperature T now becomes a "randomizing" parameter and we can set $k = 1$.

In our approach to molecular docking, we use a modified simulated annealing scheme, due to our desire to locate all low-energy local minima. We do this by using many randomly generated starting configurations and performing simulated annealing on each one. There is naturally a tradeoff between the length of each annealing run and the number of starts; we found the best result by using a large number of starting configurations with relatively short annealing runs.

Our energy calculation uses a pairwise atom calculation with terms for van der Waals interactions and electrostatics. To improve the speed of the calculation, we use the standard method of employing a distance cutoff. To avoid cutoff artifacts due to one charge in a dipole being inside the cutoff and one outside (thus giving a charge instead of a dipole interaction), partial charges are organized into groups whose total net charge are integral multiples of an electron charge, and the cutoffs are applied to whole groups. Thus, the energy calculation occurs between pairs of groups whose centers fall within the cutoff distance of each other. The total interaction energy of the ligand can be represented as a quadruple sum,

$$E = \sum_{g_l} \sum_{g_r} \sum_{i \in g_l} \sum_{j \in g_r} [E_{ij}^{vdw} + E_{ij}^{el}],$$

where the second sum is over all receptor groups g_r within the cutoff distance of the ligand group g_l. The atom pairwise terms are given by

$$E_{ij}^{vdw} = \sqrt{A_i A_j} R_{ij}^{-12} - \sqrt{B_i B_j} R_{ij}^{-6}$$

and

$$E_{ij}^{el} = \frac{q_i q_j}{R_{ij}},$$

where A_i, A_j, B_i, B_j, q_i, and q_j are the van der Waals parameters and charges for the respective atoms, and R_{ij} is the distance between the atom centers.

Our van der Waals parameters and peptide charges are based on the work of Hagler (Hagler et al., 1974; Lifson et al., 1979; Hagler et al., 1979a, 1979b) and obtained from John Moult. We use a group cutoff distance of 8 Å. Since solvent effects are not explicitly taken into account, interactions with groups having a net charge tend to be severely overestimated. To overcome this problem, we use the technique employed in the GROMOS system, neutralizing charge groups with net charge (Aqvist et al., 1985). We did this by simply subtracting an equal fraction of the net charge from each atom in the group; for example, if the group has a charge of $+1$ and has five atoms, we subtract $+0.2$ from the partial charge of each atom. This has the effect of removing the net charge from the group but leaving all higher-charge moments (dipole, quadrupole, etc.) unchanged. In particular, the hydrogen bond interactions, represented in this force-field system by a balance between dipole and van der Waals interactions, remain unaltered. Introduction of a dielectric constant, on the other hand, affects both net charge interactions and hydrogen bonds.

Basically, then, our method involves generating many random placements for the ligand and running simulated annealing on each one. However, in our first tests we encountered a fundamental technical problem: the majority of the initial positions for the ligand were found to be actually *inside* of the receptor, and the simulated annealing was found to be costly and ineffective at pushing the ligand to the outside where a good docking could be found. The problem was that the van der Waals 6–12 potential produces an extremely complex energy hypersurface when two interacting molecules physically overlap. Our solution was to introduce a precalculated grid to determine which regions of space are occupied by the receptor.

To implement the creation and use of this grid, we first decided that the most effective way to remove an overlap between the ligand and the receptor would be to have a score function that could tell us how far the ligand was from the outside of the receptor. A grid point is first determined to be either inside or outside the receptor; it is inside if it is within a specified distance from a receptor atom (heavy atoms only are used). At this point, inside points are assigned a value of one and outside points a value zero. Then, for each inside point, the nearest outside point is found and the value one is replaced by that distance. The final result is a grid that has nonnegative values indicating the distance from that point to the outside of the receptor (outside points, having value zero, are of zero distance from themselves). The actual creation of the grid is slightly more complex, due to the need to eliminate small isolated pockets (such as internal solvent pockets). This is performed at the initial stage, after points have been tentatively classified as inside or outside, by evaluating the connectivity of the outside points and reclassifying small disconnected pockets as inside.

We define a special score function for the ligand as an average, over the heavy atoms in the ligand, of the distance to the surface of the receptor, calculated by interpolating to the nearest grid point. This gives us an objective function for determining how close a ligand is to being outside the receptor; it will have a value of zero if is completely outside, and have higher values for positions further inside the receptor. The problem of producing a ligand position outside the receptor is now one of minimizing this score function. We do this by performing simulated annealing with this score function until it falls below a specific tolerance, at which point the energy-driven simulated annealing is started. We refer to this procedure as the *floating method*, since we are in essence floating the ligands to the surface of the receptor. The score function is referred to as the *floating score function*. Since the grid in essence is a shape representation of the receptor, our docking method can be viewed as a combination of shape-based and energy-based methods, where the shape-based procedure brings the ligand into the general region of interest and the energy-based procedure performs the detailed docking.

The floating grid has a secondary purpose that is employed during the energy-driven docking phase. Since the grid determines the region where the receptor is, it is useful for detecting bad steric contacts between the ligand and receptor. In the process of docking, many of the randomly generated positions for the ligand will be in bad steric contact with the receptor. By calculating the score function of the ligand, we can determine if such a contact has occurred and avoid the costly pairwise energy calculation; instead, an arbitrary, large value for the energy is assigned. This procedure gives considerable improvement in the time required to run the docking program.

The implementation of our docking method is a suite of programs called BOXSEARCH, which includes the main docking program, the program to calculate the floating grid, and a number of utility programs for comparing and clustering data. The main docking program, BS3 (BOXSEARCH, version 3), requires the user to specify the receptor and ligand coordinate files, the floating grid, the annealing schedule, the total number of runs to attempt, the floating tolerance parameter (the number that defines when a floating run is finished), and the cutoff energy that determines which dockings will be saved in the output file. In addition, the user determines which residue and atom parameter libraries are used; these determine the van der Waals parameters and charges on the ligand and receptor. BS3 will perform all the docking inside a rectangular polyhedron determined by the dimensions of the floating grid. Thus, the floating grid also serves to define the region of interest for the docking run.

To give the reader a clear picture of how BOXSEARCH works, we will outline the steps involved in setting up and running a docking calculation. The first step is to run the floating grid program FLGRID using the receptor coordinate file to generate the floating grid binary file. The user is required to specify the extent of the grid, its center, and the grid stepsize. Since these determine where the docking takes place, their choice depends on the nature of the docking problem. If we have a specific receptor site, we normally choose the grid center around the center of the site and make the extent large enough to easily accommodate the ligand. If the receptor site is not known, then we would choose the grid center near the center of the receptor molecule and choose the extent large enough to accommodate docking at any receptor point; if the receptor's largest diameter is d_R and the ligand's is d_L, then the extent of the grid must be at least $d = d_R + 2d_L$. Normally, we take the grid stepsize to be $\frac{1}{2}$ Å. The grid can be generated with the original crystallographic coordinate file, since FLGRID uses only heavy atoms and ignores all hydrogens.

The next step is to prepare the receptor and ligand coordinates for the energy calculation by adding hydrogen atoms. In our energy functions, we use a united atom approach with only the heavy atoms and polar hydrogens. The latter must be added to the coordinate files and positioned by some means. Once this is done, the residue libraries may need to be modified for the specific molecules involved in the docking. Under normal circumstances, the receptor is a protein and only minor modifications are required to the libraries to deal with incomplete residues in the crystal structure. If the ligand is a protein or peptide, then we again need only similar such modifications; however, if the ligand is nonpeptide, then we have to define the charge group structure of the ligand and assign charges to the atoms. In general there is no simple procedure for generating charges for an empirical energy function; charges cannot be assigned by semiempirical or *ab initio* orbital calculations since the van der Waals and charge parameters in the empirical energy function are fit as dependent parameters. We adopt the following approach. Since the main charge effect, having neutralized charge groups, is hydrogen bonding, we identify hydrogen-bonding sites on the ligand (donors and acceptors) and assign charges that will give reasonable energy and geometry to the hydrogen bonds formed there. For functional groups found in peptides (e.g., carbonyl and amide), we assign the same charges as are used for the peptides; for functional groups not found in peptides, we can assign charges giving reasonable hydrogen bonds with a given peptide donor/acceptor.

The final preparations involve creating a file to hold the annealing schedule and set some other miscellaneous parameters. The annealing schedule, which is stored in a separate file, includes information on the number of steps, the temperature at each step, the number of Monte Carlo runs for each step, and the maximum random rotation and translation at each step (see Table 3-1). Typically, we use larger rotations and translations at higher temperatures and smaller ones at lower temperatures. Note that the number of Monte Carlo runs at each step includes both accepted changes and rejections. The total number of independent starts must also be specified; usually this number depends on the size of the ligand and the region to be searched. There is no set formula to determine the optimal number of starts; we have used between 5,000 and 60,000, depending on the size of the ligand and the region being searched. Finally, the cutoff energy for output must be specified; final dockings with this energy or lower will be recorded in the output file. Again, this number will depend, obviously, on the size of the ligand. Since BS3 will record and save the energy range of the final dockings that fall above the output cutoff, a number of short preliminary runs with low output energy will give the user information about the energy of the final dockings and indicate an appropriate value for the cutoff energy.

Once BS3 has been run and the data produced, there are a number of utility programs for analyzing the output. The output file stores the coordinates in protein data bank (PDB) format (Bernstein et al., 1977), with successive dockings separated by TER cards. Each docking is preceded by a REMARK card that holds the final energy of the docking. The program DISTEST will perform an RMS comparison between a single set of coordinates for the ligand (stored in a separate file) and a whole output file. This is useful if we have a crystallographic structure for the complex or a position

Table 3-1. The annealing schedule.

Step #	kT (kcal/mole)	# of MC steps	Max. rotation (degrees)	Max. translation (Å)
1	10	5	18	5.0
2	8	5	18	5.0
3	6	5	18	5.0
4	4	5	18	5.0
5	2	5	18	5.0
6	1	5	18	5.0
7	0.5	10	18	5.0
8	0.25	10	18	5.0
9	0.1	50	9	2.5
10	10^{-4}	50	9	2.5

for the ligand that we have docked "by hand." Another program, CLUS-TER, is used to perform a simple cluster analysis on the output file and print the results into a new file. This program forms exclusive clusters by finding all dockings within a specified RMS tolerance of the lowest-energy one, removing them from the list, and then proceeding to the next lowest energy. This process is repeated until all dockings have been assigned to a cluster. The energy and coordinates of the lowest-energy member of each cluster is printed into a file, along with the number of members in the cluster and the average RMS distance from the lowest-energy member. Obviously, this program will reduce the size of the data file by eliminating repeated dockings in the same binding mode.

We will illustrate the effectiveness of our method by presenting the results from several docking calculations. The results presented here have been published elsewhere (Hart and Read, 1992). Our first example involves the docking of the anticancer drug methotrexate (MTX) to *E. coli* dihydrofolate reductase (DHFR). The coordinates for the receptor come from the complexed crystal structure of DHFR-MTX solved in the laboratory of J. Kraut (Bolin et al., 1982). The docking was performed with the floating grid centered at approximately the center of the observed binding site for MTX and was 25 Å per side. (The original program used two different grids, one to perform the floating and one to perform the contact exclusion. Our current version has incorporated both of these functions into one grid, the results presented here are essentially not affected by this change.) The ligands we used for the docking included the whole MTX molecule, a fragment containing only the pteridine ring and its side groups, which we call MF1, and a fragment comprising the other part of MTX, the aminobenzoyl and glutamyl groups, which we call MF2 (see Figure 3-1). The crystallographic coordinates from the complexed structure was used as the conformation for the ligands. The charges for the ligands were chosen by using the same sets of charges for the principal functional groups as are found in the peptide charge library. The nonprotonated pteridine ring nitrogens were assigned a charge of zero, which is an obvious simplification, since these could act as weak hydrogen bond acceptors.

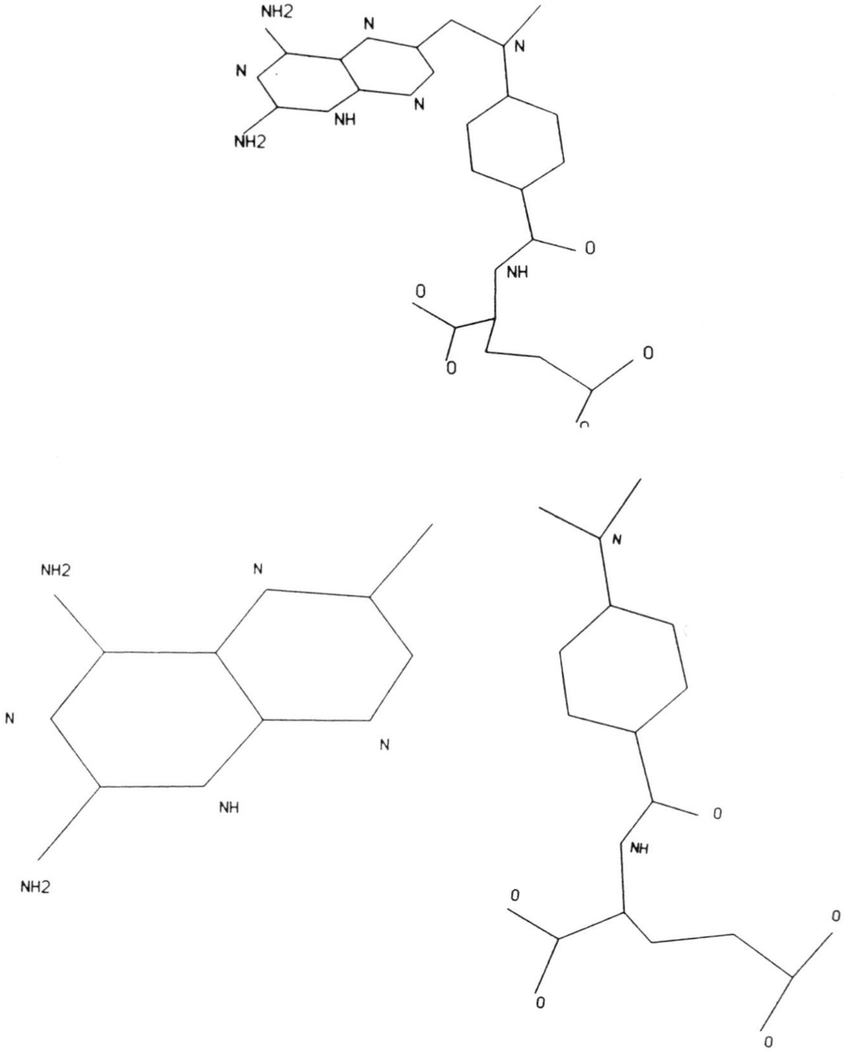

Figure 3-1. The chemical structure of the three fragments used in the methotrexate study: MTX (*top*), MF1 (*bottom left*), MF2 (*bottom right*).

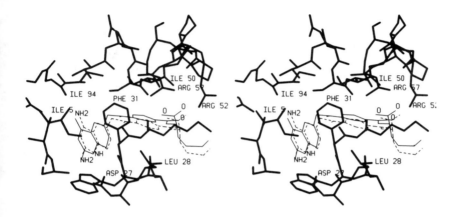

Figure 3-2. Correct docking for MTX (*dashed lines*) with the crystal structure of MTX (*thin and thick lines*).

The results for each docking run were subjected to cluster analysis and then compared to the crystallographic coordinates. Our criterion for a correct docking is that it should come within 2 Å RMS of the crystallographic structure. The results for the run are summarized in Table 3-2, which includes the ranking of the top clusters, their energy and the number of members in the cluster. As we can see, the program could reproduce the crystal structure but not as the highest ranked docking (see Figure 3-2 for structural comparison of MTX with the crystallographic structure). In particular, the correct docking for fragment MF2 placed considerably down in the docking list, indicating that the potential functions are not representing the interaction of this fragment well. One obvious problem is that the benzoyl group binds in a hydrophobic pocket. We discovered on closer examination, however, that there was considerable steric hindrance between this group and the surrounding atoms in the binding pocket; the van der Waals parameters gave too large a van der Waals radius. A likely explanation is that the spherically symmetric van der Waals interactions do not adequately represent the anisotropic nature of the electrons in the aromatic group. To overcome this problem without making major changes to our potential functions, we tried changing the van der Waals parameters on the methyl carbon type arbitrarily from $A = 1.8 \times 10^6$ and $B = 532$ to $A = 1.8 \times 10^6$ and $B = 1000$. With these new parameters, the correct docking for MTX was the highest ranked (see Table 3-3).

The docking of the two fragments MF1 and MF2 allows us to handle, at least in part, the flexibility of MTX. In Figure 3-3, we see that the fragments

Table 3-2.

Top 10 dockings of MTX to DHFR

Ranking	Energy (kcal/mole)	# members	RMS (Å)
1	−41.28	28	14.07
2	−36.26	7	10.40
3	−35.61	61	11.13
4	−34.14	9	13.85
5	−33.04	56	12.99
6	−31.87	8	0.65
7	−31.09	5	14.47
8	−28.96	4	14.35
9	−28.51	2	13.21
10	−28.42	3	13.34

Top 10 dockings of MF1 to DHFR

Ranking	Energy (kcal/mole)	# members	RMS (Å)
1	−24.18	14	12.61
2	−23.53	108	11.50
3	−23.41	36	12.65
4	−23.38	16	14.06
5	−23.03	62	10.78
6	−22.82	31	10.85
7	−22.70	35	13.53
8	−22.67	25	11.83
9	−22.61	82	11.74
10	−22.26	18	13.78
18	−21.35	14	0.45

Top 10 dockings of MF2 to DHFR

Ranking	Energy (kcal/mole)	# members	RMS (Å)
1	−29.66	6	11.86
2	−29.52	43	12.71
3	−29.40	104	10.76
4	−28.25	46	10.98
5	−28.08	4	14.30
6	−27.04	33	13.07
7	−26.64	25	14.33
8	−26.64	32	12.37
9	−26.41	42	14.78
10	−26.41	28	18.76
211	−16.74	3	1.00

Table 3-3. Top 10 dockings of MTX with alternate atom parameters.

Ranking	Energy (kcal/mole)	# members	RMS (Å)
1	−55.15	10	0.48
2	−49.67	4	14.58
3	−47.38	4	14.32
4	−42.86	17	10.97
5	−42.84	2	9.84
6	−41.79	27	12.92
7	−41.49	5	10.79
8	−41.43	5	13.51
9	−40.27	10	9.73
10	−38.85	5	11.33

do dock quite close to the crystallographic positions in MTX, although they are not low-energy members of the docking list. Thus, reassembly of the correct docking for MTX from dockings of MF1 and MF2 would involve comparing the positions of a large number of fragment pairs, one from each of the docking lists. This would require special software tools if fragment docking were to be used as a standard method of handling flexibility. However, this example does at least support the feasibility of this approach.

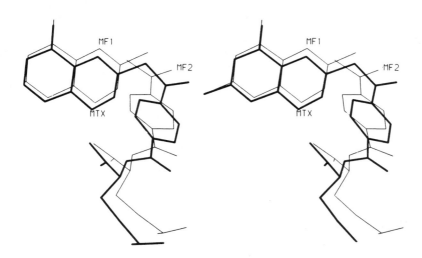

Figure 3-3. Comparison of dockings for fragments MF1 and MF2 with the crystal structure of MTX (DHFR not shown).

The second example is the docking of the third domain of the ovomu-coid inhibitor from turkey (OMTKY3) to the serine proteinase *S. griseus* proteinase B (SGPB). Both the complexed coordinates for OMTKY3-SGPB (Read et al., 1983) and the native coordinates for SGPB (A.R. Si-elecki, private communication) were available, solved in the laboratory of M.N.G. James. OMTKY3 is a small protein of 50 residues and an effective inhibitor of SGPB. Since the ligand is a protein, its charges were simply taken from our standard peptide residue charge library. The grid region was centered near the middle of the correct docking position for OMTKY3 and was 40 Å wide. Since the inhibitor is approximately 27 Å in diameter, this gives a translation search of roughly 13 Å. We performed docking to both the complexed and native SGPB structures, and found that the highest-ranked docking was within 2 Å RMS of the correct coordinates in both cases, although the docking was considerably closer using the complexed receptor (see Table 3-4). A structural comparison with the crystallographic docking can be seen in Figure 3-4. This example shows that our method can handle protein–protein docking problems.

What are the basic virtues and problems of our docking method? First, these examples show that it is basically able to find the correct dockings for both peptide and nonpeptide ligands. Second, as is seen from the MTX–DHFR studies, it is able to find most low-energy local minima, since it could find the correct docking for MF2 and MTX even with inadequate potential functions. Also, the method generates all starting points completely ran-domly, so it is totally free from user bias. The method is computationally expensive due to its use of pairwise energy calculations; on the other hand, we estimate the random search procedure is at least an order of magnitude faster than a full systematic search. The use of empirical energy calculation to generate the dockings means that the correct docking should place at a reasonably high spot on the docking list.

The two basic shortfalls of the method are the inadequacy of the po-tential functions to account for hydrophobic and solvent effects, and the inability to account for conformational changes in the molecules and errors in the coordinates. Clearly, the technique of neutralizing charge groups to deal with solvent effects is extremely crude, although the method has worked much better than might be expected. Still, it would be very desirable to include methods to handle both solvent screening and hydrophobic inter-actions. One possibility is to incorporate terms that account for changes in the amount of surface exposed to solvent on binding. Recent advances have introduced techniques for approximating the hydrophobic contribution to the binding free energy using such terms (Eisenberg and McLachlan, 1986; Still et al., 1990). Conformational changes can be of two kinds: large

Table 3-4.

Top 10 dockings of OMTKY3 to SGPB from complex

Ranking	Energy (kcal/mole)	# members	RMS (Å)
1	−68.33	7	0.41
2	−58.49	1	2.33
3	−46.20	1	3.36
4	−42.56	1	20.48
5	−42.28	2	16.96
6	−39.56	1	20.40
7	−38.99	1	18.29
8	−38.25	1	17.18
9	−37.03	1	15.65
10	−36.63	1	16.04

Top 10 dockings of OMTKY3 docking to native SGPB

Ranking	Energy (kcal/mole)	# members	RMS (Å)
1	−51.27	5	1.74
2	−47.86	4	2.18
3	−44.50	14	18.48
4	−43.02	2	16.10
5	−42.34	2	18.23
6	−42.03	9	17.75
7	−40.49	6	18.34
8	−39.64	3	11.68
9	−39.53	2	17.42
10	−39.01	3	16.97

changes in conformation of flexible portions of the ligand or receptor, and small shifts in atom position. In the latter category may also be grouped the problem of dealing with errors in the atom positions, inherent in any crystal structure. We encountered these types of problems in docking of MTX to DHFR, since a small shift in atom position would have removed the marginal steric hindrance that causes the lower ranking of the correct docking. Clearly, what is needed is a set of van der Waals parameters that are more tolerant, or "softer," than the ones we are using. Dealing with large conformational changes essentially increases the size of the search problem, since we must somehow search through the possible conformations for the optimal one. We discuss this problem in the next section, where we will evaluate a different approach to the problem of dealing with flexible ligands.

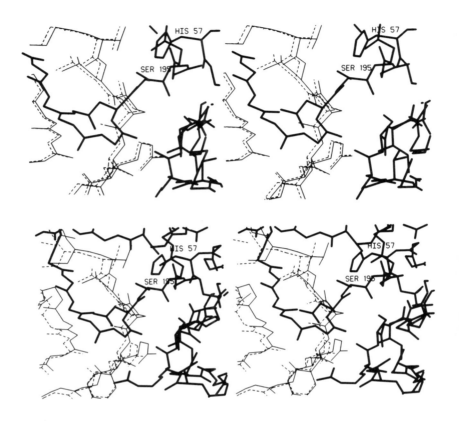

Figure 3-4. Comparison of lowest-energy dockings for OMTKY3 to complexed (*top*) and native (*bottom*) SGPB. Only active site region is shown.

4. Monte Carlo Docking of Flexible Ligands

In this section, we present a recent study handling the problem of docking flexible ligands. Basically, there are three different ways of treating flexibility: by introducing the conformational freedom during the docking process, by modeling and docking each stable conformation of the ligand independently as a rigid molecule, and by separating the molecule into rigid parts, to be docked independently and reassembled to form the correct conformaton. Each of these methods has its advantages; the first can account for the small adjustments in conformation, the second reduces the size of the search problem and performs the bonded energy calculation once for each conformation, and the third has important applications in *de novo* drug design. Our study will deal with assessing the second method, primarily because we feel that is has not been given adequate treatment elsewhere. The first method has been given detailed treatment recently by Caflisch et al. (1992).

The basic philosophy of the second approach, docking a number of stable ligand conformations as rigid fragments, is to separate the problem of dealing with the flexibility of the ligand from the actual docking. The principal advantage is that we save computational time by evaluating the internal conformations ahead of time, and the size of the docking problem is simply the size of the rigid body problem multiplied by the number of conformations we choose to represent. The primary difficulty in implementing this method is compiling and selecting the conformations to be used in the docking. Assuming torsion angles can have three stable positions, a ligand with N internal degrees of freedom might have 3^N possible conformations that must be considered, although most are either sterically or energetically unfavorable. (Naturally, this is a crude estimate. Furthermore, for large molecules we might have to consider finer torsion angles, due to the influence of nonbonded interactions.) If N is small (say $N < 4$), then we can compile the conformations "by hand" using a graphics system. However, for larger N we must have some automatic way of compiling the conformations and saving them in a format that is convenient to handle and that can be used by a docking program. Typically, we need a program to handle this task that will interface well with the docking program we are using; from a software development point of view, this task is nearly as great as the one of performing the docking.

As a test case to assess this approach, we use the example of the binding of the antibiotic chloramphenicol (CLM) to E. coli chloramphenicol acetyltransferase (CAT). The crystallographic complex structure of CLM bound to CAT was solved by A.G.W. Leslie (1990) and has been deposited in the Protein Data Bank as 3CLA. CAT makes bacteria resistant to CLM by acetylating it and making it inactive. CAT is active as a trimer; the binding site is situated on the interface of two of the subunits and consists of a fairly tight binding pocket. CLM is a small but flexible molecule, with seven freely rotatable bonds and with distinct polar and hydrophobic regions (see Figure 3-5).

To prepare CAT for docking, we first build the trimer from the coordinates for the monomer, and assigned and optimized positions for the hydrogens using the Biosym programs INSIGHTII and DISCOVER (Biosym Techologies of San Diego). Nonpolar hydrogens were removed and atoms were systematically renamed using a program written by the authors. The floating grid was generated using a probe size of 3 Å and grid-step size of $\frac{1}{2}$ Å and centered near the middle of the binding site. The sides of the grid

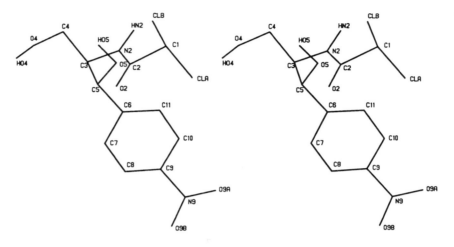

Figure 3-5. The chemical structure and atom naming of CLM.

cube were 25 Å. The assignment of charges to CLM was a problem, because the chlorine atoms, which should each hold a negative partial charge, are not commonly regarded as hydrogen bonding sites. We decided to assign charges by comparing MOPAC calculations for existing charge groups and scaling charges appropriately (calculations were performed using MOPAC version 6.0). We used this same procedure for the nitro group. For a first baseline run, assignment of positions for the polar hydrogens on CLM was based on the orientations implied by the crystal complex. The primary hydrogen bond occurs between the oxygen O4 of CLM and the ND of histidine A195 of CAT (the A indicates the residue is a member of the adjacent trimer), so we chose the position of HO4 to be optimal for this hydrogen bond. The position of HO5 was then uniquely determined, since the two hydrogens interact to form a network, with the HO5 hydrogen directed at the O4 oxygen.

We used the same procedure for all our docking experiments. Dockings consisted of 5,000 independent runs using the same annealing schedule (Table 3-1). We selected −20 kcal/mole as our energy cutoff for output, even though our initial studies indicated that we wanted mainly to look at energies below −30 kcal/mole. The reason for this is that, during the docking run, any docking that falls below the output cutoff after the annealing schedule is subjected to a further 200 low-temperature Monte Carlo steps. Thus, by choosing a higher output cutoff, we can better sample the lower-energy dockings for the ligand. We then processed the output file by removing all dockings with energy greater than −30 kcal/mole and then running CLUS-

TER to arrange the dockings into groups of increasing energy. Each group is represented by its lowest-energy member and consists of dockings falling within 2 Å RMS from that member. We can then perform RMS comparisons between these sets of coordinates and the crystallographic position of CLM, and assess the accuracy of the lowest-energy dockings.

Our first docking experiment involved docking the CLM conformation from the complexed crystal structure, i.e., docking to simply put the crystal complex back together. One problem we immediately encountered is the amount of water in the active site; the actual binding of CLM involves a considerable number of water-mediated interactions, while our calculation does not include any waters. This leads to a fair degree of "sloppiness" in the docking, with the ligand assuming a number of low-energy positions in approximately the correct binding mode but displaced along the binding grove by an angstrom or more. In principle, the water structure of native CAT (lacking bound CLM) could have been used to predict the bridging interactions, since several important solvent molecules are found in the same positions in both the CLM–CAT complex and in native CAT (Andrew Leslie, personal communication). One of the problems is that the phenol ring of CLM is not held firmly in its hydrophobic pocket because there is no hydrophobic contribution to the energy. In spite of these problems, CLM does dock successfully; the lowest-energy docking is 0.37 Å RMS from the crystallographic binding mode (see Table 3-5).

Table 3-5. Top 10 dockings of CLM to CAT.

Ranking	Energy (kcal/mole)	# members	RMS (Å)
1	−34.68	12	0.37
2	−33.55	7	7.82
3	−33.52	14	8.76
4	−32.35	5	3.15
5	−31.82	4	5.30
6	−31.49	6	11.23
7	−30.84	1	2.37
8	−30.01	9	10.51
9	−29.61	11	11.27
10	−29.54	1	4.48

Our next step was to generate a collection of stable conformations of CLM and perform docking on each one. CLM has a total of seven rotatable bonds, and, if we assume three favored positions for each torsion angle, it has potentially $3^7 = 2,187$ different possible conformations, not counting symmetry. In reality, most of these conformations will be sterically or energetically unfavorable. Since our goal is to study the feasibility of docking alternate conformations, it was not critical that we obtain every possible stable conformation of the ligand; a reasonably representative subset would be adequate. The first conformation was obtained by running energy minimization, in vacuum, on the crystallographic conformation of CLM from the complex. We then chose 10 other conformations by running molecular dynamics at a temperature of 1200 K and selecting a new conformation every 500 steps, which was then minimized. We discovered that two pairs of the 10 conformations were identical, after minimization, so we were left with the original minimized conformation plus eight from the dynamics, leaving a total of nine. Since the C1–C2 bond is fairly free to rotate, we generated an alternate conformation for this bond for each of the nine conformations by performing a 180° rotation. This gives a total of 18 conformations for our docking experiments. All minimization and dynamics were performed by the Biosym programs INSIGHTII and DISCOVER (Biosym Technologies of San Diego). We denote the original conformations as 1 Å to 9 Å, with 1 Å referring to the conformation minimized from the crystallographic structure. The alternate conformations found by rotating the C1–C2 bond are denoted 1B to 9B.

The conformations were all of comparable energy and we decided to consider them as basically equivalent, especially since the modeling was done in vacuum and the effect of solvent could easily make up the energy differences found between various conformations. Table 3-6 gives a comparison of the RMS distance from the conformation found in the crystallographic complex. Table 3-7 gives the torsion angles for three of the central rotatable bonds that serve to define the conformation. Of particular interest is that both conformations 01 and 08 are quite close to the original CLM conformation, but differ primarily in the orientation of their mobile polar hydrogens HO4 and HO5. It is of particular interest to see what the effect of changing only polar hydrogen orientation will be on the docking results.

Each conformation was stored in a separate file and the dockings were performed consecutively for each conformation using a batch command file. Aside from the different ligand coordinates, all other parameters for the runs were identical with the ones used for docking the crystallographic conformation of CLM. The total time to perform all 18 runs was approximately 13.5 hours on an SGI Crimson. We processed the data by first merging all 18 output files, removing all dockings with energy higher than −30 kcal/mole, and finally running CLUSTER on the resulting collection of dockings. By taking this approach, we would be forced to analyze the data without knowing which conformation we were dealing with. Table 3-8

Table 3-6. Comparison of CLM test conformations to conformation from complex.

	Fragment RMS (Å)
1A	1.00
1B	0.45
2A	1.81
2B	1.92
3A	2.02
3B	2.04
4A	2.02
4B	2.11
5A	1.79
5B	1.66
6A	1.89
6B	1.81
7A	1.86
7B	1.93
8A	1.08
8B	0.57
9A	2.89
9B	2.74

Table 3-7. Torsion angles of CLM test conformations and conformation from complex.

Fragment	N2–C3 (degrees)	C3–C5 (degrees)	C5–C6 (degrees)
Complex	101.7	−50.8	−89.6
1A,B	134.2	−66.7	−85.2
2A,B	146.2	60.1	−175.9
3A,B	−82.1	62.6	83.0
4A,B	135.8	61.1	82.4
5A,B	83.9	55.4	8.1
6A,B	80.9	58.6	−96.0
7A,B	135.9	59.4	−97.3
8A,B	98.9	−68.8	−85.3
9A,B	81.5	176.3	−110.7

shows a list of the top 25 clusters, with the energy, number of members of each cluster, and the RMS difference from the crystallographic binding mode. We see that the correct binding mode is predicted in the 16th cluster, with an RMS difference of less than 1 Å and with an energy very close to that found in the docking of the crystallographic conformation.

The lowest-energy docking for all the conformations had an energy of approximately −40.1 kcal/mole, which was considerably lower than the −34.7 kcal/mole for the lowest docking of the crystallographic conformation. Its conformation is fairly extended, and it binds in the opposite orientation to the observed binding mode (see Figure 3-6). In fact, this con-

Table 3-8. Top 25 dockings of all CLM conformations.

Ranking	Energy (kcal/mole)	# members	RMS (Å)
1	−40.11	10	5.72
2	−39.58	33	5.66
3	−38.90	15	7.97
4	−36.53	2	6.14
5	−35.43	7	6.41
6	−35.24	6	8.78
7	−35.22	5	3.31
8	−35.20	13	5.98
9	−35.05	11	5.25
10	−34.90	5	14.34
11	−34.82	9	10.27
12	−34.45	39	8.40
13	−34.37	4	9.12
14	−34.31	2	5.40
15	−34.16	9	8.92
16	−33.82	9	1.01
17	−33.68	4	7.57
18	−33.65	12	7.98
19	−33.61	8	6.26
20	−33.41	6	5.00
21	−33.34	11	6.70
22	−33.32	1	4.37
23	−33.30	8	5.54
24	−33.24	18	4.19
25	−32.95	1	5.40

formation and the second one in the list (with energy −39.6 kcal/mole) are in the same conformation and are in very similar docking modes, each related to the other by a 180° rotation around the long axis of the molecule, leaving the nitrophenyl and dichloride groups in the same positions. The principal interactions that give these conformations their low energy are strong polar interactions involving these groups. However, there are several problems with these predicted binding modes, primarily involving solvent effects. First, the hydrophobic phenyl group is exposed to the solvated part of the binding pocket where, in the crystal complex, several crystallographically observed water molecules form a hydrogen bonding network with the O5 hydroxyl group of CLM. Clearly, there is considerable free energy to be gained with the phenyl group in the hydrophobic pocket. Second, the O4 and O5 hydroxyl groups are near this hydrophobic pocket; in particular, this leaves these hydrogen-bond donors on CLM unsatisfied. Thus, the favorable prediction of these binding modes can be attributed to the failure of simple electrostatic calculations to account for important solvent effects.

One interesting aspect of our collection of conformations is that it con-

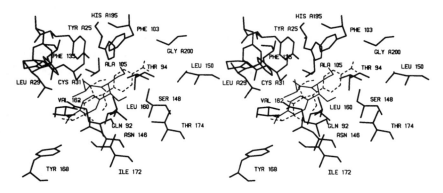

Figure 3-6. Comparison of the lowest-energy docking of CLM (*dashed lines*) with the crystal structure of CLM and CAT (*thin and thick lines*).

tains two that are similar to the crystallographic conformation and whose basic difference is the positions of the polar hydrogens HO4 and HOS. When the conformation allows the O4 and OS oxygens to be in proximity, they hydrogen-bond to one another. Thus, there are two possible ways of orienting these polar hydrogens (see Figure 3-7). The conformation with the correct positions for the polar hydrogens, conformation 1B, docks within 0.96 Å RMS of the crystal structure with an energy of −33.8 kcal/mole; the other conformation, 8B, docks 1.93 Å RMS away with an energy of −27.6 kcal/mole. Although this result comes as no surprise, it clearly shows the need to include mobility of polar hydrogens even when the heavy atoms themselves may be fairly rigid.

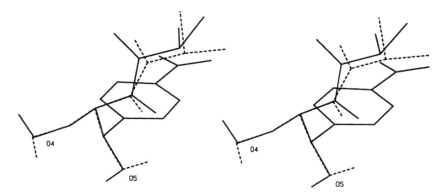

Figure 3-7. Comparison of two orientations for the hydrogens HO4 and HO5. The two conformations are 1B (*solid lines*) and 8B (*dashed lines*).

Although our study of ligand flexibility has been brief, it is clear that docking a collection of ligand conformations independently works very well in handling the problem of ligand flexibility. With the development of conformation modeling tools that interface well with the docking method, this approach seems to be a viable alternative to allowing conformational freedom during the docking process.

5. Concluding Remarks

The field of molecular docking continues to develop. It is clear that the current challenge being addressed by workers in the field is the problem of dealing with flexible ligands. While introducing conformational degrees of freedom into the docking procedure is certainly a viable approach, it is worthwhile exploring other avenues for dealing with this problem. Our short study here has illustrated the effectiveness of dealing with ligand flexibility independently of the actual docking procedure. Our study also underlines the need to develop more sophisticated energy calculation methods that are able to deal with solvent and hydrophobic effects. With other aspects of the docking problem still remaining unsolved (such as receptor flexibility, entropic effects of binding), the development of new docking methods will continue to be an active and exciting field for some time to come.

Acknowledgments. This project was supported by the Medical Research Council of Canada, the Alberta Heritage Foundation for Medical Research, and an International Research Scholars Award from the Howard Hughes Foundation (R.J.R.). We thank Andrew Leslie for generously supplying the coordinates of native chloramphenicol acetyltransferase prior to publication. We also thank Norma Duke for helpful discussions.

REFERENCES

Allen FH, Kennard O, Taylor R (1983): Systematic analysis of structural data as a research technique in organic chemistry. *Acc Chem Res* 16:146–153

Aqvist J, van Gunsteren WF, Leijonmarck M, Tapia O (1985): A molecular dynamics study of the C-terminal fragment of the L7/L12 ribosomal protein: Secondary structure motion in a 150 picosecond trajectory. *J Mol Biol* 183:461–477.

Bacon DJ, Moult J (1992): Docking by least-squares fitting of molecular surface patterns. *J Mol Biol* 225:849–858

Bernstein FC, Koetzle TF, Williams GJB, Meyer EF, Brice MD, Rodgers JR, Kennard O, Shimanouchi T, Tasumi M (1977): The protein data bank: A computer-based archival file for macromolecular structures. *J Mol Biol* 112:535–542

Böhm H-J (1992a): The computer program Ludi: A new method for the de novo design of enzyme inhibitors. *J Comput-aided Mol Design* 6:61–78

Böhm H-J (1992b): Ludi: Rule-based automatic design of new substituents for enzyme inhibitor leads. *J Comput-aided Mol Design* 6:593–606

Bolin JT, Filman DJ, Matthews DA, Hamlin RC, Kraut J (1982): Crystal structures of *Escherichia coli* and *Lactobacillus casei* dihydrofolate reductase refined at 1.7Å resolution. I. General features and binding of methotrexate. *J Biol Chem* 257:13650–13662

Caflisch A, Niederer P, Anliker M (1992): Monte Carlo docking of oligopeptides to proteins. *Proteins* 13:223–230

Connolly ML (1983a): Analytical molecular surface calculation. *J Appl Cryst* 16:548–558

Connolly ML (1983b): Solvent-accessible surfaces of proteins and nucleic acids. *Science* 221:709–713

Connolly ML (1986): Shape complementarity at the hemoglobin o1fl1 subunit interface. *Biopolymers* 25:1229–1247

DesJarlais RL, Sheridan RP, Dixon JS, Kuntz ID, Venkataraghavan R (1986): Docking flexible ligands to macromolecular receptors by molecular shape. *J Med Chem* 29:2149–2153

DesJarlais RL, Sheridan RP, Seibel GL, Dixon JS, Kuntz ID, Venkataraghavan R (1988): Using shape complementarity as an initial screen in designing ligands for a receptor binding site of known three-dimensional structure. *J Med Chem* 31:722–729

Eisenberg D, McLachlan AD (1986): Solvation energy in protein folding and binding. *Nature* 319:199–203

Gilson MK, Honig B (1988): The energetics of charge charge interactions in proteins. *Proteins* 3:32–52

Goodford PJ (1985): A computational procedure for determining energetically favorable binding sites on biologically important macromolecules. *J Med Chem* 28:849–857

Goodsell DS, Olson AJ (1990): Automated docking of substrates to proteins by simulated annealing. *Proteins* 8:195–202

Hagler AT, Dauber P, Lifson S (1979a): Consistent force field studies of intermolecular forces in hydrogen-bonded crystals. 3. The CO$\cdots$HO hydrogen bond and the analysis of the energetics and packing of carboxylic acids. *J Am Chem Soc* 101:5131–5141

Hagler AT, Huler E, Lifson S (1974): Energy functions for peptides and proteins. I. Derivation of a consistent force field including the hydrogen bond from amide crystals. *J Am Chem Soc* 96:5319–5327

Hagler AT, Lifson S, Dauber P (1979b): Consistent force field studies of intermolecular forces in hydrogen-bonded crystals. 2. A benchmark for the objective comparison of alternative force fields. *J Am Chem Soc* 101:5122–5130

Hart TN, Read RJ (1992): A multiple-start Monte-Carlo docking method. *Proteins* 13:206–222

Jiang F, Kim SH (1991): Soft docking: Matching of molecular surface cubes. *J Mol Biol* 219:79–102

Kirkpatrick S, Gelatt CD, Vecchi MP (1983): Optimization by simulated annealing. *Science* 220:671–680

Kuntz ID, Blaney JM, Oatley SJ, Langridge R, Ferrin TE (1982): A geometric approach to macromolecule-ligand interactions. *J Mol Biol* 161:269–288

Lee B, Richards FM (1971): The interpretation of protein structures: Estimation of static accessibility. *J Mol Biol* 55:379–400

Leslie AGW (1990): Refined crystal structure of type II chloramphenicol acetyltransferase at 1.75 Å resolution. *J Mol Biol* 213:167–186

Lifson S, Hagler AT, Dauber P (1979): Consistent force field studies of intermolecular forces in hydrogen-bonded crystals. 1. Carboxylic acids, amides, and the CO···H-hydrogen bonds. *J Am Chem Soc* 101:5111–5121

Metropolis N, Rosenbluth AW, Rosenbluth MN, Teller AH, Teller E (1953): Equation of state calculations by fast computing machines. *J Chem Phys* 21:1087–1092

Moon JB, Howe WJ (1991): Computer design of bioactive molecules: A method for receptor-based *de novo* ligand design. *Proteins* 11:314–328

Read RJ, Fujinaga M, Sielecki AR, James MNG (1983): Structure of the complex of *Streptomyces griseus* protease B and the third domain of the turkey ovomucoid inhibitor at 1.8 Å resolution. *Biochemistry* 22:4420–433

Shoichet BK, Kuntz ID (1991): Protein docking and complementarity. *J Mol Biol* 221:327–346

Still WC, Tempczyk A, Hawley RC, Hendrickson T (1990): Semianalytical treatment of solvation of molecular mechanics and dynamics. *J Am Chem Soc* 112:6127–6129

Wesson L, Eisenberg D (1992): Atomic solvation parameters applied to molecular dynamics of proteins in solution. *Protein Science* 1: 227–235

4

The Genetic Algorithm and Protein Tertiary Structure Prediction

Scott M. Le Grand and Kenneth M. Merz, Jr.

1. Introduction

In 1959, Anfinsen demonstrated that the primary structure (the sequence of amino acids) of a protein can uniquely determine its tertiary structure (three dimensional conformation). This implied that there must be a consistent set of rules for deriving a protein's tertiary structure from its primary structure. The search for these rules is known as the *protein folding problem*. Despite many creative attempts, these rules have not been determined (Fasman, 1989). Currently, the primary structures of approximately 40,000 proteins are known. However, only a small percentage of those proteins have known, tertiary structures. A solution to the protein folding problem will make 40,000 more tertiary structures available for immediate study by translating the DNA sequence information in the sequence databases into three-dimensional protein structures. This translation will be indispensable for the analysis of results from the Human Genome Project, *de novo* protein design, and many other areas of biotechnological research. Finally, an in-depth study of the rules of protein folding should provide vital clues to the protein folding process. The search for these rules is therefore an important objective for theoretical molecular biology.

Many theoretical efforts aimed at solving the protein folding problem have involved the optimization of a potential energy function that approximates the thermodynamic state of a protein macromolecule. These efforts are based on the assumption that the global minimum energy conformation obtained using these potential functions will correspond to the protein's tertiary structure. Since the location of the global minimum energy conformation of a protein alone under such a potential function does not necessar-

The Protein Folding Problem and Tertiary Structure Prediction
K. Merz, Jr. and S. Le Grand, Editors
© Birkhäuser Boston 1994

ily give any insight into how a protein folds, these approaches are known as *protein tertiary structure prediction.* A successful energy minimization-based protein tertiary structure prediction algorithm must satisfy two requirements. First, the conformational search technique it uses must be capable of locating the global minimum conformation of a protein under a specific potential energy function, and second, the global minimum conformation of a given protein under the potential energy function must be close to the native structure of that protein. Researchers have developed numerous algorithms for the conformational search of small polypeptides (Nemethy and Scheraga, 1990; Wilson and Cui, 1990) hydrocarbons (Anet, 1990), and proteins (Skolnick and Kolinski, 1990; Snow, 1992; Le Grand, 1993; Le Grand and Merz, 1993a, 1993b; Sun, 1993). Other groups have developed potential energy functions for molecular modeling (Weiner et al., 1986; Brooks et al., 1983, Momany et al., 1975) and tertiary structure prediction (Wilson and Doniach, 1989; Covell and Jernigan, 1990; Sippl, 1990; Crippen, 1991). Unfortunately, a combined conformational search algorithm and potential function that satisfy both of the aforementioned requirements for successful protein tertiary structure prediction have not yet been developed (Sippl, 1990; Le Grand, 1993; Le Grand and Merz, 1993b).

2. The Genetic Algorithm

The research that will be described here involves the use of the genetic algorithm to perform protein tertiary structure prediction. The genetic algorithm is an optimization technique derived from the principles of evolutionary theory (Holland, 1975; Goldberg, 1989). It has been applied to a myriad of optimization problems such as the traveling salesman problem, neural network optimization (Montana and Davis, 1989; Whitley and Hanson, 1989a, 1989b; Whitley et al., 1990; Whitley and Starkweather,1990), scheduling (Cleveland and Smith, 1989), machine learning, pattern recognition, and the solution of nonlinear equations. See Goldberg (1989) for a review of these applications.

Figure 4-1 illustrates a typical genetic algorithm as described in Grefenstette and Baker (1989). A genetic algorithm begins by encoding the k independent variables of an optimization problem as genes in a chromosome. For example, in conformational search, the genes would be the conformation determining dihedral angles of a molecule (Figure 4-2). Next, a population of N chromosomes (hereafter known as $P(t)$) is initialized with random values for each of the genes in each chromosome (Figure 4-3).

```
procedure GA
begin
    t = 0;
    initialize P(t);
    evaluate structures in P(t);
    while termination condition not satisfied do
    begin
        t = t + 1;
        select M(t) from P(t-1);
        recombine structures in M(t);
        evaluate structures in M(t);
        replace some or all of P(t-1) with M(t) to form P(t)
    end
end.
```

Figure 4-1. Flowchart of a genetic algorithm.

After this *initialization* step, the function value of the point in parameter space represented by each chromosome x is evaluated and called the chromosome's fitness $u(x)$. In conformational search, the fitness would be the potential energy of the conformation of the molecule represented by the chromosome (Figure 4-4).

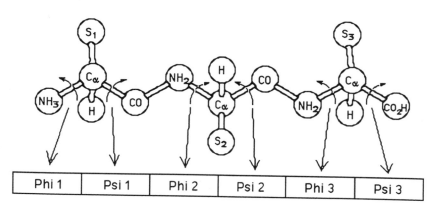

Figure 4-2. The encoding of a molecule's conformational variables as genes in a chromosome.

After initialization, a genetic algorithm cycles through rounds of *selection*, *recombination*, and *evaluation* until termination conditions are met. During the first phase, selection, the mating population $M(t)$ is selected from $P(t)$. $M(t)$ consists of one or more pairs of chromosomes known as parents. There are numerous methods of selecting $M(t)$ from $P(t)$ (Goldberg, 1989; Whitley and Hanson, 1989a). One popular method of selection is known as *proportionate selection*. In proportionate selection, a given

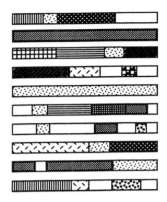

Figure 4-3. A randomly generated population of chromosomes.

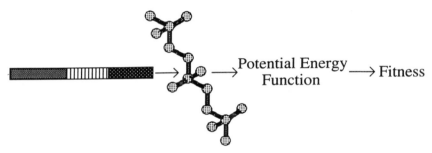

Figure 4-4. The evaluation of the fitness of a chromosome that represents a conformation of a molecule.

chromosome x is selected for inclusion into $M(t)$ with probability $p(x)$, which is proportional to the ratio of its fitness to the mean fitness of the population (Figure 4-5):

$$p(x) \approx \frac{u(x)}{\overline{u}(t)}. \tag{1}$$

During *recombination*, the genes in the pairs of parents in $M(t)$ are mixed together to produce hybrid chromosomes (hereafter to be called *children*) via the use of operators that perform processes analogous to genetic crossover and mutation. The first phase of recombination is the use of a crossover operator to create a hybrid child from each of the pairs of parents in $M(t)$. There are many crossover operators in use (Booker, 1987; Schaffer and Morishima, 1987; Sirag and Weisser, 1987; Davidor, 1989; Goldberg, 1989). The most common crossover operator, known as *simple two-point crossover*, creates a child containing all the genes from the beginning of one parent's chromosome up to a cut point, and the rest of its genes from that cut

P(t)   M(t)

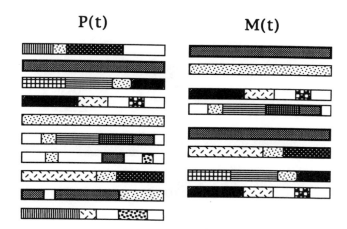

Figure 4-5. Selection of $M(t)$ from $P(t)$.

point to the end of the chromosome from the second parent (Figure 4-6[1]). A second child can be created from the genes in both parental chromosomes that are not in the first child, if desired. In conformational search, simple two-point crossover creates a child that takes one section of the molecule's dihedral angles from the first parent while the complementary section of the molecule takes its dihedral angles from the second parent. A second popular crossover operator is known as *two-point wraparound crossover*. In two-point wraparound crossover, the chromosome is treated as a ring—a child's chromosomes is created from an arc segment out of the first parent's chromosome and the complementary arc segment from the second parent's chromosome (Figure 4-6[2]). In conformational search, two-point wraparound crossover creates a child whose outer dihedral angles are from one parent and inner dihedral angles are from the other parent. The use of two-point wraparound crossover is thought to help transfer genes that are on opposite ends of the chromosome together, which would otherwise tend to be broken apart by simple two-point crossover. A third popular crossover operator is known as *uniform crossover* (Figure 4-6[3]). In uniform crossover, each gene is taken from either parent with equal probability based on the value of a random variable. In conformational search, uniform crossover would create a child whose dihedral angles were randomly selected from either parent. Like the use of two-point wraparound crossover, the use of uniform crossover also helps to solve distance-dependent crossover problems. However, it can also disrupt pairs of genes near one another that would otherwise be likely to be transferred together during crossover.

1. Two-Point Crossover

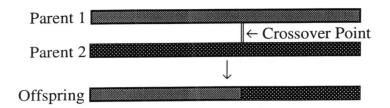

2. Wraparound Crossover

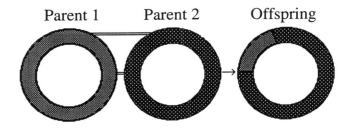

3. Uniform Crossover

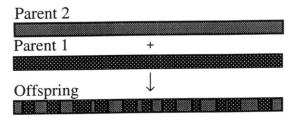

Figure 4-6. Three popular methods of crossover that are used in genetic algorithms.

The second phase of recombination is known as *mutation*. During mutation, parts of each child's chromosome are altered slightly by operators that perform processes analogous to genetic mutation (Figure 4-7). As with crossover, there are many mutation operators in use (Fogarty, 1989; Goldberg, 1989; Whitley and Hanson, 1989a). One such method is to give each of a child's genes a 3–5% chance of being changed to a random value based on the value of a random variable. In conformational search, the

Mutation

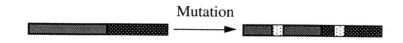

Figure 4-7. Illustration of the effect of mutation on a chromosome.

aforementioned mutation operator would randomly change the values of several dihedral angles.

During evaluation, the fitnesses of the chromosomes in $M(t)$ are evaluated in the same manner as after initialization. Once the fitnesses of the chromosomes in $M(t)$ have been calculated, they replace all or some of the members of $P(t)$ to form $P(t+1)$ during replacement. This cyclic process of selection, recombination, evaluation, and replacement repeats until user-specified termination conditions are met.

Many variations on this basic theme are possible; there are several good introductions to the subject that cover both this basic approach (Walbridge, 1989; Radcliffe and Wilson, 1990; Wayner, 1991) and many of these variations (Goldberg, 1989). Holland (1975) presents a rigorous analysis of the genetic algorithm.

3. Genetic Algorithms and Molecular Structure Prediction

The first application of the genetic algorithm to the task of protein structure prediction was that of Friedrichs and Wolynes (1989), who used a genetic algorithm to perform the conformational search of a simplified cubic lattice-based model of a protein. Each amino acid was represented solely by its C_α atom, which could exclusively occupy a single site on the lattice. A 3.8 Å lattice spacing was used to mimic the normal spacing of amino acid C_α atoms in proteins. Evolutionary fitness was measured by how well a given conformer of a protein satisfied a set of anywhere from 85 to 140 NMR-like distance constraints. The genetic algorithm also incorporated aspects of simulated annealing to improve its performance at conformational search. Their approach starts off similarly to the above described genetic algorithm by first generating N randomized conformers of a target protein and then evaluating their evolutionary fitnesses. However, the iterative section of the genetic algorithm differs significantly from the standard approach. This section first performs m mutations on each of the N conformations in the population, where a mutation consisted of rigidly rotating the section of a conformer above or below a randomly chosen site a random amount around either the x, y, or z axis. After each mutation, the conformer is checked for overlapping sites. If any such sites are found, the conformer reverts to its premutation state. Next, the evolutionary fitness is calculated and the move is either accepted or rejected using a Boltzmann-like criterion. After m mutations of each conformer are attempted, N new conformers are generated via an intriguing method of recombination. For each new conformer, a recombination site p is randomly chosen within the amino acid chain. This splits the protein into two subchains, a and b. Next, the fitnesses of all a segments in all N currently existing conformers are calculated by ignoring terms in the evolutionary fitness function not involving amino acid residues in the a chain segment. Finally, an a and b segment are chosen for the new conformer from the existing pool of N segments via proportionate selection (equation (1)). After recombination, the temperature is lowered slightly and a new round of mutations are performed upon the new population. This algorithm continues until either a conformer of optimal fitness is located or the lowest user-specified temperature is reached. Performance of this algorithm was mixed at first, and numerous techniques were used to improve upon it. The most promising of these was the utilization of a "build-up" approach, where the amino acid chain of a protein was broken into numerous segments and optimized independently. Next, a conformational search of the whole protein was attempted, utilizing the results of the previous step. This was shown to reduce computational time requirements by 40%.

Tuffery et al. (1991) applied the genetic algorithm to the *packing problem*, or the determination of amino acid side chain conformations given a fixed backbone. Their approach first centered on locating a small family of rotamers that would represent the observed conformations of amino acid side chains in proteins. Their library started with that generated by Ponder and Richards (1987), but numerous additional rotamers were located that were shown to improve the ability of the library to reproduce amino acid side-chain conformations. The resulting improved rotamer library was then used in the genetic algorithm-based conformational search of side-chain conformations in order to reduce the complexity of the search. The genetic algorithm used here is known as the "Selection-Mutation-Focusing Genetic Algorithm" and consists of five steps, four of which iterate. The first step is known as *initialization*, during which a starting population of protein conformers is generated via a probability-law-based reproduction scheme and their fitnesses are measured via the Flex potential energy function (Lavery et al., 1986). The initial probability law for reproduction of a particular side chain rotamer i is given by calculating the lowest energy state it can possibly be in a protein, $E_{i\min}$, and then calculating its probability of reproduction

p_i via a Boltzmann-like criterion. The second step is known as *selection*. During selection, protein conformers with unacceptably high energies are eliminated from the population. Next, during focusing, a cluster analysis is performed on the fraction of the population that survived selection in order to group individuals in the current population that resemble one another. During *subset transformation*, the next step, each group generated by focusing is analyzed for diversity via an entropy measure. If the entropy measure is less than a specified threshold, the most fit member of the group is retained and the rest are discarded. The groups that survive subset transformation are then used to generate a new probability law of reproduction, and new members are generated by mutation and crossover in order to replace those that were discarded. Finally, during *grouping*, the groups are assembled together into one population and selection resumes. This process iterates until grouping only generates one group. This algorithm was tested on numerous proteins and generated low-energy conformations with root mean square deviations (RMSDs) ranging from 1.43 Å to 2.44 Å relative to the native structure.

Le Grand and Merz (1993a) applied the genetic algorithm to the conformational search of several small polypeptides and to the 46-amino-acid protein crambin. Similar to the work of Tuffery et al. (1991), our work incorporated a rotamer library that was taken from Vasquez et al. (1983) in order to reduce the complexity of the search. Their genetic algorithm consists of the following steps. During initialization, a population of N conformers is generated and their evolutionary fitness is measured by their AMBER potential energy (Weiner et al., 1986). Next, during selection, a pair of parental conformers is chosen from the current population using a technique that favored selecting members of the population with high-ranking fitnesses. During recombination, the pair of parents is recombined by one of the three different forms of crossover previously illustrated in Figure 4-6. Additionally, the performance of each of these crossover methods is monitored and those that tend to create the most fit children are gradually used more often for the creation of offspring. Next, during mutation, the offspring is mutated at a frequency that is dependent on the similarity of its two parents and can range from 0 to 20%. Once the offspring is created, each amino acid is checked to insure that a rotamer existed in the rotamer library that had all of its dihedral angles within ±25 degrees of it. If no such rotamer exists, then the most similar rotamer in the rotamer library is used to replace that amino-acid residue's dihedral angles. Finally, during replacement, the member of the current population that is most similar to the offspring is located, and if the offspring is of superior fitness, it replaces it; otherwise, the offspring is discarded. The algorithm then returns to selection and continues this iterative process until either the population converges to a single conformer, no improvement is seen in the population for a specified number of iterations, or a user-specified time limit expires. This algorithm could locate the global minimum conformations of the polypeptides AAAAAAAAA, AGAGAGAGA, and YGGFM with 90% reliability in approximately 200,000 iterations. Additionally, this algorithm was also used to locate a non-native conformation of the 46-amino-acid protein crambin that had an AMBER potential energy $\sim$200 kcal below that of the gradient-minimized native structure.

Judson et al. (1992) compared the performance of the genetic algorithm, simulated annealing, and random search in the task of locating the global minimum conformations of two-dimensional polymers of various lengths. Fitness was measured via a Lennard–Jones potential. The performance of the genetic algorithm and simulated annealing was, not surprisingly, superior to that of random search. The performance of the genetic algorithm and simulated annealing was subsequently improved by incorporating simplex and conjugate gradient optimization at each iteration. No general conclusions are drawn as to whether the genetic algorithm or simulated annealing is superior for this task, and it is noted that the performance of both of these algorithms can be strongly problem-dependent.

Dandekar and Argos (1992) applied the genetic algorithm to two different protein-folding-related optimization tasks. The first task was designing a sequence to fit a prespecified zinc finger-sequence motif and the λ-repressor structural motif. Fitness for the zinc finger-sequence motif was measured by via a sequence's differences from the consensus amino acid content of all zinc finger motifs with additional terms to penalize stop codons and reward consensus matches at each amino acid position. Fitness for the λ-repressor structural motif was measured by a set of terms reflecting proper atomic burial, secondary structure, and sequence preferences. Zinc finger-like sequences were located rapidly, as were sequences that fit the λ-repressor structural motif. The second task was the *ab initio* folding of an amino acid sequence from a random conformation to a four-strand β bundle. Fitness here was measured by the number of steric clashes plus the number of residues with correct β-strand secondary structure plus a term signifying the presence of an interaction between the ends of β-strands. Later trials added a term for scatter that was measured by the sum of the distances of each atom in the protein from its center of mass. Each of these terms was then given a weight to reflect its importance. After sufficient fine tuning of the weights, the genetic algorithm was able to find the correct fold in a conformation space of 10^{24} distinct conformations. However, since this model was only applied to the folding of one motif and it also assumed perfect prediction of secondary structure, it is not of general applicability to protein tertiary structure prediction.

Judson (1992) used a genetic algorithm to evolve folding pathways for a simplified two-dimensional polymer. Evolutionary fitness was measured by folding the polymer via a sequence of genome-specified steps from an initial linear zig-zag state and then measuring its potential energy, which consisted of terms for nonbond and bonded interactions. The genetic algorithm successfully located a folding pathway for a 19-atom polymer and it took it to its lowest energy state in six folding steps.

Sun (1993) used a conformational, dictionary-based genetic algorithm and a knowledge-based fitness function (Sippl, 1990) to predict the tertiary structures of the proteins melittin (1mlt), avian pancreatic polypeptide (1ppt), and apamin (no PDB entry). A conformational dictionary, which is conceptually similar to a rotamer library, is a collection of fragments derived from a database of proteins of known structure. The conformational dictionary used in this consisted of di- to pentapeptides from 110 proteins that were less than 50% homologous to one another. Unfortunately, the conformational dictionary included both melittin and avian pancreatic polypeptide, and the method of sampling the conformational dictionary involved copying the dihedral angles of homologous pentapeptides into a conformer 10% of the time. This means that if a protein upon which tertiary structure is performed is present in the conformational dictionary, there is approximately a $20^5/(110 \times 100)$ or 300 to 1 chance of automatically sampling the native conformation of a pentapeptide. The final RMSDs for the prediction of the tertiary structures of melittin and avian pancreatic polypeptide are 1.66 Å and 1.32 Å, respectively. In contrast, the RMSD for the predicted structure of apamin, which is both significantly smaller and not present in the conformational dictionary, is 2.80 Å or 3.46 Å depending on the presence or absence of two disulfide constraints. Therefore, it is impossible to judge the quality of this genetic algorithm until it is applied to more proteins that are not present in the conformational dictionary.

Unger and Moult (1993) compared the performance of the genetic algorithm to Monte Carlo steps in the task of folding a two-dimensional lattice model of protein structure similar to that used in the work of Dill (Lau and Dill, 1989). Their work showed that the genetic algorithm-based conformational search was dramatically superior to Monte Carlo methods as it was one to two orders of magnitude faster for sequences ranging from 20 to 64 residues in length. It would be interesting to see this comparison performed with a more realistic model of protein structure.

Le Grand and Merz (1993b; Le Grand, 1993) used a modified form of their previously developed genetic algorithm (Le Grand and Merz, 1993a) and a new, knowledge-based potential function to predict the tertiary structures of the proteins melittin (2mlt), avian pancreatic polypeptide (1ppt), and crambin (1crn), as well as the tertiary structure of the four-helix bundle protein cytochrome b_{562}, where the constraint of correct α-helical secondary structure was enforced. The modifications to the original algorithm involved the use of a fragment library rather than a rotamer library. A fragment library is similar to a conformational dictionary, as used in the work of Sun (1993), except that no constraints were placed on the lengths of the fragments that could be sampled. In practice, however, no fragments were sampled that were greater than four amino acids in length. The best predicted structure of melittin had an RMSD of 1.34 Å relative to the native structure, while the best predicted structure of avian pancreatic polypeptide had an RMSD of 4.75 Å. In both cases, the fitness of the predicted structures was significantly better than that of the native structure. In the case of crambin, the predicted structure had an RMSD of 9.06 Å relative to the experimental structure. This indicated that the knowledge-based potential function probably would not work well with larger proteins. The existence of this problem was verified by predicting the tertiary structure of four α-helix bundle protein cytochrome b_{562} given the constraint of correctly predicted secondary structure. The final predicted structure was a straight, extended α-helical rod rather than a compact α-helix bundle, and was of vastly superior fitness compared to the native structure. This research showed that even if the genetic algorithm or some other conformational search technique can locate the global optimum conformations of proteins under various fitness functions, these fitness functions must also have the native structures of proteins as their global optimum conformations. This is not an easy task to achieve, and more research is needed in this area.

Bowie and Eisenberg (1991) predicted the tertiary structures of 434 repressor, a 57-residue engrailed homeodomain, and a 50-residue fragment of the B domain of protein A using a genetic algorithm-based conformational search algorithm which utilized a conformational dictionary composed of fragments ranging from 9 to 25 residues in length along with a 3D1D profile-based fragment screening algorithm (Bowie et al., 1991), and a knowledge-based six-term potential function specifically tuned to reproduce the distance matrix error (DME) of a training set of 434 repressor (1r69) conformers. The profile screening algorithm was used, instead of requiring strict homology, in order to increase the pool of allowable fragments of longer sequence length. Two attempts to fold 434 repressor located best conformers within 3.0 Å DME of the native structure; however, as mentioned above, the fitness function was biased to reproduce this native structure. Two attempts to fold the 57-residue engrailed homeodomain located best structures within 4.0 Å DME and one final structure within 2.4 Å. Finally, three attempts to fold the B-domain fragment of protein A located final conformers within 4.0 Å DME of the native structure.

4. Conclusions

The application of the genetic algorithm to the task of protein tertiary structure prediction is still in an early phase. The genetic algorithms described here, as well as the fitness functions, differ greatly. These differences eliminate the possibility of making any general conclusions about the superiority of the genetic algorithm to other search techniques, such as simulated annealing or dynamic programming. The universal elements of these genetic algorithms are the use of a population rather than a single conformer and the use of some form of crossover to create new conformers. It is not clear whether the use of crossover is superior to other approaches, but the use of a population in a genetic algorithm rather than a single conformer gives the genetic algorithm a memory that is not usually present in other search techniques. It seems likely that this will be the most important contribution to conformational search algorithms.

Finally, even if genetic algorithms or another optimization technique are capable of locating the global optimum conformations of proteins, this capability is of no use for the purpose of tertiary structure prediction unless the global optimum conformations correspond to the native states. Achieving this may turn out to be a much harder problem than conformational search.

Acknowledgments. We would like to thank Drs. James Bowie, David Eisenberg, Greg Farber, and Juliette Lecomte for stimulating conversations; Dr. David Eisenberg, The Cornell Theory Center, and The Pittsburgh Supercomputing Center for computer time; Cray Computers and the National Science Foundation for fellowship support; and the Office of Naval Research for supporting this research (N00014-90-3-4002).

REFERENCES

Anet F (1990): Inflection points and chaotic behavior in searching the conformational space of cyclononane. *J Am Chem Soc* 112:7172–7178

Anfinsen C (1959): *The Molecular Basis of Evolution.* New York: John Wiley & Sons

Booker L (1987): Improving search in genetic algorithms. In *Genetic Algorithms and Simulated Annealing*. San Mateo, CA: Morgan Kaufmann

Bowie J, Eisenberg D (1991): An evolutionary approach to folding proteins from sequence information: application to small α-helical proteins. (Unpublished results.)

Bowie J, Lüthy R, Eisenberg D (1991): A method to identify protein sequences that fold into a known three-dimensional structure. *Science* 253:164–170

Brooks B, Bruccoleri R, Olafson B, States D, Swaminathan S, Karplus M (1983): CHARMM: A program for macromolecular energy, minimization, and dynamics calculations. *J Comp Chem* 4:187–217

Cleveland G, Smith S (1989): Using genetic algorithms to schedule flow shop releases. In: *Proceedings of the Third International Conference on Genetic Algorithms*. San Mateo, CA: Morgan Kaufmann

Covell DG, Jernigan RL (1990): Conformations of folded proteins in restricted spaces. *Biochemistry* 29:3287–3294

Crippen G (1991): Prediction of protein folding from amino acid sequence over discrete conformational spaces. *Biochemistry* 30:4232–4237

Dandekar T, Argos P (1992): Potential of genetic algorithms in protein folding and protein engineering simulations. *Protein Engineering* 5:637–645

Davidor Y (1989): Analogous crossover. In: *Proceedings of the Third International Conference on Genetic Algorithms*. San Mateo, CA: Morgan Kaufmann

Fasman G (1989): Development of the prediction of protein structure. In: *Prediction of Protein Structure and the Principles of Protein Conformation*. New York: Plenum Press

Fogarty T (1989): Varying the probability of mutation in the genetic algorithm. In: *Proceedings of the Third International Conference on Genetic Algorithms*. San Mateo, CA: Morgan Kaufmann

Friedrichs MS, Wolynes PG (1989): Genetic algorithms for model bimolecular optimization problems. (Unpublished.)

Goldberg D (1989): *Genetic Algorithms in Search, Optimization, and Machine Learning*. San Mateo, CA: Addison-Wesley

Grefenstette J, Baker J (1989): How genetic algorithms work: A critical look at implicit parallelism. In: *Proceedings of the Third International Conference on Genetic Algorithms*. San Mateo, CA: Morgan Kaufmann

Holland J (1975): *Adaptation in Natural and Artificial Systems*. Ann Arbor, MI: University of Michigan Press

Judson RS (1992): Teaching polymers to fold. *J Phys Chem* 96:10102–10104

Judson RS, Colvin ME, Meza JC, Huffer A, Gutierrez D (1992): Do intelligent configuration search techniques outperform random search for large molecules? *Int J Quant Chem* 44:277–290

Lau KF, Dill KA (1989): A lattice statistical mechanics model of the conformational and sequence spaces of proteins. *Macromolecules* 22:3986–3997

Lavery R, Sklenar H, Zakrzewska K, Pullman B (1986): The flexibility of nucleic acids: (II) The calculation of internal energy and applications to mono nucleotide DNA. *J Biomol Struct Dynam* 3:989–1014

Le Grand SM, Merz KM, Jr (1993a): The application of the genetic algorithm to the minimization of potential energy functions. *J Global Opt* 3:49–66

Le Grand SM, Merz KM, Jr (1993b): The application of the genetic algorithm to protein tertiary structure prediction. *J Mol Biol* (to be submitted)

Le Grand SM (1993): Doctoral thesis. The Pennsylvania State University

Momany F, McGuire R, Burgess A, Scheraga HA (1975): Energy parameters in polypeptides. VII. Geometric parameters, partial atomic charges, nonbonded interactions, hydrogen bond interactions, and intrinsic torsional potentials for the naturally occurring amino acids. *J Phys Chem* 79(22):2361–2381

Montana D, Davis L (1989): Training feedforward neural networks using genetic algorithms. In: *Proceedings of the Third International Conference on Genetic Algorithms*. San Mateo, CA: Morgan Kaufmann

Nemethy G, Scheraga H (1990): Theoretical studies of protein conformation by means of energy computations. *FASEB Journal* 4(14):3189–3197

Ponder JW, Richards FM (1987): Tertiary templates for proteins: Use of packing criteria in the environment of allowed sequences for different structural classes. *J Mol Biol* 193:775–791

Radcliffe N, Wilson G (1990): Natural solutions give their best. *New Scientist* 126:47–50

Schaffer J, Morishima A (1987): An adaptive crossover distribution mechanism for genetic algorithms. In: *Genetic Algorithms and their Applications: Proceedings of the Second International Conference on Genetic Algorithms*. Hillsdale, NJ: Lawrence Erlbaum

Sippl MJ (1990): Calculation of conformational ensembles from potentials of mean force: an approach to the knowledge-based prediction of local structures in globular proteins. *J Mol Biol* 213:859–883

Sirag D, Weisser P (1987): Toward a unified thermodynamic genetic operator. In: *Genetic Algorithms and their Applications: Proceedings of the Second International Conference on Genetic Algorithms*. Hillsdale, NJ: Lawrence Erlbaum

Skolnick J, Kolinski A (1990): Simulations of the folding of a globular protein. *Science* 250:1121–1125

Snow ME (1992): Powerful simulated-annealing algorithm locates global minimum of protein-folding potentials from multiple starting conformations. *J Comp Chem* 13:597–584

Sun S (1993): Reduced representation model of protein structure prediction: statistical potential and genetic algorithms. *Protein Science* 2:762–785

Tuffery P, Etchebest C, Hazout S, Lavery R (1991): A new approach to the rapid determination of protein side chain conformations. *J Biomol Struct Dynam* 8(6):1267–1289

Unger R, Moult J (1993): Genetic algorithms for protein folding simulations. *J Mol Biol* 231:75–81

Vasquez M, Nemethy G, Scheraga HA (1983): Computed conformational states of the 20 naturally occurring amino acid residues and of the prototype residue a-aminobutyric acid. *Macromolecules* 16:1043–1049

Walbridge C (1989): Genetic algorithms: What computers can learn from Darwin. *Technology Review* 92:46–53

Wayner P (1991): Genetic algorithms. *Byte* 16(Jan):361–364

Weiner S, Kollman P, Nguyen D, Case D (1986): An all atom force field for simulations of proteins and nucleic acids. *J Comp Chem* 7(2):230–252

Whitley D, Hanson T (1989a): The GENITOR algorithm and selective pressure: Why rank-based allocation of reproductive trials is best. In: *Proceedings of the Third International Conference on Genetic Algorithms.* San Mateo, CA: Morgan Kaufmann

Whitley D, Hanson T (1989b): Optimizing neural nets using faster, more accurate genetic search. In: *Proceedings of the Third International Conference on Genetic Algorithms.* San Mateo, CA: Morgan Kaufmann

Whitley D, Starkweather T, Bogart C (1990): Genetic algorithms and neural networks: optimizing connections and connectivity. *Parallel Computing* 14:347–361

Whitley D, Starkweather T (1990): GENITOR II: A distributed genetic algorithm. *J Exp Theor Artif Intell* 2:189–214

Wilson S, Cui W (1990): Applications of simulated annealing to peptides. *Biopolymers* 29:225–235

Wilson C, Doniach S (1989): A Computer model to dynamically simulate protein folding: studies with crambin. *Proteins* 6:193–209

5

Conformational Search and Protein Folding

Robert E. Bruccoleri

1. Introduction

Without protein folding, there would be no life as we know it. Genetic information, stored in nucleic acids, specifies a one-dimensional sequence of amino acids that comprise protein molecules. Yet, after synthesis, protein molecules spontaneously fold into a precise three-dimensional structure. Only after the folding process has completed can proteins perform their myriad functions, such as catalysis, regulation, chemical transport, motility, structural support, etc.

The ability to accurately predict the folded form of a protein from its sequence would be very valuable for a number of reasons. First, it would realize the full potential of rapidly growing genetic knowledge. The Human Genome Project will provide the sequence of all human genes. In the absence of the three-dimensional structure of the proteins that these genes encode, this sequence information is of limited utility. The detailed understanding of the function of these proteins requires knowledge of their atomic structure.

Second, structural information provides the basis for rational drug design. If we wish to modify the operation of a protein or correct a defective one, knowledge of the structure greatly simplifies the development of pharmaceuticals to achieve these effects. In addition, the process of intermolecular recognition that takes place when a ligand binds to a protein is driven by the same energetic considerations as the intramolecular recognition that results in protein folding. A fuller understanding of protein folding will

The Protein Folding Problem and Tertiary Structure Prediction
K. Merz, Jr. and S. Le Grand, Editors
© Birkhäuser Boston 1994

lead to a better understanding of drug binding and molecular recognition in general.

Third, knowledge of the energetics of protein folding will be a great asset in the production of genetically altered proteins. For example, it would be possible to improve the expression of proteins or to increase the stability of enzymes that must operate in hostile environments.

Fourth, *de novo* design of proteins will be facilitated by the knowledge of the sequence features that control the final fold of a protein, and more importantly, the regions of sequence that can be engineered without affecting the fold.

A graphic illustration of protein folding is shown in Figure 5-1. On the left is shown the protein in linear form, where all the backbone and side-chain torsion angles have been set to the *trans* value. Its linear nature reflects the information content encoded in nucleic acids and the linear synthesis process as it takes place on the ribosome. On the right is the protein in its native form. Within a brief interval, typically seconds, after synthesis, the protein folds into this form. On the surface, the native structure of the protein would appear to be a random assemblage of atoms, but in fact the structure is a highly precise and reproducible fit of the amino acid residues in the protein sequence. The driving force behind this molecular rearrangement is the quest for stability, i.e., any molecular system under constant pressure and temperature will spontaneously reduce its Gibbs free energy and dissipate heat.

The prediction of protein structure from the amino acid sequence is a difficult problem for two fundamental reasons. First, the number of possible protein conformations is very large. Each amino acid residue has a number of rotational degrees of freedom that are accessible at physiological temperatures. Although there is substantial coupling of these degrees of freedom and there are restrictions in the range of motion, the number of possible states for proteins with hundreds of amino acids is enormous. This conformational space is orders of magnitude larger than can be enumerated on modern computers.

The second reason for the difficulty in predicting protein structure is the lack of an accurate function for calculating the Gibbs free energy of a protein in solution. The Gibbs free energy of a particular protein conformation is determined by a number of factors. Some of these factors, such as the structural geometry and packing, can be calculated accurately and quickly. Others, such as electrostatics, solvation, hydrophobicity, and entropic effects, remain a significant theoretical challenge. Further, the difference in Gibbs free energy between the native, folded form and the denatured, unfolded form of proteins is typically very small (Dill, 1990).

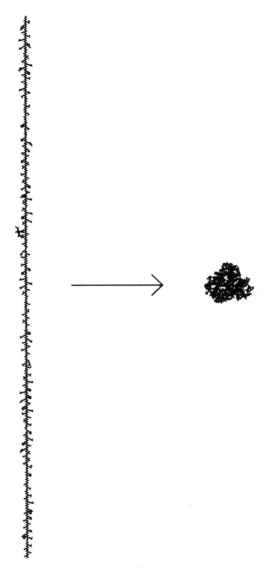

Figure 5-1. An illustration of the folding of myoglobin. On the left is a model of sperm whale myoglobin where all the backbone and rotatable side-chain torsion angles have been set to 180°, and the model was subsequently minimized with mild positional constraints to relax any gross modeling errors. This model symbolizes the linear nature of the protein structure, but is not a reflection of the unfolded form of the protein. On the right is the crystal structure of myoglobin. (Reprinted with permission of Academic Press from Phillips, 1980.)

Thus, high accuracy is needed to differentiate between the correct and incorrect structures.

The strategy that has been selected for solving the problem begins with the realization that the energy function is largely independent of exploration of the conformational space, and thus these two problems can be separately attacked. For the initial work, efforts have been focused on the structure of short segments (loops) within proteins. Such segments have the advantage that their conformational space is restricted, and therefore, one can sample their entire conformational space in a reasonable amount of computer time. The availability of a complete sampling provides a powerful test of the second part of the problem, an accurate free energy function. The complete sampling guarantees that the true minimum of the energy can be evaluated over the conformational space. Comparison of the minimum energy conformer against an experimentally determined structure represents the ultimate test of a structure prediction method. In addition, such comparisons can provide insights into improving the energy functions. In parallel with improvements to the energy function, one can also improve upon the algorithms for searching conformational space. Directed searches, genetic searching algorithms (Goldberg, 1989; Le Grand and Merz, 1994 [Chapter 4 of this volume]), and parallel searching are currently under investigation.

In this review, the progress in solving the protein folding problem along the lines of this strategy will be presented. First, the conformational search algorithm and its implementation in the program CONGEN will be described first. Then, the results of using CONGEN on a number of systems will be presented along with a comparison of the results to experimental data. Finally, improvements in the search methodology will be presented.

2. Conformational Search

In its most general form, a conformational search is just a set of nested iterations of the degrees of freedom in the system. In the initial implementation of CONGEN (Bruccoleri and Karplus, 1987), the degrees of freedom were encoded directly into the program and were capable of generating conformations for a single loop. It was clear from this prototype that the necessary operations inherent in such a search could be generalized to quasi-independent operators. These operators could then be combined in any reasonable way, and as a result a great variety of searches could be performed.

2.1 General Principles of the Conformational Search of Loops

The fundamental problem of generating loop conformations is finding a set of atomic coordinates for the backbone and side chains that satisfy all the stereochemical and steric constraints. For the sake of efficiency, it is presumed that bond lengths and bond angles are fixed, and in addition it is assumed that the peptide ω torsion angle is also planar. Under these assumptions the only degrees of freedom in the loop are the torsion angles. The conversion of these torsion angle degrees of freedom into atomic positions is straightforward, and is illustrated in Figure 5-2. Given a torsion angle defined by four atoms (A, B, C, and X), the bond length (C–X), the bond angle (B–C–X), and the coordinates of atoms (A, B, and C), we can determine the coordinates for atom X (Brooks et al., 1983).

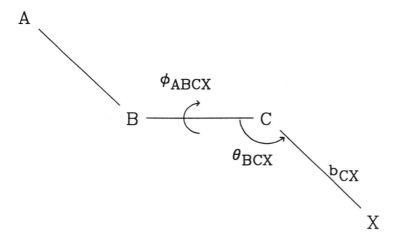

Figure 5-2. Given the positions of the three atoms A, B, and C, and the bond length b_{CX}, the bond angle θ_{BCX}, and the torsion angle ϕ_{ABCX}, the position of X can be determined. The value of the bond length may be viewed as restricting the locus of X to a sphere, the bond angle further restricts the locus to a circle, and finally the torsion locates the point on the circle. (Adapted from Bruccoleri and Karplus, 1987 and used by permission. © 1987 John Wiley & Sons, Inc.)

Given the chemical structure of proteins, the search process is divided into backbone and side-chain constructions. The backbone conformational space is normally sampled before the side chains because the chain closure condition is very restrictive. As a result, fewer samples are generated early in the process, which helps to reduce the necessary computer time.

2.1.1 Backbone construction. The generation of backbone coordinates depends heavily on the modified Gō and Scheraga chain closure algorithm

(Gō and Scheraga, 1970; Bruccoleri and Karplus, 1985). The algorithm is designed to calculate local deformations of a polymer chain, i.e., to find all possible arrangements of a polymer anchored at two fixed endpoints. Given stereochemical parameters for the construction of the polymer, and six adjustable torsion angles between the two fixed points, this algorithm will calculate values for the six torsion angles in order to perfectly connect the polymer from one endpoint to the other. In the sampling of the backbone, the use of a planar ω torsion angle reduces the number of free backbone torsion angles per residue to two, and therefore three residues are required for the application of the Gō and Scheraga algorithm. For generating conformations of loops with more than three residues, the backbone torsion angles of all but three residues are sampled, and the Gō and Scheraga procedure is used to close the backbone.

Bruccoleri and Karplus (1987) modified the Gō and Scheraga algorithm to allow small changes in the peptide bond angles (Bruccoleri and Karplus, 1985). After the first implementation of the algorithm, it was tested by deleting three residue segments in several different proteins and calling on the algorithm to reconstruct the peptide backbone. In the helices in flavodoxin (Smith et al., 1977) the algorithm failed to reconstruct a large number of these segments. An attempt was then made to reconstruct ideal α-helices where individual bond angles were perturbed by a few degrees. Many of these trials failed. Thus, it was clear that the normal variations in bond angles due to the limits of crystallographic resolution were interfering with the algorithm. By allowing the algorithm to adjust the bond angles by small amounts, typically no more than 5°, it was possible to find torsion angles that would close all three residue segments in the helices (Bruccoleri and Karplus, 1985).

The free sampling of backbone torsion angles is done with the aid of a backbone energy map. Bruccoleri and Karplus (1987) calculated the energetics of constructing the backbone for three different classes of amino acids: glycine, proline, and all the rest. This information is stored as a map (Ramachandran et al., 1963) that gives the energy as a function of discrete values ϕ, ψ, and ω, where ω can only be 0° (*cis*) or 180° (*trans*). A set of maps corresponding to grids of 60°, 30°, 15°, 10°, and 5° have been calculated; typically, a 30° sampling is sufficiently fine for good agreement.

With regard to the peptide ω angle, only the proline ω angle is normally allowed to sample *cis* values. However, CONGEN can be directed to sample *cis* ω angles for all amino acids.

The ring in proline creates special problems. The proline ring constrains the ϕ torsion to be close to −65°; any deviation from −65° distorts the ring. The minimum energy configuration of the proline ring (specifically, 1,2 dimethyl pyrrolidine) has been determined for a range of ϕ angles ($\pm 90°$) around $-65°$ using energy minimization with a constraint on ϕ, and a file has been constructed that contains these energies and the construction parameters necessary to calculate the position of C_β, C_γ, and C_δ of the proline. All of these energies are adjusted relative to a minimum ring energy equal to zero. After a chain closure is performed, any conformations that have a proline ϕ angle whose energy exceeds the minimum energy by more than the parameter, ERINGPRO, are discarded. Generally, a large value for ERINGPRO is used (50 kcal/mole) so that the chain closure algorithm does not overly restrict proline closures. The *cis–trans* peptide isomerization is handled by trying all possible combinations of *cis* and *trans* configurations. The user has complete control over which residues can be built in the *cis* isomer. Since there are only three residues involved in the chain closure, this results in no more than eight (2^3) attempts at chain closure.

There are two optimizations performed during the sampling of backbone torsions. First, whenever any atom is constructed, a check is made to see if the atom overlaps with the van der Waals radius of any other atom in the system; if so, that conformation is discarded. Second, as backbone residues are generated, CONGEN calculates the distance from the growing end back to the other fixed point. If that distance is greater than can be reached by fully extended backbone, then those conformations are discarded.

The backbone can be constructed either forward from the N-terminus or backward from the C-terminus order until only three residues remain. The N-terminus of the internal segment is anchored on the peptide nitrogen; the C-terminus is anchored on C_α. When the construction direction is from the N terminus to the C terminus, the first torsion to be sampled in a residue is the ω angle (which normally is sampled just at 180°, and can be sampled at 0° for prolines or, as an option, all the amino acids). It determines the C_α and the peptide hydrogen positions. The ϕ angle determines the position of the carbonyl carbon and C_β of the side chain; and finally, the ψ angle determines the carbonyl oxygen and peptide nitrogen of the next residue. When the construction is in the reverse direction, the ψ angle determines the peptide nitrogen; the ϕ angle determines the carbonyl carbon of the preceding residue, the peptide hydrogen, and C_β; and the ω angle determines the position of the preceding residue's C_α and carbonyl oxygen.

2.1.2 Side-chain construction.

Given a set of backbone conformations, it remains to generate a set of side-chain atom positions for each of the backbone conformations. This problem is divided into two parts, construction of individual side chains and combining results from individual side chains for all the residues.

As with the backbone atom placement, the atoms of a side chain are positioned based on free torsion angles. The side-chain torsions are processed from the backbone out as each succeeding atom requires the position of previous atoms for its placement. The sampling interval of each torsion can be either some fixed number of degrees or the period of the torsion energy. If the latter is used, and the parameters for the torsion energy specify only a single term in the Fourier series for the torsion energy, then the side-chain torsion energy is always zero.

It is common for one free torsion to generate the position of more than one atom because of side-chain branching, nonrotatable bonds, and rings. For example, although a tryptophan defined by an explicit hydrogen topology file (Brooks et al., 1983) has 11 side-chain atoms to be placed, it has only two free torsion angles. Also, some side-chain branching is symmetric, e.g., phenylalanine, and CONGEN can use such symmetry to reduce the sampling necessary.

As with the backbone construction, a search of the surrounding space is made for any constructed atom to see if there are any close contacts. However, with the side chains, there are two ways of checking for such overlaps. The first method is very simple: given the sampling of the torsion angles, each atom is constructed and checked for contacts.

The second method, van der Waals avoidance, is more time consuming, but it yields better quality structures. It is a straightforward geometrical problem to determine the range of torsion angles that will avoid constructing an atom within a given distance of other atoms in the system. As a side-chain torsion angle χ_i varies, it specifies a circular locus of points on which atoms can be constructed (Figure 5-3) (Bruccoleri and Karplus, 1987). If atoms in the vicinity of this circle are examined, the sectors of the circle that will result in the repulsive overlap of the constructed atom with its spatial neighbors can be calculated. The complement of these sectors can be used to determine values for the χ_i angles that avoid bad contacts.

The information needed for side-chain construction is stored in a side-chain topology file. It is a straightforward matter to add new amino acids to this file so that the structure of unnatural amino acids can be predicted.

Given this method for constructing individual side chains, it remains to combine side-chain conformations for all the side chains attached to a particular backbone conformer.

Since the backbone construction process provides the position of C_β, there is a strong bias to the side-chain orientation. Thus, an acceptable course of action is the generation of only one side-chain conformation for each backbone conformation. A substantial effort must be made to ensure that this one conformation has the lowest energy possible for the given

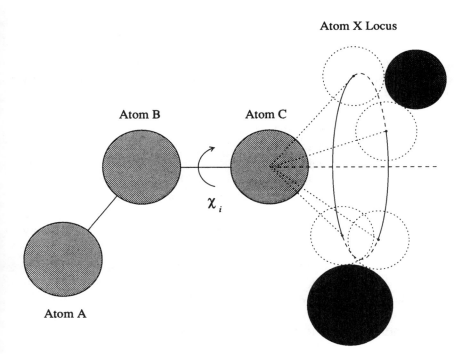

Figure 5-3. An illustration of van der Waals avoidance. The construction of atom X is based on the positions of atoms A, B, and C, the C–X bond distance, and B–C–X bond angle. Depending on the value of χ_i, the center of atom X can be located anywhere on the circle illustrated as the "Atom X Locus." Neighboring atoms, shown in dark gray, will block out parts of the circle, shown in dashed lines. The remaining part of the locus identifies the values of the torsion angle for which there are no van der Waals overlap. For illustrative purposes, the radii of the atoms in this figure are much smaller than actual values. Adapted from Bruccoleri and Novotny (1992) and used by permission. © 1992 Academic Press, Inc.

backbone. Second, because the side chains close together in sequence frequently are not close together in space, and therefore do not interact strongly, it is a reasonable approximation to treat the side chains quasi-independently. Instead of finding all combinations of side-chain atomic positions, the side chains can be processed sequentially so the time required for side-chain placement increases linearly, rather than exponentially, with the number of residues.

In order not to limit the options for using the program, five possible methods for generating side-chain positions have been implemented. Some of the methods can generate only one side-chain conformer; others can generate many. The first two methods described, ALL and FIRST, assume no quasi-independence of the side chains, whereas the others do.

The first method, named ALL, generates all possible conformations by a series of nested iterations over every side chain as described above. The second method, named FIRST, uses the same algorithm as ALL except that all the iterations terminate when the first conformation for all the side chains has been found. This method is useful for determining if a backbone conformation will accommodate the side chains when details about the side-chain energetics are not required.

The three other methods all depend on a function that evaluates the side-chain positions as they are generated so that the best ones can be selected. "Best" is defined as the conformation whose evaluation function is numerically smallest. Two evaluation functions are currently provided, one based on positional deviations, and one based on the CHARMM energy function (Brooks et al., 1983). The evaluation function based on positional deviations is present for testing CONGEN, as it provides a means for determining the limit of CONGEN's ability to generate a known structure. If coordinates are available for the side-chain atoms, this evaluation function will determine the RMS shift between a generated side chain conformation and the initial coordinates. The second evaluation function computes the CHARMM energy of the side-chain atoms.

In the first of these other methods, named INDEPENDENT, the atoms in each side chain are placed independently, with the atoms of the other side chains in the peptide being ignored; interactions with all other atoms in the system are included. The conformation that has the lowest value for the evaluation function is selected for each side chain. When the RMS evaluation function is used, this method gives the optimum conformation, though it may be sterically inappropriate. However, it should not be used when the energy is the evaluation function because it can generate conformations with large repulsive van der Waals contacts between side-chain atoms.

The COMBINATION method begins by generating a small number of the best side chain conformations for each side chain independently, as above. Then, these side-chain conformations are assembled in all possible combinations, and those combinations that do not have bad van der Waals contacts are accepted. The number of conformations saved for each side chain must be small to avoid a combinatorial explosion.

The ITERATIVE method starts with an energetically acceptable side-chain conformation for all the side chains. This conformation is generated, if possible, using the FIRST method. Starting with this conformation, all the possible positions for the side-chain atoms of the first residue are recalculated, and the conformation with the lowest energy is selected. The value of the evaluation function is also saved. This regeneration is done with all the other side-chain atoms present so that their effect can be accounted for. The process is repeated sequentially for the rest of the side chains in the gap. The process then returns to the first residue and it is repeated over each side chain until the energies of the side-chain atoms do not change or until the number of passes reaches an iteration limit. This method has the virtue that only one conformation is generated per backbone conformation, and it is an energetically reasonable one. However, if there are significant interactions between the side-chain atoms, the initial state of the side chains will bias the iterative process, and the lowest energy side-chain conformation may be missed. A test of this method is described in Section 3.5.1.

With any of the five methods described above, the CONGEN command can apply any of the minimization algorithms to the generated conformations. Minimization provides an ability to reduce the small van der Waals repulsions that are inevitable with coarse torsion grids.

2.2 Organization of a CONGEN Conformational Search

Within CONGEN, a degree of freedom signifies a computer operation applied to a group (zero or more) of atoms by either sampling a set of variables or performing an operation on existing atoms. When a conformational search is specified, the user indicates which degrees of freedom are to be sampled and also their order. The program automatically sets up a nested iteration over all of them. Only successful samples of a degree of freedom will invoke the succeeding degrees of freedom. The process can be visualized graphically in Figure 5-4.

There are two reasons for taking this abstract approach to the operation of the search. First, it allows searches of arbitrary complexity to be performed. Second, the operations inherent in sampling a degree of freedom can be separated from the process of managing the search. Such modularity greatly simplifies the implementation of the program. In addition, one can apply the methods of state-space search as developed in research into artificial intelligence (Pearl and Korf, 1987).

Currently, six "degree of freedom" operators are provided in CONGEN. Three of them deal with atomic construction using stereochemistry. The backbone degree of freedom generates the position of the peptide backbone atoms, and the chain closure degree of freedom closes a loop. Because the Gō and Scheraga (1970) procedure finds multiple solutions to the chain closure, each solution is treated as a separate sample. The side-chain degree of freedom will construct side chains onto any number of backbone residues, and depending on the method, it will generate either single samples or multiple ones.

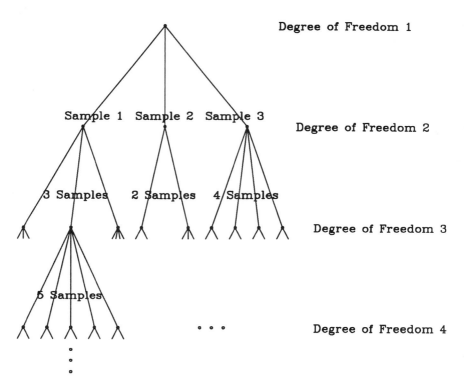

Figure 5-4. Presented is the top portion of a search tree showing how the sampling process can be represented. Each node in the tree represents a conformation. The root node of the tree, which is at the top of the figure, represents the system before the sampling of degrees of freedom has begun. Each successively deeper (meaning lower) level in the tree corresponds to a sampling of further degrees of freedom. The lines connecting nodes, or edges, represent the specific samples of degrees of freedom. The leaf nodes, which have no nodes emanating from them and which are not shown, are either completely constructed conformations or partially constructed conformations that were blocked because of the various constraints on the sampling. The completely constructed conformations will be as deep in the tree as there are degrees of freedom to be sampled. The blocked conformations will be higher up. Intermediate nodes represent partially constructed conformations with a particular sampling of the degrees of freedom above it. Constraints applied early serve to delete many more possible nodes than pruning later on in the generation process. Adapted from Bruccoleri and Karplus (1987) and used by permission. © 1987 John Wiley & Sons, Inc.

Two of the degrees of freedom are involved with input and output. The "write" degree of freedom writes a conformation to a file each time it is invoked. It can also do some limited filtering of what is written by comparing the energy of each conformer against the minimum energy seen thus far. Normally, this filter will greatly reduce the number of conformers written to a file. In all cases, this degree of freedom does not generate any atomic positions, and it always succeeds. The "read" degree of freedom can be viewed as an inverse of "write". It will read a set of conformations from a file, and then invoke succeeding degrees of freedom on each one. Conformations can be selected based on their energies, so it is possible to set up a "build-up" procedure (Pincus et al., 1982), where the best conformations from one search are used as the starting point for additional residues. Furthermore, this degree of freedom allows a user to input his own set of conformations, which can be generated by arbitrary means. This approach was used by Martin et al. (1991) to process loop conformations as found in a database.

The final degree of freedom is the "evaluate" degree of freedom. This operation is responsible for calculating either energies or root mean square (RMS) deviations. When used for energy evaluations, this degree of freedom can either calculate the energy, or it can perform minimization or dynamics on each of the conformers. When used for RMS deviations, it will compare the coordinates of the conformations against a reference coordinate set. This is used for testing the search process; in particular, to see if a search is capable of generating the original experimental coordinates.

2.3 Surface Area Rule

Because the energy function currently used in CONGEN is the *in vacuo* CHARMM potential energy, solvent effects are largely ignored. In modeling experiments with McPC 603, the conformational space of each of the loops was explored (Bruccoleri et al., 1988). In some cases, the lowest energy conformation was close to the x-ray structure. In other cases, the lowest energy conformation deviated significantly from the x-ray structure, but the next-lowest energy conformation agreed well. In all of these cases, the structures with the best agreement had the lowest accessible surface area. This made intuitive sense, as the solvent-accessible area of molecules is an approximate measure of the magnitude of solvent effects acting on the molecule; and proteins have been known to minimize their (nonpolar) solvent-exposed areas (Novotny et al., 1984, 1988). Therefore, until solvent effects are incorporated into the energy calculation, loop conformations can be selected by examining all the conformations within 2 kcal/mole of the minimum and using the one with lowest accessible surface energy. The "XCONF" command in CONGEN automates this task.

3. Applications of CONGEN to Protein Modeling

There have been a large number of applications of systematic conformational search to protein modeling. The most common type of problem is homology modeling. Since many proteins evolve through the accumulation of small mutations, there are large families of proteins that have both similar sequence and structure. The problem of predicting the structure of one protein from a homologous one is generally reduced to the prediction of the effect of the substitutions. In the case of many single substitutions, a conformational search can be used to model the replacement side chains. Where there are most extensive substitutions or short deletions and insertions, then the full loop-searching capabilities can be brought to bear.

In addition to homology modeling problems, CONGEN has also been applied to small peptide structure determination, theoretical studies of protein folding, and reconstruction of protein coordinates from C_α coordinates.

3.1 Testing on Single Loops

CONGEN was tested on a number of short segments in proteins with different secondary structures (Bruccoleri and Karplus, 1987). The testing was designed to answer two different questions. First, does the limitation in degrees of freedom to torsion angles restrict the conformational space so that native structures cannot be generated? Second, how well does the CHARMM energy function predict correct structures? Table 5-1 summarizes the results of these tests. Provided that the backbone is sampled well, the answer to the first question is that there is no limitation. As seen in the middle column of the table, conformers are usually found within 1 Å of the crystal structure. The second question is answered by the last column in Table 5-1. In many cases, single loops are predicted within 1 Å. In other cases, however, they are not. Figure 5-5 shows a comparison between the CONGEN-predicted structure and the x-ray structure for residues 41–47 in the heavy chain of the KOL Fab (Marquart et al., 1980).

3.2 Antibody Modeling

The antigen-combining sites of antibodies present ideal opportunities for experimenting with protein modeling procedures. First, the antibody structure, with its conserved β-sheeted framework and hypervariable combining site loops (antigen complementarity determining regions, or CDRs) is perfectly suited for loop-splicing experiments, both in the computer and in the laboratory (Jones et al., 1986). Figure 5-6 shows the high degree of structural homology. Second, some 15 x-ray crystallographic structures

Table 5-1. Testing conformational search on short loops.[a]

Secondary structure	Protein	Segment	Min RMS[b]	Min E RMS[c]
Helix	Flavodoxin (Smith et al., 1977)	127–131	0.610	0.896
Sheet	Plastocyanin (Colman et al., 1978)	80–84	0.860	1.150
Turn	McPC 603 (Satow et al., 1986)	L 95–99	1.349	2.660
Turn	McPC 603	L 98–102	1.074	1.436
Turn	Kol (Marquart et al., 1980)	H 41–45	0.650	1.114
Turn	Kol	H 41–46	0.685	0.789
Turn	Kol	H 41–47	0.743	0.882

[a] Tests of CONGEN using known proteins. Adapted from Bruccoleri and Karplus (1987) and used by permission. © 1987 John Wiley & Sons, Inc.
[b] This column presents the minimum RMS deviation to the crystal structure. For this calculation, the INDEPENDENT side-chain construction method was used, and side chains were evaluated based on their agreement to the crystal structure. Thus, this column shows the theoretical lower limit of agreement for conformations generated by CONGEN.
[c] Here, the ITERATIVE side-chain construction method was used; side chains were evaluated based on their CHARMM energies.

of antibody Fab fragments are currently available through the Brookhaven Protein Data Bank (Bernstein et al., 1977), and the number is steadily growing into an impressive structural base supporting both homology modeling and structural analysis of various antibody antigenic specificities. Finally, combining site model-building is often the only means of obtaining three-dimensional structural information to guide protein engineering of practically important antibodies (e.g., antitumor therapeutic immunoglobulins [Trail et al., 1993]).

The antigen-combining site resides in the antibody Fv fragment, a noncovalent dimer of heavy and light chain variable domains (VH and VL). The domains themselves consist of conserved framework regions (essentially, an antiparallel, eight-stranded β-sheet sandwich) and the six CDR loops (the L1–L3 loops in the VL domain and the H1–H3 loops in the VH domain). The loops vary in length and sequence among different antibodies, thus creating combining sites complementary to diverse antigens.

The protocol for modeling antibodies consists of three steps: matching sequence by homology, construction of the framework, and construction of the hypervariable loops.

The goal of the sequence homology step is to find the antibody of known structure whose sequence is closest to the antibody being modeled. This can be done using the standard homology matching algorithms (Needleman

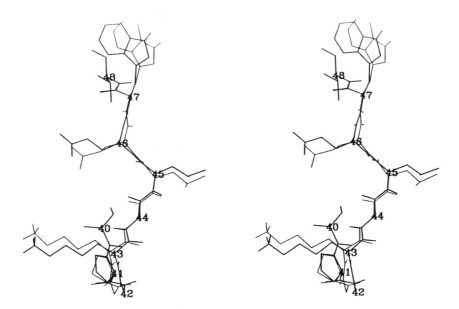

Figure 5-5. A stereo stick-figure showing the lowest energy conformation generated by CONGEN for heavy chain residues 41 to 47 in the antibody KOL (Marquart et al., 1980) (thin lines) and the x-ray conformation (thick lines). The RMS deviation between the two is 0.882 Å. Residues 40 and 48 which surround the generated segment are also included. Adapted from Bruccoleri and Karplus (1987) and used by permission. © 1987 John Wiley & Sons, Inc.

and Wunsch, 1970; Wagner and Fischer, 1974). A second factor that must be included in the choice of antibody is the length of loops that must be constructed using CONGEN. If there is a choice between two nearly homologous structures, then one should choose the one that gives the shortest difference in the hypervariable loops. Although the time will vary significantly with the number of glycine and proline residues in the loop, the local environment, and the orientation and location of the endpoints, CONGEN can perform complete searches over loops of up to about 10 residues using current computer technology.

The next step in the antibody construction protocol is building the framework. First, one must use the "splice" command in CONGEN to change the sequence of the antibody of known structure into the antibody being modeled. Second, the coordinates of the hypervariable loops of the reference structure must be deleted.

Finally, one must build the coordinates for the residues that changed in the framework. The "splice" command will preserve the backbone and C_β

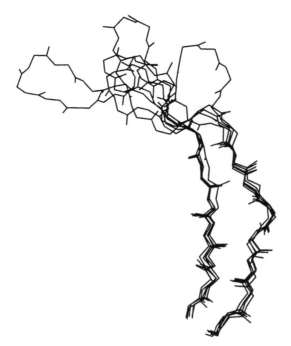

Figure 5-6. The H3 loops from five different antibodies—McPC 603 (Satow et al., 1986), New (Saul et al., 1978), Kol (Marquart et al., 1980), J539 (Suh et al., 1986), and HyHEL-5 (Sheriff et al., 1987)—are plotted along with the β-strand segments of the framework to which they are attached. The superpositions of these loops was made using the homologous residues in the entire variable domain, and therefore the quality of the fit reflects the conservation of the entire framework. From Bruccoleri and Novotny (1992) and used by permission. © 1992 Academic Press, Inc.

coordinates if the sequence substitutions are of equal length. In such cases, only a search over the side chains is necessary to build them. If a substitution requires a change in the length of the sequence, or if the substitution requires the removal or addition of a proline or glycine, then a loop search should be performed. The loop should be centered on the sequence changes, it should incorporate conserved positions from both sides of the changes, and it should be at least four residues long.

Typically, all the individual side-chain substitutions are modeled using a single side-chain degree of freedom using the iterative search method. Then each of the framework loop searches is done sequentially.

Once all the framework substitutions have been made, the hypervariable loops can be constructed. Because of computer time limitations, the searches must be performed in sequential order. Since the framework structure is known in advance, the hypervariable loops are constructed from the framework out, so that the known structure can provide a partial template for constructing the lower loops. The order used is L2, H1, L3, H2, H3, and L1.

Although the above protocol will work under ideal circumstances, Nature rarely provides ideal circumstances. However, procedures have been developed to deal with a number of problems than have arisen during antibody modeling.

One of the most common problems is the failure of CONGEN to find any conformations for a loop. A related problem is finding only high-energy conformers. In both cases, this is usually due to a small number of atoms that block the space near the endpoints of the loops because of imprecise positioning of side chains in the preceding framework construction. Usually, a visual inspection of the endpoints will reveal the problem. Alternatively, the conformational search debugging variables can be activated, and examination of the output will usually show a common residue involved in bad contacts. This second method must be used with caution, since the debugging output of the program can be voluminous.

Once the offending residue is found, the simplest solution is to incorporate it into the search of the loop itself. Because the degrees of freedom are general purpose operators, this is straightforward. Another option is to use a finer grid for the backbone residues in this region. If the space through which a loop must pass is narrow, the finer grid should help. A third option, which is less likely to solve the problem, is to change the order of searching the backbone. Depending on the residues where the chain-closure degree of freedom is applied, the conformations of the backbone will vary by a small amount.

Another common problem that arises in the search is that of loops too long to be searched in a reasonable time. The only option that has been tested is to break the search into pieces. Typically, two residues (both backbone and side chain) are sampled at a time from each end of the loop, and the resulting conformers are written to a file. Succeeding searches use a small subset consisting of the lowest energy conformers of the previous searches to extend the chain. The order of construction usually works from each end toward the center, although it is desirable to start at endpoints that are less exposed. Since less exposed endpoints have less space for sampling conformers, fewer conformers will result, and the likelihood that the correct conformer will be selected is increased.

3.2.1 McPC 603. The first published antibody modeling with CONGEN was performed on McPC 603, a phosphorylcholine-binding antibody (Satow et al., 1986). In this study, CONGEN was used to generate the conformations of each of the loops, and the results were compared against the x-ray structure as progress was made. This analysis of these conformations provided the basis for the accessible surface rule. The final results are shown in Table 5-2.

Table 5-2. Modeling of McPC 603.[a]

| | | RMS (Å) | |
Loop	Length	Total	Backbone
H1	5	1.7	0.7
H2	9	2.1	1.6
H3	8	2.9	1.1
L1	12	3.0	2.6
L2	6	1.9	1.6
L3	6	1.4	0.8
Totals	46	2.4	1.7

[a] The RMS deviations for the loops (complete and backbone only) as well as the loop lengths for the searches are given for each loop individually and for the six loops jointly. Adapted with permission from Bruccoleri et al.,1988; © 1988 Macmillan Magazines, Ltd.

3.2.2 HyHEL5. Using the protocol developed for McPC 603, a model of HyHEL5 (Sheriff et al., 1987) was constructed (Bruccoleri et al., 1988). The results are shown in Table 5-3. Because the loops in HyHEL5 were shorter than those in McPC 603, the construction was straightforward. However, two problems were noted upon comparison with the crystal structure. First, the conformer selected for the H1 loop protruded into the space occupied by the H2 loop. As a result, the H2 loop also deviated from the correct structure. This highlighted the difficulties with the sequential construction of the loops. Second, a number of the side chains deviated from the x-ray structure. This was due to the fact that the modeling was done in the absence of the lysozyme antigen, whereas the crystal structure was of the antibody, lysozyme complex. All of the deviations occurred with hydrophobic side chains that protruded to make contact with lysozyme, but were modeled to lie against the antibody.

3.2.3 The anti-digoxin antibody 26-10. Numerous monoclonal antibodies have been raised against digoxin, a cardiac glycoside (Mudgett-Hunter et al., 1985). Several properties of digoxin make it a valuable antigen in the study of antibody–antigen interactions. The antigen is quite large, and the

Table 5-3. Modeling of HyHEL5.[a]

| | | RMS (Å) | |
Loop	Length	Total	Backbone
H1	5	1.8	1.1
H2	7	3.1	2.1
H3	4	2.7	1.0
L1	5	1.8	0.6
L2	6	1.7	0.8
L3	5	4.1	1.1
Totals	32	2.4	1.5

[a] The RMS deviations for the loops (complete and backbone only) as well as the loop lengths for the searches are given for each loop individually and for the six loops jointly. (Adapted with permission from Bruccoleri et al.,1988; © 1988 Macmillan Magazines, Ltd.)

steroid moiety is nearly rigid, with only a single rotatable torsion angle, the bond to the lactone ring. Even this torsion angle is sterically hindered so that there are only two states for the rotation. There are dozens of analogs for digoxin so that one can easily probe the specificity of the interaction. Many of the antibodies against digoxin have dissociation constants in the nanomolar range or better (Mudgett-Hunter et al., 1985; Schildbach et al., 1991).

The antibody 26-10 was the first of the anti-digoxin antibodies modeled. McPC 603 was used as the reference antibody. Table 5-4 gives the definition of the loops used for the first model. Loop H1 was not modeled because the only sequence difference in the definition of the loop was H35 Glu → Asn (this turned out to be an error.)

In the first attempts at construction of the model, residues H Tyr 50 and H Trp 104 in the center of the combining site were modeled in a solvent-exposed position. It appeared from a visual inspection of the model that H Phe 32 and H Tyr 33 were interacting with H50. Since the definition for H2 started at H50, we believed that extending the search to H49 would allow for additional sampling for H50. Therefore the model was reconstructed with the extension of H2, and the side chains of residues H32 and 33 were included in the searches over H2 and H3. The resulting model had a cleft about the size of digoxin. About 10 side chains that were exposed to solvent and prominently located at the putative digoxin binding site were subjected to side-directed mutagenesis, and in almost all the cases mutations to the

Table 5-4. Modeling of 26-10.[a]

Loop	Residues	Length	RMS (Å) Total	RMS (Å) Backbone
H1[b]	Not done		5.1	3.5
H2	50–57	8	3.6	2.1
H3	98–106	9	5.5	1.0
L1	31–37	7	2.9	0.6
L2	55–66	6	5.0	0.8
L3	94–101	8	3.3	1.1
Totals			4.4	3.6

[a] The RMS deviations for the loops (complete and backbone only) as well as the loop lengths for the searches are given for each loop individually and for the six loops jointly. (Adapted from Bruccoleri (1993a) and used by permission; © 1992 Gordon and Breach Science Publishers S.A.)
[b] The deviations were calculated for residues 26–35.

selected positions resulted in alterations of antibody affinity for digoxin (Near et al., 1991; Schildbach et al., 1991, 1993).

Nevertheless, the experimental structure of 26-10 (Jeffrey et al., 1993) turned out to differ from the model that was constructed. The correct structure for 26-10 has a binding cavity lined with residues H33 Tyr, H50 Tyr, and H104 Trp as well as the framework residue, H47 Tyr.

Upon comparing the two structures in detail, it was found that the H1 loop was substantially different for residues 28–30. These residues were outside the range for the H1 loop as we had determined (Novotny et al., 1983), see Figure 5-7. As a result of this error and the sequential construction of the loops, H2 and H3 were constructed incorrectly.

The results with 26-10 clearly demonstrate the importance of determining the endpoints to be used for modeling the loops, and the need to improve upon the sequential modeling protocol. However, the fact that polypeptide segments with essentially identical sequences may differ radically in their conformations presents a major challenge to all protein modeling protocols (Kabsch and Sander, 1985).

3.2.4 Antibody 40-150 modeling. A curious mutation in the antibody 40-150 (Panka et al., 1988) has also been analyzed (Novotny et al., 1990). The mutation of H94 serine to arginine results in a thousand-fold reduction in binding constant. A subsequent deletion of the first two residues of the amino terminus of the heavy chain restores much of the binding energy.

H1 loop changes from McPC 603 to 26-10

	26	27	28	29	30
McPC 603:	GLY	PHE	THR	PHE	SER
26-10:	GLY	TYR	ILE	PHE	THR

	31	32	33	34	35
McPC 603:	ASP	PHE	TYR	MET	GLU
26-10:	ASP	PHE	TYR	MET	ASN

Figure 5-7. A comparison of the H1 hypervariable loop sequences between the initial molecule in the modeling, McPC 603, and 26-10. Sequence differences are underlined.

At first, reconstruction of the entire molecule of 40-150 was attempted. Unfortunately, the third hypervariable loop was too long to perform a successful search. CONGEN was then used to model just the environment around position H94, and it was discovered that the H arginine 94 can participate in a hydrogen-bond network with residues H Asp 101, L Arg 46, and L Asp 55. Thus, in this case, dealing with a group of polar atoms buried inside and required to satisfy their hydrogen bonding potentials, the CONGEN-generated conformations were ranked not according to their calculated *in vacuo* energy or their solvent-exposed surface but according to the strength and quality of the hydrogen bonding they participated in (i.e., the CONGEN-calculated H-bond potential). A serine in position H94 cannot participate in this network. When the two amino terminal residues are deleted, H Arg 94 becomes much more solvent exposed, and its charge would be solvated. As a result, it would not participate in the hydrogen bond network, and it would restore the network back to the state when serine was in position H94.

3.2.5 ANO2. ANO2 is an anti-dinitrophenyl antibody (Anglister et al., 1984) with very high homology to HyHEL-5 and HyHEL-10. It was modeled with CONGEN using two different starting structures (Bassolino et al., 1992). The modeling was completed prior to the solution of the x-ray structure (Brünger et al., 1991), and the two models were each compared against the x-ray structure; see Table 5-5. In both cases the RMS deviation to the heavy chain was about 2.5 Å, and the deviations to the light chain were either 2.0 Å or 2.1 Å.

Table 5-5. Modeling of ANO2.[a]

| | | RMS (Å) | |
Loop	Length	Total	Backbone
H1	9	3.9	2.4
H2	7	2.3	1.9
H3	7	6.0	4.0
L1	6	3.0	1.6
L2	7	0.9	0.9
L3	7	4.2	2.6
Totals	43	3.9	2.4

[a] The RMS deviations for the loops (complete and backbone only) as well as the loop lengths are given for each loop individually and for the six loops for Model A. The results from model B are very similar. (Adapted from Bassolino et al. (1992) with permission of Cambridge University Press.)

It is illustrative to review the protocols used for ANO2. The heavy chain of ANO2 is 73% homologous with the heavy chain of HyHEL-10, and the light chain of ANO2 is 83% homologous with the light chain of HyHEL-5. Thus, two models were built, one where the heavy chains are superimposed using a least-squares fit and the light chains are carried along, and a second model where the light chains are superimposed and the heavy chains are carried along. Both models are similar to each other except for modest variations in L3 (1.3 Å RMS), H1 (2.8 Å RMS), and H3 (2.0 Å RMS).

Some of the loop constructions were different from those encountered before. Loops L2 and H2 were nearly identical to the parent anti-lysozyme antibodies, and therefore no loop construction was performed. Loop L1 was constructed with no problems.

Loop H1 had a change from Asp to Tyr at position 27 and an alanine insertion into position 34. The loop from position 26 to 34 was too long for a single CONGEN run, so the loop was split into two overlapping searches from 26 to 30 and from 29 to 34.

Loop L3 was a seven-residue loop containing two sequential prolines. The first attempt to construct this loop failed to find any conformations. The search was then repeated with the backbone torsion grid for the prolines reduced to 15°, and low-energy conformations were obtained.

The H3 loop was also seven residues containing a proline. Using a 30° grid, no low energy conformations were found. Since there were no atoms blocking the endpoints, the search was repeated using a 15° search, and many good conformations were found.

3.3 The T Cell Receptor

The T cell receptor, coupled with proteins of the major histocompatibility complex (MHC), is the primary sensor of the cellular immune response (Ashwell and Klausner, 1990; Meuer et al., 1990). The antigen/MHC binding element is a dimer of two subunits, α and β, linked by a disulphide. There are sufficient sequence and structural similarities to strongly suggest that the T cell receptor has a variable domain with the same structure as the variable domain of antibodies (Novotny et al., 1986; Chothia et al., 1988; Claverie et al., 1989).

Novotny et al. (1991) devised a single-chain flourescein binding T cell receptor (named RFL3.8) from the variable domains of the receptor and constructed a model of it using the antibody construction protocol described above, except that the $\beta 3$ hypervariable loop was too long to be completed in reasonable time. The conformational search on this loop was truncated after seven days of computation, and the lowest energy conformer found was used in the model. In addition, the 23-residue linker between the α and β chains was modeled by using a very narrow range of ϕ, ψ torsion angles for the backbone conformations. Since the structure of the linker is expected to be disordered in solution because of its large percentage of glycines, the goal of this linker model was to generate a plausible structure.

Although experimental structural studies on the single-chain T cell receptor remain to be performed, two site-directed mutation experiments suggest that the structure is largely correct. First, after the receptor was first cloned, it was noted that it was poorly soluble. Examination of the model revealed that five hydrophobic side chains were present on the surface opposite to the combining site. In immunoglobulins, these residues were all hydrophilic. All five of these hydrophobic residues were mutated to hydrophilic residues. The resulting receptor was more soluble than the original and had nearly identical fractionation and binding properties.

Second, the model of the single-chain T cell receptor was superimposed on the anti-flourescein antibody 4-4-20 (Herron et al., 1989). The model shows a cavity whose size and relative location is similar to that of the antibody. Six residues within the cavity of the RFL3.8 T cell receptor as well as a lysine residue outside the combining site were mutated to alanine. Five of the six mutants made from residues within the cavity lost binding to flourescein, and the mutant of the external lysine had no change in binding relative to the wild type receptor (Ganju et al., 1992). Although the experience with 26-10 illustrates that the T cell receptor model could be incorrect, these experimental results are very encouraging.

3.4 Small Peptides

Conformational search has been extensively applied to the structure of small peptides (Dygert et al., 1975; Deber et al., 1976; Venkatachalam et al., 1981; Hall et al., 1982; Hall and Pavitt, 1985; Madison, 1985). CONGEN has the necessary code and topology files for D amino acids, as well as the logic for handling cyclic peptides where the residues are joined in the backbone.

This capability was used in the determination of the structure of cyclo-(D-Trp-D-Asp-L-Pro-D-Val-L-Leu) (Krystek et al., 1992), an antagonist for the endothelin ET_A receptor. The structure was determined by both NMR spectroscopy and by global energy minimization using conformational search. Since the peptide was cyclic, the modified Gō and Scheraga chain closure algorithm could be used on the backbone, and therefore the free backbone torsions could be sampled using a 10° or 15° grid. All backbone conformations were minimized by 50 steps of Adopted Basis Newton–Raphson minimization (Brooks et al., 1983).

The RMS deviation for heavy atoms between the average experimental structure and the lowest two-energy CONGEN structures were 0.25 Å and 2.12 Å. The backbone RMS deviations for these two structures were 0.22 Å and 0.37 Å. The difference between the two lowest-energy structures was the orientation of the tryptophan side chain. Otherwise, the CONGEN-determined structure was nearly identical to the experimental structure.

3.5 Side-Chain Reconstructions

Because the side-chain degree of freedom can be invoked independently of any other degree of freedom, CONGEN has been used in a number of applications where side-chain positions needed to be modeled (Novotny et al., 1984; Bruccoleri et al., 1986; Allen et al., 1987; Novotny et al., 1988; Krystek et al., 1991).

3.5.1 Test of side-chain construction. In an "experimental" test of the side-chain building capabilities of CONGEN, the backbones of an immunoglobulin VL domain (McPC 603) and myohemerythrin were stripped of their side chains and then rebuilt. Two different orders for iteration over the side chains were used: sequential from the N-terminus, or ordered by increasing distance from the center of gravity of the domain. In both cases the same result was obtained; namely, the buried core of the domain, where the side-chain packing density was the highest, was rebuilt fairly accurately, while side chains on the surface, particularly those carrying formal charges, were often placed into positions significantly different from the crystallographic positions (Novotny et al., 1988).

3.5.2 Incorrectly folded structures. An important test of any protein modeling procedure is the ability to discriminate correctly from incorrectly folded proteins. Novotny et al. (1984, 1988) constructed models of seaworm hemerythrin and the variable domain of the mouse κ light chain where the structures were clearly incorrect. Both of these proteins have the same length. The incorrect models were generated by swapping the side chains of one protein onto the backbone of the other. In the first of these studies (Novotny et al., 1984), the side chains were initially constructed using *trans* side-chain torsion angles, and then energy minimization was used to refine the side-chain positions. In the second study, the side-chain modeling was improved by using the side-chain construction operator in CONGEN to rebuild the side chains. The energies of the new models were improved, but the incorrectly folded models still had excessive nonpolar solvent-exposed side-chain surfaces.

The side-chain construction operator was tested on the native protein structures. For both proteins, side-chain atoms beyond C_β were removed, and the iterative side-chain algorithm was used to rebuild all the side chains. Although there was great variation in the individual side-chain deviations (from 0.0 Å to 6.5 Å), the largest differences were found with exposed side chains. As the energy functions used to select side-chain conformations are improved, the agreement in the interior should also improve.

3.5.3 Coiled coils. Supercoiled dimers of α-helices are a common structural motif in transcriptional factors (Kouzarides and Ziff, 1988; Landschulz et al., 1988) and fibrous proteins (Geisler and Weber, 1983). The structural parameters for the supercoiling (Crick, 1953) can be used to construct the the peptide backbone in these dimers. The side chains for a particular sequence can then be added using the side-chain construction operator from CONGEN. It is possible to optimize the dimerization by calculating the dimerization energy while varying the supercoiling parameters. Two examples of such calculations are given below.

Synthetic oligoheptapeptides having an alternating pattern of hydrophobic and hydrophilic residues (Lys-Leu-Glu-Ala-Leu-Glu-Gly)$_n$ were shown to dimerize when the repeat length was at least four. Shorter peptides showed random structure (Lau et al., 1984). The experiments were performed on peptides with repeats of one to five.

Models of the five peptides were constructed by using the supercoiling parameters of Crick and optimizing the distance between the helices and rotation angles of the helices. The peptide side chains for each of these trials was constructed with the iterative side-chain construction method of CONGEN. Using an empirical free-energy potential (Novotny et al., 1984), the free energy of helix and dimer formation was calculated, and it was found that the energy of helix and dimer formation was positive for repeat lengths of one through three, and negative for four and five.

Using the construction protocols developed for the coiled coils, models of the leucine zipper DNA-binding proteins *fos*, *jun*, and GCN4, were used to calculate the stability of various homo- and heterodimers. Recently, the x-ray structure for GCN4 was published (O'Shea et al., 1991) and the model structure for GCN4 was compared against the x-ray structure. The overall RMS deviation for a least squares superposition of all atoms was 2.70 Å, but most of this deviation is due to the hydrophilic side chains, which would be expected to be disordered in solution. The RMS deviation for the backbone atoms is 1.08 Å, and the deviation for all the side chains was 3.58 Å. The agreement for the leucine and valine side chains found in the interface was only 1.51 Å. Visually, the packing of the leucine and valine side chains in the interface is also preserved.

4. Improvements in the Search Methodology

As described above, one of the major directions in the research to predict protein folding is the methodology for conformational search. In this section, two improvements for conformational search will be described. The first is directed searching, where information generated during the search process is used to guide the process. The goal is to generate the best conformers early in the search, and thereby avoiding an exhaustive sampling of the conformational space. The second method is the implementation of the search algorithms on a parallel computer. Parallel processing promises to deliver several orders of magnitude improvement in processing power, and it is imperative that search algorithms take advantage of this technology.

4.1 Directed Searching

In the applications of CONGEN described above, the search tree (see Figure 5-4) was traversed exhaustively. As the number of degrees of freedom increases, the time necessary for this exhaustive search grows exponentially. Since the goal of the search is to find minimum energy conformations, it may not be necessary to examine every node in the tree to find the minimum. In particular, if a function of the partial conformations can provide guidance to the path leading to the lowest energy, then a search that concentrates on the branches of the tree that have lower energy conformers will likely find the minimum long before the complete tree is traversed. Alternatively, it may be useful for modeling problems just to generate low-energy structures, but not necessarily the structure of lowest energy. In artificial intelligence research, directed and heuristic search techniques are essential tools in the solution of combinatorial problems in general (Pearl and Korf, 1987).

The idea of a directed search is analogous to a concept proposed by Zwanzig (Zwanzig et al., 1992) for the resolution of Levinthal's paradox (Levinthal, 1969). Levinthal's paradox arises from the large conformational space of proteins. Proteins fold on a time scale of seconds. In that time, they can only sample around 10^{12} conformations, a number far smaller than the possible number. Zwanzig analyzed the kinetics of a simple protein folding model where bonds connecting two residues can have two states, correct or incorrect, and each bond is independent of the others. If the energy bias of a correct bond is greater than about 2 kT and the rate constant for a change to a correct bond is 1 nanosecond, then folding will take place in physiological times, namely seconds.

In order to test the concept of a directed search, it was decided to focus on a common modeling problem, the reconstruction of the peptide backbone from the C_α coordinates (Bassolino-Klimas and Bruccoleri, 1992). This problem was chosen because the RMS deviation between the generated and experimental C_α positions for a partial conformation is an excellent indicator of the fit for the entire structure. If the RMS deviation for a partial conformer is poor, then it is not possible for the entire molecule to fit. If a directed search cannot solve this reconstruction problem, then it cannot possibly be used to find minimum-energy conformers.

4.1.1 Directed search methods. In the description of search methods that follows, the term "evaluation" refers to the function that is applied to each partial conformer to assess it quality. For the problem of reconstructing the peptide backbone from the C_α coordinates, the evaluation function is the RMS deviation between the C_α coordinates in the partial conformer and the C_α coordinates in the structure that is being reconstructed. The term "expansion" is a synonym for sampling one degree of freedom on a partial conformation. The term arises from the effect in the search tree, whereby sampling of one node can result in many child nodes directly underneath.

In the simplest algorithm, best first search, the conformation with the best evaluation in the search tree is sampled before any other. The best first method has an inherent problem, however. If the experimental C_α coordinates for a residue are erroneous, the RMS deviation for conformers containing this residue will be higher than many of the conformers with fewer residues. The program will then spend all its time exploring shorter conformers until all of them have a higher RMS deviation that the conformers containing the erroneous position. Depending on the branching of the search tree at that point, this could take longer than the age of universe.

This "barrier" problem led to a second method, the deepening evaluation method (Bassolino-Klimas and Bruccoleri, 1992). In this method, the program goes through successive levels of the tree, and it chooses the best conformer at the selected level for sampling. If there are no conformers available for sampling at the selected level, then the program will move to the next level in the tree. If it hits the end of the list of degrees of freedom, then it will cycle back to the first level and continue this process until no further samplings are possible anywhere in the tree. The intent of this method is to force progress on sampling all the degrees of freedom so that complete structures are generated.

For example, when the program begins, it will sample the first degree of freedom generating a set of conformers. Next, it will evaluate this set of conformers, and select the best-fitting one for sampling the second degree of freedom. For the third sampling, the program will consider only those conformers that have been generated by sampling two degrees of freedom. If there are any, then the succeeding sampling operation will be performed on conformers that have been generated by three degrees of freedom, etc. If there are no conformers generated by two degrees of freedom, then the program will sample the next best conformer generated by sampling one degree of freedom.

When this search method was used, the program was able to generate many peptide backbones, but all of them were poor. The first few residues had good RMS deviations, but successive residues had poor RMS deviations.

Further reflection on the problem led to a successful search method, the "mix" strategy. In the "mix" strategy, the program alternates between the two strategies in its selection of which conformer to sample next. It maintains the data structures necessary to apply both search strategies, and simply switches from one to the other on each cycle of the main search loop. The effect of this strategy is to overcome the barriers that occur with the best first strategy, while ensuring that the deepening evaluation strategy is applied to partial conformers that fit the experimental data well.

4.1.2 Results of peptide backbone reconstruction. The results of using the "mix" strategy are shown in Table 5-6. In all of these cases, the complete structures were found in a single search using a 5° grid and generating approximately 5×10^7 partial conformers. The branching factor on each backbone degree of freedom was around 600 conformers. Yet, these results are comparable to other methods that use geometric information (Purisma and Scheraga, 1984), energy minimization (Correa, 1990), or protein fragments from a database (Classens et al., 1989; Reid and Thornton, 1989; Holm and Sander, 1991).

Table 5-6. Deviations for reconstructions from C_α coordinates.[a]

Protein	RMS C_α (Å)	RMS backbone (Å)
Myohemerythrin (Sheriff and Hendrickson, 1987)	0.87	0.89
Flavodoxin (Smith et al., 1977)	0.36	0.60
Concavalin A (Hardman and Ainsworth, 1972)	0.73	0.99
Triose phosphate isomerase (Banner et al., 1976)	0.46	0.71
Carboxypeptidase A (Rees and Lipscomb, 1983)	0.62	0.84
McPC 603 Heavy Chain (Satow et al., 1986)	0.30	0.50
Thioredoxin (Holmgren et al., 1975)	1.28	NA[b]
Triacylglycerol acylhydrolase (Brady et al., 1990)	1.02	NA[b]

[a] Adapted with permission from Bassolino-Klimas and Bruccoleri (1992). © 1992 John Wiley & Sons, Inc.
[b] Not available because C_α coordinates were published.

There are two types of structures used for these tests. The top half of the table contains well-refined structures that have nearly ideal geometries. The bottom half of the table contain structures for which only C_α coordinates were deposited. For *trans* peptide groups, the C_α–C_α distance should be constant, but in these structures there are significant variations in C_α–C_α distances. The standard deviation for the C_α–C_α distance in thioredoxin is ± 0.34 Å, and it is ± 0.61 Å in triacylglycerol acylhydrolase.

Despite the errors in the C_α positions, the directed search procedure was able to find good quality peptide backbones over the entire molecules. This robustness in the face of errors is very encouraging. The use of a directed search to optimize energies will depend heavily on such robustness because the energy functions are not as well behaved as the RMS deviation as an evaluation function for a directed search.

4.2 Parallel Implementation

A significant part of the strategy to solve the protein folding problem is the use of faster computers to permit greater exploration of conformational space. In the past, using faster computers typically implied upgrading to the latest computer models. However, as physical limits to sequential computational speed are being approached, use of multiple processors operating in parallel has become the dominant approach to improving computer power. Such parallelism requires modification of existing software to adapt to the new computer models.

An exhaustive conformational search is inherently adaptable to parallel computers. Each node on different branches of the search tree is independent of others, and therefore the sampling of nodes can be spread over many processors. The directed search process can also be parallelized, but the process of deciding which node to expand next depends on the contents of the search tree at the time of decision. When multiple processors are working on different parts of the search tree, the decision-making process will not have the most current information, so the parallel directed search will require additional node expansions as compared to a serial search. Hopefully, the benefit of having additional processors will outweigh the extra node expansions.

4.2.1 Implementation of parallel search on a shared-memory computer.
CONGEN was parallelized on a Silicon Graphics Power Series workstation. These workstations use a shared-memory programming model, which is the simplest model for parallelizing an existing computer code. In this model, every processor in the computer sees the same virtual address space and can access any memory location. Any changes in a memory location can be retrieved by any other processor, and there is no need to explicitly transfer data from one processor to another. Self-scheduling was also used, whereby each processor examines the state of the computation, and decides for itself what work to do next. This approach is very adaptive to the workload and is easily load-balanced.

A skeleton of the searching procedure is shown in Figure 5-8. This procedure is executed by all processors. It depends on three data structures for its operation. Each of the data structures is protected from simultaneous modification by an individual lock. The primary data structure is the search tree. The second data structure, the work queue, is a queue of nodes that are scheduled for expansion. The third data structure, the new-nodes queue, is a queue of nodes that are to be entered into the tree. These nodes represent conformations that have resulted from the sampling of a degree of freedom.

There are three possible operations that the search procedure can execute at any given time. The primary operation is the sampling of a degree of freedom, the expand operation. This operation begins by removing a node from the work queue, setting the processor's copy of the coordinates to the partial conformation in the code, and sampling the appropriate degree

```
search_nodes(space grid, coordinates)
/*
 * Perform the conformational search. This procedure executes on all processors
 * simultaneously.
 */
{
    Make local copy of coordinates;
    Make local copy of space grid;
    prev_node = search tree root;
    done = FALSE;
    while (NOT done) {
        some_work_done = FALSE;
        if (new node queue is not empty) {
            if (search tree can be locked) {
                lock the new node queue;
                add new nodes to the tree;
                clear locks;
                some_work_done = TRUE;
            }
        }
        if (work queue has many nodes for expansion)
            next_node = first node on work queue;
            expand next_node;
            prev_node = next_node;
            some_work_done = TRUE;
        }
        else if (search tree can be locked) {
            add to work queue;
            if (work queue is empty and no other nodes are being expanded) {
                done = TRUE;
            }
            else if (new nodes were added to work queue) some_work_done = TRUE;
            clear search tree lock;
        }
        if (NOT some_work_done) {
            next_node = first node on work queue;
            expand next_node;
            prev_node = next_node;
            some_work_done = TRUE;
        }
    }
}
```

Figure 5-8. A pseudocode version of the top-level search procedure in CONGEN.

of freedom. All of the new conformations generated by this sampling are entered into the new-node queue. The second possible operation is the examination of the search tree to decide which nodes are to be expanded next, i.e., added to the work queue. This process must lock the search tree, and it will add new nodes to the work queue. The final operation is the addition of new nodes into the search tree. This operation is done at the highest priority because the decision process depends on having the most current information about the current search state.

The decision between adding new nodes to the work queue and sampling nodes already on the work queue is determined by examining the length of work queue and the accessibility of the search tree. If the number of nodes in the work queue is large, then a sampling operation is performed. If the number is small, an attempt is made to lock the search tree. If that succeeds, then more nodes are added to the work queue. Otherwise, the process will take another node off the work queue and expand it.

4.2.2 Performance. The efficiency of parallelization was tested on an exhaustive search of the H3 loop of ANO2 (Anglister et al., 1984; Brünger et al., 1991; Bassolino et al., 1992); see Figure 5-9. The speedup on eight processors was 7.3-fold. The high efficiency is due to the heavy CPU time requirements for side-chain constructions, which were necessary on each of the backbone conformers generated in this search. As a result, the program spent less than 1% of its time managing the search tree and the decision-making process.

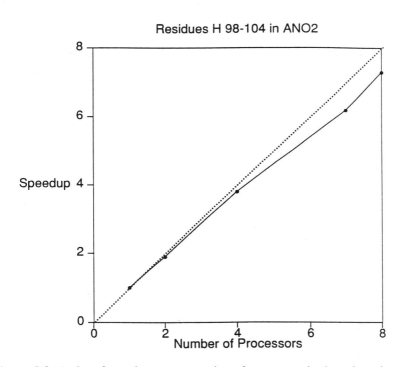

Figure 5-9. A plot of speedup versus number of processors in the exhaustive conformational search of the H3 loop in ANO2. Speedup is the ratio of elapsed time for the search on a single processor versus the elapsed time for the job with multiple processors. These runs were performed on a Silicon Graphics Iris 480/VGX workstation.

The implementation of parallelization is compatible with operation on a serial computer. The overhead for creating, maintaining, and locking the data structures described above is negligible on a uniprocessor.

5. Future Prospects

In this review, a long-term strategy has been presented for attacking one of the most significant problems in modern biochemistry, the prediction of protein folding. This strategy involves working on tractable folding problems, and using the experience gained from these to make additional progress. It is clear from the loop and antibody modeling that the search algorithms can exhaustively explore conformational space, and in many instances the CHARMM energy function coupled with the accessible surface rule can identify near-native conformations. However, greater accuracy in the ranking of conformers is still required.

Most work in the immediate future is aimed at improving the calculation of energies, and in particular, accounting for solvent effects. The Poisson–Boltzmann equation provides a more accurate calculation of electrostatic interactions in solution (Harvey, 1989) than does the simple Coulomb's law used currently. An improved method (Bruccoleri, 1993b) for solving the equation has been implemented with CONGEN, and is currently being tested. Although this equation is more accurate, its solution is much more time consuming. In addition, empirical atomic solvation parameters (Eisenberg and McLachlan, 1986) are another promising way of accounting for solvation.

Acknowledgments. I thank Drs. Edgar Haber and Jiri Novotny for their unwavering support, guidance, and insights over the years. This work was partially funded by NIH Grant PO1-HL 19259 and other grants from NIH, NSF, and ONR. I thank Phil Jeffrey for the use of data prior to publication. I am grateful to John Wiley & Sons, Inc., the American Association for the Advancement of Science, Macmillan Magazines Ltd., Academic Press, Inc., and Gordon and Breach Science Publishers, S.A., for permission to use copyrighted material. Portions of the text have been adapted from contributions to Bruccoleri and Karplus (1987), Gierasch and King (1990), Bruccoleri and Novotny (1992), and Bruccoleri (1993a).

For information about obtaining the CONGEN program, please contact Dr. Bruccoleri.

REFERENCES

Allen J, Novotny J, Martin J, Heinrich G (1987): Molecular structure of mammalian Neuropeptide Y: Analysis by molecular cloning and computer-aided comparison with crystal structure of avian homologue. *Proc Nat Acad Sci USA* 84:2532–2536

Anglister J, Frey T, McConnell HM (1984): Magnetic resonance of a monoclonal anti-spin label antibody. *Biochemistry* 23:1138–1142

Ashwell JD, Klausner RD (1990): Genetic and mutational analysis of the t-cell antigen receptor. *Annu Rev Immunol* 8:139–167

Banner DW, Bloomer AC, Petsko GA, Phillips DC, Wilson IA (1976): Atomic coordinates for Triose Phosphate Isomerase from chicken muscle. *Biochem Biophys Res Comm* 72:146–155

Bassolino D, Bruccoleri RE, Subramaniam S (1992): Modeling the antigen combining site of an anti-dinitrophenyl antibody, ANO2. *Protein Science* 1:1465–1476

Bassolino-Klimas D, Bruccoleri RE (1992): The application of a directed conformational search for generating 3-D coordinates for protein structures from α-carbon coordinates. *Proteins: Struc Funct Gen* 14:465–474

Bernstein FC, Koetzle TF, Williams GJB, Meyer EF, Brice MD, Rodgers JR, Kennard O, Shimanouchi T, Tasumi M (1977): The Protein Data Bank: A computer-based archival file for macromolecular structures. *J Mol Biol* 112:535–542

Brady L, Brzozowski AM, Derewenda ZS, Dodson EJ, Dodson GG, Tolley SP, Turkenburg JP (1990): A serine protease triad forms the catalytic center of a triacyclglycerol lipase. *Nature* 343:767–770

Brooks BR, Bruccoleri RE, Olafson BD, States DJ, Swaminathan S, Karplus M (1983): CHARMM: A program for macromolecular energy minimization and dynamics calculations. *J Comp Chem* 4:187–217

Bruccoleri RE (1993a): Application of systematic conformational search to protein modeling. *Mol Sim* 10:151–174

Bruccoleri RE (1993b): Grid positioning independence and the reduction of self-energy in the solution of the Poisson-Boltzmann equation. *J Comput Chem* 14:1417–1422

Bruccoleri RE, Haber E, Novotny J (1988): Structure of antibody hypervariable loops reproduced by a conformational search algorithm. *Nature* 335:564–568; see *Errata*, vol. 336, p. 266

Bruccoleri RE, Karplus M (1985): Chain closure with bond angle variations. *Macromolecules* 18:2767–2773

Bruccoleri RE, Karplus M (1987): Prediction of the folding of short polypeptide segments by uniform conformational sampling. *Biopolymers* 26:137–168

Bruccoleri RE, Novotny J (1992): Antibody modeling using the conformational search program, CONGEN. *Immunomethods* 1:96–106

Bruccoleri RE, Novotny J, Keck P, Cohen C (1986): Two-stranded α-helical coiled-coils of fibrous proteins. Theoretical analysis of supercoil formation. *Biophys J* 49:79–81

Brünger AT, Leahy DJ, Hynes TR, Fox RO (1991): 2.9 Å resolution structure of an anti-dinitrophenyl-spin-label monoclonal antibody Fab fragment with bound hapten. *J Mol Biol* 221:239–256

Chothia C, Boswell DR, Lesk AM (1988): The outline structure of T-cell alpha-beta receptor. *EMBO J* 7:3745–3755

Classens M, Cutsem EV, Lasters I, Wodak S (1989): Modelling the polypeptide backbone with spare parts from known protein structures. *Prot Eng* 2:335–345

Claverie JM, Prochnicka-Chalufour A, Bougeleret L (1989): Implications of a fab-like structure for the T cell receptor. *Immunol Today* 10:10–14

Colman PM, Freeman HC, Guss JM, Murata M, Norris VA, Ramshaw JAM, Venka-tappa MP (1978): X-ray crystal structure analysis of plastocyanin at 2.7 Å res-olution. *Nature* 272:319–324

Correa P (1990): The building of protein structures from α-carbon coordinates. *Proteins: Struct Funct Gen* 7:366–377

Crick FHC (1953): The Fourier transform of a coiled-coil. *Acta Cryst* 6:685–689

Deber CM, Madison V, Blout ER (1976): Why cyclic peptides? Complementary approaches to conformation. *Acc Chem Res* 9:106–113

Dill KA (1990): Dominant forces in protein folding. *Biochemistry* 29:7133–7155

Dygert M, Gō N, Scheraga HA (1975): Use of a symmetry condition to compute the conformation of Gramicidin S. *Macromolecules* 8:750–761

Eisenberg D, McLachlan AD (1986): Solvation energy in protein folding and bind-ing. *Nature* 319:199–203

Ganju RK, Smiley ST, Bajorath J, Novotny J, Reinherz E (1992): Similarity between fluorescein-specific T cell receptor and antibody in chemical details of antigen recognition. *Proc Nat Acad Sci USA* 89:11552–11556

Geisler N, Weber K (1983): Amino acid sequence data on glial fibrillary acidic protein (GFA): Implications for the subdivision of intermediate filaments into epithelial and non-epithelial members. *EMBO J* 2:2059–2063

Gierasch LM, King J (1990): *Protein Folding: Deciphering the Second Half of the Genetic Code*. Washington, DC: American Association for the Advancement of Science

Gō N, Scheraga HA (1970): Ring closure and local conformational deformations of chain molecules. *Macromolecules* 3:178–187

Goldberg DE (1989): *Genetic Algorithms in Search, Optimization, and Machine Learning*. Reading, Massachusetts: Addison-Wesley Publishing Co

Hall D, Pavitt N (1985): Conformation of cyclic analogs of enkephalin. III. Effect of varying ring size. *Biopolymers* 24:935–945

Hall D, Pavitt N, Wood MK (1982): The conformation of pithomycolide. *J Comput Chem* 3:381–384

Hardman KD, Ainsworth CF (1972): Structure of Con A at 2.4 Å resolution. *Bio-chemistry* 11:4910–4919

Harvey SC (1989): Treatment of electrostatic effects in macromolecular modeling. *Proteins: Struct Funct Gen* 5:78–92

Herron JN, Hei XM, Mason ML, Voss EW, Edmundson AB (1989): Three-dimen-sional structure of a fluorescein-Fab complex crystallized in 2-methyl-2,4-pen-tanediol. *Proteins: Struct Funct Gen* 5:271–280

Holm L, Sander C (1991): Database algorithm for generating protein backbone and side-chain coordinates from a C_α trace. *J Mol Biol* 218:183–194

Holmgren A, Söderberg B, Eklund H, Brändén C (1975): Three dimensional struc-ture of *E Coli* thioredoxin-S2 to 2.8 Å resolution. *Proc Nat Acad Sci USA* 72:2305–2309

Jeffrey PD, Strong RK, Sieker LC, Chang CY, Campbell RL, Petsko GA, Haber E, Margolies MN, Sheriff S (1993): 26-10 Fab-digoxin complex: Affinity and specificity due to surface complementarity. *Proc Nat Acad Sci USA* 90:10310–10314

Jones PT, Dear PH, Foote J, Newberger MS, Winter G (1986): Replacing the complementarity-determining regions in a human antibody with those from a mouse. *Nature* 321:522–525

Kabsch W, Sander C (1985): Identical pentapeptides with different backbones. *Nature* 317:207.

Kouzarides T, Ziff E (1988): The role of the leucine zipper in the fos-jun interaction. *Nature* 336:646–651

Krystek SR, Bruccoleri RE, Novotny J (1991): Stabilities of leucine zipper dimers estimated by an empirical free energy method. *Int J Peptide Protein Res* 38:229–236

Krystek SR Jr, Bassolino DA, Bruccoleri RE, Hunt JT, Porubcan MA, Wandler CF, Andersen NH (1992): Solution conformation of a cyclic pentapeptide endothelin antagonist: Comparison of structures obtained from constrained dynamics and conformational search. *FEBS Letters* 299:255–261

Landschulz WH, Johnson PF, McKnight SL (1988): The leucine zipper: A hypothetical structure common to a new class of DNA binding proteins. *Science* 240:1759–1764

Lau SYM, Taneja AK, Hodges RS (1984): Synthesis of a model protein of defined secondary and quaternary structure. Effect of chain length on the stabilization and formation of two-stranded α-helical coiled-coils. *J Biol Chem* 259:13253–13261

Le Grand SM, Merz, KM Jr (1994): The genetic algorithm and protein tertiary structure prediction. In *The Protein Folding Problem and Tertiary Structure Prediction*, Le Grand SM, Merz, KM Jr, eds. Boston: Birkhäuser

Levinthal C (1969): In: *Mossbauer Spectroscopy in Biological Systems.* Debrunner P, Tsibris JCM, Münck E, eds. Urbana: University of Illinois Press, pp. 22–24

Madison V (1985): Cyclic peptides revisited. *Biopolymers* 24:97–103

Marquart M, Deisenhofer J, Huber R, Palm W (1980): Crystallographic refinement and atomic models of the intact immunoglobulin molecule Kol and its antigen-binding fragment at 3.0 Å and 1.0 Å resolution. *J Mol Biol* 141:369–391

Martin ACR, Cheetham JC, Rees AR (1991): Molecular modeling of antibody combining sites. *Methods in Enzymology* 203:121–153

Meuer SC, Acuto O, Hercend T, Schlossman SF, Reinherz EL (1990): The human T cell receptor. *Annu Rev Immunol* 2:23–50

Mudgett-Hunter M, Anderson W, Haber E, Margolies MN (1985): Binding and structural diversity among high-affinity monoclonal anti-digoxin antibodies. *Mol Immunol* 22:477–488

Near RI, Bruccoleri RE, Novotny J, Hudson NW, White A, Mudgett-Hunter M (1991): The specificity properties that distinguish members of a set of homologous anti-digoxin antibodies are controlled by H chain mutations. *J Immunol* 146:627–633

Needleman SB, Wunsch CD (1970): A general method applicable to the search for similarities in the amino acid sequence of two proteins. *J Mol Biol* 48:443–453

Novotny J, Bruccoleri RE, Haber E (1990): Computer analysis of mutations that affect antibody specificity. *Proteins: Struct Funct Gen* 7:93–98

Novotny J, Bruccoleri RE, Karplus M (1984): An analysis of incorrectly folded protein models, implications for structure prediction. *J Mol Biol* 177:787–818

Novotny J, Bruccoleri RE, Newell J, Murphy D, Haber E, Karplus M (1983): Molecular anatomy of the antibody binding site. *J Biol Chem* 258:14433–14437

Novotny J, Ganju RK, Smiley ST, Hussey RE, Luther MA, Recny MA, Siliciano RF, Reinherz EL (1991): A soluble, single-chain T cell receptor fragment endowed with antigen combining properties. *Proc Nat Acad Sci USA* 88:8646–8650

Novotny J, Rashin AA, Bruccoleri RE (1988): Criteria that discriminate between native proteins and incorrectly folded models. *Proteins Struct Funct Gen* 4:19–30

Novotny J, Tonegawa S, Saito H, Kranz DM, Eisen HN (1986): Secondary, tertiary and quaternary structure of T-cell-specific immunoglobulin-like polypeptide chains. *Proc Nat Acad Sci USA* 83:742–746

O'Shea EJ, Klemm JD, Kim PS, Alber T (1991): X-ray structure of the GCN4 leucine zipper, a two-stranded, parallel coiled coil. *Science* 254:539–544

Panka DJ, Mudgett-Hunter M, Parks DR, Peterson LL, Herzenberg LA, Haber E, Margolies MN (1988): Variable region framework differences result in decreased or increased affinity of variant anti-digoxin antibodies. *Proc Nat Acad Sci USA* 85:3080–3084

Pearl J, Korf RE (1987): Search techniques. *Ann Rev Comput Sci* 2:451–467

Phillips SEV (1980): Structure and refinement of oxymyoglobin at 1.6 Å resolution. *J Mol Biol* 142:531–554

Pincus MR, Klausner RD, Scheraga HA (1982): Calculation of the three dimensional structure of the membrane-bound portion of melittin from its amino acid sequence. *Proc Nat Acad Sci USA* 79:5107–5110

Purisma EO, Scheraga HA (1984): Conversion of virtual bond chain to a complete polypeptide chain. *Biopolymers* 23:1207–1224

Ramachandran GN, Ramakrishnan C, Sasisekharan V (1963): Stereochemistry of polypeptide chain configurations. *J Mol Biol* 7:195–199

Rees DC, Lipscomb WN (1983): Crystallographic studies on Apocarboxypeptidase A at 1.54 Å resolution. *J Mol Biol* 168:367–387

Reid LS, Thornton JM (1989): Rebuilding flavodoxin from C_α coordinates: A test study. *Proteins: Struct Funct Gen* 5:170–182

Satow Y, Cohen GH, Padlan EA, Davies DR (1986): Phosphorylcholine binding immunoglobulin Fab McPC603—an X-ray diffraction study at 2.7 Å. *J Mol Biol* 190:593–604

Saul FA, Amzel LM, Poljak RJ (1978): Preliminary refinement and structural analysis of the FAB fragment from human immunoglobulin NEW at 2.0 Å resolution. *J Biol Chem* 253:585–597

Schildbach JF, Near RI, Bruccoleri RE, Haber E, Jeffrey PD, Novotny J, Margolies MN (1993): Modulation of antibody affinity by a noncontact residue: A mutagenesis and molecular modeling study. *Prot Sci* 2:206–214

Schildbach JF, Panka DJ, Parks DR, Jager GC, Novotny J, Herzenberg LA, Mudgett-Hunter M, Bruccoleri RE, Haber E, Margolies MN (1991): Altered hapten recognition by two anti-digoxin hybridoma variants due to variable region point mutations. *J Biol Chem* 266:4640–4647

Sheriff S, Hendrickson W (1987): Structure of Myohemerythrin in the azidomet state at 1.7/1.3 Å resolution. *J Mol Biol* 197:273–296

Sheriff S, Silverton EW, Padlan EA, Cohen GH, Smith-Gill SJ, Finzel BC, Davies DR (1987): The three dimensional structure of an antibody-antigen complex. *Proc Nat Acad Sci USA* 84:8075–8079

Smith WW, Burnet RM, Darling GD, Ludwig ML (1977): Structure of the semiquinone form of flavodoxin from clostridium M.P. *J Mol Biol* 117:195–226

Suh SW, Bhat TN, Navia MA, Cohen GH, Rao DN, Rudikoff S, Davies DR (1986): The galactan-binding immunoglobulin FAB J539. An x-ray diffraction study at 2.6 Å resolution. *Proteins: Struct Funct Genet* 1:74–80

Trail PA, Willner D, Lasch SJ, Henderson AJ, Hofstead S, Casazza AM, Firestone RA, Hellström I, Hellström KE (1993): Cure of xenografted human carcinomas by BR96-Doxorubicin Immunoconjugates. *Science* 261:212–215

Venkatachalam CM, Khaled MA, Sugano H, Urry DW (1981): Nuclear magnetic resonance and conformational energy calculations of repeat peptides of elastin. Conformational characterization of cyclopentadecapeptide cyclo-(L-Val-L-Pro-Gly-L-Val-Gly)$_3$. *J Am Chem Soc* 103:2372–2379

Wagner RA, Fischer MJ (1974): The string to string correction problem. *J Assoc Comput Mach* 32:168–173

Zwanzig R, Szabo A, Bagchi B (1992): Levinthal's paradox. *Proc Nat Acad Sci USA* 89:20–22

6

Building Protein Folds Using Distance Geometry: Towards a General Modeling and Prediction Method

William R. Taylor and András Aszódi

1. Introduction: From Modeling to Prediction

A known protein structure can be modified and manipulated to produce a model for another protein with which it shares some sequence similarity. Typical changes involve the substitution and reorientation of side chains and the remodeling of the main chain to accommodate possible insertions and deletions of sequences. Where the two sequences are clearly related (say, more than 50%), such changes are relatively minor and a model can easily be constructed automatically.

As the dissimilarity increases between the sequence of unknown structure and its partner of known structure, then the modeling process becomes more difficult. The relative insertions and deletions between the two structures may be so large that existing secondary structures are lost or it becomes probable that the additional segments will contain new secondary structures. Current modeling programs provide no guide for such large reorganizations which are either ignored or left to the imagination of the modeler (Pearl and Taylor, 1987).

The problem of dealing with the packing of relatively unconstrained secondary structures has been approached from a different direction. Many attempts have been made to predict the tertiary structure of a protein given only sequence data, and one of the more tractable of these approaches

The Protein Folding Problem and Tertiary Structure Prediction
K. Merz, Jr. and S. Le Grand, Editors
© Birkhäuser Boston 1994

involves prediction and subsequent packing of secondary structure (Cohen et al., 1979, 1981). These approaches have been referred to as combinatoric because they try all packing combinations allowed under a set of rules and then select those that best conform to predefined expectations.

In the past, little attempt has been made to bridge the gap between the modeling-by-homology approach and the combinatoric building approach. This has largely resulted from the independent origins of the methods. Modelers have little need to consider radically different folds, while those who make predictions do not feel justified in specifying detailed side-chain or loop interactions. However, there is also little incentive to cover the middle ground, as practical problems in this area are less common. If a clear sequence similarity exists, then a model can be built with relative ease and can be expected to be reasonably accurate—so providing a clear need for the required modeling tools. At the other extreme, the drive to develop prediction tools comes not only from the intellectual challenge of solving such a difficult problem, but also from the great need to find some structural insight into the vast quantity of sequence data that exhibits no similarity to any protein of known structure. In the middle ground, however, where there are some partial constraints provided by a known structure, there is less scope for practical application. This situation arises because protein structure is more highly conserved through evolution than is the sequence that determines it. Thus, any sequence similarity usually implies that the entire fold of the protein is specified. In other words, one commonly gets all or nothing.

This review outlines an approach to a general method of modeling based on distance geometry that spans the range from the highly constrained modeling-by-homology method to pure structure prediction, where the only constraint is that the model should look like a globular protein.

2. Scope of a General Approach

2.1 Setting a Limit to Combinatorics

Rather than develop an independent bridging method, an approach to re-solve the schism between modeling and prediction can be found by extract-ing their common features into a general method. Both problems involve the manipulation and packing of objects in three dimensions and differ only in the number of available constraints. When modeling-by-homology, these are numerous and specific, dictating both local and global structure, while when predicting structure, the constraints are few and specify only local structure with any degree of certainty.

Weak constraints allow a multitude of different solutions and a central problem is whether to investigate all solutions, sample just a few, or generate one that is, by some measure, best. Since the definition of "best" is often difficult to define, it is desirable for a prediction method to offer a variety of (if not all) solutions to the constraints that are equivalent under the uncertainties inherent in their evaluation. A well-constrained problem will, by contrast, define a unique structure and any general method must encompass this difference.

When considered further, it becomes apparent that the distinction drawn above is not fundamental but simply reflects the scale at which the problem is examined. For example, the details of side chain packing may be poorly constrained even though the overall fold of a model is fixed by homology to a known structure. This degree of latitude can easily lead to a combinatoric explosion in the number of solutions (Desmet et al., 1992), which in some circumstances may appear trivial, but if the details of a binding site were being investigated, could be critical. By contrast, in the realms of prediction, where even the overall fold is uncertain, the details of side-chain packing (indeed even the packing of loops) becomes largely irrelevant. In this situation, either the side chains are ignored, or since they are often required in some calculations, a best solution can be presented.

A decision must be made on the level of detail to which alternate solutions will be pursued, and beyond which they will simply be quenched into a unique "best" solution. This level is dictated by the radius of convergence of the method used to calculate the best solution. For example, if such a function were perfect then no combinatorics would be necessary, since the function would avoid all incorrect suboptimal solutions. A less than perfect function might be able to optimize the hydrophobic packing between secondary structures but be unable to distinguish (with its evaluation function) different folds that lead to good packing. An even worse function (typical of those currently employed in model refinement) might simply adjust local steric clashes and be unable to make any relative adjustment of the secondary structures to optimize packing.

Since few computational calculations can be worse than combinatoric in complexity, it is desirable to limit the combinatoric element of the modeling as much as possible. This places greater onus on the remaining refinement stage to converge to the correct (or at least an acceptable) solution. In the method developed below, an attempt is made to derive the final model from a very rough outline of the fold—leaving only the choice of fold (or topology) to the combinatoric stage. Various decisions have been made in previous prediction methods, some of which have generated the alignment of strands in sheets combinatorically (Cohen et al., 1980, 1981) while in more recent attempts it has been popular to specify the location of every, residue on a lattice of points (Chan and Dill, 1990; Skolnick and Kolinsky, 1990, 1991; Dill et al., 1993).

2.2 Choosing a Versatile Refinement Method

Having decided the limit of combinatoric exploration of alternate forms, it remains to adopt a refinement strategy that has both a wide radius of convergence and can incorporate the large number of specific constraints available when modeling from clear sequence homology. Conventional energy minimization or molecular dynamics methods, such as CHARMM (Brooks et al., 1983), which are often used to refine molecular models avoid this problem simply through their small radius of convergence. The physical inertia of the atoms and the lack of any long-range concerted force means that the model is unlikely to diverge much from the original template, so implicitly incorporating global constraints.

A more intermediate position can be found in the construction of models to incorporate distance constraints derived from Nuclear Magnetic Resonance (NMR) experiments. In this domain there is generally no preferred starting model but there is a sufficient number of constraints to determine an almost unique structure (data is normally collected until this is achieved). Two distinct solutions have been adopted for this type of data. One approach simply applies conventional molecular dynamics with the data-derived distances specified as pseudobonding (or target) potentials. The convergence of this simple method to a good solution is not guaranteed, as the many distances can easily become "tangled" and trapped in a local minimum. However, by satisfying local constraints first, better results can be obtained (Braun and Gō, 1985).

The alternate approach is to use the method of projective distance geometry. This method can take a full set of pairwise distances and project (or *embed*) them into three dimensions such that all the distances between the points correspond to those in the distance matrix (Kuntz et al., 1976; Crippen and Havel, 1988; Kuntz et al., 1989). If the original distances did indeed derive from a three-dimensional object, then an exact solution will be found, while if the distances contain errors, then a configuration of points will be found that best reconciles their conflicts. The method is direct in that it does not involve a kinetic approach to the final solution, giving it a distinct advantage over the preceding type of method. However, it has the great disadvantage that, although points can be given mass, distances cannot be individual weighted. This results in final configurations that can appear less than protein-like because the normally tightly constrained, bonded distances and steric repulsions are not necessarily preserved. In the treatment of NMR-derived constraints a combination of methods has been used in which a rough starting position is generated by the projective method and subsequently refined by molecular dynamics, incorporating the constraints as target distances.

The set of constraints generated by NMR experiments should be reasonably consistent with a final, three-dimensional model (assuming that the data had been collected well) and while the ensuing problems of their incorporation into a model might well present technical difficulties, no fundamental problems are posed. The situation with distances estimated by predictive methods can be expected to be much less well determined. Consider only a single force directing hydrophobic residues to pack together; if no fold is specified, then all distances between hydrophobic residues will be set to their optimal values, which will specify a distance matrix that would be highly incompatible with a three-dimensional set of points (except for trivially small sets of distances). Specifying a starting fold improves the situation somewhat by restricting the possible pairs that might form packing partners, but the situation will remain less well determined than the equivalent NMR problem.

The affinity of the prediction/modeling problem to the construction of models from NMR data suggests that the best refinement strategy might be some combination of projective distance geometry with real-space refinement. The distance geometry component is ideally suited to deal with a matrix of inconsistent constraints without becoming trapped in local minima, while the real-space component will allow strongly constrained local features, such as bumps and bonds, to be maintained. The many differences from the NMR problem, however, mean that a method cannot simply be adopted directly from that field. The development of a novel strategy, concentrating on the less familiar distance geometry component, will be described in the following section.

3. Distance Geometry

The distance matrix contains almost all geometric information (except chirality) that is needed to specify the position of the corresponding set of points in space, and the techniques employed to transform the distance data into spatial coordinates is known as *distance geometry*. The term "distance geometry" has come to refer in recent years both to the refinement of a set of target distances in three-dimensional space and also to the projective method discussed above. In this section the term will be used exclusively to refer only to the projective method.

3.1 Projection to Euclidean Space

The basic idea of the method of subspace projection (MacKay, 1983; Crippen and Havel, 1988) is as follows. Center an N-dimensional coordinate system at the centroid of the N-point ensemble, and set the base vectors $\mathbf{r}_i$ equal to the vectors pointing to the N points from the origin. At best, only $N - 1$ of these vectors will be linearly independent, and in general will not be mutually perpendicular (orthogonal) to each other, and their length will not be unity. To test for both criteria (i.e., orthogonality and normality), a symmetric $N \times N$ matrix $\mathbf{M}$, the *metric matrix*, is constructed whose elements are the scalar products of the base vectors:

$$m_{ij} = \mathbf{r}_i \cdot \mathbf{r}_j. \tag{1}$$

Two vectors are orthogonal if and only if their scalar product is zero; hence, we would wish to set all $m_{ij} = 0$, for $i \neq j$. This can be accomplished by the *diagonalization* of the metric matrix, i.e., by finding its eigenvalues and eigenvectors. The eigenvalues will be equal to the squared length of the new base vectors, and the matrix of eigenvectors will define a rotation that transforms the old nonorthogonal coordinate system into a new orthogonal one. The jth coordinate of the ith point (x_{ij}) will then be given as

$$x_{ij} = \lambda_j^{1/2} \mathbf{W}_{ij}, \tag{2}$$

where λ_j is the jth eigenvalue and $\mathbf{W}_{ij}$ is the (i, j)th element of the matrix of eigenvectors.

3.1.1 Construction of the metric matrix. The above transformation might have been just an obscure test of geometric consistency were it not for the additional valuable result that the metric matrix can be calculated from the interpoint distances alone. This allows a configuration of points to be calculated from a distance matrix with no prior knowledge of the source of the distances.

The metric matrix $\mathbf{M}$ can be constructed from the distance matrix $\mathbf{D}$ using the cosine rule, which states that in a triangle the length of a side a can be calculated given the lengths of the other two sides b, c and the angle α subtended by them:

$$a^2 = b^2 + c^2 - 2bc \cdot \cos \alpha. \tag{3}$$

Therefore, the scalar product defining the elements of $\mathbf{M}$ can be written as

$$m_{ij} = \mathbf{r}_i \cdot \mathbf{r}_j = d_{i0} \cdot d_{j0} \cdot \cos \alpha = \tfrac{1}{2}(d_{i0}^2 + d_{j0}^2 - d_{ij}^2), \tag{4}$$

where $d_{i0} = |\mathbf{r}_i|$ is the distance of the ith point from the centroid (i.e., the origin). These distances can also be calculated from the distance matrix, using Lagrange's theorem (Flory, 1969):

$$d_{i0}^2 = \frac{1}{N} \sum_{k=1}^{N} d_{ik}^2 - \frac{1}{N^2} \sum_{j<k}^{N} d_{jk}^2, \tag{5}$$

allowing the metric matrix to be specified entirely in terms of the original distance matrix.

3.1.2 Triangle inequality violations.   Before proceeding to projection into Euclidean space, the metric matrix constructed from the distance matrix by equations 4 and 5, should checked for triangle inequality violations. One of the Euclidean distance axioms states that the distances between three points i, j, k should obey the *triangle inequality*:

$$d_{ij} \le d_{ik} + d_{kj}. \tag{6}$$

Since in the current application the distances might derive from very rough (predicted) estimates, the triangle inequality axiom will often be violated. Instead of naively checking all distance triplets, which leads to an algorithm of order N^3 complexity (Havel et al., 1979), violations can be checked and removed by processing the metric matrix. From its definition (see equations 1 and 4) it follows that the cosine of the angle α_{ij} subtended by the ith and jth base vectors is given by

$$\cos \alpha_{ij} = \frac{m_{ij}}{(m_{ii}m_{jj})^{1/2}}.$$

These cosine values (calculated for all i, j pairs, where $i \ne j$) indicate triangle inequality violation whenever any value lies outside the interval -1 to $+1$. A correction can then be made by replacing m_{ij} by $\pm 0.95 \cdot (m_{ii}m_{jj})^{1/2}$ (preserving the original sign). In an analogous manner, if any of the diagonal elements m_{ii} happen to be negative, then twice the absolute value of the most negative diagonal element can be added to all diagonal elements before performing the off-diagonal checks. This "shifting," which is widely used in eigenvalue calculations (Press et al., 1986), is equivalent to the addition of the identity matrix $\mathbf{I}$ multiplied by the shift constant c to the metric matrix before proceeding to diagonalization. The operation does not affect the eigenvectors $\mathbf{x}_i$ of $\mathbf{M}$, but the eigenvalues λ_i will be shifted by c:

$$\mathbf{M} \cdot \mathbf{x}_i = \lambda_i \mathbf{x}_i, \tag{8}$$

$$(\mathbf{M} + c\mathbf{I})\mathbf{x}_i = (\lambda_i + c) \cdot \mathbf{x}_i, \tag{9}$$

as can be verified by adding $c\mathbf{x}_i$ to both sides of equation 8, giving equation 9. Consequently, the shift constant c was subtracted from the eigenvalues before the actual calculation of the coordinates (equation 2) is performed.

3.1.3 Negative eigenvalues and projection strategy. Correcting for the triangle inequality violations alone does not guarantee that the metric matrix will not have negative eigenvalues, because higher-order (e.g., tetrangle) inequality violations were not checked due to their large CPU time requirement (Easthope and Havel, 1989). Instead, the following projection strategy was adopted:

- if $M < N-1$ eigenvalues were positive, then the points were projected into an M-dimensional subspace;
- if all eigenvalues were positive, then their sum was calculated, and the M largest eigenvectors whose sum was a preset fraction (e.g., 95%) of the total were kept and used in the projection;
- projection into less than three-dimensional subspaces was not allowed.

3.2 Residue Density

The overall residue density (number of C_α atoms per unit volume) provides a global measure of compactness of the folded polypeptide molecule. The average residue density of protein molecules is about $\rho = 6 \cdot 10^{-3} \text{Å}^{-3}$ (Gregoret and Cohen, 1991). During the course of the iterations the residue density of the model chains had to be readjusted to compensate for the shrinkages accompanying projection steps. It must be kept in mind that the residue density was defined for three-dimensional space only and its value was meaningless in higher dimensions, which rendered direct adjustments in high-dimensional Euclidean spaces impossible. Consequently, for adjustments performed in distance space and immediately after projection into higher dimensions, indirect scaling methods had to be applied.

3.2.1 Density adjustment in distance space. The distribution of distances among a set of points reflects the overall density of the point set. For solid three-dimensional spheres with uniform density, the first two moments of the distribution of distances between point pairs within the sphere are

$$\overline{d_{ij}} = \tfrac{36}{35} R, \tag{10}$$

$$\overline{d_{ij}^2} = \tfrac{6}{5} R^2, \tag{11}$$

where R is the radius of the sphere (Taylor, 1993). Assuming a spherical fold, the expected radius R_{exp} of an N-residue model chain with residue density ρ in three dimensions would be

$$R_{\exp} = \left(\frac{3N}{4\pi\rho}\right)^{1/3}. \tag{12}$$

The distances were scaled so that the first two moments should be as close to the theoretical values calculated from equations 10, 11, and 12 as possible. The distance scaling factor $f_{\text{dens}, D}$ was chosen as the average of the ratios between the expected and observed values:

$$f_{\text{dens}, D} = \frac{1}{2}\left(\frac{\frac{36}{35} R_{\exp}}{\overline{d_{ij}}} + \sqrt{\frac{\frac{6}{5} R_{\exp}^2}{\overline{d_{ij}^2}}}\right). \tag{13}$$

3.2.2 Density adjustment after projection. This adjustment attempted to preserve the distribution of the pairwise C_α–C_α distances immediately after projection into greater than three-dimensional Euclidean spaces. The approach used was to find a scaling factor $f_{\text{dens},P}$ such that the squared differences between the previous distances $d_{ij}^{(\text{old})}$ and the new distances $d_{ij}^{(\text{new})}$ (calculated from the point coordinates after projection), multiplied by the factor, be minimal:

$$Q = \sum_{i,j}(d_{ij}^{(\text{old})} - f_{\text{dens},P} \cdot d_{ij}^{(\text{new})})^2 \to \min. \tag{14}$$

The solution of this simple, least-squares regression problem is

$$f_{\text{dens},P} = \frac{\sum_{i,j} d_{ij}^{(\text{old})} \cdot d_{ij}^{(\text{new})}}{\sum_{i,j}(d_{ij}^{(\text{new})})^2}. \tag{15}$$

Both the new distance matrix entries and the set of Cartesian point coordinates were then multiplied by the scaling factor $f_{\text{dens},P}$.

3.2.3 Density adjustment in three-dimensional Euclidean space. In three-dimensional Euclidean space the molecules were approximated by inertial ellipsoids (Taylor et al., 1983). The ellipsoid was centered on the centroid of the model chain, its semiaxes collinear with the principal axes of inertia. The length of the semiaxes A, B, and C were proportional to the three moments of inertia (the square root of the eigenvalues of the moment matrix). Each semiaxis was scaled by a factor $f_{0.9}$, so that the ellipsoid contained 90% of the point set. This empirical scaling was necessary to compensate for the concave crevices present on most protein surfaces. The density of an N-residue polypeptide chain was then calculated as

$$\rho = \frac{3N}{4\pi ABC} = \frac{3N}{4\pi f_{0.9}^3(\lambda_1\lambda_2\lambda_3)^{1/2}}, \tag{16}$$

where λ_1, λ_2, λ_3 were the eigenvalues of the moment matrix. This method was used to calculate the average density of the 84 proteins in our representative dataset, which was found to be $\rho = (6.3 \pm 1.3) \cdot 10^{-3}$ residues per $Å^3$, in good agreement with other studies (Gregoret and Cohen, 1991).

3.3 Handedness Correction

One of the problems with a projective distance geometry approach is that the handedness of the model is not contained in the distance matrix and is effectively randomly determined during projection. For such asymmetric objects as proteins, this is an important aspect and must be monitored and corrected when necessary. In the simple models treated here this cannot be done locally, since the full main chain is not used and the handedness of the α-carbon is, therefore, unknown. Asymmetric secondary structures (such as the α-helix), even if present, cannot provide a reliable guide, due to possible distortion in the early stages of modeling.

To overcome these problems, an algorithm was devised that, given a set of points, will find the two largest (intersecting) tetrahedra, each defined by four points from opposing octants of a Cartesian coordinate frame with its origin at the center of gravity. In practice, the points in each octant (numbered successively 1 to 8, with no diagonal connections between adjacent octants) were ranked on their radius, and combinations of remote points from even-numbered and odd-numbered octants were then scored by their tetrahedral volume. The largest even- and largest odd-numbered tetrahedra were found and the handedness of both these were checked before and after projection. A change in the handedness of both tetrahedra indicated that the coordinates should be reflected about a plane to restore their correct handedness, while if only one tetrahedron switched, then it was obvious that the model had become too distorted.

While the above algorithm can maintain a consistent handedness through alternating phases of Euclidean space and "distance space," it cannot, of course, provide any guide to the correct selection of the initial handedness for the projected model. This must come from a source external to the distance matrix, which in the current application was derived from an idealized chain fold.

4. Geometric Regularization

4.1 Bonds and Bumps

After the projection, the ensemble of the points will have changed in appearance. Generally, projection shrinks the set (for reasons discussed above) but the shape also becomes altered because information is discarded with the ignored eigenvalues and eigenvectors. Therefore a reassessment and compensation of possible distance constraint violations is again necessary after projection. For some distances, in particular steric repulsion and the virtual (C^α–C^α) bond length, this is unacceptable (if only on the grounds of visual aesthetics). To avoid such distortions, a simple geometric regularization was applied to the projected model. Any pair of sequentially adjacent points were set to the virtual bond length of 3.8 Å, while any other pairs within the bumping distance (5.5 Å) were separated to that distance. Pairs of atoms adjacent-but-one, however, were separated only when closer than 5.0 Å.

The regularization algorithm separated each of a pair of violating atoms along a line connecting their centers by a fraction of the required amount, making a new record of all positions to avoid any order dependence in its operation. The same algorithm was then reapplied to the new copy and so forth until all the ideal values were effectively attained.

4.2 Local Structure and Motifs

An equivalent regularization was applied to well-defined local structures, including secondary structure and motifs. For an atom i a local coordinate reference frame (of orthogonal unit vectors X, Y, and Z) was calculated. This was based on the two flanking positions ($i - 1$ and $i + 1$) with Z perpendicular to the plane of the triplet and X defined by the direction $C^\alpha_{i-1} \rightarrow C^\alpha_{i+1}$. Reference frames were then calculated for equivalent atoms (i) in the model and (j) in the ideal motif template. The motif frame (with atom j at its origin) was then superposed on the model frame at atom i and the same transformation applied to all atoms in the motif. Target locations are thus generated by the superposed motif for all corresponding model atoms. This approach uses the same construct as in structure comparison by the method of Taylor and Orengo (1989), allowing motifs defined by that approach to be directly incorporated.

When located at their origin (i) in the starting model (A), each vector set specifies target locations for all other atoms. A shift vector, s_{ij}, was thus defined for atom j in protein A (with atomic coordinate A_j) by the jth vector (r_{ij}) in the set associated with atom i:

$$s_{ij} = (A_j - A_i - r_{ij}). \tag{17}$$

Atoms were then shifted towards these positions using a simple algorithm described previously (Taylor, 1993). This employs a simple regularization algorithm and applying the more local shifts in the early refinement cycles, giving a new location $(\mathbf{A}'_j)$ for atom j as follows:

$$\mathbf{A}'_j = \mathbf{A}_j + |j - i|^{1/2} \, \mathbf{s}_{ij}.$$

This procedure was then applied, taking in turn each model atom with a correspondence in an ideal motif template as the superposition center.

5. Towards Protein Structure

The methods described in the preceding section are sufficiently robust to be able to take a random set of distances and massage these towards a packed chain with the correct bond geometry and packing density of a globular protein. To the trained eye, however, these "fake" proteins do not "look right." When the sequence is considered, it is apparent that they have no hydrophobic core and they also have no secondary structure. Both these missing elements represent the two most important forces directing protein folding, namely, the hydrophobic effect and hydrogen-bond formation. Respectively, they impose a global and local ordering on the chain.

5.1 Modeling Hydrophobic Packing

5.1.1 Idealized hydrophobic packing. Consider a simple model of a globular protein with a pure hydrophobic core surrounded by a pure hydrophilic shell. Any two hydrophobic residues must be closer than the diameter of the core, whereas two hydrophilic residues can be at any separation less than the diameter of the whole protein. Thus some correlation can be expected between the hydrophobicity of a pair of residues and their separation. This simple model can be made more realistic by taking the distance of a residue (α-carbon) from the center of gravity of the protein as an idealized measure of hydrophobicity, while the sum of these radii for a pair of residues (r_i and r_j) provides a crude empirical estimate (s_{ij}) of their separation:

$$s_{ij} = r_i + r_j. \tag{19}$$

This calculated distance can then be compared with the observed separation, d_{ij}, for all residue pairs (i and j) in an idealized (spherical) globular protein with residues located at the centers of close-packed spheres. The distribution of estimated and measured distances both have an approximate Gaussian form but are displaced relative to each other; however, a good fit can be obtained by simple scaling with the constraint that the expected and observed radii of gyration are equal. The latter value can be found from the pairwise distances using Lagrange's theorem (Flory, 1969).

Despite well-matched overall properties, the estimated distance is still poorly correlated with the observed distances, with the error being greatest for the more hydrophilic pairs. Thus, even in this idealized model, where hydrophobicity values exactly reflect depth in the protein, little specific packing information can be obtained.

5.1.2 Conserved hydrophobicity. The simple premise that hydrophobic residues should pack together can be refined by considering both the degree of hydrophobicity of individual amino acids and, where more than one homologous sequence is available, the degree of conservation of sequence at each aligned position.

As previously (Taylor, 1991b), for an alignment of N sequences, the hydrophobicity of a given position (f_i) was calculated as the mean over the aligned amino acid types using the scale of Levitt (1978):

$$f_i = \frac{1}{N} \sum_{j=1}^{N} H_{R_{ij}}, \tag{21}$$

where R_{ij} is the amino acid type of residue i in sequence j and H is the list of amino acid hydrophobicity values.

Conservation was measured using an amino acid relatedness table (M), specifically that of Dayhoff et al. (1978), by taking a mean over all pairs of aligned residues:

$$g_i = \frac{2}{N^2 - N} \sum_{j=1}^{N-1} \sum_{k=j+1}^{N} M_{R_{ij}, R_{ik}}. \tag{21}$$

Clearly g cannot be calculated for a single sequence. To allow the application of the following methods in this situation, a value of $g = 1$ was be adopted. A product of the measures f and g was expected to be large for positions that were both hydrophobic *and* conserved. Following the ideal model outlined above, the sum of this composite measure was used to provide a rough pairwise hydrophobic packing score (h_{ij}):

$$h_{ij} = f_i g_i + f_j g_j. \tag{22}$$

However, unlike the ideal model considered above, the nature of the tables H and M mean that h will be inversely related to distance.

5.1.3 Scaling the estimated distances. For every sequence alignment, the packing score (h) defined above will have a different range, depending on the length, number, and degree of conservation of the sequences.

The best that can be done is to scale the estimated distances to correspond with the bulk properties observed in globular proteins. The bulk property that characterizes the packing in the molecule is the residue density, ρ. This has been estimated by various means, including Voronoi polyhedra constructions (Richards, 1974), ellipsoidal approximations of shape (Taylor et al., 1983), and molecular volumes of amino acids (Teller, 1976; Creighton, 1983; Gregoret and Cohen, 1991). Considering all these approaches leads to a value for ρ between 0.006 and 0.007 residues per Å^3. For any sequence length (N) the maximum expected radius for a spherical molecule ($r_{\max}$) can then be calculated as:

$$r_{\max} = \left(\frac{3N}{4\pi\rho}\right)^{1/3}. \tag{23}$$

The conserved hydrophobic values were then scaled into the range $d_{\min} \rightarrow 2r_{\max}$, where $d_{\min}$ is the closest nonbonded approach (see Section 2.2.4). However, simply matching the largest distances would make the scaling sensitive to extreme values, and to avoid this problem the first and second moments of the distribution were matched. The properties considered were the average pairwise distance $\overline{d_{ij}}$ and the average squared distance $\overline{d_{ij}^2}$. For a sphere of uniform density, the former is slightly greater than the radius ($\overline{d_{ij}} = {}^{36}\!/_{35}\,r$) (Taylor, 1993), while the latter is simply twice the radius of gyration squared (${}^6\!/_5\,r^2$).

The estimated hydrophobic packing distance (h_{ij}^d) was calculated as

$$h_{ij}^d = ah_{ij}^{-c} - b, \tag{24}$$

where h_{ij} is calculated by equation 4 and a, b and c are parameters to be estimated under the constraint of the two moments described above.

5.2 Incorporating Hydrogen Bonds

Hydrogen-bonds (H-bonds) form between the main-chain carbonyl oxygen and hydrogen of the amide nitrogen, but none of these atoms are included in our simple (α-carbon) model. To incorporate the full atomic backbone would lead to an unacceptable increase in computation load. A simplified "virtual" H-bond that mimicked most of the expected properties was therefore constructed (Aszódi and Taylor, 1994).

The most important of these properties was the constraint that only two H-bonds can be made to each residue. Secondary to this is the observation that the peptide bond is (effectively) flat. This implies that amino and carboxy H-bonds of adjacent residues must be opposed. It was considered less important to model the polarity of the bonds, which in an α-carbon model can remain ambiguous.

The bonds were modeled by connecting the mid-points of the α-carbon virtual bonds at a distance of 4.5 Å. They were identified in the distance matrix as pairs of virtual bond midpoints that had come within a prescribed bonding radius. Ideally, a maximum number of such pairs should be selected, under the constraint that only two bonds can form per residue. This, unfortunately, is a problem of combinatorial complexity; as it is a common calculation, an approximate solution was found by ranking the candidate bonds by distance and applying single-linked clustering.

The α-carbons flanking the selected bonds were then assigned ideal separations in the distance matrix prior to projection. After projection, the bond geometry was further refined in multidimensional space towards the ideal geometry before returning again to distance space. This additional refinement improved the linearity of chains of H-bonds in a network, again using the simple iterative algorithm (described above) that was used to regularize bonds and bumps.

5.3 Local Chirality

The current model, including the features described above, is still essentially an α-carbon model and therefore does not incorporate any of the asymmetry associated with the full main chain. This implies that there is no preferential selection of one chiral form of secondary structure over the other (for example, the right-hand over the left-hand α-helix).

Such preferences might be introduced locally by favoring the occurrence of the correct virtual-bond torsion angle (Gregoret and Cohen, 1991) or on a larger scale by "flipping" the handedness of the entire helix (Saitoh et al., 1993) The former method is less effective for global structures, such as β-sheets, while the latter method would be beset by many problems if it had to be generalized. Instead we used the motif refinement algorithm described above, which can impose both local or global preferences depending on the extent of the ideal motif used (Taylor, 1993).

5.4 "Random" Proteins

Using the methods described above on random sequences of binary hydrophobicity, the solutions obtained for a particular sequence were not unique (although the major features of folding were similar) when the iteration was started with a different random initial distance matrix several times. The flexibility of the model chains (with no torsional constraints applied), the simplified model of hydrophobicity, and the lack of fine details of side chains implied that multiple solutions for the same sequence were possible.

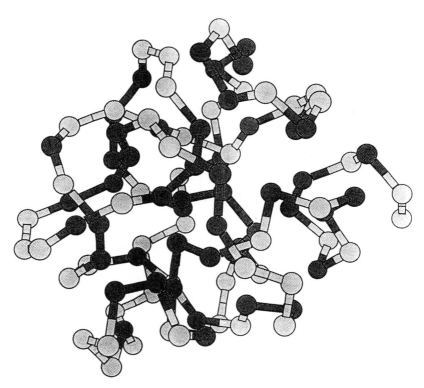

Figure 6-1. A "random" protein structure built from a random sequence of binary hydrophobic residues (*gray*) and hydrophobic residues. A globular structure has been formed with a clear hydrophobic core, but no secondary structures can be seen.

Although the non-H-bonded model chains folded into compact globules that possessed distinct, well-formed hydrophobic cores, no secondary structure was formed in these simulations (Figure 6-1). The H-bonded chains, on the other hand, *always* contained H-bond networks that were visually very similar to those found in real proteins (Figures 6-2a, 2b, and 2c). Helices were formed more easily than extended structures. Of all H-bonds, 61% were found in α-helices, π-helices contained an additional 14%, whereas antiparallel and parallel β-sheets accounted for 15% and 10% of H-bonds, respectively. The various secondary structure types occurred in almost all combinations and could be classified as "all-α proteins." β-sheets mainly occurred together with α-helices, often arranged in a familiar sandwich fashion, with the sheet lying in the middle of the molecule. Among the sheets, antiparallel hairpins were quite common, followed by mixed sheets, whereas parallel structures were relatively rare.

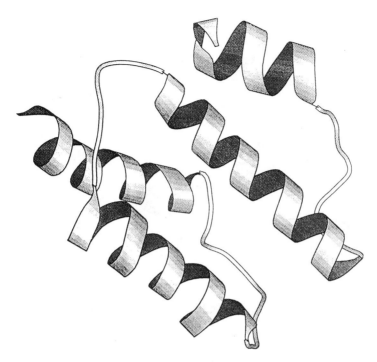

Figure 6-2a. Structures of H-bonded model chains: An "all-helical" structure, built of three α-helices and a π-helix at the chain terminus.

6. Stick Models

The methods described above are sufficient to produce structures from random sequences that embody all the familiar features of globular proteins. Unfortunately, they cannot predict the fold of a protein given only its amino acid sequence. This difficult, if not impossible, problem will be avoided by considering starting models based on simplified folds. The problem of generating suitable starting models falls into two parts. First, a realistic framework must be adopted to represent possible secondary structure locations, and second, the possible connections of these units must be explored.

6.1 Stick Architectures

Secondary structures are extended (helical) objects and, because of their linear axis, often pack in a roughly aligned manner, as in a bundle of rods. To a first approximation this allows the structure of proteins to be displayed in a very simplified way by neglecting the extended dimension and portraying only the ends of the "rods." In this representation, protein structures appear

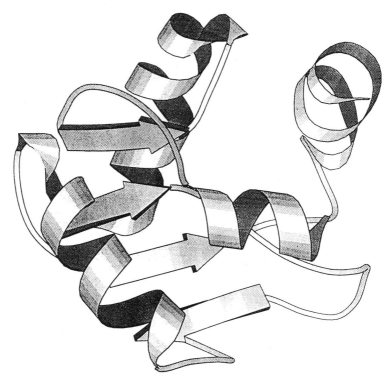

Figure 6-2b. Structures of H-bonded model chains: A mixed β-sheet built of four strands, surrounded by α-helices on both sides.

as layers of packed secondary structure. Typically, β on β (the β-sandwich class) or a β-layer between two α-layers (the alternating β/α class).

6.1.1 Alpha-helical structures. The layered structure is clear in the preceding classes because of the regularity imposed by the hydrogen-bonded β-sheets. However, this constraint is not present in the all-α class, which adopts a less regular variety of forms. A useful model for this class was devised by Murzin and Finkelstein (1988), who constructed idealized models for small globular proteins. If it is assumed that, to a first approximation, α-helices are as long as they are thick, then two helices will have N- and C-terminal endpoints that are equidistant both within a helix and between helices. This assumption of approximate symmetry allows very simple architectures to be constructed for bundles of packed helices in which all pairs of adjacent α-helices have equidistant endpoints. This constraint, combined with the adoption of an approximately spherical shape, defines a class of polyhedra that have equilateral triangles as faces and are sometimes

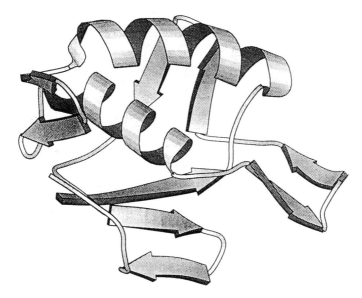

Figure 6-2c. Structures of H-bonded model chains: Three antiparallel β-hairpins and a mixed sheet, protected by two helices from one side only.

(graphically) referred to as delta-hedra. The most regular members of the class are its smallest and largest members: the tetrahedron (two helices), and at the upper end, the icosahedron (six helices; see Figure 6-3[a]).

6.1.2 Beta structures. β-strands can be incorporated into this simple framework by allowing them to lie at half the helical spacing. Endpoints were defined for a small β/α protein by connecting two hexagons that have a relative rotation of 30° (see Figure 6-3[a]). This imparts a suitable twist to the sheet (arrows) and a corresponding twist between both strands and helices. The model can also be used for stacked β proteins by neglecting the β-strands, reducing the scale by half, and taking the α-helices as β-strands (see Figure 6-3[c]). With a little distortion, both models can be extended into a general helical structure, so allowing any number of β-strands.

6.2 Transmembrane Structures

A specialized protein architecture can be found in the bundles of packed helices that typically form integral membrane proteins. Neglecting their reversed hydrophobic polarity, these helices can also be modeled using the twisted lattice of sticks described above (Taylor et al., 1994). However, this more specialized application will not be discussed in the current work.

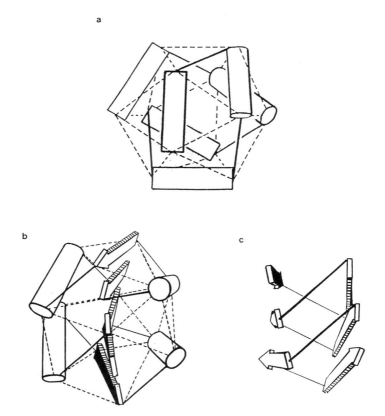

Figure 6-3. (a) By assuming α-helices have equidistant endpoints, idealized models can be constructed for small globular proteins. Six helices have 12 ends, which define an icosahedron (fewer helices define less regular polyhedra). All windings over this surface define a protein fold that can then be investigated. (b) β-strands can be incorporated into this simple framework by allowing them to lie at half the helical spacing. Endpoints were defined for a small β/α protein by connecting two hexagons that have a relative rotation of 30°. This imparts a suitable twist to the sheet (arrows) and a corresponding twist between both strands and helices. (c) The model can also be used for stacked β proteins by neglecting the β-strands, reducing the scale by half and taking the α-helices as β-strands.

6.3 Fold Combinatorics

The problem of enumerating folds becomes the simple problem of generating all tracings over an idealized framework in which the path does not cross or pass through the same point twice. The register of the sequence on the framework is set by the secondary structures, with each structure being placed on alternate edges of the polyhedron as the winding progresses. Computationally, this can be achieved by the application of a recursive routine that chooses a path from each node until there is no further secondary structure unit or a dead-end is encountered. On each of these conditions the procedure "back-tracks" to the preceding node and takes an alternative path. Exhaustive application of this procedure eventually enumerates every possible path.

Surprisingly, on the icosahedral model of six helices there are only 1264 distinct paths (less symmetric models would generate more). This can be contrasted with the alternative approach (applied to the same size of problem) of simply adding one helix onto another and allowing the fold to grow through accumulated pairwise interactions. This generates many millions of possibilities (Cohen et al., 1979), most of which infringe obvious steric constraints that are never encountered when the chain is constrained to an idealized framework.

6.3.1 Motif incorporation. The possible structures generated by an unconstrained combinatoric trace over all possible windings can be greatly reduced if a distance constraint can be placed on even one pair of structural elements. In a previous study on myoglobin (Cohen and Sternberg, 1980), the constraint implied by heme binding was imposed after relatively detailed models had been built. However, it is more cost-effective to apply any such constraints at as early a stage as possible. This might be done during the search over the tree of possibilities, and if a forbidden pairing is encountered then all remaining combinations following that node on the tree can be neglected. This "tree-pruning" strategy is most effective when the interactions being tested are sequentially local, such as the handedness of $\beta/\alpha/\beta$ units of supersecondary structure.

A general approach is simply to generate complete "stick" figures and test each pairwise interaction against a target template (which may involve any number of elements). To make such a comparison requires the target constraints—the motif—to be in an equivalent form. To facilitate this, a fragment of structure composed of secondary structures can be reduced to a series of line segments by fitting a least-squares line to each element (Taylor et al., 1983). Normally these lines would be truncated at the ends of the secondary structures that they represent (at a point normal to the first and last residues). However, the motif line segments can instead be truncated to the same length as the ideal segments. When represented in this way, the fragment or motif can be compared easily with the orientation of lines encountered in various windings over the ideal framework by any measure based on angle, distance, or both (Taylor, 1991a).

The single constraint implied by heme-binding in the globins provides a relatively weak constraint on the possible folds (Figure 6-4a); however,

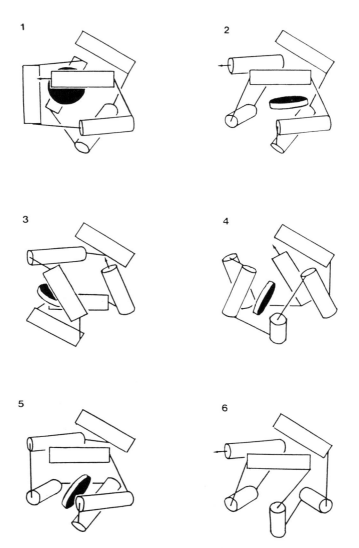

Figure 6-4a. High-scoring myoglobin-like folds are depicted on the icosahedral framework (Figure 6-3[a]). All chains begin at the topmost vertex. The C-terminus is distinguished by an arrow head and the approximate location of the heme is indicated by a disc. (1) Fold 1013 or MBN (myoglobin). (2) Fold 997; has better general helix packing than MBN. (3) Fold 420; also better general helix packing than MBN but no B/E contact. (4) Fold 1120; a well-packed, simple, meandering barrel fold but with poor loop packing. (5) Fold 465; of the same packing type as fold 1120. (6) Fold 1170; similar packing class to MBN but with no E–F hairpin (hence no heme is shown).

well-defined motifs—such as the calcium-binding EF-hand—can provide powerful constraints when used as a filter. For example, the protein parvalbumin contains six helices, two pairs of which constitute EF-hands. Applying these as a constraint (independently of each other) reduced the possible structures from over 1200 to 3. One of these corresponded to the native fold while the other two were trivial variants (see Figure 6-4b).

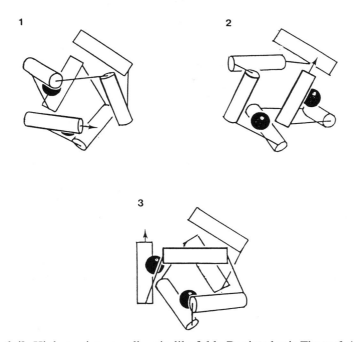

Figure 6-4b. High-scoring parvalbumin-like folds. Depicted as in Figure 6-4a except that, rather than a heme, the Ca^{++} ions are indicated as black spheres between the C–D and E–F helices. (1) Fold 1030 or CPV (parvalbumin). (2) Fold 437; differs only from CPV in the orientation of the A–B helical hairpin and scores equally. (3) Fold 868; scores slightly less than CPV and 437 and differs also in the orientation of the A–B hairpin, but this is additionally of a different longitudinal packing.

7. Protein Models

In the methodology outlined in the preceding sections, a starting structure was generated over an ideal framework resulting in a sketch of the overall fold. The following sections describe how these ideal folds can be made more protein-like using the distance-geometry modeling methods described in earlier sections.

7.1 Models from Ideal Folds

The simplest method to elaborate a "stick" model is to place atoms along each secondary structure vector at intervals appropriate to the pitch of the type of structure and allow the ideal secondary structure template, in conjuntion with the regularization algorithm outlined above, to elaborate the structure. Alternatively, it can be assumed that the hydrophobic moment of the ideal secondary structure should point towards the center of gravity of the molecule, allowing the full secondary structure to be constructed immediately. Loop residues can be similarly treated by placing these at equal intervals along the line connecting their adjacent secondary structures. These too, however, can be made slightly more realistic by introducing a radial component such that the loops describe approximate parabolic arcs.

The α-carbon models so constructed then become the initial model for the cycles of distance projection and real-space regularization during which any regular secondary structures or motifs can be refined toward their ideal form. After only ten cycles of refinement, the stick models lose their regular shapes and become good representations of their corresponding native structures—especially if motif constraints have been refined (Taylor, 1993) (see Figure 6-5).

7.2 Models from Real Folds

By substituting a real protein structure as a starting model in place of the ideal model, the same refinement methodology can be used. The effect of the ensuing refinement cycles will be to regularize secondary structures and pack hydrophobic residues together. However, if the sequence being modeled has a complete correspondence to the known structure, then little change will result, since the large number of specified distances will maintain the model close to the original structure.

Of greater interest is the application where substantial elements of the modeled protein do not correspond to the known structure. These may simply be slight variations in loop lengths but can also include new substructures that can be refined toward their predicted secondary structure, independently of the known fold (Taylor et al., 1993).

7.3 Evaluating Folds

A function is required that, given all possible windings on an ideal framework, can recognize the one that corresponds to the native fold. However, given the simplifications that are inherent in the idealized model, such a function is unlikely to be reliable, and attempts to specify it in terms of

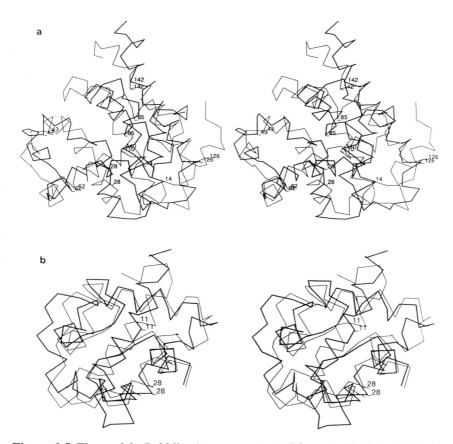

Figure 6-5. The models (bold lines) were constructed from the ideal (polyhedron) starting structure with predicted secondary structures and refined motifs. (a) Myoglobin: Equivalent residues are indicated on both structures. The largest deviation is found in the first helix, which is more deeply buried in the model. RMSd over all atoms was 7.07 Å and 3.92 Å over the core helices. (b) Parvalbumin: The molecule is viewed down the pseudo-twofold axis relating the two EF-handedness motifs. The largest deviations are in the first two helices. The first helix is oriented poorly due to the incorporation of an additional strong hydrophobic residue. The motifs are sufficiently close for their correct equivalence to be apparent. No constraints were applied between the two motifs. RMSd over all atoms was 3.98 Å, and 3.03 Å over the helices.

"stick" packing have yielded little, unless specific distance or motif constraints can be incorporated (Taylor, 1991a). The length of chain connecting secondary structures might be used as a constraint, but this is also not very effective, given the relatively small dimensions of the packed globule and the uncertainty in the secondary structure prediction. However, the fundamental problem is that the range of interactions between pairs of secondary structures is not great, as one pair of packed hydrophobic surfaces looks much like any other.

A more realistic initial step has been to apply an evaluation function to models generated from known structures. A number of methods based on empirical energy potentials allow model protein structures to be evaluated without the need to fully specify side-chain locations (Sippl, 1990). Such methods are effective at recognizing protein sequences matched— or *threaded*—onto correct homologues of known tertiary structure (Jones et al., 1993). In principle, it is only necessary to apply the method to matching a sequence against a sufficiently realistic representation of combinatorially generated structures to recognize the native fold. Two practical problems barring this simple solution are the limited generality of the empirical potentials used in evaluating different threadings and the lack of realism achieved in the models generated from "stick" structures.

Application of the distance geometry-based methods of model building described above from a starting structure based on a winding over an ideal framework can lead to model structures that are quite close to the native structures to which they correspond. The regular forms of the ideal polyhedra are initially lost by the incorporation of secondary structures of different lengths, and the subsequent "collapse" of these to form a compact hydrophobic core further obscures the regular origins of the model (Taylor, 1993). These results give some hope for success, as these models are as similar to their corresponding native structure as are homologues that can be recognized by the threading method. For example, this method can recognize the relationship of a phycocyanin sequence to a globin structure, which is a similarity that involves differences in structure that are as great as between myoglobin and its corresponding model (Figure 6-5[a]).

Preliminary results indicate that while the methods are converging, the native fold cannot yet be recognized as a unique fit. The reason for this may simply be that the models are systematically different from "real" proteins, thereby introducing an additional source of noise. Alternatively, there are many folds among the databank of "fake" proteins that, when viewed only at a less detailed level, incorporate interactions that are more similar to the globin fold than anything encountered in the databank of "real" proteins (for example, a "mirror-image" globin). The elimination of these as candidate native folds may be impossible without the full specification of (chiral) side-chain interactions.

REFERENCES

Aszódi A, Taylor WR (1994): Secondary structure formation in model polypeptide chains. *Prot Engng* (Submitted)

Braun W, Gō N (1985): Calculation of protein conformations by proton–proton distance constraints—a new efficient algorithm. *J Mol Biol* 186:611–626

Brooks BR, Bruccoleri RE, Olafson BD, States DJ, Swaminathan S, Karplus M (1983): CHARMM: A program for macromolecular energy, minimisation, and dynamics calculations. *J Comp Chem* 4:187–217

Chan HS, Dill KA (1990): Origins of structure in globular proteins. *Proc Natl Acad Sci USA* 87:6388–6392

Cohen FE, Richmond TJ, Richards FM (1979): Protein folding: Evaluation of some simple rules for the assembly of helices into tertiary structures with myoglobin as as example. *J Mol Biol* 132

Cohen F, Sternberg M (1980): On the use of chemically derived distance constraints in the prediction of protein structure with myoglobin as an example. *J Molec Biol* 137:9–22

Cohen FE, Sternberg MJE, Taylor WR (1980): Analysis and prediction of protein β-sheet structures by a combinatorial approach. *Nature* 285:378–382

Cohen FE, Sternberg MJE, Taylor WR (1981): Analysis of the tertiary structure of protein β-sheet sandwiches. *J Mol Biol* 148:253–272

Creighton TE (1983): *Proteins: Structures and Molecular Properties.* New York: Freeman

Crippen GM, Havel TF (1988): *Distance Geometry and Molecular Conformation.* Chemometrics Research Studies Press

Dayhoff MO, Schwartz RM, Orcutt BC (1978): A model of evolutionary change in proteins. In *Atlas of Protein Sequence and Structure.* Dayhoff MO, ed. Washington, DC: Nat Biomed Res Foundation, Vol. 5, Suppl. 3, pp. 345–352

Desmet J, Demaeyer M, Hazes B, Lasters I (1992): The dead-end elimination theorem and its use in protein side-chain positioning. *Nature* 356:539–542

Dill KA, Fiebig KM, Chan HS (1993): Cooperativity in protein-folding kinetics. *Proc Natl Acad Sci USA* 90:1942–1946

Easthope PL, Havel TF (1989): Computational experience with an algorithm for tetrangle inequality bound smoothing. *Bull Math Biol* 51:173–194

Flory PJ (1969): *Statistical Mechanics of Chain Molecules.* New York: Wiley-Interscience

Gregoret LM, Cohen FE (1991): Protein folding: Effect of packing density on chain conformation. *J Mol Biol* 219:109–122

Havel TM, Crippen GM, Kuntz ID (1979): Effects of distance constraints on macromolecular conformation. II. Simulation of experimental results and theoretical predictions. *Biopolymers* 18:73–81

Jones DT, Orengo CA, Taylor WR, Thornton JM (1993): A new approach to protein fold recognition. *Nature* 358:86–89

Kuntz ID, Crippen GM, Kollman PA, Kimelman D (1976): Calculation of protein tertiary structure. *J Mol Biol* 106:983–994

Kuntz ID, Thomason JF, Oshiro CM (1989): Distance geometry. *Meth Enzymology* 177:159–204

Levitt M (1978): Conformational preferences of amino acids in globular proteins. *Biochemistry* 17:4277–4285

MacKay AL (1983): The numerical geometry of biological structures. In *Computing in Biological Science*. Geisow MJ, Barrett AN, eds. Amsterdam: Elsevier Biomedical, pp 349–392

Murzin AG, Finkelstein AV (1988): General architecture of the α-helical globule. *J Mol Biol* 204:749–769

Pearl LH, Taylor WR (1987): A structural model for the retroviral proteases. *Nature* 329:351–354

Press WH, Flannery BP, Teukolsky SA, Vetterling WT (1986): *Numerical Recipes: The Art of Scientific Computing*. Cambridge: Cambridge University Press

Richards FM (1974): The interpretation of protein structures: Total volume, group volume distributions and packing density. *J Mol Biol* 82:1–14

Saitoh S, Nakai T, Nishikawa K (1993): A geometrical constraint approach for reproducing the native backbone conformation of a protein. *Proteins* 15:191–204

Sippl MJ (1990): Calculation of conformational ensembles from potentials of mean force. An approach to the knowledge-based prediction of local structures in globular proteins. *J Mol Biol* 213:859–883

Skolnick J, Kolinsky A (1990): Simulations of the folding of a globular protein. *Science* 250:1121–1125

Skolnick J, Kolinski A (1991): Dynamic Monte-Carlo simulations of a new lattice model of globular protein folding, structure and dynamics. *J Mol Biol* 221:499–531

Taylor WR (1991a): Sequence analysis: Spinning in hyperspace. *Nature* 353:388–389 (News and Views)

Taylor WR (1991b): Towards protein tertiary fold prediction using distance and motif constraints. *Prot Engng* 4:853–870

Taylor WR (1993): Protein fold refinement: Building models from idealised folds using motif constraints and multiple sequence data. *Prot Engng* 6

Taylor WR, Jones DT, Green NM (1994): A method for α-helical integral membrane protein fold prediction. *Prot Struct Funct Genet* (In press)

Taylor WR, Jones DT, Segal AW (1993): A structural model for the nucleotide binding domain of the cytochrome b_{-245} β-chain. *Protein Science*

Taylor WR, Orengo CA (1989): Protein structure alignment. *J Molec Biol* 208:1–22

Taylor WR, Thornton JM, Turnell WG (1983): A elipsoidal approximation of protein shape. *J Mol Graphics* 1:30–38

Teller DC (1976): Accessible area, packing volumes and interaction surfaces of globular proteins. *Nature* 260:729–731

7

Molecular Dynamics Studies of Protein and Peptide Folding and Unfolding

Amedeo Caflisch and Martin Karplus

1. Introduction

Proteins are fascinating. As objects in three-dimensional space, they are sometimes elegant and always complex molecules, yet they consist of only 20 different amino acid building blocks. The function of proteins is determined by their three-dimensional structure, and the majority of biological processes involve one or more protein molecules. The mechanism of the evolutionary development of specific proteins is one of the unsolved problems of biology. Most proteins are very sensitive to their environment; small temperature or pH changes can alter both their stability and their ability to function. The native structure of a protein is determined by the amino acid sequence (Anfinsen, 1972). In solution, many proteins have been shown to refold by themselves under conditions that lead to a stable native state, but *in vivo* the folding process can be very complicated and often involves other proteins, such as chaperones (Gething and Sambrook, 1992).

The time scale for protein folding in solution ranges from nanoseconds to hours. This is much too long for a steepest descent to a minimum of the potential energy surface and much too short for an exhaustive search of the conformational space. It is clear from minimization studies of folding (Nemethy and Scheraga, 1977; Levitt, 1983) and from experimental (Austin et al., 1975) and theoretical studies (Elber and Karplus, 1987; Noguti and

The Protein Folding Problem and Tertiary Structure Prediction
K. Merz, Jr. and S. Le Grand, Editors
© Birkhäuser Boston 1994

Gō, 1989a–e; Caflisch et al., 1992) of the protein conformation space that it is highly complex, with a multiminimum character that prevents folding from being the analogue of a simple, monotonic descent into a potential well. This means that barriers must be overcome in the folding process and the finding of the minimum has analogies to the complexities arising in glassy systems. Since the seminal analysis of Levinthal (1969) it has been realized that alternatives to a random search must be operative in protein folding. Although the conformation space of the denatured state is vast, there would be no search problem if each of the amino acid residues could find its correct conformation independent of the others, or if only interactions with nearest neighbors were involved. Rapid folding would be expected, as in the helix–coil transition of polymers. Protein folding has been compared to crystallization (Harrison and Durbin, 1985), but it is important to realize that it is significantly different. In the former, once a nucleus is formed, there is no problem; i.e., there are many similar sites at which subsequent molecules can coalesce independently. In protein folding, even if there were a stable nucleus, it is not clear that the protein could continue to fold by the independent condensation of residues. The fundamental distinction between the helix–coil transition or crystallization and protein folding is that long-range interactions play an essential role in determining the native state. In the limit, this implies that the conformational energy of each amino acid residue depends on all of the others in the polypeptide chain. It is this aspect of the search of the full configuration space, with its vast numbers of conformers, that leads to the Levinthal paradox.

A recent study (Zwanzig et al., 1992) argues that the Levinthal paradox can be resolved simply by introducing local main-chain propensities. They examined a bead model for the polypeptide chain that has only local interactions and found that a small bias of the main-chain propensities results in a very large reduction in the apparent first passage time to a folded state. Such an analysis misses the essential point, outlined in the previous paragraph, concerning the difference between the helix–coil transition, which corresponds to the case studied by Zwanzig et al., and the cooperative transition of a protein.

Naively, the best way to find out how a protein folds is to start with a model for the denatured state of a protein and to simulate the transition to the native state. The simulation itself would solve the Levinthal paradox if the model used were sufficiently detailed and accurate. An approach that should work, in principle, is to use an atom-based model for the potential energy function and to follow a molecular dynamics trajectory from one or more denatured conformers to the native state in the presence of the appropriate solvent. With the available methodology and computing power, such a simulation would require approximately 10^{11} hours (or 10^7 years) for a 100-residue protein where the experimental transition to the folded state takes place in about 1 sec. With the teraflop supercomputers that are on the horizon (Deng et al., 1990), such simulations would still take 10^3 years, although it would be possible to examine the faster portions of the folding transition by such a direct method. Clearly, when such simulations become possible they should be done, even though it is likely that a very large fraction of the trajectory would be uninteresting and that a very large amount of human time would be required to interpret the results.

The practical difficulties in doing such brute force simulations has led to several types of theoretical approaches to the dynamics of protein folding. One approach is based on the complete atomic model, but biases the system to speed up the folding transition by several orders of magnitude (as in simulated annealing with NMR distance constraints), or instead studies unfolding under special conditions (e.g., use of a forcing potential, high temperature) that reduce the time involved from microseconds to picoseconds. Related studies deal with small protein fragments (e.g., α-helices, β-turns) for which the conformation space is sufficiently small so that full searches can be accomplished and/or transitions of interest occur on a manageable time scale. Alternatively, the entire polypeptide is considered, but a simplified description is introduced. One approach is to approximate each amino acid residue by one or two quasi particles and to use the C_α pseudo-dihedral angles as the only variables. The problem can be further reduced by restricting the quasi particles to move on a lattice (e.g., a cubic lattice) and using Monte Carlo procedures to follow the folding process. Such lattice calculations have the advantage over the physically more correct ("off-lattice") atomic models in that the total number of conformations is reduced, and a large fraction of all the conformations can be examined in a given time (Shakhnovich and Gutin, 1990; Shakhnovich et al., 1991).

In addition to methods that are based on following the dynamics of a polypeptide chain with varying levels of detail, there exist a range of models for protein folding that are phenomenological in character (Karplus and Weaver, 1976, 1979). These models provide a qualitative description of the folding process, often including a conceptual approach to the solution of the Levinthal paradox. In this review we concern ourselves primarily with molecular dynamics simulations of peptides and proteins that are being used to analyze elements of protein folding and unfolding. Such simulations constitute an approach to the protein folding problem that has been made possible relatively recently by improved simulation technology and the increase in speed of the available computers. A review of the theory of protein folding that considers other aspects of the problems is given by Karplus and Shakhnovich (1992).

2. Fragment Studies

An approach that can be used to obtain information concerning the detailed kinetics of protein folding is to study peptide fragments that can be treated by detailed simulations. A number of such studies have been made, and more are in progress. To illustrate the possibilities, we shall describe several of them here and outline the more important results.

Czerminski and Elber (1989, 1990) did vacuum simulations of a blocked alanine tripeptide, which is the shortest unit that can form an i to $i + 4$ helical hydrogen bond. The simplicity of this system and the absence of solvent permitted them to make a full analysis of the multiminimum nature of the potential surface. Minima were found with the three peptide units in the neighborhood of the local minima for (ϕ, ψ) equal to $(-60, -60;$ α-helix), $(-120, 120;$ β-sheet), $(-60, 60)$ and $(60, -60)$. A total of 138 minima were found in the energy range up to 6.3 kcal/mol, relative to the lowest minimum. This is to be compared with $4^{3.5}$, or 128, combinations of the most probable (ϕ, ψ) values. The 6,216 reaction paths between the minima were investigated. For direct paths connecting two minima without an intermediate, the barriers were found to be in the range of 0.5 to 5.5 kcal/mol, with the higher energies most common. Most of the direct paths involved changes in only one dihedral angle, with two thirds of the "flips" involving ψ and one third involving ϕ. Whether indirect paths led to significantly lower barriers between pairs of minima was not discussed. A master equation approach was used to study the dynamics, with the kinetic constants estimated from transition state theory. This led to lifetimes for the various minima in the picosecond to nanosecond range at 400 K.

The "folding/unfolding" transition of a blocked alanine dipeptide was investigated recently (Lazaridis et al., 1991). The reaction path between the "folded" configuration $(\phi_1, \psi_1; \phi_2, \psi_2)$ equal to $(-72, -57; -81, -31)$ and an "extended" conformation $(-83, 129; -83, 129)$ was determined *in vacuo* by a free-energy simulation method with a dielectric constant of 50 to mimic the effect of an aqueous solvent. Two paths were found that are essentially mirror images of each other with ψ_1 and ψ_2 undergoing successive changes (ϕ_1 changes very little); an intermediate was found where either ψ_1 or ψ_2 had undergone a transition. The calculated barriers were in the range of 1.5 to 2 kcal/mol. In a free-energy simulation including an explicit model for the water molecules, similar results were obtained.

The major difference in the latter is that the extended configuration with hydrogen bonds to water is about 3.4 kcal/mol more stable than the "turn," which is similar to an earlier value (Tobias et al., 1990); in the model with a dielectric constant of 50, the extended conformer was less stable by about 1 kcal/mol. The form of the potential of mean force in the presence of water is similar to that in the absence of water with an intermediate and with barriers of 2.3 and 2.8 kcal/mol.

Case et al. (personal communication) have studied the dynamics of tetrapeptides for which there are NMR data indicating that the turn conformation makes a significant contribution to the equilibrium population in aqueous solution. In a 5 ns simulation of the peptide Ala-Pro-Gly-Asp, they found transitions between "turns" with 1–3 and 1–4 bifurcated $C=O \cdots H-N$ hydrogen bonds and essentially extended conformations on a time scale of about 500 ps. It was noted that the transitions had diffusive elements, suggesting that a simple transition state description might not be valid. This corresponds to the results of stochastic dynamics simulations of the β-strand to coil transition (Yapa et al., 1992).

A study (Robert and Karplus, unpublished results) of the folding of the peptide $(Ala)_2(Asp)(Ala)_6$ from an extended chain to an α-helix *in vacuo* (dielectric constant of 1) and in aqueous solution suggest similar folding mechanisms, but significantly different time scales. *In vacuo*, a structure formed rapidly (3 ps), in which the carboxyl group of the Asp interacted with three NH groups. This remained stable for about 80 ps, when a fourth NH group moved in and this initiated rapid (in a few ps) helix formation. The simulation in aqueous solution required about 200 ps to form the three NH/Asp configurations, and helix initiation required on the order of 600 ps.

Tirado-Rives and Jorgensen (1991) used a molecular dynamics simulation to study a 15-residue ribonuclease A S-peptide analogue at $5°C$ and $75°C$ in the presence of an explicit water model. They found that the peptide was stable for 300 ps at $5°C$, while it unfolded in less than 500 ps at $75°C$. Although experiments show that the peptide is more stable at low temperature (in agreement with the calculation), it is clear that equilibrium was not reached in the dynamics, since the observed helix–coil equilibrium constant is near unity at $5°C$. In the two higher-temperature simulations, unfolding occurred; in one, the transition took about 100 ps and in the other about 350 ps. Examination of the simulation suggested that an important element of the unfolding transition is the replacement of an α-helical (i to $i + 4$) hydrogen bond by water hydrogen bonds through an intermediate involving a 3_{10} (i to $i + 3$) or reverse turn hydrogen bond. This is in accord with the x-ray studies of Sundaralingam and Sekharudu (1989), who observed that when a water molecule inserts into an α-helix, a reverse turn is frequently formed; see also Karle et al. (1990). Details of the dihedral angle transitions are not given in the paper, so no conclusion can be drawn about their rates. Also, it is important to recognize that in a simulation of helix unfolding, nothing is learned about the rate-determining helix-initiation step.

In related work, Soman et al. (1991) have simulated the isolated helix H (residues 132 to 149) of myoglobin for 1 ns at 300 K. They found that the helix unfolded progressively from the C to the N terminus. Tight (i to $i + 3$) turns (3_{10} helices) were intermediates in the unfolding and often appeared prior to aqueous solvation. Insertion of H_2O occured in some cases, but not all. Often a single water molecule was inserted between the C=O and NH, but sometimes several water molecules were involved in transient intermediates. The transition were found to have diffusive elements with a time scale on the order of 100 ps. Considerable detail about the unfolding process is given in the paper.

A 100 ps molecular dynamics simulation of hydrated decaalanine (initially in a canonical right-handed conformation) at 300 K was performed by DiCapua et al. (1990). The distances between carbonyl oxygen and the amide nitrogen in the i to $i + 4$ hydrogen bonds were monitored. During the trajectory the helix was stable, except for the 5 NH–OC 1 hydrogen bond, which separated at about 15 ps (for 10 ps) and at ca. 70 ps for the remainder of the simulation. Analysis of the orientation of the water revealed that a bridge structure (i.e., water acting as donor for the CO group of residue 1 and acceptor for the NH of residue 5) was formed after about 70 ps but was not present during the transient destabilization, although a water was inserted and acted mainly as a hydrogen-bond donor. The helix destabilization coupled to water insertion is in agreement with simulations described above and the experimental results of Sundaralingam and Sekharudu (1989).

The helix denaturation of a polyalanine peptide (13 residues) was simulated both *in vacuo* and in aqueous solution with an explicit water model by Daggett and Levitt (1992a). They performed several simulations, lasting 200 ps each, at temperatures ranging from 298 to 473 K. The helical state was stable *in vacuo* at all temperatures, whereas in solution the helix denatured at higher temperatures. The role of water in destabilizing and denaturing the helix was investigated. Daggett and Levitt concluded that the type of water insertion proposed by Sundaralingam and Sekharudu (1989) was not the mechanism of helix unfolding. Instead they found that the helices tended to open up due to high-temperature fluctuation, after which the polar C=O and NH groups of the helix interacted with water molecules.

Although the simulations of Daggett and Levitt (1992a) were too short to reach equilibrium, the free energy was estimated from the average waiting time for the helix→coil and coil→helix transition for each residue (a fast event treated as if it were at equilibrium) and then averaged over all residues during the last 100 ps of each trajectory. It was found that in solution the free energy change for the helix→coil transition decreased with increasing temperature and transitions to coil conformations were unfavorable below 423 K. *In vacuo*, such transitions were unfavorable at all temperatures and the free energy was nearly temperature independent. Moreover, the free energy of activation for the single residue transition was found to increase with increasing temperature in all cases except for the helix→coil transition in solution.

Another type of study of peptide fragments makes use of free energy simulation methods for estimating the change of stability when amino acids are altered. An example is provided by molecular dynamics simulations, which were used to estimate the change in the conformational stability of proline-replaced polyalanine α-helices (Yun et al., 1991). An unfavorable effect was found for proline replacements in the middle of the helix, whereas a substitution at the first position of the helical N-terminus yielded a favorable contribution to the free energy of folding. In a related study, good agreement with experimental results was obtained for the differences in stability of oligoalanine peptides with single amino acid replacements in the middle of the chain (Hermans et al., 1992). Although this type of work is not concerned with the dynamics of folding, it is nevertheless of interest in the present context because it may be useful in interpreting the stability of intermediates or protein fragments involved in folding mechanisms, such as the diffusion collision model.

A number of simulations of protein fragments have reached longer time scales than those possible in the all-atom plus solvent simulations described above. The method uses a potential of mean force surface that implicitly includes the solvent and employs stochastic dynamics to introduce the dynamic effects of the solvent. To further simplify the problem, each residue is represented by a single interaction center ("atom") located at the centroid of the corresponding side chain, and the centers are linked by "virtual" bonds (Levitt, 1976; McCammon et al., 1980). For the potential energy of interaction between the residues, assumed to be Val in an α-helical simulation (McCammon et al., 1980), a set of energy parameters obtained by averaging over the side-chain orientations was used (Levitt, 1976). Terms that approximate solvation and the stabilization energy of helix formation were included. The diffusive motion of the chain "atoms" expected in water was simulated by using a stochastic dynamics algorithm based upon Brownian dynamics. Starting from an all-helical conformation, the dynamics of several residues at the end of a 15-residue chain was monitored in a number of independent 12.5 ns simulations at 298 K. The mobility of the terminal residues was quite large, with an approximate rate constant of 10^9 per sec for the transition between coil and helix states. This mobility decreased for residues further into the chain. Unwinding of an interior residue required simultaneous displacements of several residues in the coil, so larger solvent frictional forces were involved. The coil region did not move as rigid body. Instead, the torsional motions of the chain were correlated so as to minimize dissipative effects. Such concerted behavior is not consistent with the conventional idea that successive transitions occur independently. Analysis of the chain diffusion tensor showed that the frequent occurrence of correlated transitions resulted from the relatively small frictional forces associated with such motions (Pear et al., 1981). Further, the correlated nature of the torsional transition suggests that unwinding occurs in a relatively localized fashion and that a limiting value of about 10^7 per sec would be reached for the interior of the helix.

A similar model has been used recently to study β-sheet to coil transitions in a β-hairpin (Yapa et al., 1992). From an analysis of ten 90 ns simulations, it was shown that rate constants for the transition between the coil and strand state are on the order of 10^{10} to 10^{11} per sec. This high rate occurs even though the adiabatic potential has barriers on the order of 3 kcal/mol. Unlike the α-helix results, the transition rate constant decreases only slowly as one goes from the end of the strand toward the interior. Also, nonterminal residues are sometimes found in the coil state, while the end residues form a regular sheet. This behavior may be related to the occurrence of β-bulges.

A corresponding approach has been applied to the behavior of structural motifs of proteins. An example is a simulation of two α-helices connected by a coil segment (Lee et al., 1987). This simulation served to examine a possible elementary step in protein folding, as described by the diffusion-collision model (Karplus and Weaver, 1976), i.e., the coalescence of a pair of helices. The system considered in the simulations is a 24-residue peptide in which the first and last eight residues formed an α-helix and the intervening eight residues are initially in a random-coil conformation. Twelve trajectories were generated with a total time of 820 ns. The exact lengths of the individual trajectories were not important because the parameters were chosen so that a stable, coalesced structure does not form. Instead, the system folds and unfolds many times during the simulation, and rate constants for the coalescence and dissociation reactions could be determined. The values obtained are on the order of 10^8 per sec. For the unfolded system, there is a strong bias in the connecting loop towards shorter distances, with a maximum in the radial distribution function near 27 Å, even though the fully extended conformation has a length of 45 Å. This is due primarily to the entropic contribution, since the intervening chain has many more allowed conformations at intermediate lengths. Such a trend has been observed experimentally in a study of the end-to-end distances in a series of oligopeptides (Haas et al., 1978). This model calculation of coalescence used stable helices to reduce the time required. In a real system, helix to coil transitions would be coupled to the collisional association events. Given the estimates of the helix to coil transition rates, such a study should be possible by combining a stochastic dynamics simulation with a kinetic model for the helix–coil transition.

3. Simplified Protein Simulations

Simulation of folding for a complete protein represent a much more complex problem than the fragment studies just described. Early work in this area (Levitt and Warshel, 1975) used a C_α-type model for the protein and coupled energy minimization with thermalization *in vacuo* in an attempt to fold bovine pancreatic trypsin inhibitor (BPTI), a 58-residue protein. Conceptually, such an approach is of interest; if the potential surface, or better, potential of mean force surface of a protein, were such that a relatively simple procedure can find the minimum, there would be no Levinthal paradox. The limiting case would be a single potential well connecting the extended coil conformations to the native structure. Clearly that is not what is found for real proteins, although an empirical potential can be modified to obtain a relatively well behaved surface that leads to a folded structure. In the potential used for the BPTI folding, for example, biases for turns and extended strands were introduced in the regions that have such conformations in the native structure (Levitt and Warshel, 1975; Hagler and Honig, 1978). With such a potential, one out of five runs that began with the terminal α-helix present and involved 600 cycles of minimization and thermalization had features of the native form, although the root mean square (RMS) difference was about 6 Å, as compared with 3.4 Å, the best value possible for the simplified model.

Another way of modifying the potential to achieve rapid folding is to introduce distance constraints. These have been used to simplify the energy minimization problem and, more recently, as a way of employing molecular dynamics with a simulated annealing protocol for structure determination by NMR (Brünger and Karplus, 1991). In a trial study of the protein crambin with NOE interproton distance constraints (Brünger et al., 1986), it was found that the native structure is achieved from an extended conformation in a few picoseconds of high-temperature molecular dynamics, a time on the order of 10^{12} times faster than that observed in solution. This decrease in time appears to involve two aspects of the constraints. The first is that stable secondary structural elements are formed in 1 psec, instead of approximately 10 msec, and the second aspect is that the need to search a large portion of the conformation space is eliminated by the long-range constraints (in terms of distance along the chain rather than physical distance). It was shown that if the secondary structure is not formed before collapse, misfolded structures that are local minima can result. This is consistent with the diffusion–collision and related phenomenological description of protein folding (Karplus and Weaver, 1979; Kim and Baldwin, 1990).

A BPTI fragment that contains the 30,51 S–S bond and has a large portion of the β-sheet and the C-terminal α-helix has been studied experimentally by Oas and Kim (1988); they have shown that the system is marginally stable with a structure close to that of native BPTI. When this construct was simulated at 500 K (Robert and Karplus, unpublished results), rapid denaturation occurred. A corresponding simulation at 500 K with the explicit inclusion of solvent showed the beginnings of denaturation over a time scale of 200 psec. The C-terminal α-helix began to unfold and water molecules replaced the terminal hydrogen bonds of the β-sheet. The behavior corresponds to the fragment studies cited previously, in term of the slowing of the time scale relative to vacuum simulations and the explicit role of water–protein interactions at a certain stage of denaturation.

4. High-Temperature Unfolding Simulations of Proteins

Recently, a number of high-temperature molecular dynamics simulations of protein denaturation have appeared, and more are in progress. The simulation of protein unfolding is of interest for several reasons. First, the utilization of the x-ray structure of the native protein as the starting point provides initial conditions corresponding to a known, physically stable state. The choice of an unfolded or denatured structure for the simulation of protein folding would be more difficult because of the lack of experimental information and the multiplicity of possible starting structures. Second, it is known that protein folding is cooperative in that the transition between the denatured and native state is of the all-or-none type at equilibrium (Karplus and Shakhnovich, 1992). There is evidence that the rate-limiting step (i.e., the main free-energy barrier to folding and unfolding) is near the native state in many proteins (Creighton, 1988). Thus, the investigation of the initial phases of protein unfolding can help to elucidate the cooperative nature of protein folding, as well as of unfolding. Third, a number of detailed experimental studies of the protein unfolding transition have been made; an example is given in the barnase results described below. Finally, it has been shown by experiments that several proteins have stable, partially folded intermediates under mild denaturing conditions (Dolgikh et al., 1981; Ptitsyn et al., 1990; for reviews, see Kuwajima, 1989; Ptitsyn, 1992). These intermediates are characterized by native-like secondary structure and a native-like tertiary fold. However, their tertiary structure is less rigid than in the native state, probably due to the absence of the tight packing of side chains that occurs in the protein interior (Ptitsyn, 1992). The term "molten globule" has been introduced for such compact states (Ohgushi and Wada, 1983), which appear to be intermediates on the folding pathway in some cases. There is evidence that the main energy barrier in protein folding is encountered after the formation of the molten globule state, i.e., between the molten globule and the native state. It is likely that this part of the folding process is mirrored by the presently available simulations of the unfolding process.

Vacuum molecular dynamics simulations with a polar hydrogen model for reduced BPTI were performed by heating the system from 300 to 1000 K over a time of 450 psec (Brady et al., unpublished results). Such a simulation of denaturation is in accord with the thermodynamic analysis of Shakhnovich and Finkelstein (1989) in which the first step of denaturation is regarded as a "vacuum" process and the protein expands to a molten, globule-like state with a transition temperature of the order of 500 K. A relatively sharp transition in the RMS deviation from the native structure was found at about 500 K; the RMS increased from about 2 Å to 4 Å. After this, the RMS increased slowly until there was a second transition close to 1000 K, in which the RMS reached about 7 Å. The first transition corresponds to a loss of tertiary structure with preservation of secondary structure; i.e., the β-sheet and α-helix are present but reoriented relative to each other. In the second transition all the secondary structure is lost, with the α-helix disappearing before the β-sheet. The loss of the secondary structures is accompanied by the formation of new hydrogen bonds, including "C5 ring"-like structures, a local minimum on the vacuum dipeptide map. The final system is a random, coil-like globule, whose radius of gyration is only slightly larger that that of native BPTI.

Fan et al. (1991) have utilized high-temperature vacuum molecular dynamics to study the molten globule state of α-lactalbumin. Some native-like secondary structure was recently shown to be present in the molten-globule state of guinea pig α-lactalbumin by NMR hydrogen exchange experiments (Baum et al., 1989; Dobson et al., 1991). A number of main-chain NH groups were observed to be highly protected from solvent exchange in the molten globule state; these were found to occur mainly in segments that are helical in the native structure. In particular, two major helical regions, including residues 23–34 (helix B) and residues 89–96 (helix C), were identified as stable by these experiments. To investigate the stability of α-helices B and C in α-lactalbumin, Fan et al. performed two sets of simulations (with a dielectric constant of $4r$, r being the distance in Å between interacting atoms); one was at a temperature of 500 K (lasting 50 ps) and the other at 1000 K (lasting 20 ps). Starting coordinates were model-built from the x-ray structure of the homologous baboon α-lactalbumin (Acharya et al., 1989) since no x-ray coordinates are available for the guinea pig protein. Fan et al. imposed harmonic constraints on the helices in one simulation and on the hydrophobic core in another, for each value of the temperature. Upper and lower bounds for the distances of all pairs of constrained atoms were introduced by adding or subtracting 0.2 Å from the distances in the native structure. They found that constraining the helices does not stabilize the hydrophobic core, whereas constraining the hydrophobic core does stabilize the α-helices. The hydrophobic contribution to the free energy of unfolding was estimated from the change in the solvent accessible surface area during the simulation. It was found that this contribution was more favorable for the structures obtained with hydrophobic constraints than those obtained with the helix constraints. As expected, conformations characterized by a more compact hydrophobic core have a better hydrophobic free energy. Fan et al. do not give quantitative description of the behavior of the radius of gyration and of RMS deviation from the x-ray structure during the simulation. It is difficult to determine, therefore, whether what they observe corresponds to the molten globule state. Nevertheless, their results on the coupling of helix stability with the existence of a hydrophobic core is of considerable interest.

Daggett and Levitt (1992b) investigated the unfolding of reduced bovine pancreatic trypsin inhibitor (BPTI) by high-temperature molecular dynamics. BPTI was solvated in a box of water molecules and periodic boundary conditions were applied. Simulations were performed at constant temperature by coupling to an external bath (Berendsen et al., 1984). The system was brought to the target temperature by scaling the velocities. Five simulations (lasting up to 550 ps) were performed: native BPTI at 298 K and 423 K and fully reduced BPTI at 298 K, 423 K, and 498 K; in the latter, all three S–S bonds were not present. Properties of BPTI as a function of temperature and the state of the disulfide bonds were averaged over the entire simulation. Since no detailed description of the behavior of such properties as a function of time is given in the paper, it is difficult to interpret the results, e.g., to determine whether the simulation reached a stable unfolding intermediate. The main conclusion presented by Daggett and Levitt (1992b) is that reduced BPTI at high temperature was relatively compact but had a larger radius of gyration than native BPTI (a volume increase of 5 to 25%), while there was little change at 423 K in the oxidized form; presumably a higher temperature would have been required to denature the native protein on the simulation time scale. Reduced BPTI at high temperature showed increased main-chain and side-chain mobility over that of the native protein. Since the average number of water molecules inside reduced BPTI was observed to be independent of structure, the expansion was not caused by solvent penetration. Thus, the present results are rather similar to those observed in the vacuum simulation of Brady et al. (unpublished) described above. However, it is important to note that the authors used a reduced density in their constant volume, high-temperature simulations. This is likely to be the reason for the lack of water penetration, which is in disagreement with experiment (Ptitsyn, 1992) and with the barnase simulations described below. Moreover, when a low-temperature (298 K) simulation with normal water density was performed for the unfolded reduced structure, there was a significant increase of protein–water interactions over that in the native state. Because of the lack of experimental results on BPTI kinetic intermediates, the authors compared the BPTI simulation with data for other proteins.

Mark and van Gunsteren (1992) have reported a simulation of the thermal denaturation of hen egg white lysozyme. As in Daggett and Levitt (1992b), the enzyme was solvated in a box of water molecules with periodic boundary conditions and the temperature was kept constant by coupling to an external heat bath. They performed a control run of 550 ps at 300 K and a run at 500 K, which was branched after 85 ps by cooling the system to 320 K over 5 ps. The simulation at 320 K was continued for 190 ps, while the 500 K simulation was stopped after 180 ps. They monitored the radius of gyration, the RMS deviation of the $C\alpha$ atoms from the starting x-ray structure, and changes in the main-chain hydrogen bonding pattern as a function of simulation time. At 500 K, the radius of gyration increased steadily for about 90 ps and then remained effectively constant for 40 ps before rising again more rapidly than before. A similar behavior was observed for the C_α RMS deviation from the x-ray structure. The simulation branched from the intermediate state showed an approximately constant radius of gyration at 320 K, a marked increase in solvent penetration (the authors report a 50% increase in solvent accessible surface area but do not discuss the penetration of the water molecules) and a high degree of native-like secondary structure. Hydrogen exchange NMR experiments (Miranker et al., 1991) have shown that lysozyme has two distinct folding domains, which correspond to the two structural domains (McCammon et al., 1976; Janin and Wodak, 1983). No difference between the two domains was observed in the simulation. Van Gunsteren (private communication) has argued that this result is not inconsistent with the hydrogen exchange experiments, i.e., that rapid fluctuations in accessibility can explain the NMR results. However, this argument is incorrect because it is based on a misunderstanding of the mechanism of hydrogen exchange, which, for the lysozyme, measures the equilibrium between the unfolded (exposed NH) and folded (protected NH) structures, rather than fluctuations in the latter. From the number of helical main-chain hydrogen bonds as function of time, Mark and van Gunsteren (1992) concluded that the formation of secondary structure occurs after the collapse of the peptide chain. This is in sharp contrast with experimental results based on far-UV ellipticity measurements, which demonstrate that there is formation of secondary structure at an early stage of folding (Gilmanshin and Ptitsyn, 1987; Kuwajima et al., 1987; Goldberg et al., 1990).

Recently, molecular dynamics simulations lasting 500 ps or more of apomyoglobin in aqueous solution were carried out by Brooks (1992) at 312 K, and by Tirado-Rives and Jorgensen (1993) at 298 K and at 358 K (pH values of 6 and 4 were used at the high temperature). The studies used different force fields; the CHARMM polar hydrogen model (Brooks et al., 1983) was used by Brooks (1992) and the AMBER-OPLS all-atom model by Tirado-Rives and Jorgensen (1993); the same water model, TIP3P (Jorgensen et al., 1983), was used by the two groups. The two room-temperature simulations (298 K and 312 K) address the native structure of apomyoglobin. In both simulations helices A, E, G, and H were stable. However, for the remaining helices, there are differences. Brooks (1992) found that the other helices also existed during the entire simulation, with B as stable as A, E, G, and H, while C, D, and F were more mobile and moved into the heme cavity. By contrast, Tirado-Rives and Jorgensen (1992) found that helix B unfolded in part and helices C, D, and F unfolded completely in a relatively short time. Both results are consistent with the experimental findings of Hougson et al. (1990, 1991) in that helices A, G, and H constitute a stable subdomain that retains substantial helicity upon partial unfolding in environments of decreasing pH; as to the other helices, the experiments do not clearly distinguish between the two simulations. Brooks (1992) found that the all-atom RMS deviation (RMSD) from the crystal structure reached a plateau value after about 60 ps. Different behavior of the RMSD for the main-chain heavy atoms relative to the crystal structure were found at 358 K by Tirado-Rives and Jorgensen (1993); in the 358 K run at pH = 4, the RMSD increased continuously until the end of the simulation (500 ps), whereas in the 358 K run at pH = 6 it reached a plateau value after the first 350 ps. It is surprising that in the 298 K control run (pH value of 6.0) the RMSD did not reach a plateau value. Tirado-Rives and Jorgensen state that the helical content averaged over the last 50 ps of both trajectories at 358 K was in agreement with CD estimates (Houghson et al., 1990). No discussion of the role of water is given in either paper.

High-temperature molecular dynamics simulations have been performed to study the unfolding of barnase, a ribonuclease excreted from *Bacillus amyloliquefaciens* for which a large body of experimental work on folding and unfolding is available from Fersht and co-workers. Barnase is a monomeric enzyme (110 residues) that is of particular interest as a folding model, as suggested by Fersht (for a review, see Fersht, 1993). Both its crystal structures (Mauguen et al., 1982; Baudet and Janin, 1991) and its structure in solution are known (Bycroft et al., 1991); there are no disulfide linkages to constrain the unfolded state and the three prolines in barnase are all trans. The structure of barnase consists of three α-helices and a five-stranded β-sheet that are stabilized by three hydrophobic cores in the native structure (Figure 7-1). Transition states and pathways of barnase folding have been investigated in detail by protein engineering (Matouschek et al., 1989; Fersht et al., 1992a; Matouschek et al., 1992a; Meiering et al., 1992; Serrano et al., 1992a, 1992b) and NMR hydrogen-exchange trapping experiments (Bycroft et al., 1990; Matouschek et al., 1992b). The two techniques, which yield complementary results (Fersht et al., 1992b; Serrano et al., 1992c), have shown that the rate-determining step for both folding and unfolding involves the crossing of a free-energy barrier near the native state.

Solvated barnase denaturation simulations started with the x-ray structure and were performed with a deformable boundary potential (Brooks and Karplus, 1983, 1989); the system consisted of 9009 solvent atoms and 1091 protein atoms. Two denaturation simulation were performed at 600 K (A600, 120 ps; R600, 150 ps) and a 300 K control trajectory was run for 250 ps. The 600 K temperature was used to speed up the unfolding transition by at least six orders of magnitude relative to the experimental denaturation temperature of approximately 327 K; the activation energy for unfolding is 20 kcal/mol (Kellis et al., 1989; Matouschek et al., 1990). The A600 simulation was started after 4 ps of simulation at 300 K, while the R600 run was initiated after 100 ps of simulation at 300 K. The two different initial structures were chosen to evaluate the effect of the initial conditions on the overall unfolding behavior. The A600 simulation was branched after 90 ps dynamics, by cooling the system to 300 K; it was continued for 160 ps; the name B300 is used to designate this simulation. The

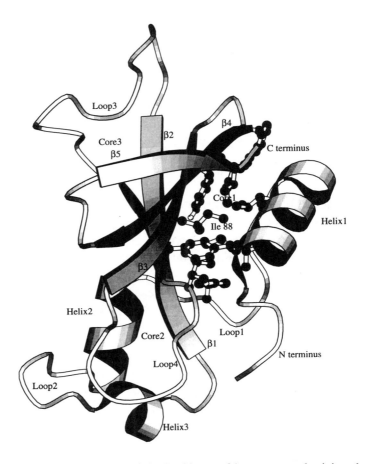

Figure 7-1. Schematic picture of the backbone of barnase, emphasizing the secondary structure elements. Side chains of hydrophobic $core_1$ are plotted in a ball-and-stick representation. The secondary structural elements include the following residues: N-terminus (1–5), $helix_1$ (6–18), $loop_1$ (19–25), $helix_2$ (26–34), $loop_2$ (35–40), $helix_3$ (41–45), type II β-turn (46–49), $strand_1$ (50–55), $loop_3$ (56–69), $strand_2$ (70–75), $loop_4$ (76–84), $strand_3$ (85–90), type I β-turn (91–94), $strand_4$ (95–100), type III' β-turn (101–104), $strand_5$ (105–108), C-terminus (109–110). The atomic coordinates of the protein were kindly provided by Dr. A. Cameron and Professor G. Dodson of the University of York. Figure made with the Molscript program (Kraulis, 1991).

temperature of the system was controlled by weak coupling to an external bath (Berendsen et al., 1984).

In addition, two molecular dynamics runs starting with the x-ray structure and lasting 250 ps were performed at 600 K (V600) and 800 K (V800) without explicit solvent molecules (*in vacuo*) starting from the x-ray structure; a distance dependent dielectric constant was used, as in the simulation of Fan et al. (1991). During the V600 *in vacuo* run, the radius of gyration (R_g) decreased to about 13.0 Å in the first 50 ps (R_g of the minimized x-ray structure is 13.6 Å), and was approximately constant during the rest of the simulation. The heavy atom RMS positional deviation (RMSD) from the x-ray structure increased to about 4.5 Å in the first 60 ps and did not change significantly after that. In V800, R_g increased to about 14.3 Å during the first 10 ps; it then decreased to a value that oscillated between 13.5 and 14.0 Å in the last 50 ps, while the heavy atom RMSD from the x-ray structure reached a plateau value of 9 Å after about 170 ps. The structures obtained at the end of the V600 and V800 runs are characterized by a tight packing of the loops and a large amount of distortion in the secondary structure elements. At the end of V600 the major α-helix (helix$_1$) of barnase preserves about 40% of its helical content and packs against the stable β-sheet. In V800, helix$_1$ unfolded during the first 50 ps; most of the β-sheet was disrupted in the first 50 ps (apart from strands 3 and 4, which lasted until 120 ps). The *in vacuo* simulations are briefly compared with the simulations that included explicit water molecules at the end of this section.

Plots of R_g and RMSD as a function of time for the simulations with explicit solvent are given in Figures 7-2a and 7-2b. In the control simulation at 300 K, R_g showed a very small increase (the average R_g is 13.7 Å), which was mainly due to the rearrangements of surface hydrophilic side chains, and the heavy atom RMSD from the x-ray structure is 1.9 Å (the main-chain atom RMSD is 1.5 Å) during the last 50 ps. The overall conformation, hydrophobic core compactness (Figure 7-5a–c), and secondary structure elements are found to be stable at room temperature. Furthermore, an average of 0, 2, and 1 water molecules were present in core$_1$ (within 7 Å from the core center), core$_2$ (within 6 Å), and core$_3$ (within 6 Å), respectively. Experimental evidence indicates that core$_2$ contains three water molecules, whereas core$_1$ and core$_3$ are fully inaccessible to solvent in the x-ray structure (A. Cameron and K. Henrick, unpublished results; Serrano et al., 1992c).

In A600 and R600, R_g started to increase only after 30 ps, while the RMSD increased immediately. This suggests that the protein was undergoing a local conformational search to find convenient pathway(s) for denaturation. In the A600 simulation, R_g increased between 30 and 45 ps to about 14.2 Å; this was followed by the formation of an intermediate state, while in R600, R_g continuously increased during the first 70 ps. In A600, the intermediate state lasted for about 18 ps; the average R_g of the 45–63 ps

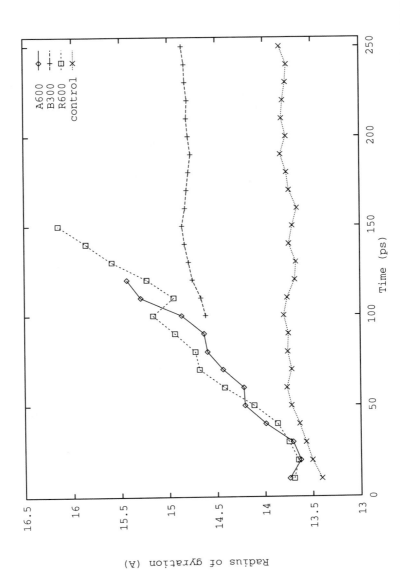

Figure 7-2a. Radius of gyration as a function of simulation time averaged over 10 ps intervals: A600 (*solid line*), B300 (*dashed line*), R600 (*broken line*), control run at 300 K (*dotted line*).

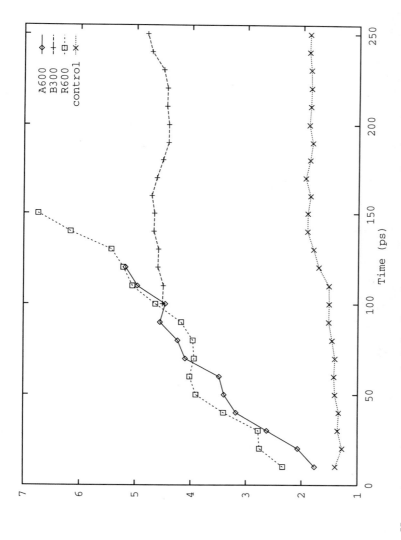

Figure 7-2b. Heavy atom root mean square deviation from the x-ray structure as a function of simulation time averaged over 10 ps intervals: A600 (*solid line*), B300 (*dashed line*), R600 (*broken line*), control run at 300 K (*dotted line*).

structures was 14.2 Å. This intermediate was characterized by a structure similar to the native one and a partially solvated $core_2$ (solvation of the cores is discussed in the next section). In A600 and R600, another intermediate state was formed after about 80 and 70 ps, respectively. It was characterized by a relatively small distortion in the secondary structural elements, a fully solvated $core_2$, and partially solvated $core_1$ and $core_3$ (see Figure 7-1 for notation). During the B300 simulation (started after 90 ps of A600 dynamics), R_g remained approximately constant at an average value of 14.8 Å; assuming a spherical geometry, the volume increase from the minimized x-ray structure amounts to 29%. Fersht and co-workers have shown that there exists an intermediate on both the folding and unfolding pathway of barnase (Matouschek et al., 1992a). It has some of the properties of B300 (e.g., unfolded N-terminus, $loop_1$, $loop_2$, and $loop_4$, distorted secondary structure elements, solvated $core_2$, and weakened hydrophobic interactions in $core_1$ and $core_3$) but no R_g measurement is available. Measurements of the α-lactalbumin molten globule by ultracentrifugation (Kronman et al., 1967) and by quasielastic light scattering (Gast et al., 1986) show a volume increase of about 30%.

In the high-temperature simulations, there were many structural changes (most of which are common to A600 and R600), but the overall fold was conserved (Figures 7-3a and 7-3b), as it appears to be in the experimentally observed intermediate (Matouschek et al., 1992a). In A600, the N-terminus, $loop_1$, and $loop_2$ began to unfold during the first 30 ps. This was followed by partial denaturation of the hydrophobic cores; $core_2$ denatured relatively rapidly, followed by $core_1$, $core_3$, and $loop_3$. The same sequential order for the hydrophobic core solvation was seen in R600. Solvation of hydrophobic $core_1$ was coupled with a large distortion of both $helix_1$ and the edge strands of the β-sheet. $Helix_1$ and $helix_2$ lost about 50% of the native α-helical hydrogen bonds in the A600 and R600 simulations; $helix_3$ unfolded after about 20 ps. In the β-sheet, about half of the native interstrand hydrogen bonds had disappeared after 100 ps in both trajectories; in R600 the β-sheet was almost fully solvated in the last 30 ps (120 to 150 ps), while in the last 20 ps of A600 (100 to 120 ps), there were about 50% of the native interstrand hydrogen bonds. In the B300 simulation, the percentages of native interstrand hydrogen bonds at the edges of the sheet (strands 1–2 and 4–5), and at the center of the sheet (strands 2–3 and 3–4), stabilized at about 50% and 70%, respectively. Interactions with water molecules replaced many of the helical and β-sheet hydrogen bonds (see Section 5).

The accessibility of $core_1$ to water was coupled with the relative motion of $helix_1$ and the β-sheet (see Figures 7-3a and 7-3b); in A600 they began to move apart at about 30 ps and their separation was continuous during the

30–75 ps period; between 75 and 100 ps there was a small closing movement in accord with a decrease of water in the core, followed by expansion for the remainder of the simulation (see strand$_3$ residues 85–90 and helix$_1$ residues 6–18 in Figure 7-3a). During the R600 simulation an essentially continuous opening motion was seen (Figure 7-3b). In B300, the relative orientation of helix$_1$ and the β-sheet was stable, the steady state of partial solvation of core$_1$ was coupled with a nearly constant number of hydrogen bonds in helix$_1$ and the central part of the β-sheet (strands 2–3 and 3–4). The structure at the end of B300 was similar to the conformation of A600 after 90 ps. The all-atom (backbone, side chain) RMSD after optimal superposition between these two structures was 2.32 Å (1.89 Å, 2.68 Å). This is to be compared with the deviations from the native structure; they were 4.55 Å (4.15 Å, 4.91 Å) after 90 ps of A600 and 4.94 Å (4.25 Å, 5.36 Å) at the end of B300.

5. Solvent Role in the Unfolding Transition

The results from the simulations described in the previous section provide considerable insights into the nature of the unfolding transition of globular proteins. An important point is that water penetration plays an essential role in the disruption of secondary structure and of hydrophobic clusters. This is in disagreement with theoretical arguments, which have suggested that water is not involved in the transition from the native to the molten globule state (Shakhnovich and Finkelstein, 1989). It also disagrees with the results of Daggett and Levitt (1992b), who used a low water density in their high-temperature simulations. However, it is in accord with measurements of the partial specific volume of the molten globule and the heat capacity change in the transition from the native to the molten globule state (Ptitsyn, 1992). It has been concluded that there are several hundred water molecules in the interior of the molten globule state of α-lactalbumin. In what follows, we summarize the available results. Much of the information (all of it for hydrophobic cores and β-sheets) comes from the barnase simulation outlined above.

The picture that emerges is that of a stepwise but partly cooperative unfolding phenomenon from the native to a compact globule state. By this we mean that the initial break-up of secondary structure is simultaneous with and coupled to the denaturation of the hydrophobic cores that are involved in the stabilization of the tertiary structure. During the transition there is a limited expansion of the protein and the overall fold is preserved,

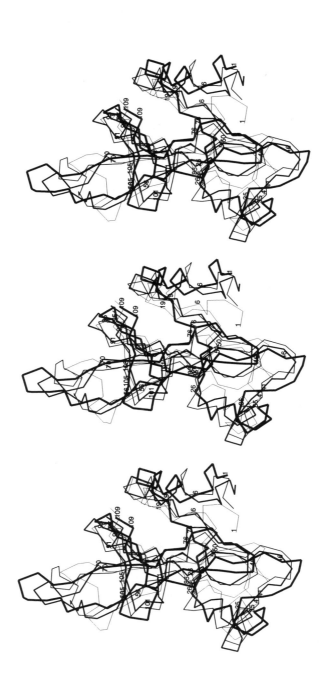

Figure 7-3a. Stereo drawing of the barnase C_α atoms to illustrate the relative helix$_1$/β-sheet motion during A600. 1 ps (*thin line and labels*), 70 ps (*medium line*), 90 ps (*thicker line*), last ps (*thick lines and labels*). (In all stereofigures, the view is cross-eyed for the left pair of images and wall-eyed for the right pair).

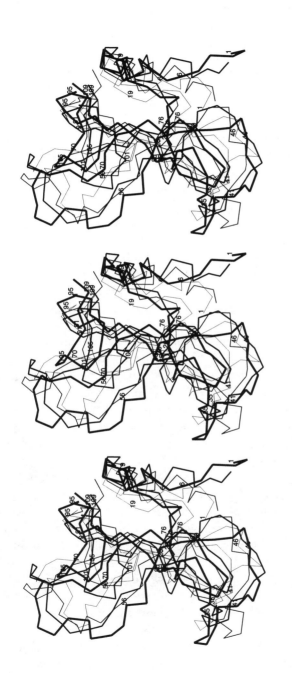

Figure 7-3b. Stereo drawing of the barnase C_α atoms to illustrate the relative helix$_1$/β-sheet motion during R600. 1 ps (*thin line and labels*), 70 ps (*medium line*), 90 ps (*thicker line*), last ps (*thick lines and labels*). (In all stereofigures, the view is cross-eyed for the left pair of images and wall-eyed for the right pair).

though the loops and secondary structural elements are present in a looser and significantly distorted form. In all of this, entropic effects due to the elevated temperature may play a role, i.e., at high temperature a flexible polypeptide chain with its polar groups participating in hydrogen bonds to water has a lower free energy than a more rigid helix with enthalpically stronger intrahelical hydrogen bonds. This effect may be amplified by the use of the room-temperature water density.

For α-helices and β-sheets the denaturation process shows similar characteristics. The essential factor is that water molecules compete with the existing hydrogen bonds of the secondary structure. Simulations of isolated helices (see Section 2 on fragment studies) and helices as parts of proteins (apomyoglobin and barnase) have shown that the water molecules attack primarily the more accessible carbonyl groups. A water molecule that has formed a hydrogen bond to a carbonyl oxygen can then insert into the helix and replace the helical C=O$\cdots$NH hydrogen bond. Of course not every water molecule that makes a hydrogen bond to a carbonyl group inserts directly. Moreover, some helical hydrogen bonds are broken without simultaneous water insertion. For β-sheets, the process involves waters that also interact primarily with carbonyl oxygens but now the waters insert to replace interstrand hydrogen bonds.

As a detailed example, we consider the A600 simulation results for helix$_1$ of barnase. The C-terminal hydrogen bond was lost after about 30 ps and did not reform. The three N-terminal hydrogen bonds were broken after about 50 ps; they reformed in the last part of the simulation. Reformation of the N-terminal hydrogen bond in helix$_1$ (at 90 ps and 110 ps) occurred through a 3_{10} hydrogen bond; this phenomenon was observed during molecular dynamics simulations of an α-helical analogue of the ribonuclease A S-peptide (Tirado-Rives and Jorgensen, 1991) and is in accord with the x-ray studies of Sundaralingam and Sekharudu (1989). The helical hydrogen bonds between 15NH$\cdots$CO11 and 16NH$\cdots$CO12 were broken after about 70 ps; they were converted into a 3_{10} form in the last 20 ps of A600. A water molecule accepted from the NH group of residue 15 and donated to the CO of residue 11 during the entire B300 simulation. The NH group of residues 12 and 13 converted from the 3_{10}- to the α-type in the first half of the B300 simulation, while the NH of residue 14 underwent the opposite conversion after about 80 ps. The 11NH$\cdots$CO7 and 17NH$\cdots$CO13 helical hydrogen bonds and the 16NH$\cdots$CO13 forming a 3_{10} hydrogen bond were stable during B300.

The barnase β-sheet is composed of five antiparallel β-strands in consecutive order (Figure 7-1) and provides a clear example of the role of water in its high-temperature denaturation. In its native state, the sheet has a regular structure, apart from a β-bulge in strand$_1$ between residue 53 and 54; it has a twist of about 90 degrees (from strand$_1$ to strand$_5$; Figure 7-1). Solvent insertion began at the edges of the β-sheet (see Figures 7-4a and 7-4b). The C-terminal part of strand$_1$ (near the β-bulge) was rapidly solvated in both A600 and R600 (Figure 7-4a); this is likely to originate from the instability of this irregular structure. At 70 ps several solvent molecules were located between strands and participate in hydrogen bonds with the main-chain NH and CO groups as donor, acceptor or both. In R600, hydrogen bonds between strands 1 and 2 came apart at about 70 ps (in A600 they separated after about 100 ps). This was followed by separation (accompanied by water insertion) of strands 3 and 4 at 120 ps, strands 2 and 3 at 130 ps, and strands 4 and 5 at 150 ps; in A600, except for strands 1 and 2, one or more interstrand hydrogen bonds remained until the end of the simulation.

To focus on the nature of the motions of water molecules interacting with the β-sheet during R600, six waters that spent more than 30 ps within 3 Å of any main-chain atom of the β-sheet are labeled (from A to F) in Figures 7-4a and 7-4b. From 60 to 110 ps, water A belonged to a cluster of solvent molecules interacting with the β-bulge part of strand$_1$. At 70 ps (bottom stereopicture of Figure 7-4a), it donated a bifurcated hydrogen bond to the CO group of strand$_1$ residues 52 and 53; at 90 ps (top stereopicture of Figure 7-4b), it acted as acceptor for another water, which donated a bifurcated hydrogen bond to the same CO groups. From 50 to 110 ps, water B was inserted between strands 1 and 2; it donated to the CO group of residue 73 in strand$_2$. At 110 ps, it also accepted from the NH groups of residue 53 (strand$_1$) and residue 73 (strand$_2$), acting as an interstrand bridging element. It then moved to the C-terminal part of strand$_3$, where it donated to the CO group of residue 90 at 150 ps. Water C interacted with the N-terminal part of strand$_1$ from 50 to 100 ps; it then moved away from the β-sheet and came closer again at 130 ps for about 5 ps. Water D solvated the N-terminal part of strand$_4$ during the 90–130 ps period. Waters E and F inserted between the C-terminal part of strand$_4$ and the N-terminal part of strand$_5$ during the 70–130 ps period.

The role of water in disruption of hydrophobic cores is more complex. Water molecules penetrate in a number of ways, all of which tend to involve the preservation of their hydrogen bonds. This may be achieved by waters hydrogen bonding to polar groups (e.g., OH of Tyr residues) and to charged groups that enter the hydrophobic core. Alternatively, the water molecules form hydrogen-bonded chains or clusters around hydrophobic residues. Also, some waters may also be involved in breaking up the secondary structural elements that are held together by the hydrophobic cores. Any

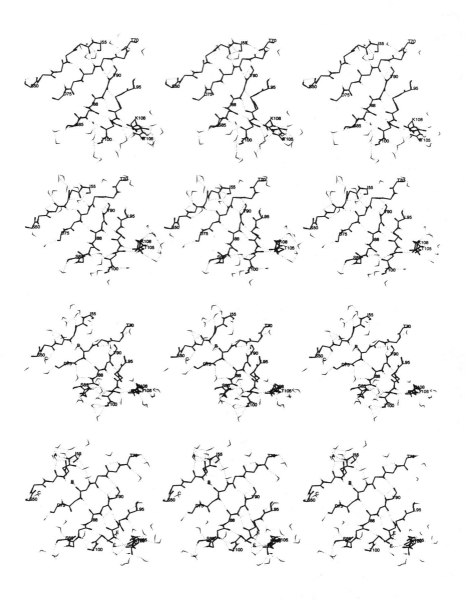

Figure 7-4a. Stereoview of water penetration into the β-sheet during R600. Main-chain N and O atoms are thick, hydrogens are thin, and hydrogen bonds are dotted; water molecules within 3 Å of any main-chain atom of the β-sheet are shown. Water molecules labeled from A to F are discussed in the text. From top to bottom: 10, 30, 50, 70 ps.

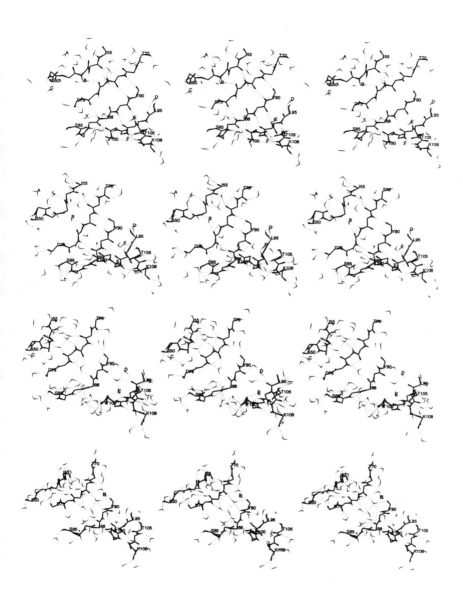

Figure 7-4b. Stereoview of water penetration into the β-sheet during R600. Main-chain N and O atoms are thick, hydrogens are thin, and hydrogen bonds are dotted; water molecules within 3 Å of any main-chain atom of the β-sheet are shown. Water molecules labeled from A to F are discussed in the text. From top to bottom: 90, 110, 130, 150 ps.

such water penetration is coupled to opening up of the hydrophobic core by relative motion of the secondary structural elements. In barnase, for example, the solvent-exposed surface area of the core increases as more water molecules penetrate into the core. Figure 7-5a–c shows the time dependence of the solvent-accessible surface area of the side chains of the principal hydrophobic core of barnase. Increase in $core_1$ surface area (Figure 7-5a) and water influx (not shown) are nearly simultaneous and begin at about 35 ps in A600 and 25 ps in R600. A maximum of 16 water molecules has penetrated at 82 and 84 ps in A600 (average of 12 for the 67–86 ps period); most of them are expelled again so that at 90 and 98 ps only 7 water molecules are located within 7 Å of its center (average of 10 for the 89–98 ps period). In the last 10 ps of A600 (from 111 to 120 ps) the number of water molecules in $core_1$ increases again to an average of 17. The water molecules penetrate mainly in the two directions parallel to the $helix_1$ axis. Furthermore, hydrophobic residues at the edges of the core (Leu 20, Tyr 24, Tyr 90, Trp 94, and Ile 109) were solvated first, whereas residues in the center of the core (Leu 14, Ala 74, Ile 88, and Ile 96) were more compact, in agreement with protein engineering results (Serrano et al., 1992c). The solvation of the main hydrophobic core was coupled to both the relative motion of $helix_1$ and the β-sheet and the distortion of the secondary structure elements. During the B300 simulation, the solvent accessible surface area of $core_1$ was nearly constant (average of 431 Å^2; 14 water molecules within 7 Å). This steady state of solvation of the core was correlated with the nearly constant number of hydrogen bonds in the secondary structure elements.

Solvation of $core_2$ began after about 10 ps in A600 and R600 (Figure 7-5b). It was coupled to the rearrangement of the subdomain consisting of $helix_2$, $loop_2$, and $helix_3$ which unfolded along with the β-turn formed by residues 46–49 (Figures 7-3a and 7-3b). Apart from a small closing movement of $core_2$ during the 20–50 ps period in R600, the number of water molecules increased continuously during the first 100 ps; a plateau value was reached after about 100 ps. $Core_3$ was more stable than either $core_2$ or $core_1$. It began to unfold after about 60 ps and 70 ps in the R600 and A600 trajectories, respectively. The late unfolding of $core_3$ and $loop_3$, relative to $core_2$, is consistent with experimental results (Serrano et al., 1992c).

During the *in vacuo* simulation V600 the solvent-accessible surface area of the cores showed an oscillating behavior around average values of 461 Å^2, 390 Å^2, and 162 Å^2, for $core_1$, $core_2$, and $core_3$, respectively. In V800, opening of the cores began in the first 10 ps ($core_2$ and $core_1$) or 100 ps ($core_3$); the solvent-accessible surface area reached a plateau value

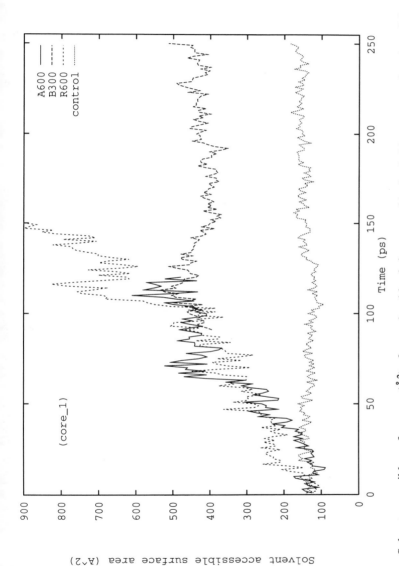

Figure 7-5a. Solvent-accessible surface area (Å²) of nonpolar side-chain atoms of hydrophobic core₁ as a function of time (ps). The Lee and Richards (1971) algorithm (CHARMM implementation) and a probe sphere of 1.4 Å radius were utilized; A600 (*solid line*), B300 (*dashed line*), R600 (*broken line*), control run at 300 K (*dotted line*).

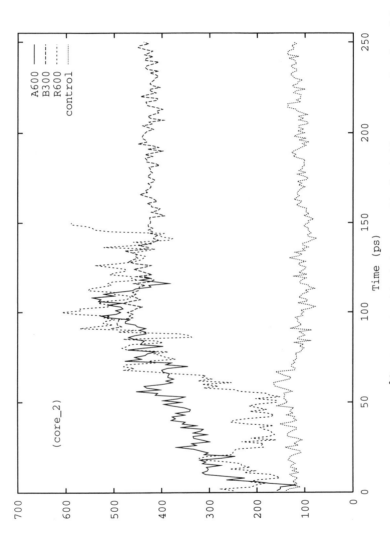

Figure 7-5b. Solvent-accessible surface area (Å^2) of nonpolar side-chain atoms of hydrophobic core$_2$ as a function of time (ps). The Lee and Richards (1971) algorithm (CHARMM implementation) and a probe sphere of 1.4 Å radius were utilized; A600 (*solid line*), B300 (*dashed line*), R600 (*broken line*), control run at 300 K (*dotted line*).

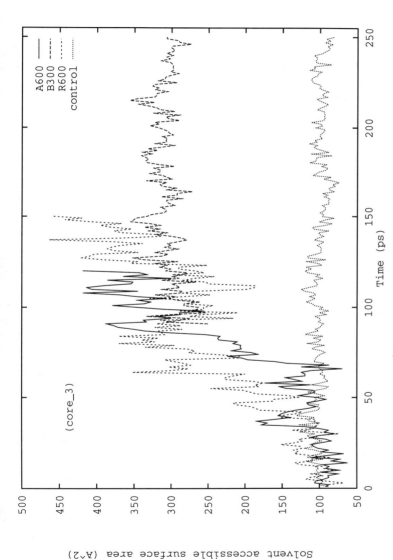

Figure 7-5c. Solvent-accessible surface area ($Å^2$) of nonpolar side-chain atoms of hydrophobic $core_3$ as a function of time (ps). The Lee and Richards (1971) algorithm (CHARMM implementation) and a probe sphere of 1.4 Å radius were utilized; A600 (*solid line*), B300 (*dashed line*), R600 (*broken line*), control run at 300 K (*dotted line*).

after 40 ps, 50 ps, and 150 ps, for $core_2$, $core_1$, and $core_3$, respectively; the sequential order was the same as in A600 and R600.

6. Conclusions

At the present stage of computational hardware, folding studies are limited to small fragments or to simplified models of proteins. The results that have become available are particularly useful as a source of parameters for phenomenological descriptions of protein folding, such as the diffusion–collision model. Much more can be done at an all-atom level of detail to simulate the dynamics of the early stages of unfolding from the native state to a compact organized globule. Here the results of a number of recent simulations focus on the important role in the denaturation process of explicit water molecules, rather than the general dielectric effect of aqueous solvent. Both for secondary structural elements (α-helices and β-sheets), and for hydrophobic cores, the entrance of water molecules that are hydrogen bonded to each other or to the protein polar groups are an essential part of the denaturation process. Some detailed examples, particularly from simulations of barnase, have been presented to illustrate the dynamic role of water molecules in protein unfolding.

Much remains to be done to obtain a full understanding of protein unfolding and to extend that knowledge to the folding process. All simulations of unfolding of entire proteins, in contrast to fragments, were performed at unrealistically high temperatures to increase the rate so that it falls within the nanosecond time scale of present day simulations. Most of these high-temperature simulations used explicit models of water molecules at a density corresponding to that appropriate for room temperature; it is likely that this also contributes to increasing the unfolding rate. These special conditions will have to be investigated further to determine whether significant artifacts are introduced. Also, alternative approaches to the denaturation process should be examined. Finally, the question of how best to use the information obtained from unfolding studies for analyzing the folding process has yet to be determined.

Acknowledgments. The work was supported in part by a grant from the National Institutes of Health and a gift from Molecular Simulations, Inc. A.C. was supported by the Schweizerischer Nationalfonds (Swiss National Science Foundation).

REFERENCES

Acharya KR, Straut DI, Walker NPC, Lewis M, Phillips DC (1989): Refined structure of baboon α-lactalbumin at 1.7 Å resolution. Comparison with c-type lysozyme. *J Mol Biol* 208:99–127

Anfinsen CB (1972): The formation and stabilization of protein structure. *Biochem J* 128:737–749

Austin RH, Beeson KW, Eisenstein L, Frauenfelder H, Gunsalus IC (1975): Dynamics of ligand binding to myoglobin. *Biochemistry* 14:5355–5373

Baudet S, Janin J (1991): Crystal structure of a barnase-d(GpC) complex at 1.9 Å resolution. *J Mol Biol* 219:123–132

Baum J, Dobson CM, Evans PA, Hanly C (1989): Characterization of a partly folded protein by nmr methods: Studies on the molten globule state of guinea pig α-lactalbumin. *Biochemistry* 28:7–13

Berendsen HJC, Postma JPM, van Gunsteren WF, DiNola A, Haak JR (1984): Molecular dynamics with coupling to an external bath. *J Chem Phys* 81:3684–3690

Brooks BR, Bruccoleri RE, Olafson BD, States DJ, Swaminathan S, Karplus M (1983): CHARMM : A program for macromolecular energy, minimization, and dynamics calculations. *J Comput Chem.* 4:187–217

Brooks CL III, Karplus M (1983): Deformable stochastic boundaries in molecular dynamics. *J Chem Phys* 79:6312–6325

Brooks CL III, Karplus M (1989): Solvent effects on protein motion and protein effects on solvent motion. *J Mol Biol* 208:159–181

Brooks CL III (1992): Characterization of "native" apomyoglobin by molecular dynamics simulation. *J Mol Biol* 227:375–380

Brünger A, Clore GM, Gronenborn AM, Karplus M (1986): Three-dimensional structure of proteins determined by molecular dynamics with interproton distance restraints: Application to Crambin. *Proc Natl Acad Sci USA* 83:3801–3805

Brünger A, Karplus M (1991): Molecular dynamics simulations with experimental restraints. *Acc Chem Res* 24:54–61

Bycroft M, Matouschek A, Kellis JT, Serrano L, Fersht AR (1990): Detection and characterization of a folding intermediate in barnase by NMR. *Nature* 346:488–490

Bycroft M, Ludvigsen S, Fersht AR, Poulsen FM (1991): Determination of the three-dimensional solution structure of barnase using nuclear magnetic resonance spectroscopy. *Biochemistry* 30:8697–8701

Caflisch A, Niederer P, Anliker M (1992): Monte Carlo minimization with thermalization for global optimization of polypeptide conformations in Cartesian coordinate space. *Proteins: Structure, Function and Genetics* 14:102–109

Creighton TE (1988): Toward a better understanding of protein folding pathways. *Proc Natl Acad Sci USA* 85:5082–5086

Czerminski R, Elber R (1989): Reaction path study of conformational transitions and helix formation in a tetrapeptide. *Proc Natl Acad Sci USA* 86:6963–6967

Czerminski R, Elber R (1990): Reaction path study of conformational transitions in flexible systems: Applications to peptides. *J Chem Phys* 92:5580–5601

Daggett V, Levitt M (1992a): Molecular dynamics simulations of helix denaturation. *J Mol Biol* 223:1121–1138

Daggett V, Levitt M (1992b): A model of the molten globule state from molecular dynamics simulations. *Proc Natl Acad Sci USA* 89:5142–5146

Deng Y, Glimm J, Sharp DH (1990): Los Alamos Report, pages LA-UR-90–4340

DiCapua FM, Swaminathan S, Beveridge DL (1990): Theoretical evidence for destabilization of an α-helix by water insertation: Molecular dynamics of hydrated decaalanine. *J Am Chem Soc* 112:6768–6771

Dobson CM, Hanley C, Radford SE, Baum JA, Evans PA (1991): In *Conformations and Forces in Protein Folding*. Nall BT, Dill KA, eds. pages 175–181

Dolgikh DA, Gilmanshin RI, Brazhnikov EV, Bychkova VE, Semisotnov GV, Venyaminov SY, Ptitsyn OB (1981): A-Lactalbumin: Compact state with fluctuating tertiary structure. *FEBS Letters* 136:311–315

Elber R, Karplus M (1987): Multiple conformational states of proteins: A molecular dynamics analysis of myoglobin. *Science* 235:318–321

Fan P, Kominos D, Kitchen DB, Levy RM, Baum J (1991): Stabilization of α-helical secondary structure during high-temperature molecular-dynamics simulations of α-lactalbumin. *Chemical Physics* 158:295–301

Fersht AR, Matouschek A, Serrano L (1992a): The folding of an enzyme. I: Theory of protein engineering analysis of stability and pathway of protein folding. *J Mol Biol* 224:771–782

Fersht AR, Matouschek A, Sancho J, Serrano L, Vuilleumier S (1992b): Pathway of protein folding. *Faraday Discuss* 93:183–193

Fersht AR (1993): Protein folding and stability: The pathway of folding of barnase. *FEBS letters* 325:5–16

Gast K, Zirwer D, Welfle H, Bychkova VE, Ptitsyn OB (1986): Quasielastic light scattering from human α-lactalbumin: Comparison of molecular dimensions in native and 'molten globule' state. *Int J Biol Macromol* 8:231–236

Gething M-J, Sambrook J (1992): Protein folding in the cell. *Nature* 355:33–45

Gilmanshin RI, Ptitsyn OB (1987): An early intermediate of refolding α-lactalbumin forms within 20 ms. *FEBS Letters* 223:327–329

Goldberg ME, Semisotnov GV, Friguet B, Kuwajima K, Ptitsyn OB, Sugai S (1990): An early immunoreactive folding intermediate of the tryptophan synthase β_2 subunit is a 'molten globule.' *FEBS Letters* 263:51–56

Haas E, Katchalski-Katzir E, Steinberg IZ (1978): Brownian motion of the ends of oligopeptide chains in solution as estimated by energy transfer between the chain ends. *Biopolymers* 17:11–31

Hagler AT, Honig B (1978): On the formation of protein tertiary structure on a computer. *Proc Natl Acad Sci USA* 75:554–558

Harrison S, Durbin R (1985): Is there a single pathway for the folding of a polypeptide chain? *Proc Natl Acad Sci USA* 82:4028–4030

Hermans J, Anderson A, Yun RH (1992): Differential helix propensity of small apolar sidechains studied by molecular dynamics simulations. *Biochemistry* 31:5646–5653

Houghson FM, Wright PE, Baldwin RL (1990): Structural characterization of a partly folded apolyoglobin intermediate. *Science* 249:1544–1548

Houghson FM, Barrick D, Baldwin RL (1991): Probing the stability of partly folded apomyoglobin intermediate by site-directed mutagenesis. *Biochemistry* 30:4113–4118

Janin J, Wodak S (1983): Structural domains in proteins and their role in the dynamics of protein function. *Prog Biophys Mol Biol* 42:21–78

Jorgensen WL, Chandrasekhar J, Madura J, Impey RW, Klein ML (1983): Comparison of simple potential functions for simulating liquid water. *J Chem Phys* 79:926–935

Karle IL, Flippen-Anderson JL, Uma K, Balaram P (1990): Apolar peptide models for conformational heterogeneity, hydration, and packing of polypeptide helices: Crystal structure of hepta- and octapeptides containing α-aminoisobutyric acid. *Proteins: Structure, Function and Genetics* 7:62–73

Karplus M, Weaver DL (1976): Protein folding dynamics. *Nature* 260:404–406

Karplus M, Weaver DL (1979): Diffusion-collision model for protein folding. *Biopolymers* 18:1421–1437

Karplus M, Shakhnovich E (1992): Protein folding: Theoretical studies of thermodynamics and dynamics. In *Protein Folding*. Creighton TE, ed. New York: WH Freeman

Kellis JT Jr, Nyber K, Fersht AR (1989): Energetics of complementary side-chain packing in a protein hydrophobic core. *Biochemistry* 28:4914–4922

Kim PS, Baldwin RL (1990): Intermediates in the folding reactions of small proteins. *Annual Review of Biochemistry* 59:631–660

Kraulis P (1991): Molscript, a program to produce both detailed and schematic plots of protein structures. *J Appl Crystallogr* 24:946–950

Kronman MJ, Holmes LG, Robbins FM (1967): Inter and intramolecular interactions of α-lactalbumin. VIII The alkaline conformational change. *Biochim Biophys Acta* 133:46–55

Kuwajima K, Yamaya H, Miwa S, Sugai S, Nagamura T (1987): Rapid formation of secondary structure framework in protein folding studied by stopped-flow circular dichroism. *FEBS Letters* 221:115–118

Kuwajima K (1989): The molten globule state as a clue for understanding the folding and cooperativity of globular-protein structure. *Proteins: Structure, Function and Genetics* 6:87–103

Lazaridis T, Tobias DJ, Brooks CL III, Paulaitis ME (1991): Reaction path and free energy profiles for conformational transitions: An internal coordinate approach. *J Chem Phys* 95:7612–7625

Lee S, Karplus M, Bashford D, Weaver DL (1987): Brownian dynamics simulation of protein folding: A study of the diffusion-collision model. *Biopolymers* 26:481–506

Lee B, Richards FM (1971): The interpretation of protein structures: Estimation of static accessibility. *J Mol Biol* 55:379–400

Levinthal C (1969): In *Mössbauer Spectroscopy in Biological Systems*, DeGennes P et al., eds. Urbana, IL: University of Illinois Press. Proceedings of a meeting held at Allerton House, Monticello, IL

Levitt M, Warshel A (1975): Computer simulation of protein folding. *Nature* 253: 694–698

Levitt M (1976): A simplified representation of protein conformations for rapid simulation of protein folding. *J Mol Biol* 104:59–107

Levitt M (1983): Protein folding by restrained energy minimization and molecular dynamics. *J Mol Biol* 170:723–764

Mark AE, van Gunsteren WF (1992): Simulation of the thermal denaturation of hen egg white lysozyme: Trapping the molten globule state. *Biochemistry* 31:7745–7748

Matouschek A, Kellis JT Jr, Serrano L, Fersht AR (1989): Mapping the transition state and pathway of protein folding by protein engineering. *Nature* 340:122–126

Matouschek A, Kellis JT Jr, Serrano L, Bycroft M, Fersht AR (1990): Transient folding intermediates characterized by protein engineering. *Nature* 346:440–445

Matouschek A, Serrano L, Fersht AR (1992a): The folding of an enzyme. IV: Structure of an intermediate in the refolding of barnase analysed by a protein engineering procedure. *J Mol Biol* 224:819–835

Matouschek A, Serrano L, Meiering EM, Bycroft M, Fersht AR (1992b): The folding of an enzyme. VH/H exchange—nuclear magnetic resonance studies on the folding pathway of barnase: Complementarity to and agreement with protein engineering studies. *J Mol Biol* 224:837–845

Mauguen Y, Hartley RW, Dodson EJ, Dodson GG, Bricogne G, Chothia C, Jack A (1982): Molecular structure of a new family of ribonuclease. *Nature* 297:162–164

McCammon JA, Gelin BR, Karplus M, Wolynes PG (1976): The hinge-bending motion in lysozyme. *Nature* 262:325–326

McCammon JA, Northrup SH, Karplus M, Levy RM (1980): Helix-coil transitions in a simple polypeptide model. *Biopolymers* 19:2033–2045

Meiering EM, Serrano L, Fersht AR (1992): Effect of active site residues in barnase on activity and stability. *J Mol Biol* 225:585–589

Miranker A, Radford SE, Karplus M, Dobson CM (1991): Demonstration by NMR of folding domains in lysozyme. *Nature* 349:633–636

Nemethy G, Scheraga HA (1977): Protein folding. *Quart Rev Biophys* 10:239–352

Noguti T, Gō N (1989a): Structural basis of hierarchical multiple substates of a protein. I: Introduction. *Proteins: Structure, Function and Genetics* 5:97–103

Noguti T, Gō N (1989b): Structural basis of hierarchical multiple substates of a protein. II: Monte Carlo simulation of native thermal fluctuations and energy minimization. *Proteins: Structure, Function and Genetics* 5:104–112

Noguti T, Gō N (1989c): Structural basis of hierarchical multiple substates of a protein. III: Side chain and main chain local conformations. *Proteins: Structure, Function and Genetics* 5:113–124

Noguti T, Gō N (1989d): Structural basis of hierarchical multiple substates of a protein. IV: Rearrangements in atom packing and local deformations. *Proteins: Structure, Function and Genetics* 5:125–131

Noguti T, Gō N (1989e): Structural basis of hierarchical multiple substates of a protein. V: Nonlocal deformations. *Proteins: Structure, Function and Genetics* 5:132–138

Oas TG, Kim PS (1988): A peptide model of a protein folding intermediate. *Nature* 336:42–48

Ohgushi M, Wada A (1983): 'Molten globul state': A compact form of globular proteins with mobile side-chains. *FEBS Letters* 614:21–24

Pear MR, Northrup SH, McCammon JA, Karplus M, Levy RM (1981): Correlated helix–coil transitions in polypeptides. *Biopolymers* 20:629–632

Ptitsyn OB, Pain RH, Semisotnov GV, Zerovnik E, Razgulyaev OI (1990): Evidence for a molten globule state as a general intermediate in protein folding. *FEBS Letters* 262:20–24

Ptitsyn OB (1992): The molten globule state. In *Protein Folding*, Creighton TE, ed. New York: WH Freeman, pages 243–300

Serrano L, Kellis JT Jr, Cann P, Matouschek A, Fersht AR (1992a): The folding of an enzyme. II Substructure of barnase and the contribution of different interactions to protein stability. *J Mol Biol* 224:783–804

Serrano L, Matouschek A, Fersht AR (1992b): The folding of an enzyme. III: Structure of the transition state for unfolding of barnase analysed by a protein engineering procedure. *J Mol Biol* 224:805–818

Serrano L, Matouschek A, Fersht AR (1992c): The folding of an enzyme. VI: The folding pathway of barnase: Comparison with theoretical models. *J Mol Biol* 224:847–859

Shakhnovich EI, Finkelstein AV (1989): Theory of cooperative transitions in protein molecules. I. Why denaturation of a globular protein is a first order phase transition. *Biopolymers* 28:1667–1680

Shakhnovich EI, Gutin A (1990): Enumeration of all compact conformations of copolymers with random sequence of links. *J Chem Phys* 93:5967–5971

Shakhnovich EI, Farztdinov G, Gutin A, Karplus M (1991): Protein folding bottlenecks: A lattice Monte Carlo simulation. *Physical Review Letters* 67:1665–1668

Soman KU, Karimi A, Case DA (1991): Unfolding of an α-helix in water. *Biopolymers* 31:1351–1361

Sundaralingam M, Sekharudu YC (1989): Water-inserted α-helical segments implicate reverse turns as folding intermediates. *Science* 244:1333–1337

Tirado-Rives J, Jorgensen WL (1991): Molecular dynamics simulations of the unfolding of an α-helical analogue of ribonuclease A S-peptide in water. *Biochemistry* 30:3864–3871

Tirado-Rives J, Jorgensen WL (1993): Molecular dynamics simulations of the unfolding of apomyoglobin in water. *Biochemistry* 32:4175–4184

Tobias DJ, Sneddon SF, Brooks CL III (1990): Reverse turns in blocked dipeptides are intrinsically unstable in water. *J Mol Biol* 216:783–796

Yapa K, Weaver DL, Karplus M (1992): β-sheet coil transitions in a simple polypeptide model. *Proteins: Structure, Function and Genetics* 12:237–265

Yun RH, Anderson A, Hermans A (1991): Proline in α-helix: Stability and conformation studied by dynamics simulation. *Proteins: Structure, Function and Genetics* 10:219–228

Zwanzig R, Szabo A, Bagchi B (1992): Levinthal's paradox. *Proc Natl Acad Sci USA* 89:20–22

8

Contact Potential for Global Identification of Correct Protein Folding

Gordon M. Crippen and Vladimir N. Maiorov

1. Introduction

The classical protein folding problem is to predict the three-dimensional (3D) conformation of a protein given only its amino acid sequence. In the thermodynamic approach one attempts to simulate this by choosing some kind of potential function of conformation and then searching for the conformation(s) having the global minimum of this function. There are two interrelated problems that need to be solved in order to ensure the calculations are feasible and to get correct answers: How to represent conformations, and how to construct the function? The standard approach to computational conformational analysis, molecular mechanics, represents each atom in the molecule as a point in 3D space, so that atoms can move in a continuous fashion by smoothly changing their x, y, and z coordinates. Sometimes the atomic Cartesian coordinates are calculated from specified internal coordinates (bond lengths, bond angles, and torsion angles), but in any case conformational movements are smooth and continuous. Potential functions of conformation are generally chosen as long sums of two-, three-, and four-atom interactions, where each term is a continuous function of atomic coordinates, often having some physical significance. The adjustable parameters within these terms are subsequently varied so as to reproduce some selection of experimentally observed conformations, crystal structures, known bond-rotation barriers, enthalpies of sublimation,

The Protein Folding Problem and Tertiary Structure Prediction
K. Merz, Jr. and S. Le Grand, Editors
© Birkhäuser Boston 1994

vibrational frequencies, etc. The intent is that the function should simulate the enthalpy of a molecule at room temperature (or sometimes at $0°$ K), and entropic effects, such as solvation and conformational variability at room temperature, must be simulated by lengthy molecular dynamics calculations using this same function. However, even verification of the function with respect to energetically determined phenomena is typically limited. In particular, achieving agreement with experimentally observed conformations means that the function should have a substantial local minimum relatively close to the experimental value. Any relevant potential function having interatomic attractions and repulsions has a number of local minima that increases rapidly with the number of atoms. Thus, molecular mechanics potential functions are generally tuned to agree well with the experimentally observed conformations of a selection of molecules in the vicinity of the conformational space of those same conformations. However, in a more global sense, there may be other much deeper local minima corresponding to very different conformations. The importance of this problem became most obvious after discovering that typical atom–atom potential functions cannot generally favor the native fold over a deliberately misfolded structure, given exactly the same amino acid sequence for both (Novotny et al., 1984, 1988). Crippen and co-workers have devised potential functions having better global properties (Crippen and Snow, 1990; Seetharamulu and Crippen, 1991) in the sense that the functions are intended to simulate the free energy of a solvated polypeptide chain insofar as the global minimum should be near the crystal structure for a training set of small proteins. This is a qualitatively different parameter-adjustment task, made extremely difficult because the number of local minima apparently increases exponentially with the size of the molecule (Crippen, 1975). Since the molecule is represented in terms of interacting particles positioned by smoothly varying Cartesian coordinates, and the potential is a continuous function of these coordinates, even finding a local minimum requires converging on it with a nonlinear minimization algorithm, and after that, one must verify that the near-native minimum is indeed the very lowest of all local minima.

One way of making the problem more tractable is to convert from a continuous conformational space to a (mathematically speaking) *finite discrete* one. Instead of allowing atomic coordinates to vary continuously so that important conformations correspond to local minima of the potential over the continuous space of all conformations, it would be much more convenient if there were just a finite (though possibly long) list of allowed conformations a protein could take on. Then, if one of these is designated to be the native conformation, the potential is merely required to have a lower value for it than for any other conformation in the list. Verification of the requirement amounts simply to evaluating the function for each conformation in the finite conformational space. There are two plausible ways to construct such a discrete space. The traditional way in theoretical polymer chemistry is to model a protein as a walk on some sort of lattice, so that for short chains *all* conformations can be examined, subject to the constraints of certain allowed interresidue bond lengths and bond angles. The second way is to use pieces of the known crystal protein 3D structure as examples of how polypeptide chains can fold. The advantage of the latter is that the conformations are obviously much more realistic, but since there are only a limited set of protein crystal structures, not all possible conformations would be explored.

The exploration of discrete conformational space has been reported in the literature for some time. For instance, Covell and Jernigan (1990) looked at all conformations available to a few proteins corresponding to Hamilton walks on certain lattices and showed that a function of interresidue contacts previously derived from a survey of protein crystal structures (Miyazawa and Jernigan, 1975) could rank the native conformation among the best 2%. Lau and Dill (1989) found that only certain kinds of amino acid sequences can produce a unique global minimum of a particular extremely simple contact potential for two-dimensional (2D) square-lattice walks. Sippl and co-workers (Hendlich et al., 1990; Sippl, 1990) constructed a potential of mean force for the interactions among C^β atoms from a survey of protein crystal structures that tended to prefer the native conformation of several proteins over some thousands of alternatives, but not in all cases.

We treat a related but fundamentally different question: Given the finite set of all possible conformations and given one of these as the native structure, can one devise a potential function that clearly favors the native sequence versus 3D structure correspondence over all the rest? If so, does this function have any predictive power? What restrictions are there on the types of such functions and the kinds of conformations that can be globally favored? In other words, we consider a restricted version that one might call the *multiple choice "recognition problem"*: Given the amino acid sequence of a protein and a large selection of globular conformations that includes the correct native fold, choose the one native conformation. The need to discriminate correct versus incorrect 3D structure also naturally arises in attempting to predict a protein's conformation by homology modeling, where there may be several different ways to arrange variable loops. Other possible applications could be an assessment of alternative conformations of a protein derived according to NMR data if the latter are not sufficient (Pastore et al., 1991), or choosing between different chain tracings through the electron density in the early stages of determining a protein's x-ray crystal structure.

A number of different researchers have suggested various criteria for the recognition problem, such as the number of hydrophobic contacts (Bryant and Amzel, 1987). Novotny and co-workers (Novotny et al., 1984, 1988) analyzed the accessible surface area in terms of its polar/apolar ratio and the distribution of this ratio for different amino acid side chains, as well as atomic packing and empirical energy and free-energy functions. Chiche and co-workers related solvation free energy (Eisenberg and McLachlan, 1986) to the correctness of a protein fold using the observed, approximately linear dependence of the solvation energy on the protein chain length (Chiche et al., 1990). Almost the same problem was also attacked under a different formulation as an assessment of sequence-versus-3D fold correspondence (Jones et al., 1992; Luethy et al., 1992). For example, in the 3D profile approach (Luethy et al., 1992), the ability to discriminate between the correct and an incorrect fold for seven different proteins was made possible by taking into account solvent accessibility, polarity, and local type of secondary structure, and then judging from their relative scores calculated on that basis, and from the general relation between the scores of correct crystal structures and their chain lengths. These authors were also able to detect an incorrectly folded segment in an otherwise correct structure. However, the goal has been to recognize the correct fold as better in some sense when only limited numbers of alternatives are provided. We believe it is much more difficult to favor the native fold over larger numbers of alternatives. In the approach described here, we follow the line of building a function of interresidue contacts that prefers the native conformation over absolutely all possible alternatives. An attempt to solve the multiple recognition problem was undertaken earlier by Sippl and co-workers (Hendlich et al., 1990) by potentials of mean force, where the same scheme of generating the alternative conformations was used as in our earlier work (Crippen, 1991). In our approach the general correctness of concept is proven for the case of a model 2D square lattice, and then it is tailored to handle real protein structures. As a result we are able to account for all the proteins we examined by learning to identify the kinds of native conformations that can be treated this way and by correctly dealing with homologous proteins.

2. Two-dimensional Square Lattice Model

Let us represent a protein having n residues as a self-avoiding walk of $n - 1$ steps on a very large, two-dimensional, square lattice, given the lattice spacing is unity. Therefore, sequentially adjacent residues are distance 1 apart. Each occupied lattice point has a sequence number $1 \leq i \leq n$ and a type $1 \leq t_i \leq 20$, corresponding to the 20 naturally occurring amino acids.

We define a contact to be if adjacent lattice points are occupied by two residues that are not sequentially adjacent: $d_{ij} \leq d_c = 1$ and $|i - j| > 1$. Since the walks are self-avoiding, a contact between residues i and j implies $|i - j| \geq 3$. Two conformations, c and c', of the same protein are defined as different if corresponding residues (i.e., having the same sequence number) cannot be superimposed by rigid body motions (translations, rotations, and mirror inversions).

Let S_c be the set of contacts occurring in conformation c. Identify the contacts by sequence separations, type of residue with lower sequence number, and type of residue with higher sequence number. Since the same residue type may occur more than once in the protein, the list of contacts may have more than one entry with the same description. We design a potential function (mimicking the free energy of a real protein as a function of conformation) that depends only on the contacts: $E(c) = E(S_c)$ assuming that only two-body interactions are significant. In particular, let the effects of contacts be simply additive:

$$E(c) = \sum_{k \in S_c} f(k) \tag{1}$$

where contact function f takes into account the types of residues involved, their sequence separation, and even which residue is higher in sequence number.

For n residues on an $m \times m$ lattice, where $m \gg n$, there may be n_{rr} adjacent occupied pairs of points (residue–residue contacts plus $n - 1$ sequentially adjacent residues pairs), n_{rs} residue–solvent contacts, and n_{ss} solvent–solvent contacts. Then

$$\begin{aligned} n_{\text{rr}} &= n - 1 + x \\ n_{\text{rs}} &= 2n + 2 - 2x \\ n_{\text{ss}} &= 2(m - 1)m - 3n - 1 + x, \end{aligned} \tag{2}$$

where always $x \geq 0$; $x = 0$, in particular, for a fully extended conformation; and $x \gg 0$ for globular conformations. Since all three types of contacts can be expressed in terms of a single parameter, it is sufficient to take into explicit account only the residue–residue contacts for defining E, even though solvation of real proteins makes considerable contribution to their free energy.

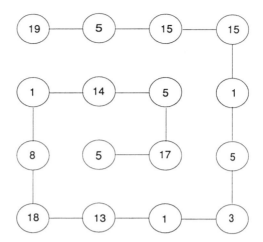

Figure 8-1a. Some 16-point lattice configurations are shown here and in Figures 8-1b–e. This figure is the arbitrarily chosen "native" conformation showing the arbitrarily chosen sequence of residue types. Alternative rigid conformations are shown in 8-1b and 8-1c.

We define a conformation c to be *rigid* if there is no other conformation $c' \neq c$ such that $S_c = S_{c'}$. For cutoff distance $d_c = 1$, that implies that rigid conformations tend to be rather compact. In Figures 8-1a–e, conformations 8-1a, 8-1b, 8-1c, and 8-1e are all rigid, but conformation 8-1d is not, because there are two ways to place its last residue and the list of contacts would be exactly the same for both. One way to distinguish the two alternatives for conformation 8-1d would be that in one conformation, one solvent "molecule" would be in contact simultaneously with residue sequence numbers 15 and 16, while in the other conformation, another solvent molecule would be in contact with 6 and 16. However, we have assumed that the potential function depends only on a sum of two-body contacts. Another way to distinguish the two alternatives would be to raise the distance cutoff $d_c > \sqrt{2}$. Similarly for $d_c = 1$, the lattice helix analogue 8-1e is rigid, but an extended chain is not. Raising the cutoff to $\sqrt{2} < d_c < 2$ allows the extended chain to be rigid. Since in real polypeptides a long isolated α-helix can be stable while a single extended strand is not, we keep the cutoff $d_c = 1$. If we require the native conformation c_{nat} to be the *unique* global minimum of E, i.e., $E(c_{\text{nat}}) < E(c)$ for all $c \neq c_{\text{nat}}$, then the native must be a rigid conformation.

For a 16-residue chain on a square lattice, there are 802,075 distinct conformations, not counting their symmetry-related counterparts. Is it possible to build a potential E of the form in Equation (1) such that the arbitrarily

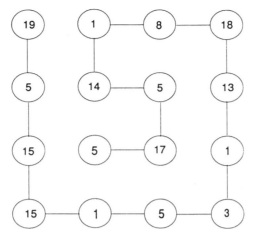

Figure 8-1b. A 16-point lattice configuration. An alternative rigid conformations is shown here.

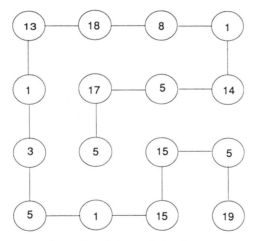

Figure 8-1c. One more alternative rigid conformation.

selected "native" conformation (Figure 8-1a) with its arbitrarily chosen amino acid sequence can be the unique global minimum of E? To eliminate any ambiguities about uniqueness, we define f in Equation (1) to be an integer-valued function of the contacts. Our contact function f first groups contacts into a number of ranges of sequence separations, where the first range includes separations of three and perhaps more, and the last range covers all separations greater than the upper limit of the previous range. Then for each range, residue types are grouped into a number of mutually

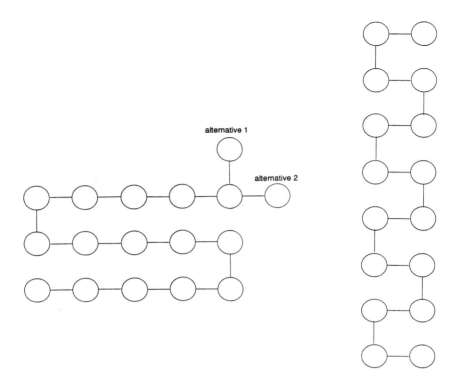

Figure 8-1d, 1e. The conformation 8-1d (*left*) is not rigid, insofar as two alternative placements of its last residue are possible. The 8-1e (*right*) is the lattice analog of a helix.

exclusive and comprehensive classes. Finally, each range has a table of integers ϵ, so that $f(k) = \epsilon_{ij}$, where the contact k involves a residue of type class i in contact with a sequentially higher residue of type class j. The direct way to determine f is to carry out a branch-and-bound, depth-first search over the number of ranges, the minimal sequence separation of each range, the number of classes in each range, the assignment of residue types to the classes in each range, whether the interaction table should be symmetric, and finally what integer values should be put into the tables, starting with small absolute values. There are two checks along the way that keep the combinatorial explosion manageable. The first is that a given choice of ranges and their sequence separations may be eliminated if it fails to distinguish between the native and some alternative set of contacts even when residue types are fully separated into 20 classes in each range. The second check is that the chosen distribution of residues into classes in each range must distinguish between the native and each alternative. For instance, exactly the same nine contacts occur in the conformations of Figures 8-1a and 8-1b if there is only one range, i.e., the potential does not depend on sequence separations. The first solution found by the branch-and-bound search happens to be Table 8-1, which might be simply paraphrased as "don't count hairpin turn contacts, but otherwise make as many contacts as possible, although contacts involving residue type 19 are not as favorable as others." Although this potential function is sufficient to make conformation a the global minimum over 802,075 alternatives, it can be viewed as having been determined by looking at only three of them! Alternative 1 is just conformation 8-1a with the sequence reversed; 2 is conformation 8-1b; and 3 is conformation 8-1c. Then in the search, 1 and 2 are active in setting up the classification of ranges and residues type classes, and 1 and 3 are active in determining the interaction tables. Of course the potential can be viewed as being determined by other sets of alternatives just as there are many choices of basis in linear programming), but the important thing is that at least one set of critical alternatives is so small.

Table 8-1. Contact potential function that favors the lattice configuration in Figure 8-1a.[a]

Separation	Class	ϵ_{ij}
3	{all}	(0)
4–∞	$1 = \{1 - 18, 20\}$ $2 = \{19\}$	$\begin{pmatrix} -2 & -1 \\ -1 & 0 \end{pmatrix}$

[a] Classification of contacts is by sequence separation and by subsets of numeric residues type; the third column gives the matrix of empirically determined interaction parameters for contacts between residues of various classes.

Outside of the rigidity of conformer 8-1a, there is nothing very special about it, and one could produce potential functions that uniquely favor various other conformations with other sequences, or indeed, that simultaneously favor the respective native conformation of more than one "protein." The helical conformation 8-1e, for example, is uniquely favored by

$$\text{separation} = 3, \quad \text{class } 1 = \{\text{all}\}, \quad \epsilon = (-1);$$
$$\text{separation} = 4 - \infty, \quad \text{class } 1 = \{\text{all}\}, \quad \epsilon = (0). \tag{3}$$

There is a simple procedure to automatically generate a small set of conformations that includes those critical for determining the potential function. Let every residue in the given protein be a node in a graph, and connect two nodes whenever they correspond to sequentially adjacent (i.e., bonded) residues or to residues in contact. Then, any Hamilton walk on this graph corresponds to a reassignment of residue sequence numbers and hence residue types to the nodes, and all edges not traversed in the walk are contacts. Because contacts are distinguished only on the basis of sequence separation and the two residue types, two different walks may produce the same set of contacts. Thus, conformation 8-1a gives rise to 552 different walks on a 4×4 lattice, of which there only 69 different sets of contacts, and these drive the combinatorial search to the same potential function as before. If conformation 8-1e is viewed as a homopolymer, there are 116 walks, among which there are only 16 distinct sets of contacts, resulting in the same potential as before.

3. Protein Structures

3.1 Selection of Protein Chains and Generating Alternative Protein Conformations

The total set of protein crystal structures we considered were the 109 polypeptide chains in the October 15, 1990 release of the Brookhaven Protein Data Bank (PDB) (Abola et al., 1987) with coordinates of N, C^α, C', C^β, and O atoms, and no obvious chain breaks in the middle. Disordered or unresolved residues at the N- or C-termini were not included in the polypeptide chains we considered here. For brevity, we will refer to these chains by their PDB code and the chain identifier in the PDB file (e.g., 3ins.A is the A chain of insulin). The full name of each protein can be found in Table 8-2. If there were alternate conformations, only the first of them was used. Generally, we included only the accurately determined structures (2.5 Å nominal resolution), although some lower-resolution structures having no interior chain breaks were included in this study, sometimes to increase the number of alternative conformations we could generate, and sometimes to increase the number of short protein chains considered. We also included two other PDB entries that technically did not fulfill the 2.5 Å resolution criterion: 1bds is a structure determined from NMR data by distance geometry calculations having unknown accuracy, and 1hvp is a hypothetical conformation built by homology modeling. In the final analysis, these two caused no special problems. The 109 protein structures ranged from 21 residues for the shorter chain 3ins.A to 498 residues for 8cat.A. However, we used only the smallest 86 chains as reference structures because these all had 255 or fewer residues. The limit of 255 is due to the database packing scheme we used, where each contact in each alternative encodes its sequence separation in one 8-bit byte. Even so, our total database of all contacts for all 691,165 alternatives of all the reference proteins required a few hundred megabytes of storage. Thus, the 23 largest structures (2cab, 4rhv.1, 1cse.E, 1pyp, 1rhd, 2cyp, 1abp, 5cpa, 4tln, 1pfk.A, 1hmg.A, 6ldh, 4ape, 4mdh.A, 3gpd.G, 7api.A, 8adh, 1phh, 3pgk, 1cts, 3grs, 2taa.A, and 8cat.A) were used only for building alternative conformations.

Instead of using walks on a 3D structure lattice to generate a finite set of alternative conformations, the whole list of involved proteins can be used in a "threading" procedure (Hendlich et al., 1990; Crippen, 1991). For the amino acid sequence corresponding to a protein and consisting of N_{res} residues, all fragments of 3D structures of all other proteins (with length not less than N_{res}) were employed as frames to position the sequence. For example, crambin (1crn) has 46 residues, while rubredoxin (3rxn) has 52. Using the amino acid sequence of 1crn applied to the coordinates of 3rxn residues 1–46, 2–47, ..., 7–52 produces seven plausible alternative conformations of crambin. This procedure is certainly a better way than lattice walks to naturally imitate lengthy lists of alternative conformations that have to compete with native structure, because of their authentically "protein-like" spatial arrangements of atoms.

The 86 reference structures are shown in Table 8-2 and also listed below in Table 8-6 in order of chain length. In addition, 19 of these reference structures have one or more homologous structures, by which we denote other crystal structures of proteins having the same chain length and strong sequence identity or the same proteins in different crystal environments and/or complexed with different ligands. These were not used in any training set and served only to assess the quality and predictive power of the deduced contact potentials. In all, there are 95 homologous structures, as listed in Table 8-3, along with their corresponding 19 reference structures. As in our previous work, we derived alternative conformations for each reference structure from all larger references and from the 23 very large structures. Consequently the smallest references had more alternatives (16,521 for 3ins.A), but even the largest, 4rhv.2, had 2,127.

3.2 Compactness

Our aim was to construct a function of the interresidue contacts such that each reference structure has a lower function value than any of its alternatives, just as the native conformation of a real protein has a lower free energy than any kinetically accessible alternative conformation. Although we make no claim that the function we determine in this work resembles the real free energy, we will loosely refer to our function as the *contact energy*. Our preliminary studies indicated that some proteins are particularly

Table 8-2. *Part 1 of 4.* List of the reference proteins sorted by PDB code.

PDB code	Resolution $(\text{Å})^a$	No. residues	Chain ID^b	Title and source
155c	2.5	121		Cytochrome c550, P. denitrificans
1abp	2.4	306		L-arabinose-binding protein, E. coli
1acx	2.0	108		Actinoxanthin, A. globisporus
1bds	–	43		Sea anemone antihypertensive antiviral protein
1bp2	1.7	123		Bovine pancreatic phospholipase A2
1cc5	2.5	83		Cytochrome c5, Azotobacter
1ccr	1.5	111		Cytochrome c, rice
1crn	1.5	46		Crambin, Abyssinian cabbage
1cse	1.2	63	I	Eglin C (complexed with subtilisin carlsberg)
		274	E	Subtilisin carlsberg (complexed with eglin C)
1ctf	1.7	68		L7/L12 5OS ribosomal protein (C-terminal domain), E. coli
1cts	2.7	437		Pig citrate synthase
1cy3	2.5	118		Cytochrome c3, D. desulfuricans
1ecd	1.4	136		Hemoglobin (erythrocrvorin, deoxy), C. thummi thummi
1est	2.5	240		Porcine tosyl-elastase
1fdx	2.0	54		Ferredoxin, P. aerogenes
1fxl	2.0	147		Flavodoxin, D. vulgaris
1gcn	3.0	29		Porcine glucagon
1gcr	1.6	174		Calf γ-II crystallin
1hip	2.0	85		High potential iron protein (oxidized), C. vinosum
1hmg	3.0	175	B	Haemagglutinin, influenza virvs
		328	A	
1hmq	2.0	113		Hemerythrin (met), sipunculid worm
1hoe	2.0	74		α-Amylase inhibitor, S. tendae
1hvp	–	99		Retrovirus HIV-1 protease
1lh4	2.0	153		Leghemoglobin (deoxy), yellow lupin
1lyz	2.0	129		Hen egg white lysozyme
1lz1	1.5	130		Human lysozyme
1mba	1.6	146		Sea hare myoglobin
1mbd	1.4	153		Sperm whale myoglobin

^{a} A "–" sign denotes NMR (1bds) and model (1hvp) protein structures for which the notion of the resolution is not applicable.
^{b} In the case of more than one chain in a PDB file, the chain identifiers are given.

Table 8-2. *Part 2 of 4.* List of the reference proteins sorted by PDB code.

PDB code	Resolution (Å)	No. residues	Chain ID[a]	Title and source
1paz	1.55	120		Pseudoazurin (oxidized, Cu^{++}), A. faecalis
1pcy	1.6	99		Plastocyanin (Cu^{++}), poplar
1pfk	2.4	320		Phosphoiructokinase, E. coli
1phh	2.3	394		P-hydroxybenzoate hydroxylase, P. fluorescens
1pp2	2.5	122		Calcium-free phospholipase A-2, rattlesnake
1ppt	1.37	36		Avian pancreatic polypeptide
1pyp	3.0	280		Yeast pyrophosphatase
1rei	2.0	107		Human Bence-Jones immunoglobulin variable portion
1rhd	2.5	293		Bovine rhodanese
1rn3	1.45	124		Bovine ribonuclease A
1sn3	1.8	65		Scorpion neurotoxin, variant 3
1tim	2.5	247		Chicken triose phosphate isomerase
1wrp	2.2	102		Bacterial TRP repressor
2abx	2.5	74		α-bungarotoxin, braided krait venom
2act	1.7	218		Actinidin, kiwifruit
2alp	1.7	198		Alpha-lytic protease, L. enzymogenes
2aza	1.8	129		Azurin (oxidized), A. denitrificans
2b5c	2.0	85		Bovine cytochrome b5
2c2c	2.0	112		Cytochrome c2 (oxidized), R. rubrum
2cab	2.0	256		Human carbonic anhydrase (form B)
2ccy	1.67	127		Cytochrome c', R. molischianum
2cdv	1.8	107		Cytochrome c3, D. vulgaris
2cna	2.0	237		Concanavalin A, jack bean
2cyp	1.7	293		Yeast cytochrome c peroxidase
2fb4	1.9	216	L	Human immunoglobulin light chain
2gn5	2.3	87		Bacteriophage gene 5 DNA binding protein
2hhb	1.74	141	A	Human hemoglobin (deoxy)
		146	B	
2lhb	2.0	149		Sea lamprey hemoglobin V (cyano, Met)
2lzm	1.7	164		T4 phage lysozyme
2mlt	2.0	26		Bee melittin

[a] In the case of more than one chain in a PDB file, the chain identifiers are given.

Table 8-2. *Part 3 of 4.* List of the reference proteins sorted by PDB code.

PDB code	Resolution (Å)	No. residues	Chain ID[a]	Title and source
2ovo	1.5	56		Ovomucoid (third domain), pheasant
2pab	1.8	114		Human prealbumin
2pka	2.05	80	A	Porcine kallikrein A
		152	B	
2rhe	1.6	114		Human lambda immunoglobulin variable domain (Bence-Iones)
2sga	1.5	181		Proteinase A, S. griseus
2sns	1.5	141		Staphylococcal nuclease
2sod	2.0	151		Bovine Cu, Zn superoxide dismutase
2ssi	2.6	107		Streptomices subtilisin inhibitor
2stv	2.50	184		Tobacco necrosis virvs coat protein
2taa	3.0	478		Taka-amylase A, A. oryzae
351c	1.6	82		Cytochrome c551 (oxidized), P. aeruginosa
3adk	2.1	194		Porcine adenylate kinase
3ebx	1.4	62		Sea snake erabutoxin B
3fab	2.0	207	L	Human lambda immunoglobulin FAB'
		219	H	
3fxc	2.5	98		Ferredoxin, S. platensis
3fxn	1.9	138		Flavodoxin (oxidized), clostridium
3gap	2.5	208		Catabolite gene activator protein, E. coli
3gpd	3.5	334		Human D-glyceraldehyde-3-phosphate dehydrogenase
3grs	1.54	461		Human glutathione reductase
3icb	2.3	75		Bovine calcium binding protein
3ins	1.5	21	A	Pig insulin
		30	B	
3pgk	2.5	415		Yeast phosphoglycerate kinase
3rp2	1.9	224		Rat mast cell protease
4ape	2.1	330		Endothiapepsin, fungal
4dir	1.7	159		Dihydrofolate reductase, E. coli
4fd1	1.9	106		Azotobacter ferredoxin
4mdh	2.5	333		Porcine cytoplasmic malate dehydrogenase
4pti	1.5	58		Bovine pancreatic trypsin inhibitor

[a] In the case of more than one chain in a PDB file, the chain identifiers are given.

Table 8-2. *Part 4 of 4.* List of the reference proteins sorted by PDB code.

PDB code	Resolution (Å)	No. residues	Chain ID[a]	Title and source
4rhv	3.0	40	4	Human rhinovirvs 14 coat protein
		236	3	
		255	2	
		273	1	
4sbv	2.8	199		Southern bean mosaic virvs coat protein
4tln	2.3	316		Bacterial thermolysin
5cpa	1.54	307		Bovine carboxypeptidase A alpha
5cpv	1.6	108		Carp calcium-binding parvalbumin B
5cyt	1.5	103		Cytochrome c (reduced), tuna
5rxn	1.20	54		Rubredoxin (oxidized, Fe^{+++}), Clostridium
6ldh	2.0	329		Dogfish lactate dehydrogenase
7api	3.0	36	B	Human modified α-1-antitrypsin
		339	A	
8adh	2.4	374		Horse apo-liver alcohol dehydrogenase
8cat	2.5	498		Bovine catalase
9pap	1.65	212		Papain (Cys-25 oxidized), papaya
9wga	1.8	170		Wheat germ agglutinin

[a] In the case of more than one chain in a PDB file, the chain identifiers are given.

difficult to bring into agreement with our goal, perhaps because they are not adequately compact or globular, certainly necessary conditions for lattice models of proteins (Crippen, 1991). For example, if a polypeptide chain crystallizes as a dimer with many interchain contacts, it is unreasonable to use the coordinates of a monomer in isolation as a reference structure in developing our energy function, because the contacts that stabilize the conformation would not be included in our calculations.

In order to develop a quantitative criterion for deciding the suitability of a structure for use as a reference, and generally in order to distinguish between compact and noncompact structures, we examined two functions of a conformer's radius of gyration, r_g, and number of contacts, N_c: (i) the ratio e_g of the radius of gyration of the native structure to the minimal radius of gyration $r_g(\text{min})$ over the set of all its alternatives

$$e_g = r_g/r_g(\text{min}) \tag{4}$$

and (ii) the ratio of the maximal number of contacts for all alternatives,

Table 8-3. List of the 19 reference proteins used in the present work with their 95 homologues.[a]

No. residues	PDB codes and chain identifiers Reference[a]	List of homologous[d]	RMSD (Å)[c]
58	4pti	5pti	0.59
62	3ebx	5ebx[b]	0.15
82	351c	451c	0.03
99	1pcy	2pcy 3pcy 4pcy 5pcy 6pcy	0.12
106	4fd1	1fd2 2fd2	0.20
108	5cpv	1cdp 4cpv	0.27
112	2c2c	3c2c	0.09
124	1rn3	5rsa 6rsa 7rsa	0.15
129	1lyz	1lzt 2lym 2lyz 2lz2 2lzt 3lym [3–8]lyz 1lym.A	0.35
136	1ecd	1eca 1ecn 1eco	0.06
138	3fxn	4fxn	0.21
146	1mba	2mba 3mba 4mba	0.21
153	1lh4	1lh[1–3] 1lh[5–7] 2lh[1–7]	0.12
153	1mbd	5mbn 1mb5 1mbc 1mbn 1mbo 4mbn	0.41
159	4dfr.A	7dfr	0.74
164	2lzm	1l[01–02] 1l[04–10] 1l[12–25] 1l[27–35] 1lyd 3lzm	0.12
170	9wga.A	1wgc.A 2wgc.A 7wga.A	0.21
212	9pap	1ppd	0.18
237	2cna	3cna	0.74

[a] The following reference proteins having homologies were excluded from the reduced training set: 1pcy, 4df1, 5cpv, 1lyz, 3fxn, 1mba, 1mbd, 4dfr.A, 2lzm, 9pap, 2cna (see Table 8-6).
[b] Neurotoxin B, 1nxb was excluded from the list because of noticeable compactness criteria violation: $e_g = 1.17$ and $e_N = 1.54$, despite being homologous to the reference 3ebx.
[c] C^α-based distance RMSD (Equation (12)), averaged over all the homologous structures.
[d] Digits in brackets mean the whole range of numbers, e.g., [3–8]lyz is 3lyz, 4lyz, ..., 8lyz.

$N_c(\text{max})$, to the number of contacts for the native structure

$$e_N = N_c(\text{max})/N_c. \qquad (5)$$

Here N_c corresponds to the discrete form of the contact function, as described below. In the beginning of our study we used the directly observed minimal radius of gyration and maximal of the number of contacts taken over the whole list of alternatives for a protein structure of the given chain length. Later we used the empirically determined dependences of $r_g(\text{min})$ and $N_c(\text{max})$ on polypeptide chain length derived by examining these characteristics over all reference/alternative data sets. We found by linear regression over all alternatives of all reference structures that the minimal radius of gyration depends on the number of amino acid residues N_{res} as follows:

$$r_g(\text{min}) = -1.26 + 2.79(N_{res})^{1/3} \qquad (6)$$

with correlation coefficient of 0.997. Another way to estimate the minimal possible radius of gyration as a function of N_{res} is to model a globular protein as an ellipsoid of rotation (Damaschun et al., 1969) with mean partial volume of 134 Å^3 per residue. Then the minimal radius of gyration is achieved at unit eccentricity, i.e., spherical shape, giving the same functional form as Equation (6), but changing the coefficients from -1.26 and 2.79 to 0 and 2.46, respectively. The $r_g(\text{min})$ values resulting from the two functions differed by less than 6% over the range of N_{res} considered, but the ellipsoid model curve fitted the data slightly worse. Similarly, we found that the maximal number of contacts fitted the linear regression equation

$$N_c(\text{max}) = -53.17 + 4.25N_{res} \qquad (7)$$

with correlation coefficient 0.992. Note that the slope value of 4.25 indirectly bears out the correctness of the cutoff distances described below for specifying contacts; we really have something like the first coordination sphere for each residue in a contact.

We found that the position of a given protein structure on the e_g versus e_N diagram (Figure 8-2) adequately reflects the degree and nature of its compactness. One might point out that two types of noncompactness occurred in different protein structures: One of them is characterized by noticeably larger values of r_g compared to its minimal value, and the second type is marked by a definitely smaller number of contacts N_c compared to the maximum possible for a polypeptide chain of the given length. These two types of noncompactness may occur separately, e.g., high radius of gyration for 2mlt.A ($e_N = 1.28$, $e_g = 1.73$) and 1hmg.B ($e_N = 1.47$, $e_g = 1.92$); or low number of contacts for 2gn5 ($e_N = 1.95$, $e_g = 1.25$) and 1cy3 ($e_N = 1.98$, $e_g = 1.23$); and simultaneously, e.g., for 1gcn ($e_N = 2.26$, $e_g = 2.09$) and 7api.B ($e_N = 2.50$, $e_g = 1.77$), as shown in Figure 8-2. Clearly most of the protein structures are rather compact, being clustered in the lower left part of the diagram, while 17 proteins obviously

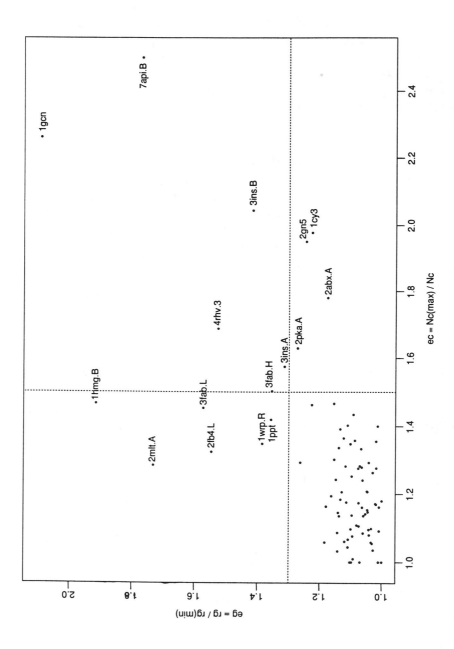

have noncompact conformations. We chose

$$e_N < 1.5 \quad \text{and} \quad e_g < 1.3 \tag{8}$$

as the requirements for compactness. We realize that the distribution in Figure 8-2 is fairly continuous throughout the diagram, and therefore these limits are somewhat arbitrary. However, we employed them in this work because such a differentiation helped us determine the desired energy function, and it is also in good agreement with visual inspections of the protein folds. Note that while we required the reference structures to be compact according to this definition, the alternative conformations had no such constraint. In fact, 5 to 20% of the alternatives for each reference protein turned out to be compact.

In order to use Equation (8) for a particular protein structure having chain length N_{res}, one needs to know $r_g(\text{min})$ and $N_c(\text{max})$. The direct way is to generate the many thousands of alternative structures and calculate r_g and N_c for each. Not only is this tedious, but for large N_{res}, there are sometimes substantial deviations from the very regular trend shown for smaller proteins. The reason is that the number of alternatives decreases as the chain length increases, simply because we are dealing with a fixed number of proteins from which to generate alternatives. Consequently, inasmuch as we have established the accurate relations given in Equations (6) and (7), we could use them in the calculations to quickly obtain $r_g(\text{min})$ and $N_c(\text{max})$.

4. Method

4.1 Contact Potential Derivation

We evaluated conformations according to the interresidue contacts formed. The exact definition of a contact we continue to use (see Table 8-4) is designed to be applicable even if the sequence of a given conformation is changed. We considered only the backbone N, C', and O atoms plus the side chain C^β, even building in an artificial C^β if the original residue is Gly. Then

---

Figure 8-2. Compactness diagram (see text) used to characterize 86 reference protein chains in terms of radius of gyration and number of contacts. The 16 noncompact structures beyond the limits $e_N < 1.50$ and $e_g < 1.30$ (marked by dotted lines) are indicated by their corresponding PDB codes and chain identifiers. One more extremely noncompact structure, 4rhv.4, is off scale at $e_N = 4.52$ and $e_g = 2.54$. *See figure on preceding page.*

a backbone–backbone contact was counted whenever $d(O, N) < 3.2\,\text{Å}$ and $d(C, N) > 3.9\,\text{Å}$; a backbone–side chain contact required $d(N$ or O, $C^\beta) < 5.0\,\text{Å}$ and no other atom between the interacting pair closer than $1.4\,\text{Å}$ to the line segment joining them; and a side chain–side chain contact requires $d(C^\beta, C^\beta) < 9.0\,\text{Å}$ and similarly no interfering atom between them. Interactions had to be between residues differing by at least three in sequence. Backbone atoms involved in contacts were ascribed to residue type Gly, but side-chain atoms corresponded to their correct residue types.

Table 8-4. Boundary parameters of the discrete and continuous contact functions (see text).

		Cutoff distances $(\text{Å})^b$		
		Discrete	Continuous	
From atom	To atom (or line)	$L = U$	L	U
N	O	3.20	3.20	3.20
N (see [a])	C	3.90	3.60	3.60
N or O	C^β	5.00	2.50	5.50
C^β	C^β	9.00	6.50	9.50
any atom	line joining two contact atoms	1.41	1.41	1.41

[a] Because the definition of a backbone–backbone contact requires a short N–O distance but a long N–C distance in order to stipulate a roughly linear hydrogen bond, the sense of these limits is reversed, compared to Equations (15) and (16).
[b] L and U denote the lower and upper distance cutoffs of the contact function, Equations (15), (16), and (17).

We have assumed the contact potential function E for a given protein conformation is a sum of the values ϵ assigned to the individual contacts:

$$E = \sum_{\substack{\text{contact residues} \\ i \text{ and } j}} \epsilon\big(\text{class } (i), \text{class } (j), |i - j|\big) \qquad (9)$$

where the ϵ's depended on the same very detailed standard classification according to sequence separation and residue-type classes proposed before (Crippen, 1991) (Table 8-5). This classification is a plausible one that groups together helix formers as opposed to helix breakers for short-range (i.e., sequence separation ≤ 4) interactions, and hydrophobic against hydrophilic residues for long-range interactions. We assumed the importance of a contact does not depend on which residue is higher in sequence, so the interaction matrices in Table 8-7 are all symmetric, and

Table 8-5. The standard classification of amino acid type contributions to the contact potential.[a]

Separation	Class
3	$1 = \{G\}, 2 = \{ALICMF\}, 3 = \{VHS\}, 4 = \{P\}$ $5 = \{RDEQ\}, 6 = \{TKN\}, 7 = \{YW\}$
4	$1 = \{G\}, 2 = \{ALICMF\}, 3 = \{VHS\}, 4 = \{P\}$ $5 = \{RDEQ\}, 6 = \{TKN\}, 7 = \{YW\}$
5–7	$1 = \{G\}, 2 = \{AV\}, 3 = \{LICMF\}, 4 = \{YHWST\},$ $5 = \{KR\}, 6 = \{P\}, 7 = \{DNEQ\}$
8–∞	$1 = \{G\}, 2 = \{AV\}, 3 = \{LICMF\}, 4 = \{YHWST\},$ $5 = \{KR\}, 6 = \{P\}, 7 = \{DNEQ\}$

[a] Amino acid residue contacts are grouped by sequence separation and by subsets of types of residue indicated by the single-letter residue code.

there are a total of 112 parameters to adjust (4 separation ranges, each having $7 \times (7 + 1)/2 = 28$ interaction parameters among 7 classes of amino acid).

At the initial stage of the work (Crippen, 1991) it was required only that

$$E(\text{alternative}) \geq E(\text{reference}) \tag{10}$$

for each reference and each respective kth alternative. Later (Maiorov and Crippen, 1992) we demanded that strict inequality hold by a margin T_k for the kth alternative given by

$$T_k = q D_k. \tag{11}$$

Here q is an empirically adjusted coefficient (see below), and D_k is the C^α root mean square distance deviation (RMSD) (Nishikawa and Ooi, 1972; Levitt, 1976), between the reference and the kth alternative structures:

$$D^2 = (1/N_{\text{pairs}}) \sum_{i<j} (d'_{ij} - d_{ij})^2 \tag{12}$$

where d_{ij} and d'_{ij} are the distances between the ith and jth C^α atoms in the reference and kth alternative structures, respectively, and N_{pairs} is a number of interatomic distances: $N_{\text{pairs}} = N_{\text{res}}(N_{\text{res}} - 1)/2$. Generally speaking, one might also use the optimal rigid-body superposition coordinate-based RMSD instead of that in Equation (12). (See, for example, McLachlan's work [McLachlan, 1979] for a concise classification of the approaches for optimal superposition and respective RMSD calculations.) In this case the exact measure of structural difference was unimportant, so we chose the more easily calculated distance RMSD of Equation (12). Thus for a given reference structure and its kth alternative, we required

$$E(k\text{th alternative}) - E \geq T_k. \tag{13}$$

The underlying idea here is to make the energy of an alternative lie above that of the corresponding reference structure by at least some minimal margin that increases linearly with their conformational difference. Test computations showed that choosing a very small positive value for q reduced Equation (13) to approximately Equation (10), made the set of inequalities easier to solve, led to very similar energies for the reference structure and some of its alternatives, and left no room on the energy scale between the reference structure and the lowest alternative for the homologous proteins, which were expected to scatter in this range. On the other hand, we empirically learned that too large a q caused a marked increase in the computing time required to find a solution for the set of inequalities. A reasonable compromise was $q = 3$, the value used throughout this work. Although our potential function is required to have the free energy-like property of favoring the native conformation, Equation (13) has no relation to physical energy or temperature scales. Therefore the units for E (and q) are arbitrary. It turned out that the method to solve homogeneous sets of inequalities (Jurs, 1986), as in Equation (10), could be applied to sets of inhomogeneous inequalities, as in Equation (13), and was therefore used in all that follows. In addition, we have also corrected some inconsistencies in the use of the contact function reported earlier (Maiorov and Crippen, 1992).

The procedure for determining the ϵ's included the following three steps. *Step 1*: Directly solving the entire set of 690,000 linear inequalities of the form in Equation (13) was hopelessly slow. On the other hand, at the solution, only a relatively small number of inequalities were active, as shown earlier (Crippen, 1991), particularly those inequalities arising from the more challenging compact alternatives. Moreover, some of the noncompact alternatives might become active as well. However, it was practically impossible to identify those of both kinds (compact and noncompact) that appeared to be active. Therefore we simply selected the first 49 alternatives for each reference no matter whether they obeyed the compactness criteria of Equation (8) or not. In the optimization procedure all the starting values were set to the arbitrary value of -0.1, and the ϵ's rapidly converged to a set of first-approximation values. *Step 2*: Next we "combed" through the full list of alternatives to each reference for any alternative that violated Equation (13). Adding these to the previous list of inequalities increased the size of the problem only slightly, and the first approximation ϵ's were a good starting point for calculating the second approximation. Actually this was a very efficient way to extract all alternatives that were essential from the contact energy difference viewpoint and were missed at the first step. The clever selection of alternatives in the first two steps was the key to being able to handle much larger sets of inequalities than before. *Step 3*: It was found that sometimes a third step of refinement of the potentials is required because some alternatives that satisfied Equation (13) before Step 2 did not at the end of the step. The remedy was to return to the basic set of inequalities in the first step, repeat the combing, and produce a third set of ϵ's from the second approximation. Our experience showed that sometimes one more iteration might be needed, because additional combing might also reveal some weakly violating alternatives. However, the total number of such alternatives was found to be very small, and moreover additional optimization steps improved almost nothing compared to the three-step optimization protocol. Therefore, for contact potential calculations we did not use extra optimization beyond the described three-step protocol.

4.2 Discrete and Continuous Forms of the Contact Function

The procedure described above was applied to deduce contact potentials for a training set consisting of all 69 compact proteins from the reference list, excluding all homologous structures (see Table 8-3). (Incidentally, note that Table 8-3 does not list neurotoxin B 1nxb [Tsernoglou et al., 1978] as homologous to erabutoxins 3ebx [Smith et al., 1988] and 5ebx [Corfield et al., 1989], in spite of strong sequence similarity, because of its noticeable shape distortion: $e_g = 1.17$ and $e_N = 1.54$.) However, we subsequently found that on rare occasions, the resulting contact energy E for some of the homologous structures was greater than that of the lowest alternative. For example, for the reference bovine pancreatic trypsin inhibitor crystal structure 4pti (Marquart et al., 1983), there is the homologous 5pti (Wlodawer et al., 1984) differing in RMSD by only 0.59 Å, yet $E(5pti)$ was an appreciable 26.7 arbitrary units greater than $E(4pti)$ and 9.8 above E of the lowest alternative. Since the assignment of "reference" and "homologous" structures was absolutely arbitrary, this outcome ought to be considered a violation of Equation (13). Although this happened to be the only violation of this kind, we were compelled to eliminate it.

The difficulty arose from the "all-or-nothing" definition of a contact, as described above. We could rewrite Equation (9) as

$$E = \sum_{\substack{\text{contacts} \\ i,j}} V(d_{ij}, U) \, \epsilon_{ij}, \tag{14}$$

where V is the value of a contact depending on d_{ij}, the relevant interatomic distance, and the cutoff value U. The discrete contact function applied earlier (Crippen, 1991) has

$$V(d_{ij}, U) = \begin{cases} 1 & \text{if } d_{ij} \leq U, \\ 0 & \text{otherwise.} \end{cases} \tag{15}$$

Even slight changes in interatomic distances between two homologous structures may cause significantly different lists of contacts. The solution to which we later came (Maiorov and Crippen, 1992) was to use a continuous contact function where V becomes a smooth sigmoidal function of d_{ij}, going from 1 below a lower cutoff distance L to 0 above an upper cutoff U:

$$V(d_{ij}, U, L) = \begin{cases} \frac{(d_{ij}-U)^2(2d_{ij}-3L+U)}{(U-L)^3} & \text{if } L \leq d_{ij} \leq U, \\ 1 & \text{if } d_{ij} < L, \\ 0 & d_{ij} > U. \end{cases} \tag{16}$$

In fact, Equation (15) is a limiting case of Equation (16) when $U = L$. For contacts involving side-chain atoms, we reserved an option to include the effect of possible interfering atoms k near the line segment joining the interacting atoms i and j by defining the modified contact strength V_m to be

$$V_m(d_{ij}, U, L) = V(d_{ij}, U, L) \prod_k (1 - V(d_{ijk}, U', L')), \tag{17}$$

where d_{ijk} is the distance from atom k to the line segment joining atoms i and j, and U', L' are the respective upper and lower limits.

In order to determine suitable cutoff values for the continuous contact definition, we chose a limited training set of reference structures, their alternatives, and their homologous structures, namely 4pti (12,701 alternatives and 5pti) (Marquart et al., 1983; Wlodawer et al., 1984), 3ebx (12,316 alternatives and 5ebx) (Smith et al., 1988; Corfield et al., 1989), and 351c (10,483 alternatives and 451c) (Matsuura et al., 1982). Then the cutoffs were adjusted so that each reference and its homologous structure spanned a small range of energies, while there was a large increase in energy going from the highest homologous structure to the lowest alternative. This is the only role the homologous structures played in the fitting because otherwise Table 8-3 makes it clear that homologous structures are extremely similar to their corresponding reference structures (from 0.06 to 0.74 Å RMSD), making their energies so easy to fit they were not needed in the training sets. It was found that continuity of the contact function is of critical importance only for contacts involving side-chain C^β's, while the contact function form for other types of contacts may remain discrete, as shown in Table 8-4. Similarly, we also used only the discrete form of the contact function term responsible for possible interfering atoms near the line segment joining the interacting pair of atoms (Equation (17)), as indicated by the last line of Table 8-4 where the continuous $U = L = 1.41$ Å. In what follows, we will refer to this hybrid form of the function as the "continuous contact function" and to the old version as the "discrete" one. Note that in determining the ϵ's, the very approximate first step of the procedure uses the discrete form of contact function, while the more accurate continuous form was employed in the following two steps.

Except for treating the homologous structures, there is not a big difference between the discrete and the continuous contact functions. For example, the relationship between the discrete number of contacts and the "effective number of contacts," defined to be the sum of all contact values (Equation (16)) for the conformer according to the continuous form of the contact function, is quite linear (N_c(discrete) $= 0.84 N_c$(continuous) $+ 1.99$) and has a correlation coefficient of 0.998.

5. Correct Folding Identification

5.1 Complete Training Set (CTS)

The first question we dealt with was whether we could satisfy Equation (13) even by including in the training set all 69 compact structures having readable chain length less than 256. As stated before, for each reference structure we selected the first 49 alternatives, which produced a set of 3,381 inequalities for the whole training set. After 1,224 iterations of optimization, the solution for the first step was found and used as the start for the next step. On the second step, 3,847 constraints were added as described above, making altogether 7,228 inequalities. The optimization converged to a solution after 7,501 iterations. Finally, on the third step only 67 new constraints were added to the 3,381 from the first step (for a total of 3,448 inequalities), and 3,101 iterations completed the procedure. Checking the final potential against the whole data base showed perfect agreement with Equation (13) for all 69 references and all their alternatives. Thus the 7,295 ($= 49 \times 69 + 3,847 + 67$) constraints used in the CTS were sufficient to predict correctly 530,062 constraints from a total of 73 proteins (including four noncompact structures, 3fab.L, 2fb4.L, 3fab.H, and 4rhv.3, which were excluded from the CTS), for an average "predictive significance" of $(530,062 - 7,295)/7,295 = 71.7$. Also all the 95 homologous structures in Table 8-3 had contact energies less than the lowest alternative of the corresponding reference structure.

5.2 Reduced Training Set (RTS)

Having seen that it was possible to fit all the proteins, we next tried to reduce the training set, seeking to determine the minimum number of proteins
necessary to deduce a potential that could make a prediction of the same
quality. Thus, we excluded from the training set the 32 structures (Table 8-
6) having reference energies in the CTS potential more than (a somewhat
arbitrary chosen margin of) 90 units below the lowest alternative, and were
successful in finding a solution. The information about interresidue interactions contained in the remaining 37 structures (whose alternatives made
a total of $6,047 = 49 \times 37 + 4,168 + 66$ constraints) was sufficient to
make correct predictions for exactly the same proteins as before with the
complete training set. In this case the average "predictive significance"
of a constraint in this calculation is somewhat better than with the CTS:
$530,062/6,047 = 87.7$.

The values of the resulting RTS parameters seem to have clear physical
meaning (Table 8-7). For example, for sequence separations of 8 residues
and more (fourth separation range) the largest positive (i.e., unfavorable)
values are observed for interactions between pairs of positively charged
side chains of Lys and/or Arg residues (group 5 and group 5 $\epsilon = 7.48$) or
between pairs of negatively charged/polar Asp, Asn, Glu and Gln residues
(group 7 and group 7 $\epsilon = 2.21$), in agreement with the obvious electrostatic
repulsion between side chains having like charges. On the other hand, the
largest negative interaction parameters are for pairs of the hydrophobic
residues Leu, Ile, Cys, Met, and Phe (group 3 and group 3 $\epsilon = -9.22$)
or for these hydrophobic residues and nonpolar side chains of Ala and Val
(group 2 and group 3 $\epsilon = -6.45$), thus reflecting the tendency of these
residues to form favorable hydrophobic interactions with each other.

In general, there was an apparent correlation in contact energies of
native structures and their chain lengths found for 69 compact reference
structures (Figure 8-3) that fits

$$E(\text{native}) = 53.43 - 4.39 N_{\text{res}} \tag{18}$$

with a correlation coefficient of -0.932. This relation may be helpful in
predicting the conformation of a novel protein sequence. If the lowest
proposed conformation of a protein still gives an energy well above the
value expected for such a chain length, then the correct native conformation
probably has not yet been suggested.

Note that all the smallest structures in the list (the first seven in Table 8-6 and Figure 8-3) are noncompact and violate the fitting condition,
Equation (13). However, only 6 of the remaining 10 noncompact structures

Table 8-6. *Part 1 of 3.* Contact energy and contact energy difference for the 86 reference protein structures and their 95 homologues calculated with RTS potential.

PDB code[a]	No. res	No. alts	Contact energy (arbitrary units)[b] ref.	alt.	diff.	Homologous No.	min.	max.	diff.[c]
* 3ins.A	21	16521	−45.7	−92.8	−47.0				
* 2mlt.A	26	15980	−26.4	−103.7	−77.3				
* 1gcn	29	15658	13.9	−118.5	−132.3				
* 3ins.B	30	15551	−49.5	−125.3	−75.8				
* 1ppt	36	14920	−4.1	−94.6	−90.5				
* 7api.B	36	14919	−82.2	−114.7	−32.5				
* 4rhv.4	40	14506	−4.0	−119.6	−115.5				
1bds	43	14199	−149.2	−134.1	15.2				
1crn	46	13895	−180.8	−161.9	18.9				
1fdx	54	13094	−218.7	−197.8	20.9				
5rxn	54	13093	−172.8	−137.9	35.0				
2ovo	56	12896	−184.2	−166.8	17.4				
4pti	58	12701	−200.9	−183.3	17.6	1	−193.4	−193.4	10.1
3ebx	62	12316	−196.2	−174.8	21.4	1	−197.8	−197.8	23.0
1cse.I	63	12220	−163.0	−142.2	20.8				
1sn3	65	12031	−190.8	−173.8	17.1				
- 1ctf	68	11751	−321.6	−200.6	121.0				
1hoe	74	11198	−227.2	−203.9	23.3				
* 2abx.A	74	11197	−191.8	−237.1	−45.2				
- 3icb	75	11106	−360.2	−262.2	98.0				
* 2pka.A	80	10660	−209.9	−231.7	−21.8				
351c	82	10483	−249.0	−217.6	31.4	1	−244.8	−244.8	27.2
1cc5	83	10395	−273.6	−252.2	21.4				
1hip	85	10222	−258.4	−233.6	24.8				
2b5c	85	10221	−304.1	−251.0	53.1				
* 2gn5	87	10052	−216.3	−269.8	−53.5				
3fxc	98	9138	−398.4	−370.9	27.4				
1hvp.A	99	9055	−358.3	−333.1	25.2				
- 1pcy	99	9054	−505.0	−322.1	182.9	5	−526.0	−499.3	177.2

[a] The noncompact structures are marked by asterisks, while those excluded from the list for the RTS potential derivation are marked by "-" sign.

[b] "Ref." is the contact energy of the reference structure; "alt." refers to the lowest contact energy over all the alternatives; and "diff." is the difference between the energy of the reference structure and the lowest energy over all of the alternatives.

[c] "Min." and "max." are the minimal and maximal contact energies over the list of homologues corresponding to a given reference structure (see Table 8-3). "Diff." is the difference between the contact energy of the highest homologous structure and the lowest alternative.

Table 8-6. *Part 2 of 3.* Contact energy and contact energy difference for the 86 reference protein structures and their 95 homologues calculated with RTS potential.

PDB code[a]	No. res	No. alts	ref.	alt.	diff.	No.	min.	max.	diff.[c]
				Contact energy (arbitrary units)[b]		Homologous			
* 1wrp.R	102	8813	−347.3	−352.7	−5.5				
5cyt.R	103	8733	−407.5	−390.4	17.1				
- 4fd1	106	8498	−383.3	−291.7	91.6	2	−419.2	−399.9	108.1
1rei.A	107	8420	−380.8	−362.2	18.6				
2cdv	107	8419	−275.4	−252.4	22.9				
2ssi	107	8418	−375.0	−337.0	38.0				
- 1acx	108	8343	−396.9	−293.2	103.8				
- 5cpv	108	8342	−605.5	−386.8	218.7	2	−609.2	−601.2	214.4
1ccr	111	8125	−356.8	−279.6	77.1				
2c2c	112	8053	−363.2	−325.0	38.2	1	−367.7	−367.7	42.7
1hmq.A	113	7982	−385.8	−358.9	26.9				
2pab.A	114	7912	−463.8	−394.2	69.6				
2rhe	114	7911	−393.1	−389.7	3.4				
* 1cy3	118	7642	−180.8	−364.9	−183.2				
- 1paz	120	7509	−566.0	−348.3	217.7				
155c	121	7443	−340.9	−317.4	23.5				
1pp2.R	122	7378	−389.4	−365.7	23.6				
1bp2	123	7314	−466.1	−399.0	67.1				
1rn3	124	7251	−391.0	−345.7	45.3	3	−415.9	−371.4	25.7
- 2ccy.A	127	7067	−489.0	−396.6	92.5				
- 1lyz	129	6946	−652.0	−557.8	94.2	13	−697.8	−623.7	65.9
- 2aza.A	129	6945	−529.2	−389.3	139.9				
- 1lz1	130	6886	−689.1	−351.0	338.2				
1ecd	136	6543	−568.4	−517.3	51.1	3	−564.5	−548.3	31.0
- 3fxn	138	6430	−817.5	−505.7	311.8	1	−815.8	−815.8	310.1
2hhb.A	141	6264	−572.3	−528.0	44.3				
2sns	141	6263	−426.5	−343.6	82.9				
- 1mba	146	5997	−722.5	−527.2	195.3	3	−732.0	−709.5	182.4
2hhb.B	146	59	−596.3	−524.2	72.1				

[a] The noncompact structures are marked by asterisks, while those excluded from the list for the RTS potential derivation are marked by "-" sign.

[b] "Ref." is the contact energy of the reference structure; "alt." refers to the lowest contact energy over all the alternatives; and "diff." is the difference between the energy of the reference structure and the lowest energy over all of the alternatives.

[c] "Min." and "max." are the minimal and maximal contact energies over the list of homologues corresponding to a given reference structure (see Table 8-3). "Diff." is the difference between the contact energy of the highest homologous structure and the lowest alternative.

Table 8-6. *Part 3 of 3.* Contact energy and contact energy difference for the 86 reference protein structures and their 95 homologues calculated with RTS potential.

PDB code[a]	No. res	No. alts	Contact energy (arbitrary units)[b]						
			ref.	alt.	diff.	No.	min.	max.	diff.[c]
- 1fx1	147	5944	−689.3	−452.2	237.0				
- 21hb	149	5843	−767.2	−512.5	254.7				
- 2sod.O	151	5744	−636.3	−422.2	214.1				
2pka.B	152	5695	−442.1	−410.2	31.9				
1lh4	153	5647	−582.0	−504.8	77.2	13	−611.5	−560.2	55.5
- 1mbd	153	5646	−765.6	−519.3	246.3	6	−775.5	−709.1	189.8
- 4dfr.A	159	5375	−650.6	−456.4	194.2	1	−612.1	−612.1	155.7
- 2lzm	164	5154	−846.0	−484.7	361.2	34	−890.9	−803.1	318.4
9wga.A	170	4895	−479.4	−449.0	30.4	3	−493.4	−459.5	10.5
- 1gcr	174	4726	−766.3	−436.5	329.9				
* 1hmg.B	175	4684	−281.2	−512.3	−231.1				
- 2sga	181	4443	−757.8	−569.8	188.0				
- 2stv	184	4325	−709.8	−592.0	117.9				
- 3adk	194	3944	−674.3	−513.1	161.1				
- 2alp	198	3795	−757.9	−645.1	112.7				
4sbv.A	199	3758	−760.3	−716.2	44.1				
* 3fab.L	207	3477	−618.2	−553.4	64.8				
- 3gap.A	208	3442	−829.1	−697.1	132.0				
- 9pap	212	3309	−799.8	−475.3	324.5	1	−797.4	−797.4	322.1
* 2fb4.L	216	3180	−659.3	−461.6	197.7				
- 2act	218	3117	−881.9	−562.1	319.8				
*_3fab.H	219	3086	−690.0	−644.0	46.0				
- 3rp2.A	224	2940	−1022.2	−762.4	259.8				
* 4rhv.3	236	2603	−870.6	−703.4	167.2				
- 2cna	237	2575	−1040.8	−643.6	397.2	1	−975.7	−975.7	332.1
- 1est	240	2496	−997.8	−643.2	354.6				
- 1tim.A	247	2320	−1064.4	−661.7	402.7				
- 4rhv.2	255	2127	−1026.6	−776.6	250.0				

[a] The noncompact structures are marked by asterisks, while those excluded from the list for the RTS potential derivation are marked by "-" sign.

[b] "Ref." is the contact energy of the reference structure; "alt." refers to the lowest contact energy over all the alternatives; and "diff." is the difference between the energy of the reference structure and the lowest energy over all of the alternatives.

[c] "Min." and "max." are the minimal and maximal contact energies over the list of homologues corresponding to a given reference structure (see Table 8-3). "Diff." is the difference between the contact energy of the highest homologous structure and the lowest alternative.

Table 8-7. Contact potentials that satisfy the 73 proteins and corresponding 95 homologous structures. Potentials are deduced with only 37 proteins—the reduced training set (RTS).

			Separation 3				
	G	AL1CMF	VHS	P	RDEQ	TKN	YW
1	0.61463						
2	−2.31789	2.21881					
3	5.72213	−3.23617	1.12215				
4	9.46103	−5.31923	−1.99999	−0.13026			
5	−4.51548	2.54359	−1.80098	−1.19346	−5.11983		
6	−1.31957	−1.98296	2.74735	0.77238	−4.79110	−6.85097	
7	3.43496	1.62466	−3.27015	−4.73947	−8.83572	−2.34533	−0.32983

			Separation 4				
	G	AL1CMF	VHS	P	RDEQ	TKN	YW
1	−1.24276						
2	−1.69222	4.50557					
3	−2.86872	−6.50263	0.40535				
4	11.70225	5.86310	0.28909	0.41115			
5	3.50425	−0.62710	4.44767	−1.04092	−0.08432		
6	0.86780	−1.27039	4.38732	3.54864	1.18010	−5.74981	
7	6.08202	−1.06606	−1.13969	2.10415	1.72003	−10.68984	−4.21411

			Separation 5–7				
	G	AV	L1CMF	YHWST	KR	P	DNEQ
1	−1.51236						
2	1.33242	0.64862					
3	−0.87751	−0.00247	−6.15579				
4	−4.55054	−1.11002	−3.93129	−1.01535			
5	0.53346	4.52234	−6.54672	6.21924	0.92650		
6	−0.16157	−0.47477	3.18217	0.20033	−4.13299	−0.06414	
7	−6.81042	3.65731	−4.25397	−3.68618	3.54672	1.27479	−0.43524

			Separation $\geq$ 8				
	G	AV	LICMF	YHWST	KR	P	DNEQ
1	−0.46815						
2	−1.70775	−4.03101					
3	1.46479	−6.44646	−9.22260				
4	−1.85291	−1.19226	−3.63516	0.31521			
5	−2.46188	0.58900	−0.98564	0.54491	7.48431		
6	2.69578	−0.46241	0.24442	1.05796	−1.77739	4.85059	
7	−1.75095	3.14822	2.57605	−0.24511	−0.70244	6.14219	2.21398

of larger size have violations. The energy margin between reference and lowest alternative is generally substantial (Table 8-6) except for the reference Bence-Jones protein 2rhe (Furey et al., 1983), which has energy only 3.1 units below an alternative derived from the related FAB-protein 2fb4.L (Marquart et al., 1980) with RMSD = 1.17 Å. This is an unusual situation where 2fb4.L is a close homologue of 2rhe. The next closest alternative has RMSD 10.4 Å and energy 94.0 above the reference (Figure 8-4a). Otherwise Figure 8-4a is typical of the energy distribution for all the compact proteins. Another example of contact energy versus RMSD distribution is that for wheat germ agglutinin 9wga.A (Wright, 1990) shown in Figure 8-4b. All alternatives are well above the native. Also presented in the figure are three homologous structures (Table 8-3) not taken into account during RTS potential derivation. Nevertheless, all of them are also below the lowest alternative, although in one case by only a little.

5.3 Tests of Significance of a Classification

To test the significance of the contact potential classification scheme used (Table 8-5) we attempted to deduce three additional potentials from the same 37 proteins of the RTS. We used the same computational protocol as described above, only with three different classification schemes.

The first test used the best (i.e., fewest adjustable ϵ's) contact classification Table 8-8 found before (Crippen, 1991) which was derived in the course of iterative reassignment of starting classification (Table 8-5), and which favors 45 reference structures over all their alternatives (see Table 8-8). We were unable to locate a reasonable solution for a new training set, RTS, even after 32,000 iterations of optimization of the protocol described above. Perhaps a solution of the quality of our CTS and RTS potentials could be found after much more computing effort, but it seems unlikely.

In the second test, we used the same sequence separation classifications and the same number of residue classes in each as before in the CTS and RTS potentials (Table 8-5), but with random assignment of residues types to classes. The first step succeeded, but the second step failed by exceeding our limit of 15,000 on the number of constraints while "combing" the 32nd protein of the 37 in the training set. Presumably even with a much greater limit on constraints, the calculation would fail to find a solution at great computational expense. Apparently, the classification scheme of Table 8-5 is not only in general agreement with conventional wisdom about residue type similarities, but the particular classification is more important than merely the number of adjustable ϵ's.

The third test was to interchange the residue classifications for the first and fourth separation ranges in Table 8-5, and then attempt to satisfy

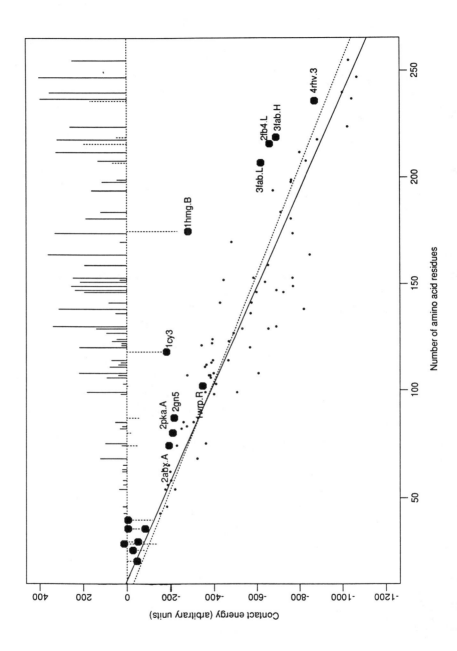

Table 8-8. The classification of amino acid type contributions to a contact potential found before (Crippen, 1991) that favors 45 reference protein structures over all their alternatives.

Separation	Class	ϵ_{ij}			
3	{all}	−0.008			
4	{all}	0.004			
5–7	{all}	0.021			
8–∞	1 = {GYHSRNE}	−0.123	−0.074	−0.054	0.123
	2 = {AV}	−0.074	0.123	−0.317	0.156
	3 = {LICMF}	−0.054	−0.317	−0.263	−0.010
	4 = {PWTKDQ}	0.123	0.156	−0.010	−0.004

all the inequalities. We were surprised that this calculation succeeded, resulting in what we refer to below as the T3 parameters. With these we are able to correctly predict the same proteins as with the RTS parameters. The distribution of the contact energies of reference structures versus the number of residues with the T3 potential is approximately the same, with slightly different linear regression coefficients: the intercept is 25.29, the slope is −4.49, and the correlation coefficient is −0.918, compared to 14.87, −3.97, and −0.920, respectively in Equation (18). The root mean square difference between contact energies of compact reference structures calculated with RTS and T3 is only 67.4 arbitrary units, compared to the 1100 units for the total range of reference energies.

We didn't try to conduct the systematic search for simpler classification schemes carried out in our earlier work (Crippen, 1991) because here we treat many more proteins, vastly more alternative conformations, and finally

---

Figure 8-3. Plot of contact energy versus number of residues of 86 reference structures. Sixty-nine compact and 17 noncompact structures are shown by points and filled circles, respectively, the latter being marked by their PDB code. The PDB markers for the seven smallest structures, 3ins.A, 2mlt.A, 1gcn, 3ins.B, 1ppt, 7api.B, and 4rhv.4, are omitted for clarity. The plot also shows contact energy differences between the lowest-energy alternative and the reference structure for each protein as lines from the zero energy level toward the respective value for the 69 compact (solid lines) and the 17 noncompact (dotted lines) proteins (upper part). Linear regression straight line (solid) for the set of compact reference structures is characterized by slope −4.37, intercept 47.17, and correlation coefficient −0.932. For comparison, the linear regression line (dotted) is calculated with the T3 contact potential (see text) with slope −3.97, intercept 14.87, and correlation coefficient −0.920. *See figure on preceding page.*

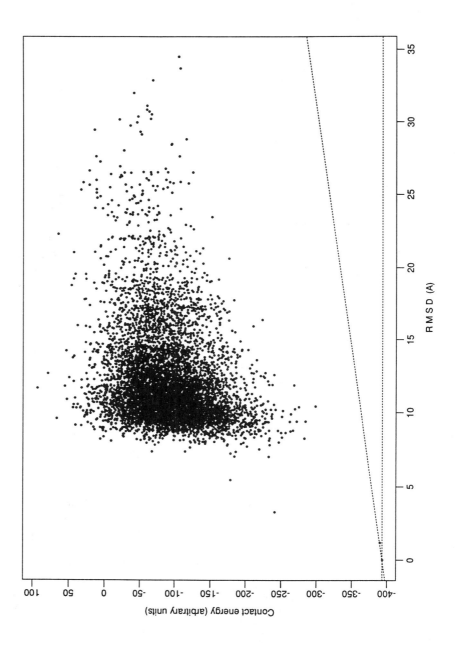

we demand in Equation (13) not merely that the reference have energy less than or equal to each alternative, but that there be a substantial margin. Solving inhomogeneous inequalities is qualitatively different from, and more time consuming than, solving homogeneous ones.

The RTS potentials were also checked against a representative sampling of Figure 8-5 crystal structures with fewer than 256 residues that had been added to the Protein Data Bank after the release of October 15, 1990 that we exclusively used: trypsin inhibitor II 2eti (Chiche et al., 1989), 28 residues, human growth factor 4tgf (Kline et al., 1990), 50 residues, FK506 binding protein 1fkf (Michnick et al., 1991; van Duyne et al., 1991), 107 residues, and T-cell surface glycoprotein 1cd4 (Ryu et al., 1990), 173 residues. These were not included in any way in the determination of the RTS potential. Using the same list of protein structures as before, the alternatives for each of these structures were generated (Figures 8-5a–d). Only the structure of 4tgf growth factor (13,495 alternatives), which is clearly not compact (parameters $e_g = 1.38$ and $e_N = 1.51$ exceed the corresponding threshold values of 1.30 and 1.50), has a number of alternatives with contact energy less than the reference structure (Figure 8-5b). The three others, 2eti (15,766 alternatives), 1fkf (8,421 alternatives), and 1cd4 (4,769 alternatives) (Figures 8-5a, 5c and 5d, respectively) demonstrate obvious satisfaction of Equation (13) for all alternatives generated. While large proteins are relatively easy to fit, partly because of the relatively smaller number of alternatives, the trypsin inhibitor 2eti is a remarkably small structure with quite a large number of alternatives that we can nonetheless successfully predict because it obeys our requirements for compactness and many internal contacts.

6. Discussion

It is interesting to compare our results with that of Sippl's group (Hendlich et al., 1990), the most similar work outside our group that we are aware of. They obtained many different potentials of mean force for C^β–C^β interactions only by surveying a database of 101 separate chains in protein x-ray

---

Figure 8-4a. Plots of the contact energy calculated with RTS potential versus distance RMSD for reference crystal structures: 2rhe, Bence-Jones protein (Furey et al., 1983) (114 residues, 7,911 alternatives). The alternative closest to the reference structure (RMSD =1.17 Å) came from the crystal structure of FAB-protein 2fb4.L (Marquart et al., 1980); cf. Figure 8-4b. *See figure on preceding page.*

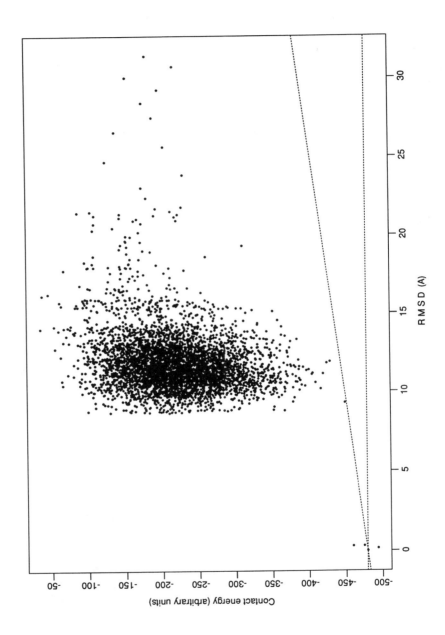

crystal structures (see Table 3 in Hendlich et al., 1990). To make predictions of the folding for one protein, they would remove it from the database and use the remaining 100 to derive the effective energy of interaction as a function of distance between side chains, broken down into 15 sequence-separation classes, and for each of these, all 210 residue-pair type classes. Then they generated a set of alternative conformations in the same way we do, and compared their calculated energies of the native versus all its alternatives. One view of their potentials is that they consist of $210 \times 15 = 3150$ different histograms as a function of C^β–C^β distance, while we have only $4 \times 7 \times (7 + 1)/2 = 112$ ϵ's (Tables 8-5, 8-7). However, we adjusted our ϵ's empirically to satisfy a large number of inequalities, while their histograms are not adjustable parameters. In return, we reach much greater predictive power: A training set of 37 compact proteins invariably favors the native structure of 73 proteins over all alternative conformations. By way of comparison, their table 7 lists 53 proteins that we would consider compact, ranging between 21 and 199 residues in length. Of these, the two corresponding potentials of mean force (denoted in their work as potentials "S" and "A"), derived apparently from 100 crystal structures in each case, could favor the native over the alternatives in both the S and A cases only for 34 compact proteins and 2 noncompact ones.

We believe there are three reasons for the superior predictive power of our contact potential. First, we derive our ϵ's by comparing the native conformation with misfolded alternatives, rather than surveying only native conformations for a potential of mean force. Training the potential must involve examples of both correctly and incorrectly folded proteins. Secondly, we find it crucial to deal only with compact native conformations, as judged both from the radius of gyration and from the relative number of contacts. We cannot account for the crystal structure of an isolated noncompact polypeptide chain when its conformation is stabilized by intermolecular contacts in the crystal, and we suspect this has led to some

---

Figure 8-4b. Plots of the contact energy calculated with RTS potential versus distance RMSD for reference crystal structures: 9wga.A, wheat germ agglutinin (Wright, 1990) (170 residues, 4,895 alternatives). Three homologous structures having fairly small RMSD to reference 9wga.A structure are also shown (Table 8-3), and reveal some scattering of the contact energies. Nonetheless, the lowest contact energy alternative (about 9.5 Å RMSD apart) is higher than that of highest homologous, and was not taken into account during the contact potential derivation. Contact energy level of the native reference structure and the threshold margin of $3 \times$ RMSD set by Equations (11) and (13) are marked by dotted lines. *See figure on preceding page.*

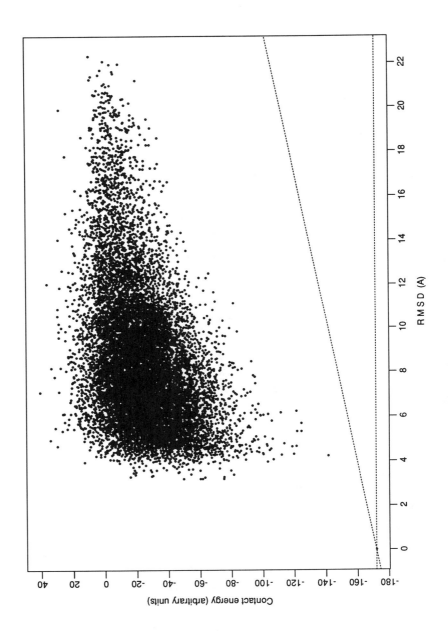

of the difficulties experienced by Sippl's group, since they do not discriminate between compact and noncompact native conformations. Thirdly, they considered only side chain–side chain interactions, but we found that backbone–backbone and especially backbone–side chain interactions are important (see the "G" columns in Table 8-7).

A priori one might assume the classification scheme for residue–residue interactions is very important. The assumption is based on common knowledge about grouping together helix formers versus helix breakers for short-range interactions, while grouping according to hydrophobicity for medium and long-range interactions. Indeed, this is the line of reasoning that led to the classification scheme in Table 8-5 (Crippen, 1991), and subsequently was used to produce the CTS and RTS potentials. However, our classification is certainly not unique, as demonstrated by the success of the test with T3 potential derivation. Although the RTS potential is very powerful in its ability to satisfy a half million inequalities, other equally good potentials could be found, possibly involving fewer parameters and possibly having even better predictive power.

We have been fortunate that some kinds of trouble simply do not arise. For example, although 1hvp.A is not an experimentally determined structure, but was rather postulated by homology modeling (Weber et al., 1989), it nonetheless can be easily accounted for by our contact function, whereas it could not be predicted by Sippl and co-workers (Hendlich et al., 1990). As another example, we utterly disregard any ligands or prosthetic groups in proteins, even large ones covalently attached to the polypeptide chain. Even so, we observe no correlation between the presence or absence of prosthetic groups and the quality of our predictions, just as long as the native is compact according to our criteria in Equation (8).

One might suspect that compact structures always have lower energy. On the contrary, we observed that the range of contact energies for compact and noncompact alternatives calculated with the RTS potential are approximately the same. Both types of alternatives may be found among the very best (which are of course always higher than the reference structure energy) and the very worst. This means in particular that one must consider both types of alternatives when forming the set of inequalities of Equation (13), rather than only compact ones.

Figure 8-5a. Plot of the contact energy calculated with RTS potential versus distance RMSD for crystal structures not used in the work before RTS potential derivation: trypsin inhibitor 2eti (Chiche et al., 1989) (28 residues, 15,766 alternatives). The notations are the same as in Figure 8-4. *See figure on preceding page.*

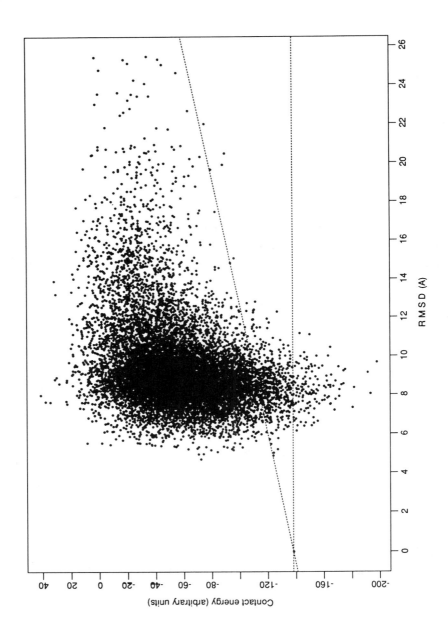

The major improvement of our latest work (Maiorov and Crippen, 1992) over earlier efforts (Crippen, 1991), is the introduction of a substantial margin in Equation (13) that becomes small when the RMSD becomes small, coupled with a continuous contact function that guarantees a small difference in contacts for a small RMSD. For example, the reference structure of cytochrome c 5cyt.R (Takano and Dickerson, 1981) has an alternative derived from rice embryo cytochrome c 1ccr (Ochi et al., 1983) with RMSD = 0.47 Å, 2rhe has an alternative from 2fb4.L with RMSD = 1.17 Å, or hen egg white lysozyme 1lyz (Diamond, 1974) has an alternative from human lysozyme 11z1 (Blake et al., 1983) with RMSD = 1.93 Å. Then quite naturally, all these "homologous" alternatives have energies only slightly above the corresponding reference structure's but still satisfy Equation (13) by a small margin.

It is encouraging that the "novel" folding pattern found in the recently determined NMR and x-ray crystal structures of 1fkf, FK506 binding protein, seems not to be new from the viewpoint of interatomic contact arrangements, given that we can correctly predict it on the basis of the reduced training set of 37 old protein structures. This finding allows one to hope that only minor readjustments of the contact potential will generally be required to keep a high level of predictive power as more and more proteins are considered.

In spite of the hopeful results we have obtained so far, there are two special cases we will treat in the future. First, sequence homologues (see Table 8-3) are correctly treated in the analysis without even being employed in the derivation of the potential, but we have not paid special attention to proteins having very similar conformation but low sequence identity. Instead, such pairs of proteins were used only in a general fashion to generate alternative conformations for each other. Presumably we have to demand that the two different sequences applied to essentially the same conformation should produce very similar contact energies. The second case is that of noncompact native proteins, which we have so far simply excluded

---

Figure 8-5b. Plot of the contact energy calculated with RTS potential versus distance RMSD for crystal structures not used in the work before RTS potential derivation: human growth factor 4tgf (Kline et al., 1990) (50 residues, 13,495 alternatives). Only this 4tgf structure of the four comprising Figure 8-5 was estimated to be noncompact (see text), and therefore a considerable number of alternatives have contact energy less than that of the reference noncompact structure. The other three (8-5a, 8-5c, and 8-5d) perfectly satisfy the constraints of Equation (13), even though they were not used in any way for deriving the contact potential. The notations are the same as in Figure 8-4. *See figure on preceding page.*

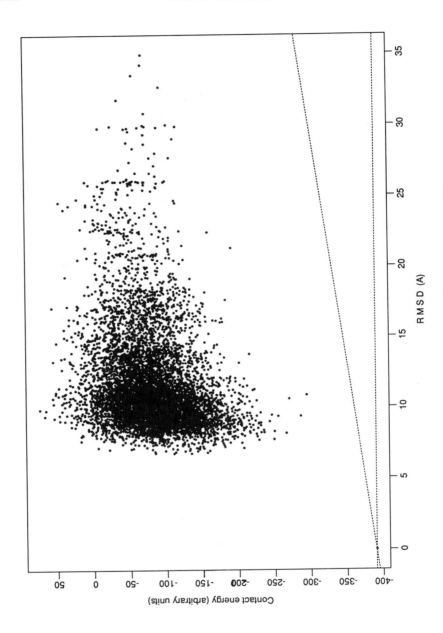

from the derivation of our potential as well as its testing. We find that for such a reference structure there are generally many alternatives having substantially lower contact energies. However, there are apparently very few examples of protein crystal structures where a polypeptide chain fails our compactness test without having significant interactions with neighboring chains. It is likely that such crystal structures have to be considered in the crystal environment of multimeric aggregates of polypeptide chains such that the multimer is compact.

The results presented here suggest that the problem of identifying the correct foldout of a large but discrete set of alternatives is basically solved. Given such a powerful tool for identifying the native fold, the most promising next goal is to implement a method to suggest possible "native" folds for a given amino acid sequence when the correct answer is not known and when it is not just a segment out of an already determined crystal structure. A future use of these contact potentials would be for predicting the conformation of a protein having a known sequence, but no known 3D structure (crystal, or NMR, or modeled by homology, whatever). As long as there is some 3D structure of a larger protein where a contiguous chain segment of the right length approximately adopts the native structure of the novel protein, it is very straightforward to find this segment, and the search would not require much from the computational viewpoint. Note that one needs no assumptions about sequence homology. Any completely unrelated protein of known 3D structure is an eligible source of alternative conformations, and all you have to do is choose the conformation or conformations with lowest potential. Given that there seems to be a limited number of structural motifs (Chothia, 1992), having a correct one in the Protein Data Bank is not unlikely. On the other hand, suppose the novel protein is so fortunate as to have a crystal structure for a closely homologous protein. It is still likely that there will be no completely contiguous portion of the known that matches the native of the novel protein, simply because of the usual insertions and deletions one sees on the surfaces of globular proteins. This means that a general prediction program built around one of these contact potentials would have to set up a correspondence between

Figure 8-5c. Plot of the contact energy calculated with RTS potential versus distance RMSD for crystal structures not used in the work before RTS potential derivation: FKSO6 binding protein 1fkf (van Duyne et al., 1991) (107 residues, 8,421 alternatives); this perfectly satisfies the constraints of Equation (13), even though they were not used in any way for deriving the contact potential. The notations are the same as in Figure 8-4. *See figure on preceding page.*

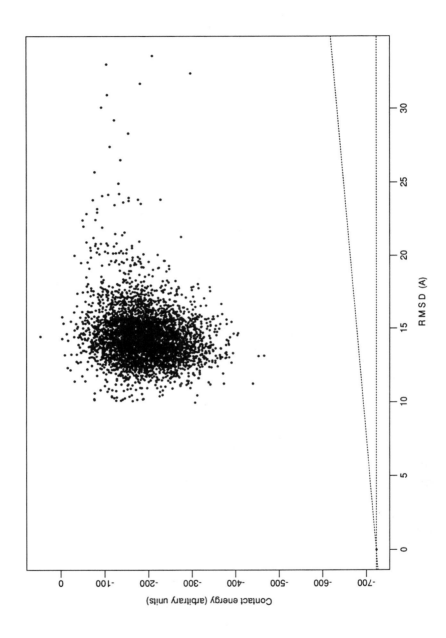

the novel sequence and the known structure that was not necessarily just a displacement of contiguous segments but must also allow for insertions and deletions.

Acknowledgments. This work was supported by grants from the National Institutes of Health (GM37123) and the National Institute of Drug Abuse (DA06746).

REFERENCES

Abola EE, Bernstein FC, Bryant SH, Koetzle TF, Weng J (1987): Protein data bank. In *Crystallographic Databases—Information Content, Software Systems, Scientific Applications* (Allen FN, Bergerhoff G, Sievers R, eds.). pp. 107–132, Data Commission of the International Union of Crystallography, Bonn-Cambridge-Chester

Blake CCF, Pulford WCA, Artymiuk PJ (1983): X-ray studies of water in crystals of lysozome. *J Mol Biol* 167:693–723

Bryant SH, Amzel LM (1987): Correctly folded proteins make twice as many hydrophobic contacts. *Int J Peptide Prot Res* 29:46–52

Chiche L, Gaboriaud C, Heitz A, Mornon JP, Castro C, Kollman PA (1989): Use of restrained molecular dynamics in water to determine three-dimensional protein structure: Prediction of the three-dimensional structure of Ecballium elaterium trypsin inhibitor II. *Proteins* 6:405–417

Chiche L, Grigoret LM, Cohen FE, Kollman PA (1990): Protein model structure evaluation using the solvation free energy of folding. *Proc Natl Acad Sci USA* 87:3240–3243

Chothia C (1992): One thousand protein families for the molecular biologist. *Nature* 357:543–544

Corfield PWR, Lee TJ, Low BW (1989): The crystal structure of erabutoxin *A* at 2.0 Å resolution. *J Biol Chem.* 264:9239–9242

Covell DG, Jernigan RL (1990): Conformations of folded proteins in restricted spaces. *Biochemistry* 29:3287–3294

Crippen GM (1975): Global optimization and polypeptide conformation. *J Comput. Chem* 18:224–231

Figure 8-5d. Plot of the contact energy calculated with RTS potential versus distance RMSD for crystal structures not used in the work before RTS potential derivation: T-cell surface glycoprotein 1cd4 (Ryu et al., 1990) (173 residues, 4,769 alternatives). It perfectly satisfies the constraints of Equation (13), even though they were not used in any way for deriving the contact potential. The notations are the same as in Figure 8-4. *See figure on preceding page.*

Crippen GM (1991): Prediction of protein folding from amino acid sequence over discrete conformation spaces. *Biochemistry* 30:4232–4237

Crippen GM, Snow ME (1990): A 1.8 Å resolution potential function for protein folding. *Biopolymers* 29:1479–1489

Damaschun G, Mueller JJ, Puerschel HV, Sommer G (1969): Berechnung der Form kolloider Teilchen aus Röntgen-Kleinwinkeldiagrammen. *Monatshefte für Chemie* 100:1701–1714

Diamond R (1974): Real-space refinement of the structure of hen egg-white lysozome. *J Mol Biol* 82:371–391

Eisenberg D, McLachlan AD (1986): Solvation energy in protein folding and binding. *Nature* 319:199–203

Furey W, Wang BC, Yoo CS, Sax M (1983): Structure of a novel Bence-Jones protein (Rhe) fragment at 1.6 Å resolution. *J Mol Biol* 167:661–692

Hendlich M, Lackner P, Weitckus S, Floeckner H, Froschauer R, Gottsbacher K, Casari G, Sippl MJ (1990): Identification of native protein folds amongst a large number of incorrect models. *J Mol Biol* 216:167–180

Jones DT, Taylor WR, Thornton JM (1992): A new approach to protein fold recognition. *Nature* 358:86–89

Jurs PC (1986): *Computer Software Applications in Chemistry.* New York: John Wiley, pp. 198–199

Kline TP, Brown FK, Brown SC, Jeffs PW, Kopple KD, Mueller L (1990): Solution structures of human transforming growth factor α derived from ^{1}H NMR data. *Biochemistry* 29:7805–7813

Lau KF, Dill KA (1989): A lattice statistical mechanics model of the conformational and sequence spaces of proteins. *Macromolecules* 22:3986–3997

Levitt M (1976): A simplified representation of protein conformations for rapid simulation of protein folding. *J Mol Biol* 104:59–107

Luethy R, Bowie JU, Eisenberg D (1992): Assessment of protein models with three-dimensional profiles. *Nature* 356:83–85

Maiorov VN, Crippen GM (1992): Contact potential that recognizes the correct folding of globular proteins. *J Mol Biol* 227:876–888

Marquart M, Deisenhofer J, Huber R, Palm W (1980): Crystallographic refinement and atomic models of the intact immunoglobulin molecule *Kol* and its antigen-binding fragment at 3.0 Å and 1.9 Å resolution. *J Mol Biol* 141:369–391

Marquart M, Walter J, Deisenhofer J, Bode W, Huber R (1983): The geometry of the reactive site and of the peptide groups in trypsin, trypsinogen and its complexes with inhibitors. *Acta Crytallogr* B39:480–482

Matsuura Y, Takano T, Dickerson RE (1982): Structure of cytochrome c_{551} from *Pseudomonas. Aeroginosa* refined at 1.6 Å resolution and comparison of the two redox forms. *J Mol Biol* 156:389–409

McLachlan AD (1979): Gene duplications in the structural evolution of chymotrypsin. *J Mol Biol* 128:49–79

Michnick SW, Rosen MK, Wandless TJ, Karplus M, Schreiber SL (1991): Solution structure of *FKBP*, a rotamase enzyme and receptor for FK506 and rapamycin. *Science* 252:836–839

Miyazawa S, Jernigan RL (1985): Estimation of interresidue contact energies from protein crystal structures: Quasi-chemical approximation. *Macromolecules* 18: 534–552

Nishikawa K, Ooi T (1972): Tertiary structure of protein. II. Freedom of dihedral angles and energy calculations. *J Phys Soc Japan* 32:1338–1347

Novotny J, Bruccoleri R, Karplus M (1984): An analysis of incorrectly folded protein models. *J Mol Biol* 177:787–818

Novotny J, Rashin AA, Bruccoleri RE (1988): Criteria that discriminate between native proteins and incorrectly folded models. *Proteins: Struct, Funct, Genet* 4:19–30

Ochi H, Hata Y, Tanaka N, Kakudo M, Sakurai T, Aihara S, Morita Y (1983): Structure of rice ferricytochrome *c* at 2.0 Å resolution. *J Mol Biol* 166:407–418

Pastore A, Atkinson RA, Saudek V, Williams RJP (1991): Topological mirror images in protein structure computation: An underestimated problem. *Proteins* 10:22–32

Ryu SE, Kwong PD, Truneh A, Porter TG, Arthos J, Rosenberg M, Dai X, Xuong NH, Axel R, Sweet RW, Hendrickson WA (1990): Crystal structure of an HIV-binding fragment of human *CD4*. *Nature* 348:419–426

Seetharamulu P, Crippen GM (1991): A potential function for protein folding. *J Math Chem* 6:91–110

Sippl MJ (1990): Calculation of conformational ensembles from potentials of mean force. *J Mol Biol* 213:859–883

Smith JL, Corfield PWR, Hendrickson WA, Low BW (1988): Refinement at 1.4 Å resolution of a model of erabutoxin *B*. Treatment of ordered solvent and discrete disorder. *Acta Crystallogr* A44:357–359

Takano T, Dickerson RE (1981): Conformation change of cytochrome C. I. Ferrocytochrome C structure refined at 1.5 Å resolution. *J Mol Biol* 153:79–94

Tsernoglou D, Petsko GA, Hudson RA (1978): Structure and function of snake venom curarimimetic neurotoxins. *Mol Pharmacol* 14:710–721

van Duyne GD, Standaert RF, Karplus PA, Schreiber SL, Clardy J (1991): Atomic structure of *FKBP-FK506*, an immunophilin-immunosuppressant complex. *Science* 252:839–842

Weber IT, Miller M, Jaskolski M, Leis J, Skalka AM, Wlodawer A (1989): Molecular modeling of the HIV-1 protease and its substrate binding site. *Science* 243:928–931

Wlodawer A, Walter J, Huber R, Sjolin L (1984): Structure of bovine pancreatic trypsin inhibitor. Results of joint neutron and X-ray refinement of crystal form II. *J Mol Biol* 180:301–329

Wright CS (1990): 2.2 Å resolution structure analysis of two refined *N*-acetylneuraminyllactose wheat germ agglutinin isolectin complexes. *J Mol Biol* 215:635–651

9

Neural Networks for Molecular Sequence Classification

Cathy H. Wu

1. Introduction

Nucleic acid and protein sequences contain a wealth of information of interest to molecular biologists, since the genome forms the blueprint of the cell. Currently, a database search for sequence similarities represents the most direct computational approach to decipher the codes connecting molecular sequences with protein structure and function (Doolittle, 1990). If the unknown protein is related to one of known structure/function, inferences based on the known structure/function and the degree of the relationship can provide the most reliable clues to the nature of the unknown protein. This technique has proved successful and has led to new understanding in a wide variety of biological studies (Boswell and Lesk, 1988).

There exist good algorithms and mature software for database searches and sequence analyses (Gribskov and Devereux, 1991). Sequence comparison algorithms based on dynamic programming (Needleman and Wunsch, 1970) have emerged as the most sensitive methods, but have a high computational cost—of order N^2 with respect to sequence length N. The FastA program (Pearson and Lipman, 1988) identifies related proteins rapidly using a lookup table to locate sequence identities (Lipman and Pearson, 1985). The QuickSearch method (Devereux, 1988) provides an even faster but less sensitive search against the database that is represented with a sparse hash table. A BLAST approach (Altschul et al., 1990), which directly approximates alignments that optimize a measure of local similarity, also permits fast sequence comparisons. Recently, a BLAZE search program

The Protein Folding Problem and Tertiary Structure Prediction
K. Merz, Jr. and S. Le Grand, Editors
© Birkhäuser Boston 1994

that involves full dynamic programming algorithm of Smith and Waterman (1981) has been implemented on a massively parallel computer (Brutlag et al., 1992). However, due to the advance of genetic engineering technology and the advent of the human genome project, the molecular sequence data has been accumulating at an accelerating rate. This makes the database search computationally intensive and ever more forbidding, even with the rapid advance of new search tools. It is, therefore, desirable to develop methods whose search time is not constrained by the database size.

A classification method can be used as an alternative approach to the database search problem with several advantages: (1) speed, because the search time grows linearly with the number of sequence classes (families), instead of the number of sequence entries; (2) sensitivity, because the search is based on information of a homologous family, instead of any sequence alone; and (3) automated family assignment. We have developed a new method for sequence classification using back-propagation neural networks (Wu et al., 1991, 1992; Wu, 1993). In addition, two other sequence classification methods have been devised. One uses a multivariant statistical technique (van Heel, 1991), the other uses a binary similarity comparison followed by an unsupervised learning procedure (Harris et al., 1992). All three classification methods are very fast, thus applicable to the large sequence databases. The major difference between these approaches is that the classification neural network is based on "supervised" learning, whereas the other two are "unsupervised." The supervised learning can be performed using a training set compiled from any existing second generation database and can be used to classify new sequences into the database according to the predefined organization scheme of the database. The unsupervised system, on the other hand, defines its own family clusters and can be used to generate new, second-generation databases.

The neural network technique has its origins in efforts to produce a computer model of the information processing that takes place in the nervous system (Rumelhart and McClelland, 1986). One can simply view a neural network as a massively parallel computational device, composed of a large number of simple processing units (neurons). The neurons communicate through a rich set of interconnections with variable strengths (weights), in which the learned information is stored. Artificial neural networks with back-propagation currently represent the most popular learning paradigm, and have been successfully used to perform a variety of input-output mapping tasks for recognition, generalization, and classification (Dayhoff, 1990). In fact, neural networks can approximate linear and nonlinear discriminant analysis with much stronger capability of class separation (Gallinari et al., 1988; Asoh and Otsu, 1990; Webb and Lowe, 1990). As a technique for computational analysis, neural network technology has been applied to many studies involving the sequence data analysis (please see Hirst and Sternberg, 1992, for a recent review). Back-propagation networks have been used to predict protein secondary structure (Qian and Sejnowski, 1988; Holley and Karplus, 1989; Kneller et al., 1990) and tertiary structure (Bohr et al., 1990; Chen, 1993; Liebman, 1993; Xin et al., 1993), to distinguish ribosomal binding sites from nonbinding sites (Stormo et al., 1982) and encoding regions from noncoding sequences (Lapedes et al., 1990; Uberbacher and Mural, 1991), and to predict bacterial promoter sequences (Demeler and Zhou, 1991; Horton and Kanehisa, 1992; O'Neill, 1992).

This chapter reviews the design and implementation of the neural network sequence classification systems, analyzes factors affecting system performance, and describes system applications.

2. System Design

The neural network system was designed to embed class information from molecular databases and used as an associative memory to classify unknown sequences (Figure 9-1). There are three major design issues: (1) the input/output mapping, (2) the sequence encoding schema, and (3) the neural network architecture.

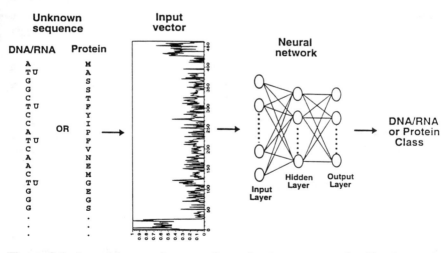

Figure 9-1. A neural network system for molecular sequence classification. The molecular sequences are first converted by a sequence-encoding schema into neural net input vectors. The neural network then classifies them into predefined classes according to sequence information embedded in the neural interconnections after network training.

The neural system is designed to classify new (unknown) sequences into predefined (known) classes. In other words, it would map molecular sequences (input) into sequence classes (output). Two neural systems have been developed using "second generation" molecular databases as the training sets. These databases are termed second generation (Pabo, 1987) because they are organized according to biological principles, or more specifically, within these databases the sequence entries are grouped into classes based on sequence similarities or other properties. The first neural system, the Protein Classification Artificial Neural System (ProCANS), is trained with the annotated PIR database (Barker et al., 1992) and classifies protein sequences into PIR superfamilies. The second system is the Nucleic Acid Classification Artificial Neural System (NACANS) that classifies DNA/RNA sequences. NACANS has two separate neural networks. The RDP network is trained with the Ribosomal RNA Database Project (RDP) database (Olsen et al., 1992), and maps ribosomal RNA sequences into phylogenetic classes (Woese, 1987). The FES network is developed with the intron/exon sequences of the FES database (Konopka and Owens, 1990), and discriminates intron/exon sequences.

The sequence-encoding schema is used to convert molecular sequences (character strings) into input vectors (numbers) of the neural network classifier (Figure 9-1). There are two different approaches for the sequence encoding. One can either use the sequence data directly, as in most neural network applications of molecular sequence analysis, or use the sequence data indirectly, as did Uberbacher and Mural (1991). Where sequence data is encoded directly, most studies (e.g., Qian and Sejnowski, 1988; Lapedes et al., 1990) use an indicator vector to represent each molecular residue in the sequence string. That is, a vector of 20 input units (among which 19 have a value of 0, and one has a value of 1) to represent an amino acid, and a vector of four units (three values are 0 and one is 1) for a nucleotide. This representation preserves the sequentiality of the residues along the sequence string. It is not, however, suitable for sequence classifications where long and varied-length sequences are to be compared.

The key element of the sequence encoding schema presented here is a hashing function, called the n-gram extraction method, that was originally used by Cherkassky and Vassilas (1989) for associative database retrieval. The n-gram method extracts and counts the occurrences of patterns of n consecutive residues (i.e., a sliding window of size n) from a sequence string. The counts of the n-gram patterns are then converted into real-valued input vectors for the neural network, with each unit of the vector representing an n-gram pattern (Figure 9-2). The value of each neuron is the product of the pattern count and weight, scaled to fall between 0 and 1. The count is the number of times the n-gram pattern appears in the sequence string. The weight of each amino acid is the measure of its frequency of occurrence in nature. The size of the input vector for each n-gram extraction is m^n, where m is the size of the alphabet.

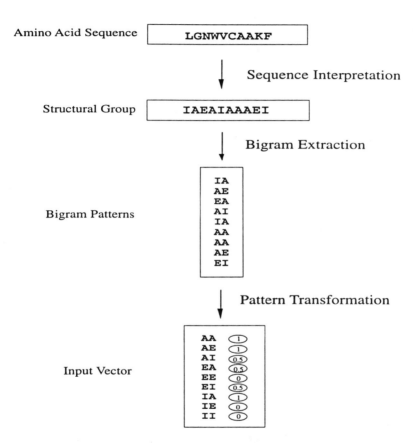

Figure 9-2. The sequence-encoding schema of the neural classification system. The encoding involves three steps: (1) sequence interpretation, during which each sequence string is converted into strings of different alphabet sets; (2) n-gram extraction, when all different n-gram patterns are extracted from the sequence; and (3) pattern transformation, when the occurrence of each n-gram pattern is counted and converted into a real-valued input vector of the neural network.

The concept of the n-gram method is similar to that of the k-tuple method. The latter has been successfully applied to sequence analyses, such as the comparison of k-tuple locations for database search (Lipman and Pearson, 1985) and the statistical analysis of k-tuple patterns/frequencies for sequence discrimination and analysis (Karlin et al., 1989; Claverie et al., 1990). The n-gram input representation satisfies the basic coding assumption so that encodings of similar sequences are represented by "close" vectors and vectors representing sequences of different classes are very different (Figure 9-3). The n-gram method has several advantages: (1) it maps sequences of different lengths into input vectors of the same length; (2) it provides certain representation invariance with respect to residue insertion and deletion; and (3) it is independent from the *a priori* recognition of certain specific patterns.

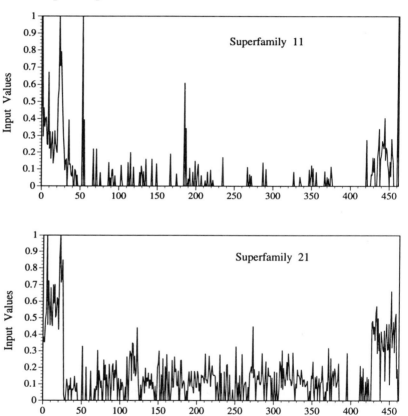

Figure 9-3. The input vectors derived from the n-gram encoding method. The value plotted for each of the 462 input units (input vector positions) is the median value of all training patterns for superfamilies 11 and 21, respectively.

The original sequence string can be attached with various biological meanings and represented by different alphabet sets in the encoding. The alphabet sets used for protein sequences include amino acids (20 letters), exchange group (6 letters), structural group (3 letters), and hydrophobicity group (2 letters). The six exchange groups are derived from the PAM (accepted point mutation) matrix, which represents the conservative replacements found in homologous sequences through evolution (Dayhoff, 1972). The two hydrophobicity groups, hydrophobic and hydrophilic, are allocated according to the hydrophobicity scale of the amino acid. The three structural groups, ambivalent, external, and internal, are assigned based on the structural feature of the amino acid (Karlin et al., 1989). The alphabet sets can be easily expanded later to attach more biological meaning to a sequence, including the representation of secondary and supersecondary structural objects. This structural representation can be derived from the protein structure database, Brookhaven Protein Data Bank. The alphabets for nucleic acid sequences include the four-letter AT(U)GC, the two-letter RY for purine and pyrimidine, and the two-letter SW for strong and weak hydrogen binding.

The major drawback of the n-gram method is that the size of the input vector tends to be large. This indicates that the size of the weight matrix (i.e., the number of neural interconnections) would also be large because the weight matrix size equals n, where $n = $ input size $\times$ hidden size $+$ hidden size $\times$ output size. This prohibits the use of even larger n-gram sizes, e.g., the trigrams of amino acids would require 20^3, or 8000, input units. Furthermore, accepted statistical techniques and current trends in neural networks favor minimal architecture (with fewer neurons and interconnections) for its better generalization capability (Le Cun et al., 1989). To address this problem, we have attempted different approaches to reduce the size of n-gram vectors.

Recently, we have developed an alternative sequence encoding method by adopting the Latent Semantic Indexing (LSI) analysis (Deerwester et al., 1990) used in the field of information retrieval and information filtering. The LSI approach is to take advantage of implicit, high-order structure in the association of terms with documents in order to improve the detection of relevant documents on the bases of terms used. The particular technique used is the Singular-Value Decomposition method (SVD), in which a large "term-by-document" matrix is decomposed into a set of factors from which the original matrix can be approximated by linear combination (Figure 9-4).

In the present study, the term-by-document matrix is replaced by the "term-by-sequence" matrix, where "terms" are the n-grams. For example, an 8000×894 matrix can be used to represent the term vectors of 894 protein sequences, with each term vector containing the 8000 a3 n-grams (i.e., trigrams of amino acids). The large sparse term-by-sequence matrix would

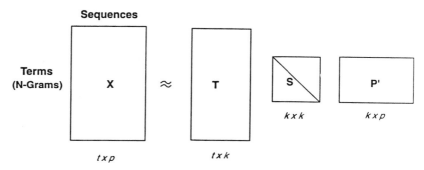

Figure 9-4. The singular-value decomposition (SVD) of a "term-by-sequence" matrix. The original term-by-sequence matrix (X) is approximated using the k largest singular values and their corresponding singular vectors. S is the diagonal matrix of singular values. T and P, both having orthogonal, unit-length columns, are the matrices of the left and right singular vectors, respectively; t and p are the numbers of rows and columns of X, respectively; m is the rank of X ($m \leq \min(t, d)$), whereas k is the chosen number of dimensions in the reduced model ($k \leq m$).

be decomposed into singular triplets, i.e., the singular (s) values, and the left and right singular vectors (Figure 9-4). With k being the chosen number of dimensions in the reduced model, the right s-vectors corresponding to the k largest s-values are then used as the input vectors for neural networks. In this example, if the right s-vectors corresponding to the 100 largest s-values are used, then the size of the input vector would be reduced from 8000 to 100. Figure 9-5 plots the right s-vectors corresponding to the 20 largest s-values computed from the a3 term-by-sequence matrix. While the s-vectors of sequences within the same family are similar, the s-vectors of different superfamilies (i.e., superfamilies 1 vs. 21) are very different. Therefore, as with the n-gram sequence encoding method, the SVD method also satisfies the basic coding assumption.

3. Neural Network Architecture

The neural networks used in this research are three-layered, feed-forward networks (Figure 9-1) that employ a back-propagation learning algorithm (Rumelhart and McClelland, 1986). (Please see Wu et al., 1992, for a more detailed description of the neural network model.) In the three-layered architecture, the input layer is used to represent sequence data, the hidden layer to capture information in nonlinear parameters, and the output layer to represent sequence classes. The networks are feed-forward because all the neurons (processing units) are fully interconnected to all neurons in the

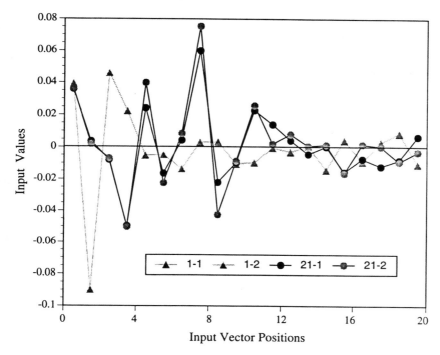

Figure 9-5. The input vectors derived from the SVD encoding method for a 8000 × 894 "term-by-sequence" matrix. The right singular vectors corresponding to the 20 largest singular values are plotted; 1-1, 1-2, 21-1, and 21-2 represents the first and second sequence entries of superfamily 1, and the first and second sequence entries of superfamily 21, respectively.

next layer. A feed-forward calculation is used to determine the output of each neuron by applying a nonlinear activation (squashing) function to its net input. The back-propagation learning algorithm applies the generalized delta rule to recursively calculate the error signals and adjust the weights of the neural interconnections. The error function that back-propagation minimizes is a sum square error function, which is a function of weights; and back-propagation is really just a gradient descent applied to this function. The weights are iteratively modified for every training pattern being read in until the system is converged to the tolerance (error threshold) or until a fixed upper limit on the epochs is reached. The error threshold is a user-defined value of root mean square error, which is the square root of the sum square error calculated from all patterns across the entire training file.

It has been known that back-propagation networks do not scale up very well from small systems to large ones. A collection of small networks, however, has been demonstrated to be an effective alternative to large back-propagation networks (Kimoto et al., 1990; Lendaris and Harb, 1990). A modular network architecture that involves multiple independent network modules is used to embed the large database. In ProCANS, each separate module, termed database module, is used to train one or a few protein functional groups of the PIR database (Figure 9-6).

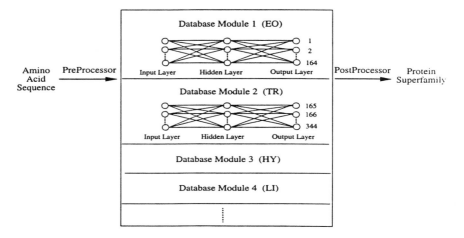

Figure 9-6. The neural network architecture of ProCANS. A database module is used to train one or a few protein functional groups containing many superfamilies. Each module is a three-layered, feed-forward, back-propagation neural network. During the training phase, each module is trained separately. During the prediction phase, the unknown sequences are classified on all modules with classification scores combined.

Presently, ProCANS has four modules, namely EO, TR, HY, and LI. They are developed with seven-protein functional groups, consisting of the first 690 superfamilies and 2724 entries of the annotated PIR database (PIR1, Release 32.0, March 31, 1992). The EO module is trained with the electron transfer proteins and the oxidoreductases (superfamilies 1–164), TR module with the transferases (superfamilies 165–344), HY module with the hydrolases (superfamilies 345–536), and the LI module with lyases, iso-merases, and ligases (superfamilies 537–690). During the training phase, each database module is trained separately using the sequences of known superfamilies (i.e., training patterns). During the prediction phase, the un-known sequences (prediction patterns) are classified on all modules with classification scores combined.

The size of the input layer (i.e., number of input units) is dictated by the sequence encoding schema chosen. In the n-gram encoding method, the size is m^n, where m is the size of the alphabet. In the SVD encoding method, the size is the number of dimensions (k) chosen in the reduced model. The output layer size is determined by the number of classes represented in the network, with each output unit representing one sequence class. The hidden size is determined heuristically, usually a number between input and output sizes. The networks are trained using weight matrices initialized with random weights ranging from -0.3 to 0.3. Other network parameters included the learning factor of 0.3, momentum term of 0.2, a constant bias term of -1.0, error threshold of 0.01, and training epochs (iterations) of 800.

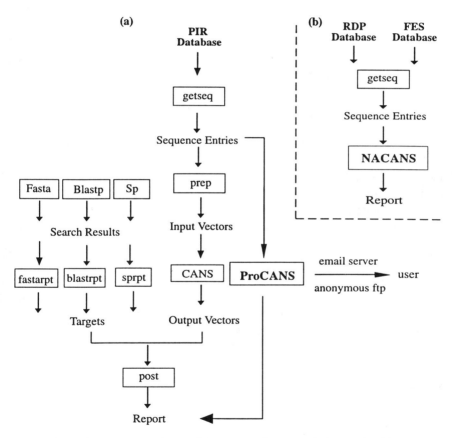

Figure 9-7. The structure charts of the (a) ProCANS and (b) NACANS systems. Both systems have three components, preprocessor (prep), neural network classifier (CANS), and postprocessor (post). The target files of unclassified sequences are generated by other database search tools, and are used to evaluate neural network results.

4. System Implementation

The system software has three components: a preprocessor program (prep) to create from input sequence files the training and prediction patterns, a neural network program (CANS) to classify input patterns, and a postprocessor program (post) to summarize classification results (Figure 9-7). The programs have been implemented on the Cray Y-MP8/864 supercomputer of the University of Texas System Center for High Performance Computing.

The preprocessor has two modules, one for the n-gram extraction, the other for the SVD computation. In the n-gram encoding, these vectors are directly used as neural network input vectors. In the SVD encoding, the n-gram vectors are further reduced into right singular vectors using a SVD program. The program, which was developed by Michael Berry, employs a single-vector Lanczos method (Berry, 1992). The right s-vectors are then processed before input into the neural network such that the component values are scaled between 0 and 1.

The neural network program is a set of programs and subroutines intended to be a tool for developing neural network applications. An interface program serves as an automated, neural network generator using user-defined parameters, including network configurations (number of neurons for each layer), network parameters (learning factor, momentum term, bias term, error threshold, and training epochs), and weight matrix options (initialization with random weights, reading from a weight file, or prediction with trained weights). The back-propagation program module contains fully vectorized FORTRAN codes for fast execution.

During the postprocessing of ProCANS, the superfamily numbers of unclassified sequence entries are identified by three database search programs running on the Cray, namely, fasta (Pearson and Lipman, 1988), blastp (Altschul et al., 1990) and sp (Davison, 1990). They are used as target values for evaluating the classification accuracy of the neural system (Figure 9-7).

5. Data Sets

5.1 ProCANS

Six data sets (Table 9-1) compiled from PIR1 (containing annotated and classified entries) and PIR2 (containing unclassified entries) are used to evaluate the system performance. The first two data sets divide the 2724 PIR1 entries into disjoint training and prediction sets. The prediction patterns for data sets 1 and 2 are compiled using every second and third entry

Table 9-1. Data sets for ProCANS training and prediction.

Data Set	Network Config. (Input × Hidden × Output)	Total # of Interconnections	Training		Prediction	
			Database	# Patterns	Database	# Patterns
1	462 × 200 × 164,180,192,154	507,600	PIR1	1604	PIR1	1120
2	462 × 200 × 164,180,192,154	507,600	PIR1	2073	PIR1	651
3	462 × 200 × 164,180,192,154	507,600	PIR1	2724	PIR2	482
4	462 × 20 × 18	9,600	PIR1	588	PIR1	284
5	462 × 20 × 18	9,600	PIR1	872	PIR2	31
6	462 × 10 × 10	4,720	PIR1	1438	PIR2	301

from superfamilies with more than one or two entries, respectively. Data set three uses all 2724 PIR1 entries for training, and 482 unclassified PIR2 entries for prediction. The first three data sets are trained and predicted on the four database modules (Table 9-1 and Figure 9-6). The fourth and fifth data sets are subsets of data sets 2 and 3, respectively, with only superfamilies of more than 20 entries. Since among the 690 superfamilies only 18 contain more than 20 sequences, a smaller network that has 18 output units and 20 hidden units is used to train and predict these data sets. The last data set, compiled from the 10 largest superfamilies among all 2,601 superfamilies, is trained and predicted on a neural network with 10 hidden units and 10 output units. Each of the 10 superfamilies contains more than 60 entries.

5.2 NACANS

The data set used for the RDP network contains 473 entries in 28 phylogenetic classes of the RDP database (Table 9-2). Among the 473 16S ribosomal RNA sequences, 316 are used for training, and the remaining 157 sequences for prediction. The FES network is developed with 3515 exon/intron sequences of the FES database. The sequences are compiled from different organisms and are trained on four separate network modules, one each for mammalian, invertebrate, plant, and viral sequences (Table 9-2). About two-thirds of the sequences are used for training and the remaining one third for prediction.

Table 9-2. Data sets for NACANS training and prediction.

Network	Data Set	Number of Patterns		
		Total	Training	Prediction
RDP	16S rRNA	473	316	157
FES	Mammal	647	432	215
	Invertebrate	996	664	332
	Plant	979	652	327
	Virus	893	595	298

6. System Performance

The predictive accuracy is expressed with three terms: the total number of correct patterns (true positives), the total number of incorrect patterns (false positives), and the total number of unidentified patterns (false negatives). The sensitivity is the percentage of total correct patterns, the specificity is 1 minus the percentage of total incorrect patterns. A sequence entry is considered to be accurately classified if its classification matches the target value (the known class number of the entry) with a classification score above the threshold (i.e., the cutoff value). The classification score ranges from 1.0 for perfect match to 0.0 for no match. In ProCANS, the predictive accuracy is measured at two stringencies. The high stringency selects the best fit (the superfamily with highest score) with a threshold 0.9. The low stringency condition for the first three data sets (with 154 to 192 superfamilies) is the best five fits (the superfamilies with five highest scores) with a threshold of 0.01. The low stringency condition for the second three data sets (with 10 or 18 superfamilies) is the best fit with a threshold of 0.01. In NACANS, the predictive accuracy is measured with best fit only at a threshold of 0.01.

6.1 ProCANS Performance

Twenty-five encoding methods were tested for ProCANS, among which ae12 encoding was the best (Wu et al., 1991, 1992). The result presented below is based on the ae12 encoding, whose input vector is concatenated from vectors representing four separate n-grams, namely, a1 (monograms of amino acids), e1 (monograms of exchange groups), a2 (bigrams of amino acids), and e2 (bigrams of exchange groups). The vector has 462 units, which is the sum of the four vector sizes (i.e., $20 + 6 + 400 + 36$).

The prediction results are summarized in Table 9-3. After 800 iterations, close to 100% of all training patterns are trained, at which time the

Table 9-3. The predictive accuracy of ProCANS.

Data Set	Patterns Trained (%)	Accuracy at Low Stringency (%)			Accuracy at High String. (%)		
		Correct	Incorrect	Unidentified	Corr.	Incorr.	Unident.
1	98.88	89.64	10.36	0.00	63.66	0.45	35.89
2	98.89	91.73	6.62	1.65	68.20	0.00	31.80
3	99.56	81.57	17.32	1.11	47.13	0.41	52.46
4	100.00	100.00	0.00	0.00	91.20	0.00	8.88
5	100.00	*93.55	6.45	0.00	70.97	0.00	29.03
6	100.00	*98.99	1.01	0.00	93.58	0.00	6.42

* The results are obtained by counting only first-fits as correct patterns. The sensitivity (% correct patterns) is 100% for data sets 5 and 6 if first three fits are counted.

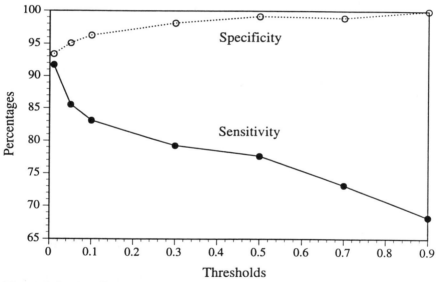

Figure 9-8. The effects of threshold on the sensitivity and specificity of ProCANS classification. The threshold is the cutoff value of the classification scores, ranging between 0.0 and 1.0. The sensitivity is the percentage of total correct patterns, the specificity is 1 minus the percentage of total incorrect patterns. The plot is based on data set 2 in Table 9-1.

weight matrices are used to classify the prediction patterns. At the low stringency, the sensitivity is between 81.6 to 100%. At the high stringency, on the other hand, a close to 100% specificity is achieved (i.e., very few false identifications), although the sensitivity is reduced. Figure 9-8 illustrates the effect of threshold on the sensitivity and specificity using data

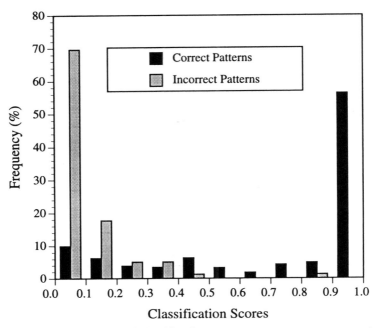

Figure 9-9. Distribution of classification scores among patterns correctly and incorrectly classified by ProCANS. The plot is based on data set 3 in Table 9-1.

set 2 as an example. As would be expected, at lower threshold values, more superfamilies are identified, which results in higher sensitivity (more true positives), but a lower specificity (more false positives). The classification scores for data set 3 are plotted to observe the distribution of the scores among the correctly identified patterns and the incorrectly identified patterns (Figure 9-9). The plot shows clearly that close to 60% of the correct patterns have a score of at least 0.9, whereas about 70% of the incorrect patterns have a score of 0.1 or less.

A detailed analysis of the misclassified sequence patterns reveals three important factors affecting classification accuracy: the size of the superfamily, the sequence length, and the degree of similarity. It is observed that the superfamily size is inversely correlated with the misclassification rate (Figure 9-10). There are two possible explanations. First, a large superfamily has more training patterns, and thus is more likely to be matched by the query sequence (it has been noted that the neural network can classify an unknown sequence accurately if it has a high degree of similarity to any training patterns). Second, the network can generalize better from a large number of training patterns. The latter feature of the network is explored by using data set 6 (see discussion for Figure 9-12, next paragraph). The

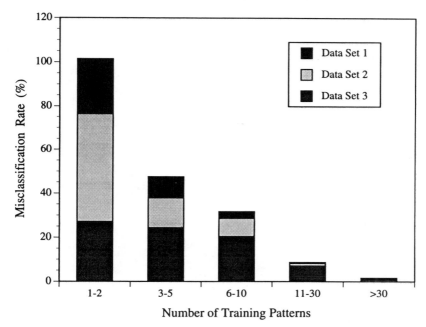

Figure 9-10. The effect of superfamily size on the classification accuracy. The results are obtained from the first three data sets in Table 9-1. The misclassification rate is the percentage of incorrect patterns per total number of prediction patterns for each superfamily size group. The superfamily size is expressed with the number of training patterns.

sequence length also affects the predictive accuracy. Generally, a sequence can be correctly classified if its length is at least 20% of the original length, although some sequences as short as 10% are classified (Figure 9-11). That the third prediction set has a lower accuracy is mainly due to the large number of sequences belonging to single-membered or double-membered superfamilies, and sequences of small fragments. Indeed, the sensitivity approaches 100% in data sets 4, 5, and 6, which contain only large super-families.

The degree of similarity is the most important factor affecting the sensitivity of any database search method. Sequence alignment often fails when similarity drops to less than 25 to 30% sequence identity. Figure 9-12 shows that the neural network classification may fail for single-membered superfamilies when similarity is less than 40%. However, with large super-families in data set 6, the classification score is still very high for similarities below 40%, indicating the network has gained some generalization capability from the large number of training patterns.

More unclassified sequences from PIR2 will be prepared to look into

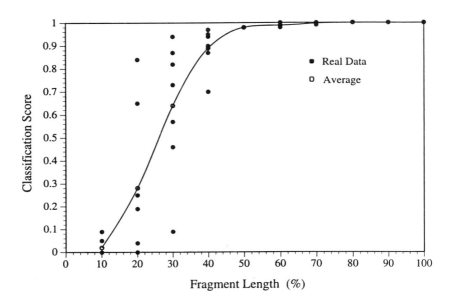

Figure 9-11. The effect of fragment length on the classification accuracy. The original sequences of seven single-membered superfamilies are truncated from the N-terminus to lengths ranging from 90 to 10% of the original length. The classification scores range from 1.0 for perfect match to 0.0 for no match. The plot is curve-fitted through the average values.

the "twilight zone" of less than 30% sequence identity. The sensitivity of ProCANS is expected to increase with the continuing accumulation of sequence entries available for training. Moreover, the accuracy is evaluated using the result of other search methods as the standard (i.e., it is considered to *mis*classify if it does not match the result of others). In a real situation, however, ProCANS may outperform others in some cases. For example, sequence entry A24459 was classified by ProCANS as cytochrome c (with a high score), but by blastp as complement C2. The alignment showed high similarities between the first halves of A24459 and several cytochrome c proteins, which may reflect a distant evolutionary relationship (personal communication with Dr. Barker of PIR).

Preliminary studies have been conducted for the SVD encoding method and compared with the ae12 n-gram encoding (Table 9-4). The data set used has 894 PIR1 protein sequences classified into 164 superfamilies. Among the sequences, 659 are used for training, and the remaining 235 for prediction. The SVD result shown is the SVD computation for ae123 n-gram vectors, which are concatenated from the a1, a2 and a3 vectors (for the monograms, bigrams and trigrams of amino acids) and e1, e2 and e3

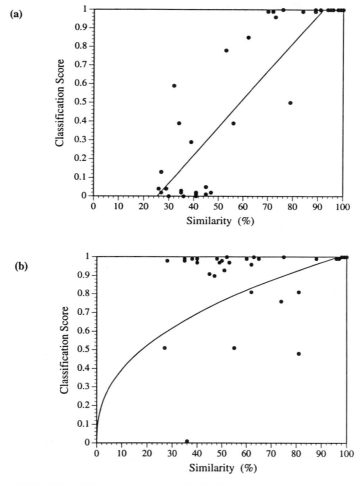

Figure 9-12. The effect of sequence similarity on classification accuracy with (a) single-membered superfamilies, and (b) superfamilies of more than 60 members. The similarity is expressed with the percentage of sequence identity for the sequence that is best aligned with the prediction sequence. The results for (a) and (b) are obtained from data sets 3 and 6 in Table 9-1, respectively.

vectors (the monograms, bigrams and trigrams of exchange groups). The ae123 n-gram extraction of the 894 proteins generates an 8678×894 term-by-sequence matrix. After the SVD computation, a reduced model of 50 dimensions is used to reduce the size of input vectors from 8678 to 50. The result shows that the sizes of the input vectors and the weight matrix can be reduced by SVD without reducing the predictive accuracy. When the input vectors from the n-gram method and the SVD method are combined,

Table 9-4. Comparisons of the *n*-gram and SVD sequence encoding methods for ProCANS.

Encoding method	Network configuration	Number of connections	Trained patterns (%)	Predictive accuracy (%)
N-gram (ae12)*	$462 \times 200 \times 164$	125,200	96.05	94.89
SVD (ae123)	$50 \times 100 \times 164$	21,400	97.57	94.47
N-gram + SVD	$512 \times 100 \times 164$	67,600	97.88	97.02

* The code used in parentheses represents the type of *n*-gram vectors (please see text).

the classification accuracy is improved to 97.02% (note: the 94.89% has been a performance ceiling for this data set using *n*-gram encoding) (Table 9-4). Conceivably, the improvement results from additional sequence information embedded in the a3 *n*-grams. It would be difficult to input the a3 *n*-gram vector directly to the neural network without a reduction—it would be too large (with 8000 units), and the vector would be too sparse (too many zeros) for the neural network to be trained effectively.

The CPU time for both training and classification is directly proportional to the number of neural interconnections and the number of patterns. The training time is also directly related to the number of iterations. The time for training the large database modules ranges from five to seven Cray CPU hours (using a single processor). It takes a total of 47 seconds to classify 482 prediction patterns on all four modules, which averages to approximately 0.1 CPU second per pattern. The classification time on a full-scale system embedded with all known superfamilies is estimated to be about four times that of the present system (less than 0.5 CPU seconds), because the full system is about four times larger. The classification time compares favorably with other database search methods. Presently, it is about an order of magnitude faster than other database search methods. However, with the continuing rapid accumulation of database entries, the saving in time will become increasingly significant because the search time of ProCANS is proportional to the number of superfamilies, whereas that of other methods is proportional to the number of entries. The neural network can be updated for new PIR releases simply by training with new entries, which would require only a small fraction of the time needed for a complete retraining of the entire network.

6.2 NACANS Performance

Among several tested, the best *n*-gram encoding method for the RDP net-

Table 9-5. The predictive accuracy of NACANS.

Data sets	Encoding methods	Network configuration	Number of connections	Trained patterns (%)	Predictive accuracy (%)
RDP	N-gram (drs5)*	1088×80×28	89,280	100.00	96.18
RDP	SVD (d7)	80×50×28	5,400	100.00	100.00
FES-Mammal	N-gram (d5rs6)	1150×50×2	57,600	100.00	95.81
FES-Invertebrate				100.00	90.96
FES-Plant				100.00	90.83
FES-Virus				100.00	95.30
FES-Mammal	N-gram + SVD	114×20×2	2,320	100.00	92.56
FES-Invertebrate	(d3) (d6)			100.00	93.37
FES-Plant				100.00	92.02
FES-Virus				99.16	92.59

* The code used in parenthesis represents the type of n-gram vectors (please see text).

work is drs5, which concatenates d5, r5 and s5, pentagram patterns of AUGC, RY (purine, pyrimidine), and SW (strong, weak hydrogen-binding) alphabets. A predictive accuracy of 96.2% is achieved counting only first fit at a cutoff classification score of 0.01 (Table 9-5). The same data set is also processed by the SVD method as a comparison. The n-gram extraction used is d7 (heptagrams of AUGC alphabet), which generates a term-by-sequence matrix of size 16384 (4^7 n-grams) × 473 (RNA sequences). A reduced dimension of 80 is chosen after SVD computation. The result shows that with the SVD encoding, classification accuracy can be improved (up to 100%) even with a much smaller network architecture (Table 9-5). Significantly, the information embedded in d7 n-grams alone is sufficient for 100% classification. This information would be very difficult to capture without the SVD reduction, however, due to its size (16,384 units).

The best encoding method for the FES network is d5rs6, which is a concatenation of d5 (pentagrams of ATGC alphabet), r6 (hexagrams of RY alphabet) and s6 (hexagrams of SW alphabet). The predictive accuracies range between 90.8 and 95.8% for the four data sets, counting only first fit at the threshold of 0.01 (Table 9-5). Similar accuracies, ranging between 92.02 and 93.37%, are obtained by using a much smaller network that employs a combined n-gram and SVD encoding. In this encoding, input vectors are generated by concatenating the d3 (trigrams of ATGC alphabet) n-gram vectors (with 4^3 or 64 units) with the SVD vectors (with 50 units) reduced from the d6 (hexagrams of ATGC) n-gram vectors.

7. System Applications

The major applications of the classification neural networks are rapid sequence annotation and automated family assignment. ProCANS is an alternative database search method. When doing database searches, the researcher may wish to get a quick list of closely related proteins, or he may wish to study distant and subtle relationships. ProCANS can be used as a filter program for other database search methods to minimize the time required to find relatively close relationships. As with other search methods, the major task for superfamily identification is to distinguish true positives from false positives. With the present system, a close to 100% specificity can be achieved at a high threshold of 0.9, with a more than 50% sensitivity. Therefore, one can use ProCANS to screen a large number of unknown protein sequences and give true identifications to more than half of the query sequences quickly. ProCANS can then be run at a lower threshold and classify another 30 to 40% unknown sequences into a reduced search space. The saving in search time will become increasingly significant due to the accelerating growth of the sequence databases. A second version of ProCANS, ProCANS II, is being developed using the Blocks database (Henikoff and Henikoff, 1991), which lists all sequences of the motif region(s) of a protein family. Since the Blocks database is compiled based on the protein groups of the ProSite database (Bairoch, 1992), the system would map protein sequences into ProSite groups. This neural system is aiming at sensitive protein classification and detection of distant relationships by applying motif information, because the presence of motifs often indicate similar function or structure even in the absence of overall sequence similarity.

The neural classification system can also be used to automate family assignment. An automated classification tool is especially important for the organization of database according to family relationships and for handling the influx of new data in a timely manner. Among all the entries in the PIR database, only approximately a quarter of them are classified and placed in PIR1 (9,633 entries in release 32). ProCANS will be used for superfamily identification of the sequences in PIR2 (14,207 preliminary entries) and PIR3 (16,458 unverified entries). The tool is generally applicable to any second-generation databases that are developed according to family relationships, because the neural network employs a "supervised" learning algorithm. The results of the two neural systems indicate that the sequence encoding methods apply equally well to both the protein and nucleic acid sequences, although the former has a 20-letter alphabet and the latter has a 4-letter alphabet. Preliminary studies have been conducted to classify DNA sequences (containing both protein-encoding regions and intervening sequences) directly into PIR superfamilies with satisfactory results. It is, therefore, possible to develop a gene classification system that can classify indiscriminately sequenced DNA fragments.

The neural networks used in this research deviate from accepted statistical techniques and from current trends in neural networks, which favor minimal architectures (Barton, 1991; Le Cun et al., 1989). The number of weights trained in the ProCANS networks exceeds by one to two orders of magnitude the number of training samples. The smallest network used in this study, which has 4,720 weights (interconnections) and 1,438 training patterns, appears to have gained a better generalization capability compared to other networks used (see the discussion for Figure 9-12). Indeed, the sensitivity approaches 100% in data sets that contain only large superfamilies (i.e., superfamilies with a large number of training patterns) (Table 9-3). A partial protein classification system can be developed using only large superfamilies and becomes immediately useful. (The superfamilies that have five or more sequence entries account for about 20% of all PIR superfamilies.) While the sensitivity of the neural systems is expected to increase with the continuing accumulation of sequence entries available for training, minimal architecture will be adopted to improve network generalization. The SVD computation, which has reduced the sizes of input vector and weight matrix significantly, is one approach toward this goal.

The neural system has two products: a "neural database" that consists of a set of weight matrices that embed family information in the neural interconnections after iterative training, and a system software that utilizes the neural database for rapid sequence classification. The neural database and the system software have been ported to other computer platforms, including an Intel iPSC hypercube and a PC, for speedy on-line protein classification. The system will be distributed to the research community using an anonymous ftp and an electronic mail server.

Acknowledgments. This work is supported by the University Research and Development Grant Program of Cray Research, Inc. The author also wishes to acknowledge the computer system support of the University of Texas System Center for High Performance Computing.

REFERENCES

Altschul SF, Gish W, Miller W, Myers EW, Lipman DJ (1990): Basic local alignment search tool. *J Mol Biol* 215:403–410

Asoh H, Otsu N (1990): An approximation of nonlinear discriminant analysis by multilayer neural networks. *Proc Intn'l Joint Conf on Neural Networks* (June) III:211–216

Bairoch A (1992): PROSITE: A dictionary of sites and patterns in proteins. *Nucl Acids Res* (Suppl) 20: 2013–2018

Barker WC, George DG, Mewes H-W, Tsugita A (1992): The PIR-international protein sequence database. *Nuc Acids Res* (Suppl) 20: 2023–2026

Barton (1991): A matrix method for optimizing a neural network. *Neural Computation* 3:450–459

Berry MW (1992): Large-scale sparse singular value computations. *Int J Supercomputer Applications* 6:13–49

Bohr H, Bohr J, Brunak S, Cotterill RMJ, Fredholm H, Lautrup B, Peterson SB (1990): A novel approach to prediction of the 3-dimensional structures of protein backbones by neural networks. *FEBS Letters* 261:43–46

Boswell DR, Lesk AM (1988): Sequence comparison and alignment: The measurement and interpretation of sequence similarity. In: *Computational Molecular Biology: Sources and Methods for Sequence Analysis*, pp. 161–178. Lesk AM, ed. New York: Oxford University Press

Brutlag DL, Dautricourt J-P, Fier RDJ, Moxon B, Stamm R (1992): BLAZE™: An implementation of the Smith–Waterman sequence comparison algorithm on a massively parallel computer. *Extended Abstracts of the 2nd International Workshop of Open Problems in Computational Molecular Biology:* 60–68

Chen S (1993): Characterization and learning of protein conformations. *Proc. 2nd International Conference on Bioinformatics, Supercomputing and Complex Genome Analysis* :391–399

Cherkassky V, Vassilas N (1989): Performance of back propagation networks for associative database retrieval. *Proc Int Joint Conf on Neural Networks* I:77–83

Claverie J-M, Sauvaget I, Bougueleret L (1990): K-tuple frequency analysis: From intron/exon discrimination to T-cell epitope mapping. In: *Molecular Evolution: Computer Analysis of Proteins and Nucleic Acid Sequences, Methods in Enzymology,* Vol. 183. Doolittle RF, ed. New York: Academic Press. pp. 237–252

Davison DB (1990): Sequence searching on supercomputers. In: *Computers and DNA, Santa Fe Institute Studies in the Sciences of Complexity*. Bell G, Marr TG, eds. Reading, MA: Addison-Wesley, pp. 93–97

Dayhoff MO (ed.) (1972): *Atlas of Protein Sequence and Structure,* Volume 5. National Biomedical Research Foundation, Washington, D.C.

Dayhoff J (1990): *Neural Network Architectures, An Introduction.* New York: Van Nostrand Reinhold

Deerwester S, Dumais ST, Furnas, Landaur TK, Harshman R (1990): Indexing by latent semantic analysis. *J Amer Soc for Information Science* 41:391–407

Demeler B, Zhou G (1991): Neural network optimization for E. coli promoter prediction. *Nuc Acids Res* 19:1593–1599

Devereux J (1988): A rapid method for identifying sequences in large nucleotide sequence databases. Ph.D. Thesis, University of Wisconsin

Doolittle RF (1990): Searching through sequence databases. In: *Molecular Evolution: Computer Analysis of Proteins and Nucleic Acid Sequences, Methods in Enzymology, Vol. 183*. Doolittle RF, ed. New York: Academic Press. pp. 99–110

Gallinari P, Thiria S and Soulie FF (1988): Multilayer perceptrons and data analysis. *Proc Intn'l Joint Conf on Neural Networks* I:391–399

Gribskov M, Devereux J (eds.) (1991): *Sequence Analysis Primer.* New York: Stockton Press

Harris N, Hunter L, States D (1992): Megaclassification: Discovering motifs in massive datastreams. *Proceedings of 10th National Conference on Artificial Intelligence*, AAAI Press

Henikoff S, Henikoff JG (1991): Automated assembly of protein blocks for database searching. *Nuc Acid Res* 19:6565–6572

Hirst JD, Sternberg MJE (1992): Prediction of structural and functional features of protein and nucleic acid sequences by artificial neural networks. *Biochemistry* 31:7211–7218

Holley LH, Karplus M (1989): Protein secondary structure prediction with a neural network. *Proc Natl Acad Sci USA* 86:152–156

Horton PB, Kanehisa M (1992): An assessment of neural network and statistical approaches for prediction of E. coli promoter sites. *Nuc Acids Res* 20:4331–4338

Karlin S, Ost F, Blaisdell BE (1989): Patterns in DNA and amino acid sequences and their statistical significance. In: *Mathematical Methods for DNA Sequences*, Waterman MS, ed. Boca Raton, FL: CRC Press, Inc. pp. 133–157

Kimoto T, Asakawa K, Yoda M, Takeoka M (1990): Stock market prediction system with modular neural networks. *Proc Int Joint Conf on Neural Networks* (June) I:1–6

Kneller DG, Cohen FE, Langridge R (1990): Improvements in protein secondary structure prediction by an enhanced neural network. *J Mol Biol* 214:171–182

Konopka AK, Owens J (1990): Non-continuous patterns and compositional complexity of nucleic acid sequences. In: *Computers and DNA, SFI Studies in the Sciences of Complexity*, Vol. VII. Bell G, Marr T, eds. Addison-Wesley. pp. 147–155

Lapedes A, Barnes C, Burks C, Farber R, Sirotkin K (1990): Application of neural networks and other machine learning algorithms to DNA sequence analysis. In: *Computers and DNA, SFI Studies in the Sciences of Complexity*, Vol. VII. Bell G, Marr T, eds. Addison-Wesley. pp. 157–182

Le Cun Y, Boser B, Denker JS, Henderson D, Howard RE, Hubbard W, Jeckel LD (1989): Backpropagation applied to handwritten zip code recognition. *Neural Computation* 1:541–551

Lendaris GG, Harb IA (1990): Improved generalization in ANNs via use of conceptual graphs: A character recognition task as an example case. *Proc Intn'l Joint Conf on Neural Networks* (June) I:551–556

Liebman MN (1993): Application of neural networks to the analysis of structure and function in biologically active macromolecules. *Proc 2nd International Conference on Bioinformatics, Supercomputing and Complex Genome Analysis* :331–347

Lipman DJ, Pearson WR (1985): Rapid and sensitive protein similarity searches. *Science* 277:1435–1441

Needleman SB, Wunsch CD (1970): A general method applicable to the search for similarities in the amino acid sequences of two proteins. *J Mol Biol* 48:443–453

Olsen GJ, Overbeek R, Larsen N, Marsh TL, McCaughey MJ, Maciukenas MA, Kuan W-M, Macke, TJ, Xing Y, Woese CR (1992): The ribosomal RNA database project. *Nuc Acids Res* (Suppl) 20:2199–2200

O'Neill MC (1992): Escherichia coli promoters: Neural networks develop distinct descriptions in learning to search for promoters of different spacing classes. *Nuc Acids Res* 20:3471–3477

Pabo CO (1987): New generation databases for molecular biology. *Nature* 327: 467

Pearson WR, Lipman DJ (1988): Improved tools for biological sequence comparisons. *Proc Nat Acad Sci* 85:2444–2448

Qian N, Sejnowski TJ (1988): Predicting the secondary structure of globular proteins using neural network models. *J Mol Biol* 202:865–884

Rumelhart DE, McClelland JL (eds.) (1986): *Parallel Distributed Processing: Explorations in the Microstructure of Cognition. Volume 1: Foundations.* MIT Press.

Smith TF, Waterman M (1981): Identification of common molecular subsequences. *J Mol Biol* 147:195–197

Stormo GD, Schneider TD, Gold L, Ehrenfeucht A (1982): Use of the 'Perceptron' algorithm to distinguish translation initiation sites in *E. coli. Nuc Acids Res* 10:2997–3011

Uberbacher EC, Mural RJ (1991): Locating protein-coding regions in human DNA sequences by a multiple sensor-neural network approach. *Proc Natl Acad Sci USA* 88:11261–11265

van Heel M (1991): A new family of powerful multivariant statistical sequence analysis techniques. *J Mol Biol* 220:877–887

Webb AR, Lowe D (1990): The optimized internal representation of multilayered classifier networks performs nonlinear discriminant analysis. *Neural Networks* 3:367–375

Woese CR (1987): Bacterial evolution. *Microbiological Reviews* 51:221–271

Wu CH, Ermongkonchai A, Chang TC (1991): Protein classification using a neural network protein database (NNPDB) system. *Proceedings of the Analysis of Neural Network Applications Conference* :29–41

Wu CH, Whitson G, McLarty J, Ermongkonchai A, Chang T (1992): Protein classification artificial neural system. *Protein Science* 1:667–677

Wu CH (1993): Classification neural networks for rapid sequence annotation and automated database organization. *Computers & Chemistry* 17:219–227

Xin Y, Carmeli T, Liebman M, Wilcox GL (1993): Use of the backpropagation neural network algorithm for prediction of protein folding patterns. *Proc. 2nd International Conference on Bioinformatics, Supercomputing and Complex Genome Analysis* :359–375

10

The "Dead-End Elimination" Theorem: A New Approach to the Side-Chain Packing Problem

Johan Desmet, Marc De Maeyer, and Ignace Lasters

The prediction of a protein's tertiary structure is still computationally infeasible, mainly because of the huge number of *a priori* possible global conformations. The prediction of side-chain conformations that are attached to a given, fixed main-chain structure is considered as a less complex but still important subproblem with applications to, for instance, homology modeling (Lee and Subbiah, 1991). Nevertheless, the determination of the global minimum energy conformation (GMEC) of a set of protein side chains was up to now still limited to small, densely packed units (Ponder and Richards, 1987; Lee and Levitt, 1991).

Recently, we have developed an algorithm that allows the identification of the most probable conformation (the GMEC) of a *large* set of protein side chains (Desmet et al., 1992). The program makes use of the side-chain rotamer concept (Ponder and Richards, 1987) and is based on an original mathematical theorem, which imposes a suitable condition to identify rotamers that cannot be members of the global minimum energy conformation. Iterative elimination of such rotamers reduces the combinatorial modeling problem to computationally feasible dimensions.

The algorithm has been successfully applied to the repositioning of the side chains of two lysozymes and to the mutation of these proteins into each other (homology modeling). Also, the modeling of the insulin dimer and bovine pancreatic trypsin inhibitor (BPTI) is discussed. The results have been compared with the known crystallographic structures. The correctness

The Protein Folding Problem and Tertiary Structure Prediction
K. Merz, Jr. and S. Le Grand, Editors
© Birkhäuser Boston 1994

of the predicted side-chain orientations has been correlated with the various types of amino acid residues as well as with the solvent accessibility of each side chain in the x-ray structure. In addition, reliable conclusions could be drawn about the influence of an inexact main-chain structure, a limited rotamer library and an approximate energy function on the results of a particular modeling experiment.

1. Introduction

In one of the first issues of *Protein Engineering* Dill (1987) presents an analysis of the way the protein folding problem has been set about so far. He distinguishes essentially two basic strategies. The first method is called the "holistic" approach, which studies the protein and its properties as a whole. "The focus is on the general, nonspecific, global averaged properties of principally two states: the native and unfolded." The *global* stability of a protein, for example, has been dissected in terms of the hydrogen bond, the influence of the ionic strength, the hydrophobic effect, etc. (Dill, 1987). From a historical point of view this method preceded the second way of tackling the folding problem, the "surgical" approach. The latter predominantly concentrates on *local* properties of a protein molecule such as specific interactions, ligand binding, and catalysis. These properties are manipulated by surgical removal or transplantation using protein engineering techniques, both *in vitro* and *in computro*.

In our opinion neither of these approaches alone will ever be capable of solving the folding problem. It is true of course that each of these methods has been helpful in trying to identify pieces of the folding puzzle. However, with respect to protein folding, the holistic method focuses too little on the specific, local, and determining factors. The surgical approach usually and perhaps necessarily leaves out the properties of the protein in the unfolded state or denies the importance of entropy for a protein's structure and function. Moreover, the latter thermodynamic parameter is a typical holistic characteristic of the molecule as a whole and it cannot be accounted as a sum of local contributions.

Regarding the model-building of proteins and their properties, one is thrown on crystallographically determined structures, the graphical representation of which is of fascinating clarity but deceptive rigidity. Mesmerized by this, supported by the success of engineering techniques, and equipped with ever more powerful computers, the model builder has focused on local interactions. Proteins were mutated *in computro*. Inhibitors were developed on a graphics screen. The surgical method has reigned supreme.

But the net outcome of all this has not (yet) redeemed the expectations, as was anticipated by Blow (1983).

The reasons behind the relatively modest results of theoretical structure and function modeling with respect to practice are probably three-fold. The main problem is the *complexity* of a protein chain that is composed of hundreds of amino acid residues and that can, in principle, adopt myriad global structures (Levinthal, 1968). A second problem, and perhaps a consequence of the former, is the supersimplified energy function that is used to calculate internal interactions. This is a semiempirical, parameter-based *potential* energy function, not *free* energy function. The third problem is typically a surgical one: interactions are modeled *locally*. For instance, mutations are often considered acceptable on the mere condition that hydrogen bonds are saturated and van der Waals close contacts are avoided. The basic tool for this type of work is energy minimization. But the latter is bound to generate conformations that are trapped in local minimum wells. Even if conformational space is searched in a more systematic way, it is by no means guaranteed that the global energy minimum is reached.

With respect to the modeling of side-chain conformations, we have recently succeeded in finding a method that *does* obtain the global minimum potential energy conformation, given some restrictions that will be discussed in this work (Desmet et al., 1992). The method makes use of the well-known side-chain rotamer concept (Janin et al., 1978; James and Sielecki, 1983; McGregor et al., 1987; Ponder and Richards, 1987) and is based on an original and simple mathematical theorem that allows rapid identification of side-chain conformations that cannot be members of the global minimum energy structure. Such rotamers are qualified as "dead-ending" since *any* further combination with other rotamers will inevitably lead to a global conformation that is different from the minimum energy structure. The theorem is implemented in the form of an algorithm, the core of which is an elimination criterion: the *Dead-End Elimination (DEE) criterion.*

In a typical experiment on a set of residues to be modeled, all of the latter's appearances are successively submitted to the DEE criterion. It has been found that the elimination of incompatible rotamers renders the elimination criterion more restrictive in a succeeding cycle, resulting in additional eliminations. In practice, it has turned out that an iterative application of the DEE criterion leads to a *quasi-exponential* decrease of the *logarithm* of conformational space as a function of the iteration cycles. In general, for a set of 100–200 residue side chains to be positioned, the conformational space often settles at a number that is sufficiently small to analyze by means of "brute force" combinatorial analysis.

Hitherto, this method is the only one that is capable of finding the absolute, global minimum energy conformation for a relatively extended set of protein side chains. Alternative methods, such as Monte Carlo simulations (Holm and Sander, 1991; Lee and Levitt, 1991), "genetic algorithms" (Tuffery et al., 1991), or "rule-based" algorithms (Summers and Karplus, 1989) succeed perfectly well in predicting "viable" structures, although they cannot determine the lowest-energy structure with mathematical certainty. But what about the necessity of being able to identify the absolute minimum-energy structure? First there is the *fundamental* importance. It is generally accepted that the absolute minimum coincides with the most probable conformation. This has to be considered as the best conformation one ever *can* predict, given the applied potential energy function. Second, the *practical* use immediately follows from the previous. Of course, reliable conclusions can only be drawn from duplicate structures. But there is more. In proportion as the "best" structure deviates more from the "true" (say, crystallographic) structure, one is forced to conclude that the applied potential energy function (or the rotamer library; see below) satisfies less. Thus, the DEE algorithm provides a feedback tool for the integrity of the energy model. Also, it should be possible to derive more accurate energy parameters by using this method.

The third aspect is both of fundamental and practical nature. Proteins *de facto* behave "holistically." Their global properties exceed the sum of local contributions. One would therefore be tempted to conclude that a correct global structure cannot be achieved by local modeling. However, this is not nearly proven. Although the fundamental question about the limits to locality has not yet come to a definite answer, there are at least some particular cases in which locality *can* be delimited. (1) Proteins can often be subdivided into mutually independent domains with no or little cooperativity between them (Privalov, 1989). (2) It has been found that proteins are often remarkably tolerant with respect to point mutations (Schellman et al., 1981; Bowie et al., 1990; Reidhaar-Olson and Sauer, 1990). (3) (De)stabilization by multiple substitutions is often additive, but sometimes it is not (Shortle et al., 1988). It is likely that the DEE method may be helpful to pinpoint the edges of locality. It cannot become trapped into local energy minima and *therefore it automatically takes into account progressive "domino effects."* It must therefore be possible to study induced conformational changes that propagate throughout the structure.

It is true, of course, that for the time being the main chain is kept fixed and this itself drastically restricts the available conformational space. However, in the near future some important improvements may be expected with respect to main-chain adaptations. Also, we have studied the effect of "inexact" main-chain structure on the reliability of side-chain predictions. In this work we describe the repositioning of the side chains of the insulin dimer, bovine pancreatic trypsin inhibitor (BPTI) and hen and human lysozyme on the original main-chain template. Also, both lysozymes were transformed into each other in a homology modeling experiment (thus using an imperfect backbone). The modeling results have been compared with the crystallographically determined structures. We believe that these experiments provide sufficient data to draw reliable conclusions about the usefulness of the method's prerequisites, i.e. the backbone quality, the rotamer library, and the potential energy function.

2. The Dead-End Elimination (DEE) Theorem

2.1 Definitions

2.1.1 The rotamer library.   Statistical analysis of crystallographically determined protein structures has shown that the side-chain dihedral angles are not uniformly distributed (Janin et al., 1978; Bhat et al., 1979; Benedetti et al., 1983; James and Sielecki, 1983; Ponder and Richards, 1987; Holm and Sander, 1991; Tuffery et al., 1991). In general, these torsion angles are clustered into gauche + (+60°), gauche − (−60°), and trans (180°) populations. (The notations follow the recommendations of the IUPAC Commission on Macromolecular Nomenclature [1979] and conform to those of Benedetti et al. [1983] and Ponder and Richards [1987].) The aromatic residues Phe and Tyr show a marked preference for a χ_2-angle near 90°. The asymmetric aromatics His and Trp can adopt the +90° or −90° conformation. The orientation of the amide plain of Asn and Gln is more vaguely determined. Yet, a slight preference is observed for a 0° or 180° conformation.

We have opted for the rotamer library of Ponder and Richards (1987), which is based on the analysis of 2273 amino acid residues from 19 well-resolved and refined structures. Not all physically possible rotamers are present in the original library. We have therefore created the absent rotamers ourselves by generating the standard gauche and trans combinations with the present partial rotamers, followed by a test on self-overlap. As a result, our rotamer library contains 212 elements, distributed over 17 amino acid types with rotatable side chains. The rotamer library is shown in Table 10-1 in an encoded form.

Our algorithm allows to extend the library by taking a user-defined number of steps around the standard angles. The size of the steps is usually

Table 10-1. *The rotamer library.* For each of the 17 amino acids having rotatable side chains, the possible rotamers are indicated in an encoded form. A rotamer is fully defined by four dihedral angles (not taking into account a fifth dihedral for lysine), which are listed in four consecutive columns, corresponding to χ_1-χ_2-χ_3-χ_4. Codes: −, gauche − (−60°); +, gauche + (+60°); t, trans; 0, ≈ 0°; P, ≈ 90°; p, ≈ −90°; *, ≈ 60° (for methyl groups); ".", no dihedral. The full (noncoded) rotamer library contains, apart from the dihedral angles expressed in degrees, also the standard deviation on each angle (Ponder and Richards, 1987), which determines the size of the steps that are eventually taken around the standard angles.

Column 1

Residue	χ_1	χ_2	χ_3	χ_4
Val	−	*	*	.
	+	*	*	.
	t	*	*	.
Leu	−	t	*	*
	+	+	*	*
	t	+	*	*
	t	t	*	*
Ile	−	−	*	*
	−	t	*	*
	+	t	*	*
	t	+	*	*
	t	t	*	*
Ser	−	−	.	.
	−	+	.	.
	−	t	.	.
	+	−	.	.
	+	+	.	.
	+	t	.	.
	t	−	.	.
	t	+	.	.
	t	t	.	.
Thr	−	−	.	.
	−	+	.	.
	−	t	.	.
	+	−	.	.
	+	+	.	.
	+	t	.	.
	t	−	.	.
	t	+	.	.
	t	t	.	.
Csh	−	−	.	.
	−	+	.	.
	−	t	.	.
	+	−	.	.
	+	+	.	.
	+	t	.	.
	t	−	.	.
	t	+	.	.
	t	t	.	.
Phe	−	P	.	.
	−	0	.	.
	+	P	.	.
	t	P	.	.

Column 2

Residue	χ_1	χ_2	χ_3	χ_4
Tyr	−	P	.	.
	−	0	.	.
	+	P	.	.
	t	P	.	.
Trp	−	P	.	.
	−	p	.	.
	+	P	.	.
	+	p	.	.
	t	P	.	.
	t	p	.	.
His	−	P	.	.
	−	p	.	.
	+	P	.	.
	+	p	.	.
	t	P	.	.
	t	p	.	.
Asp	−	0	.	.
	+	0	.	.
	t	0	.	.
Asn	−	0	.	.
	−	t	.	.
	+	0	.	.
	+	t	.	.
	t	0	.	.
	t	t	.	.
Glu	−	−	0	.
	−	+	0	.
	−	t	0	.
	+	−	0	.
	+	t	0	.
	t	t	0	.
Gln	−	−	0	.
	−	−	t	.
	−	t	0	.
	−	t	t	.
	+	t	0	.
	+	t	t	.
	t	+	0	.
	t	+	t	.
	t	t	0	.
	t	t	t	.

Column 3

Residue	χ_1	χ_2	χ_3	χ_4
Met	−	−	−	*
	−	−	t	*
	−	+	+	*
	−	+	t	*
	−	t	−	*
	−	t	+	*
	−	t	t	*
	+	−	−	*
	+	−	t	*
	+	+	+	*
	+	+	t	*
	+	t	−	*
	+	t	+	*
	+	t	t	*
	t	−	−	*
	t	−	t	*
	t	+	+	*
	t	+	t	*
	t	t	−	*
	t	t	+	*
	t	t	t	*
Lys	−	−	−	−
	−	−	−	t
	−	−	t	−
	−	−	t	+
	−	−	t	t
	−	+	+	+
	−	+	+	t
	−	+	t	−
	−	+	t	+
	−	+	t	t
	−	t	−	−
	−	t	−	t
	−	t	+	+
	−	t	+	t
	−	t	t	−
	−	t	t	+
	−	t	t	t
	+	−	−	−
	+	−	−	t
	+	−	t	−
	+	−	t	+
	+	−	t	t
	+	+	+	+

Column 4

Residue	χ_1	χ_2	χ_3	χ_4
	+	+	+	t
	+	+	t	−
	+	+	t	+
	+	+	t	t
	+	t	−	−
	+	t	−	t
	+	t	+	+
	+	t	+	t
	+	t	t	−
	+	t	t	+
	+	t	t	t
	t	−	−	−
	t	−	−	t
	t	−	t	−
	t	−	t	+
	t	−	t	t
	t	+	+	+
	t	+	t	−
	t	+	t	+
	t	+	t	t
	t	t	−	−
	t	t	−	t
	t	t	+	+
	t	t	+	t
	t	t	t	−
	t	t	t	+
	t	t	t	t
	−	−	−	−
	−	−	−	t
	−	−	t	−
	−	−	t	+
	−	−	t	t
	−	+	+	+
	−	+	+	t
	−	+	t	−
	−	+	t	+
Arg	−	+	t	t
	−	t	−	−
	−	t	−	t
	−	t	+	+
	−	t	+	t
	−	t	t	−
	−	t	t	t
	+	−	−	−
	+	−	−	t
	+	−	t	−
	+	−	t	t
	+	+	+	+

Column 5

χ_1	χ_2	χ_3	χ_4
−	t	t	t
+	−	−	−
+	−	−	t
+	−	t	−
+	−	t	+
+	−	t	t
+	+	−	−
+	+	−	t
+	+	+	+
+	+	+	t
+	+	t	−
+	+	t	t
+	t	−	−
+	t	−	t
+	t	+	+
+	t	+	t
+	t	t	−
+	t	t	+
+	t	t	t
t	−	−	t
t	−	t	−
t	−	t	+
t	−	t	t
t	+	−	−
t	+	−	t
t	+	+	+
t	+	+	t
t	+	t	−
t	+	t	+
t	+	t	t
t	t	−	−
t	t	−	t
t	t	+	+
t	t	+	t
t	t	t	−
t	t	t	+
t	t	t	t

Legend:

+/−	≈ +/−60°
t	≈ 180°
0	≈ 0°
P	≈ 90°
p	≈ -90°

a user-set fraction of the standard deviation known for each individual dihedral angle. This is a valuable tool, especially for bulky, aromatic side chains that easily make bad contacts even in their correct orientations. It will be obvious, however, that increasing the number of rotamers imposes a severe load on the performance of the algorithm.

2.1.2 The potential energy function. The function that is used for the calculation of the potential energy of (part of) a protein in a given conformation is the following (Wodak et al., 1986):

$$
\begin{aligned}
E = \sum_{\text{bonds}} k_b (b - b_0)^2 + \sum_{\text{angles}} k_\theta (\theta - \theta_0)^2 \\
+ \sum_{\text{torsions}} k_\varphi \big(1 + \cos(n\varphi - \delta)\big) + \sum_{\substack{\text{nonbonded} \\ \text{pairs}}} \left(\frac{A}{r_{ij}^{12}} - \frac{B}{r_{ij}^6} \right) \\
+ \sum_{\text{H-bonds}} \left(\frac{A'}{r_{ij}^{12}} - \frac{B'}{r_{ij}^{10}} \right) + \sum_{\text{charges}} \frac{q_i q_j}{r_{ij}^2}.
\end{aligned}
\tag{1}
$$

The function comprises the usual terms for bond stretching, bond-angle bending, a periodic function for torsion angles, a Lennard–Jones potential for nonbonded atom pairs, a 10–12 potential for hydrogen bonds, and a coulombic function for charged atoms. Note that the dielectric constant is set equal to r_{ij}, the distance between the atoms i and j (Warshel and Levitt, 1976). This energy model is used by the BRUGEL modeling package (Delhaise et al., 1984), in which the present algorithm has been developed. The energy parameters used by BRUGEL are derived from the CHARMM library (Brooks et al., 1983). All modeling experiments described in this work were carried out in the presence of explicit hydrogens.

2.1.3 The template. The Dead-End Elimination algorithm searches for the global energy minimum of a user-defined set of rotatable side chains. The search occurs in the presence of a collection of atoms that are kept fixed, which is called the *template*. The latter is by default the sum of (1) main-chain atoms, (2) C_β-atoms, (3) possible ligands (water molecules, metal ions, substrates, haem groups, etc.), (4) interacting proteins (e.g., monomers not modeled from a multimer), (5) side-chain atoms of residues that one does not want to include in the modeling set. The template structure is thus the immediate environment of the side chains that are subject to a modeling experiment.

2.2 Mathematical Analysis of the Dead-End Elimination Theorem

2.2.1 *Theorem.* The Dead-End Elimination (DEE) theorem relies on a mathematical expression, the DEE criterion, which is a screening tool for rotamers that cannot be members of the global minimum energy conformation (GMEC). The theorem itself states that *a particular rotamer i_r is not compatible with the GMEC if for the same residue i an alternative rotamer i_t exists for which the following inequality holds true*:

$$E(i_r) + \sum_j \min_s E(i_r j_s) > E(i_t) + \sum_j \max_s E(i_t j_s), \text{ where } i \neq j. \quad (2)$$

In this expression i and j are numbers associated with particular residues; i_r, i_t, and j_s are rotamers known in the rotamer library for the corresponding residue type. The function $E(i_r)$ denotes the interaction energy of side-chain atoms with the surrounding template plus its own self-energy. This will below be called the *inherent energy* of a rotamer. $E(i_r j_s)$ is the non-bonded pairwise interaction energy between two rotamers. The functions $\min_s$ and $\max_s$ are evaluated by searching respectively the lowest and the highest value for their argument by cycling over s.

It is obvious that rotamers that satisfy the DEE criterion may simply be discarded from further consideration in searching for the GMEC.

2.2.2 *Proof of the DEE theorem.* The potential energy of a protein consisting of a set of side chains i appearing in a rotameric state r and being attached to a predefined template structure can be expressed as follows:

$$E_{\text{global}} = E_{\text{template}} + \sum_i E(i_r) + \sum_i \sum_j E(i_r, j_s), \quad \text{where } i < j. \quad (3)$$

Here, E_{template} is the template self-energy; the other functions and parameters have the same meaning as in expression (2).

By definition of the GMEC any given global structure must necessarily have a potential energy that is higher than or equal to that of the GMEC itself. We explicitly take into account cases in which the GMEC has degenerated. Then, several global conformations, though different, are termed "GMEC." We can therefore write

$$E_{\text{global}} \geq E_{\text{GMEC}}. \quad (4)$$

The right-hand side of this inequality can be written out using expression (3). Doing so, the rotamers that are part of the GMEC are marked with the subscript "g." Meanwhile the energetic terms corresponding to a well-defined residue of interest i (the residue that is submitted to a DEE test at a given moment) are isolated:

$$E_{\text{GMEC}} = E_{\text{template}} + E(i_g) + \sum_j E(i_g j_g)$$
$$+ \sum_j E(j_g) + \sum_j \sum_k E(j_g k_g), \quad \text{where } j, k \neq i; \, j < k. \tag{5}$$

If the very residue i adopts an undefined rotameric state t, eventually different from the GMEC conformation g, then the protein assumes a global "perturbed" GMEC with the following energy:

$$E_{\text{global}} = E_{\text{template}} + E(i_t) + \sum_j E(i_t j_g)$$
$$+ \sum_j E(j_g) + \sum_j \sum_k E(j_g k_g), \quad \text{where } j, k \neq i, \, j < k. \tag{6}$$

Substitution of equations (5) and (6) into (4) and elimination of identical terms in both sides gives

$$E(i_t) + \sum_j E(i_t j_g) \geq E(i_g) + \sum_j E(i_g j_g), \quad \text{where } i \neq j. \tag{7}$$

Since the maximum of a function always exceeds or equals any individual value of the given function, we have, for all i, j $(i \neq j)$:

$$\max_s E(i_t j_s) \geq E(i_t j_g) \tag{8}$$

and, oppositely for the minimum,

$$\min_s E(i_g j_s) \leq E(i_g j_g). \tag{9}$$

Substitution of (8) and (9) into (7) yields the inequality

$$E(i_t) + \sum_j \max_s E(i_t j_s) \geq$$
$$E(i_g) + \sum_j \min_s E(i_g j_s), \quad \text{where } i \neq j. \tag{10}$$

Now, any rotamer that satisfies the DEE criterion, expression (2), must, in view of the general truth of expression (10), also obey the following inequality:

$$E(i_r) + \sum_j \min_s E(i_r j_s) >$$
$$E(i_g) + \sum_j \min_s E(i_g j_s); \quad \text{where } i \neq j. \tag{11}$$

This can only be the case if $i_r \neq i_g$, or in other words, the rotamer i_r must be different from the GMEC rotamer. This proves that expression (2) is a valid elimination criterion.

2.2.3 The extended DEE theorem (EDEE theorem). The DEE theorem as discussed in previous sections does not impose any restrictions on the concept of a rotamer. A practical definition of a rotamer within the scope of the theorem could be *a well-defined collection of atoms in a given configuration that interact with the template and with other collections.* The atoms are not necessarily connected. The only limitation is that two collections of atoms behave mutually independently, i.e. the conformation of one such rotamer should not be altered by changing the conformation of another. This definition leaves room for some creativity. It means that the idea of a rotamer may be stretched to *the side-chain conformation of an unlimited number of single residues*. In what follows we will confine ourselves to "double residues" and "double rotamers." The extension toward multiple residues and rotamers is perfectly analogous.

The inherent energy of a double rotamer $[i_r j_s]$, symbolized as $\mathcal{E}([i_r j_s])$, is defined as

$$\mathcal{E}([i_r j_s]) = E(i_r) + E(j_s) + E(i_r j_s), \quad \text{where } i \neq j. \tag{12}$$

Note that $E(i_r j_s)$, the interaction energy between the individual rotamers, has now become part of the inherent energy of the double rotamer.

The interaction between a double rotamer and another, e.g., single rotamer k_t can be expressed as

$$\mathcal{E}([i_r j_s]k_t) = E(i_r k_t) + E(j_s k_t), \quad \text{where } i \neq j; \ i, j \neq k. \tag{13}$$

These definitions facilitate the formulation of an extended variant of the DEE theorem (the EDEE theorem) by analogy to the original one. The EDEE theorem states that *a particular double rotamer $[i_r j_s]$ is not compatible with the GMEC if for the same double residue $[i \ j]$ an alternative double rotamer $[i_u j_v]$ exists for which the following inequality holds true*:

$$\mathcal{E}([i_r j_s]) + \sum_k \min_t \mathcal{E}([i_r j_s]k_t) > \mathcal{E}([i_u j_v]) + \sum_k \max_t \mathcal{E}([i_u j_v]k_t), \tag{14}$$

where $i \neq j; \ i, j \neq k$.

The mathematical proof of this theorem runs analogous to that of the elimination theorem for single rotamers and will therefore be truncated.

The right-hand side of expression (4) can be expanded while isolating all terms concerning the double residue of interest $[i, j]$:

$$E_{\text{GMEC}} = E_{\text{template}} + E(i_g) + E(j_g) + E(i_g j_g) + \sum_k E(k_g) \\ + \sum_k \left(E(i_g k_g) + E(j_g k_g) \right) + \sum_k \sum_l E(k_g l_g), \tag{15}$$

where $i \neq j; \ i, j \neq k, l; \ k < l$. Substitution of the equations (12) and (13) into (15) gives

$$E_{\text{GMEC}} = E_{\text{template}} + \mathcal{E}([i_g j_g]) + \sum_k \mathcal{E}([i_g j_g]k_g) \\ + \sum_k E(k_g) + \sum_k \sum_l E(k_g l_g), \tag{16}$$

where $i \neq j$; $i, j \neq k, l$; $k < l$. If the double residue $[i, j]$ adopts an undefined conformation $[i_u j_v]$, then the energy becomes

$$
\begin{aligned}
E_{\text{global}} = E_{\text{template}} &+ \mathcal{E}([i_u j_v]) + \sum_k \mathcal{E}([i_u j_v]k_g) \\
&+ \sum_k E(k_g) + \sum_k \sum_l E(k_g l_g),
\end{aligned}
\tag{17}
$$

where $i \neq j$; $i, j \neq k, l$; $k < l$. The only essential difference between the equations (5) and (6) on the one hand and (16) and (17) on the other hand is that in the former couple of equations the residue of interest is single in nature while in the latter it is double. An identical strategy can now be followed to show that the following inequality is generally true:

$$
\mathcal{E}([i_u j_v]) + \sum_k \max_t \mathcal{E}([i_u j_v]k_t) \geq \mathcal{E}([i_g j_g]) + \sum_k \min_t \mathcal{E}([i_g j_g]k_t),
\tag{18}
$$

where $i \neq j$, $i, j \neq k$. Consequently, each double rotamer $[i_r j_s]$ that answers the EDEE criterion, expression (14), will also obey the following inequality:

$$
\mathcal{E}([i_r j_s]) + \sum_k \min_t \mathcal{E}([i_r j_s]k_t) > \mathcal{E}([i_g j_g]) + \sum_k \min_t \mathcal{E}([i_g j_g]k_t),
\tag{19}
$$

where $i \neq j$; $i, j \neq k$. This is only possible if $[i_r j_s]$ is different from $[i_g j_g]$, which proves that the EDEE criterion (14) is a valid elimination criterion.

3. Implementation of the (E)DEE Theorem in the BRUGEL Package

The algorithm has been incorporated into the BRUGEL modeling package (Delhaise et al., 1984). It makes optimal usage of earlier developed routines, such as those for the rotation around dihedral angles, the calculation of potential energy, and the dynamic allocation of arrays. The DEE algorithm itself is designed in a modular way (see Figure 10-1). There are five major routines (stages A to E). The program flow is driven by a coordination routine that enables the user to skip or repeat certain calculations.

3.1 Initialization and Creation of an Extended Rotamer Library (Stage A)

As stated before, it is possible to let the program automatically generate an extended version of the standard rotamer library by taking a preset number of steps of preset size around (i.e., above and below) the standard angles. For the time being, a particular number of steps applies to a particular dihedral angle (χ_1, χ_2, χ_3, or χ_4) of *all* rotamers of a given residue *type*. For instance, if the user chooses two steps on χ_1 and χ_2 for Trp, then *all* Trp-residues

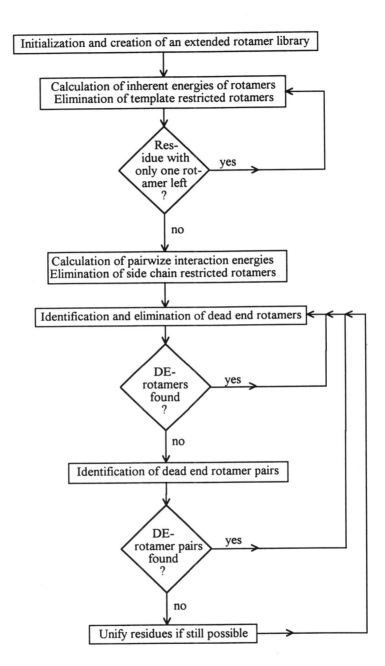

Figure 10-1. *Flow chart of the DEE algorithm.* The details of each subroutine are discussed in Section 3. The flow chart is given for the "pairwise unification method" (see Section 3.5).

in the molecule will initially have 5×5 times the number of Trp-rotamers in the standard library, thus $25 \times 6 = 150$ rotamers. It is obvious that this tool should be handled very carefully. If the number of residues to be modeled is small (e.g., less than 40), we normally take two steps on χ_1 and χ_2 for all aromatic residues. This has been done in a modeling experiment on BPTI (see Section 4.3). However, in the experiments on lysozymes (see Section 4.4), about 100 residues are modeled at once and therefore we could not afford to take steps on dihedrals.

3.2 Elimination of Template Restricted Rotamers (Stage B)

The interaction energies between each rotamer of each residue to be modeled and the template is calculated and compared with a user-set threshold value (usually 30 kcal/mol). Rotamers that exceed this threshold are considered incompatible with the template and are therefore eliminated. If for a given residue only one rotamer passes the template test, this residue is fixed and it becomes part of the template. Then, by default, this stage is repeated until no residues become uniquely determined. In the event of residues having *no* template-compatible rotamers at all, this is signalized to the user and the threshold energy is doubled until at least one rotamer can be preserved.

3.3 Elimination of Side-Chain Restricted Rotamers (Stage C)

The pairwise interaction energies between all possible rotamer pairs are calculated. Then all rotamers are tested for an inevitable clash with some other residue, e.g., rotamers that have a minimal interaction energy (with the rotamers of some neighboring residue), which is still above a certain threshold value (usually 30 kcal/mol) are discarded from further consideration.

3.4 Application of the (E)DEE Theorem (Stage D)

3.4.1 Application to single rotamers (DEE criterion). From this stage on, no time-consuming energy calculations are required anymore. The values are now readily available to be used in searching for dead-end rotamers based on the DEE criterion, expression (2). This enables very fast screening of all remaining rotamers. From practical considerations, a more applicable formulation of the dead-end criterion is the following:

$$E(i_r) + \sum_j \min_s E(i_r j_s) >$$
$$\min_n \{ E(i_n) + \sum_j \max_s E(i_n j_s) \}, \quad \text{where } i \neq j. \tag{20}$$

Indeed, if the rotamer i_r is dead-ending on the basis of a comparison with some other rotamer i_t, then it must certainly be dead-ending relative to the rotamer i_n, which gives rise to the lowest value possible of the right-hand side of the original DEE criterion (2). Conversely, if i_r is not dead-ending when compared with i_n, it will not be in a dead-end situation relative to any other rotameric state for residue i. That the minimization of the right-hand side of the DEE criterion is permitted follows from the fact that the DEE theorem imposes no restriction on the rotamer i_t. Moreover, this procedure renders a stronger and more applicable elimination criterion. In practice, one can make an ordered list of the "worst" interaction energies for all rotamers of residue i and use the best of these as a threshold value for the possible elimination of rotamers i_r.

By default, the DEE routine is executed in an iterative way. If in a given cycle dead-end rotamers are identified (and eliminated) then they will be absent in a next cycle and this will quantitatively alter the values of both sides of expression (20). Consequently, in a succeeding cycle additional dead-end rotamers may be found. The iteration ends when no more dead-end rotamers are found in one cycle.

3.4.2 Application to double rotamers (EDEE criterion). Analogously to single rotamers, now all possible double rotamers are checked for eventual GMEC incompatibility (expression (14) with the right-hand side minimized over u and v). Since double rotamers cannot be eliminated at this stage (only a list of single rotamers is kept in memory), dead-end double rotamers are flagged. This routine is usually *not* repeated to exhaustion for computational reasons. Instead, after having executed it once, the program passes again to the single rotamer elimination routine.

The additional information that is included in the flagged dead-end rotamer pairs must be treated with great care when evaluating single rotamers. The fact of a double rotamer $[i_r j_s]$ being dead-ending means that either i_r or j_s or both are not GMEC compatible. Therefore, pairs that have been flagged may *not* be discarded from the $\sum_j \max_s E(i_t j_s)$ computation when applying the DEE criterion (2) to single rotamers, and thus the extra information is useless. However, it can be proved that dead-end rotamer pairs *may indeed* be ignored when evaluating the $\sum_j \min_s E(i_r j_s)$ terms (Lasters and Desmet, 1993). This possibly augments the left-hand side of the DEE criterion, which increases the probability of single rotamers to be dead-ending.

The high performance of the algorithm mainly results from the iterative application of the DEE theorem to single and to double rotamers, alternatively. The *logarithm* of the total number of possible global conformations

($\log N$) when plotted as a function of the number of iterations typically shows a *quasi-exponential* decrease. When modeling small proteins of up to 100 residues, using the standard rotamer library, the above-described routines usually reduce the conformational space to less than 10^{10} global combinations. Occasionally, even the absolute GMEC is found. If not, the program automatically proceeds to the next stage.

3.5 Further Unification to Super-Rotamers (Stage E)

In this stage rotamers are actually unified and treated *as if they were single rotamers*. A number of unification methods and criteria have been tested of which we will discuss two.

At first, we used an "add on" mechanism, i.e., in a particular addition step the rotamers of an added residue are combined with all rotameric combinations from a previous step. This leads to growing clusters of residues that are treated as one "super-residue." Exactly which residue is selected to be added to an existing cluster is determined as follows. First it is investigated which residues have a nonzero interaction energy with others in some rotamer combination. Interacting residues are collected in clusters. Next, the residues in each cluster are arranged in decreasing order of potential interaction with former residues. Finally, each cluster is submitted to the "add on" routine, which is in fact a conventional combinatorial "tree search" routine, though truncated by the generalized DEE theorem. This procedure has proved to be most useful but it requires user-defined threshold values to discriminate between "interacting" and "noninteracting" residues.

We have also developed a "pairwise unification" procedure in which single rotamers are actually unified to pairs (doublets), then to quadruplets, octuplets, etc. Non-unified residues are collected in a "rest residue." The unification criterion is based on a maximal "average interaction energy," A, between two (super-)residues i and j:

$$A(i, j) = \frac{\sum_{r=1}^{n_i} \sum_{s=1}^{n_j} |E(i_r j_s)|}{n_i \times n_j};$$

(21)

n_i and n_j are the number of (super-)rotamers for the (super-)residues i and j, respectively. At each unification level the DEE theorem for *single* rotamers is iteratively applied to all super-rotamers until exhaustion and then the EDEE theorem is executed once on all possible *pairs* of doublets (or quadruplets, octuplets, etc.). This procedure ends as soon as the absolute GMEC is identified or if no unification is still possible. In the latter case there is only one unified super-residue and a rest residue so that evaluation of the pairwise rotamer combinations with the EDEE criterion must lead to the absolute GMEC.

4. Applications of the DEE Algorithm

4.1 Introduction

In Section 2 it has been proven that rotamers that satisfy the DEE criterion cannot be members of the searched GMEC. However, only practice can show that such rotamers do exist. At the very birth of the theorem it was not at all clear how effective it would be. Taylor (1992) and others (among whom were we ourselves) initially were very skeptical. It was therefore quite a surprise that one of our first modeling experiments (the determination of the GMEC of the insulin dimer; see Section 4.2) succeeded by the *mere application of the DEE theorem.*

The "weak point" in the DEE criterion (expression 2) resides in the computation of the $\sum_j \max_s E(i_t j_s)$ terms. It means that the algorithm does its very best to turn environmental residues j such that they interact in the worst possible manner with the rotamer i_t. However, by minimizing the right hand side of the criterion over all rotamers of the residue i (expression 20), one retrieves the rotamer that best knows to *avoid* bad interactions. From intuitive reasoning one would expect that flexible, solvent-directed residues are far more clever in this than bulky buried aromatics. It was therefore a big surprise to see that elimination proceeded *almost* as fast for core residues as for solvent-exposed ones. Nevertheless, if residues are encountered that are relatively immune to dead-end elimination, they are often clustered in well-packed, buried regions (results not shown).

Until now we have performed numerous small- and large-scale modeling experiments of which we will discuss some representative examples. First we will review one of the first experiments, the repositioning of the insulin dimer side chains (Desmet et al., 1992). This experiment was performed using the "cluster addition mechanism" (see Section 3.5) and an extended rotamer library with 363 rotamers. Then we will discuss the modeling of the small protein BPTI using the same approach but with an even larger rotamer library of 636 elements. Finally an extensive discussion will be dedicated to the modeling of hen and human lysozyme. These experiments were carried out with the standard library and using the "pairwise unification method." Both proteins have also been transformed into each other and the results of these homology modeling experiments were correlated with putative determinants for correct prediction such as the rotamer library, the energy function, the template accuracy, crystallographic resolution and refinement, the residue type, and solvent exposure of side chains.

4.2 Modeling of Insulin

The insulin dimer (PDB code 3INS, "PDB" = Brookhaven Protein Data Bank, Bernstein et al., 1977) consists of 102 residues in all, 76 of which have rotatable side chains. The two Zn^{2+}-atoms were present, the water molecules were not. The modeling was done in the presence of all H-atoms. For this experiment we used an extended version of the standard rotamer library (Table 10-1), i.e., one step of a size equal to the standard deviation was taken on χ_1 and χ_2 of all aromatic side chains. This way, the number of rotamers for Phe and Tyr amounts to 36, and for Trp and His it is 54. For the 76 residues to be modeled the initial number of possible rotamer combinations, N, equals 2.7×10^{76}.

Two separate runs were performed. The first experiment was executed as described in Section 3. The threshold energy for template and side-chain incompatible rotamers (stages B and C, respectively) was set at 30 kcal/mol. This reduced N to 1.6×10^{58} (stage B) and then to 6×10^{53} (stage C, see Figure 10-2). In a second experiment the former routines were turned off and the program was started directly at the DEE routines (stages D). The latter require no threshold values and thus it could be checked whether in the first experiment GMEC rotamers had been eliminated due to too low threshold values. (It turned out not to be the case; see below.)

The iterative application of the DEE theorem (on single rotamers) until exhaustion, followed by one cycle of the EDEE theorem (on pairs) and then back to the singles, and so on again until exhaustion, finally reduced N to only 10800 (first experiment) and 7200 (second one). Next, the "cluster addition procedure" was followed, which led to single and identical global structures in both experiments. Since in the second run no user-defined thresholds were needed, and since the DEE algorithm can never eliminate GMEC-compatible rotamers, the final structure must be the GMEC. The energy calculations took 1.8 hr (first run) and 5.6 hr (second run) of CPU time on an IRIS 4D/310 computer. It should be noted that the results of these calculations can be stored so that they can immediately be retrieved from disk in later experiments. Therefore the computing time of the pure DEE calculations is of more relevance. The latter took only 3 and 14 CPU-minutes in the first and second run, respectively. It is these values that should be compared with the performance of alternative methods. For instance, the simulated annealing routines in Monte Carlo procedures require of the order of *hours* (Lee and Subbiah, 1991). However, far more

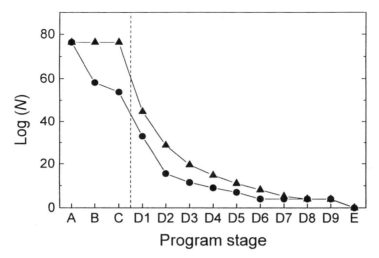

Figure 10-2. *Program performance for the modeling of insulin.* The evolution of the logarithm of the total number of possible rotamer combinations (N) is shown as a function of the program stage. Program stages (A–E) correspond to those as described in Section 3. The numbers correspond to the iteration cycle. The lower curve (containing filled circles) shows $\log(N)$ when the complete algorithm is executed. The upper curve (containing filled triangles) shows $\log(N)$ when only (E)DEE routines are applied. The vertical line separates non-DEE from DEE routines.

important than computational speed is the fact that the DEE method yields a global minimum, which cannot be guaranteed by stochastic techniques such as Monte Carlo simulations (Taylor, 1992).

In spite of the fact that the insulin structure is a GMEC, not all side chains were correctly predicted. Residues are considered to be correctly positioned if *both* their modeled χ_1 *and* χ_2 angles can be attributed to the same dihedral class as observed in the x-ray structure. This criterion is more logical but also more restrictive than evaluating the dihedral angles separately as done by Summers and Karplus (1989). With our criterion, 55 out of the 76 modeled residues could be termed correct (72%). The average difference between the correctly modeled dihedral angles and those observed in the crystal structure were: $\Delta\chi_1 = 8.9° \pm 7.0°$ and $\Delta\chi_2 = 12.8° \pm 9.2°$. The resulting structure had not been energy minimized.

A marked correlation was observed between the correctness of the prediction of a given residue and its extent of burying (Desmet et al., 1992, figure 3). The 29 residues that have a solvent accessible surface area (ASA) of up to 10% of their maximal ASA were positioned with an accuracy of 93%. From the 47 exposed residues (with an ASA above 10%; Miller et al., 1987) 28 were correct (60%). The lower score for solvent-directed residues can be attributed to a lower packing density and to the absence of solvent parameters in the energy function. The drop in prediction quality versus solvent accessibility is observed in *all* our modeling experiments.

4.3 Modeling of BPTI

Bovine pancreatic trypsin inhibitor (BPTI, PDB-code: 4PTI) is a small, monomeric protein of 58 residues, 36 of which are nontrivial to model. The crystallographic resolution is 1.5 Å and the structure has been refined to an R-factor of 16.2%. These conditions make BPTI an ideal test case. Due to the small number of residues with rotatable side chains we could afford to drastically extend the rotamer library. Concretely, we have repositioned the 36 nontrivial residues using the standard rotamer library but with two steps both on χ_1 and χ_2 for all aromatic residues. Thus, for Phe and Tyr there were $5 \times 5 \times 4 = 100$ rotamers and for Trp and His there are $5 \times 5 \times 6 = 150$. The results of this experiment (A) have been compared with those of the same experiment (B) without taking steps.

Table 10-2 shows the results of the BPTI-modeling. In experiment A (fine-rotamer library) 26 out of 36 residues (72%) were correctly positioned (χ_1, χ_2-criterion). In experiment B (coarse library), the total score is identical but differences are observed on the level of individual residues.

Table 10-2. *Modeling of BPTI.* The results of two side-chain placement experiments (A and B) on bovine pancreatic trypsin inhibitor (BPTI) are given as a function of the percentual solvent accessibility of the residues (% ASA, "accessible surface area"). Experiment A is executed with an extended version of the standard rotamer library (see text for more details). In experiment B the standard rotamer library was used as such. The percentual ASA of a side chain is defined as the ratio of the ASA in the crystal structure versus the maximal ASA of the same side chain in an extended conformation in a model tripeptide.

% ASA	# residues	# correct exp. A	# correct exp. B	% correct exp. A	% correct exp. B
0-10	4	4	3	100	75
10-20	6	5	5	83	83
20-40	13	9	10	69	77
> 40	13	8	8	62	62
total	36	26	26	72	72

Based on the exposure in the crystal structure, the residues were then divided into four classes of solvent accessibility. In the group of 0–20% ASA almost all residues are correct (exp. A: 90%; exp. B: 80%). Again it is seen that the prediction quality gradually decreases with increasing solvent exposure. The exposed residues, say those having an ASA higher than 20%, are predicted with an accuracy of 65% and 69% in the experiments A and B, respectively.

Among the buried residues, Phe 4 (17% ASA) forms an interesting exception. In the crystal structure, the aromatic plane is packed onto the positively charged guanidine group of Arg 42. However, when modeling, such a type of residue-specific interaction is absent in the potential energy function. As a result, the typical crystal orientation of both residues is not favored in the modeling experiments and both Phe 3 and Arg 42 (64% ASA) were incorrectly modeled in experiment A as well as in B. Also, the modeled Phe 3 is 41% solvent accessible. The lack of solvent parameters in the energy function may also account for a great deal to the misplacement of this residue.

Phe 33 (9.8% ASA) is correct when using the fine rotamer library (exp. A) and wrong when using the coarse library (exp. B). The modeling of this residue nicely illustrates what effects can influence the orientation of bulky and buried aromatic residues. In experiment B, the vicinal residue Phe 22 is fixed in the first cycle of the program stage B (see Section 3.2). It means that for Phe 22 there is only one rotamer in the library that is compatible with the template at that position. That very rotamer may well be considered as correct in experiment B ($\Delta\chi_1$ with crystal structure = 3.2°, $\Delta\chi_2 = 10.4°$). Upon fixation of residue Phe 22, its side-chain atoms are transferred to the template in the second cycle of stage B. Then its presence prohibits a correct orientation of Phe 33. For this residue the library rotamer that is the closest to the crystal structure has a $\Delta\chi_1 = 9.4°$ and a $\Delta\chi_2 = 23.8°$. This means that this Phe is well, but not exactly, represented in the rotamer library. Anyway, the deviation for the closest library rotamer of Phe 22 and Phe 33 from the crystal structure is sufficient to make the fixed Phe 22 prohibit Phe 33 to adopt its best conformation. This kind of "domino effect" is nonexistent when using a less coarse rotamer library. Then, for Phe 22, $\Delta\chi_1 = 3°$ and $\Delta\chi_2 = 3.7°$ and for Phe 33, $\Delta\chi_1 = 0.4°$ and $\Delta\chi_2 = 9.7°$. Thus, in this experiment both crystal rotamers are much better represented in the library and as a consequence they do not hamper each other and are therefore correctly positioned. This kind of induced domino effect is seen often in our experiments and it is typical for core residues where the conformations of vicinal side chains are strongly coupled (Reid and Thornton, 1989). It illustrates the necessity of a fine and representative rotamer library. Of course, the price to pay on the combinatorial level is very high. Whereas in experiment A there are initially 3×10^{48} possible global structures, in experiment B there are "only" 2×10^{36}.

4.4 Modeling of Lysozymes

4.4.1 Repositioning of the side chains of hen and human lysozyme. As a test case for the performance of the algorithm with respect to homology modeling, we have chosen two reasonably sized, well-resolved and refined proteins, being hen and human lysozyme. Their crystallographic properties are given in Table 10-3. Both molecules possess 95 residues with rotatable side chains. In order to have reference structures to compare the homology modeling results with, these residues were first rebuilt on the original template.

Table 10-3. *Crystallographic properties of hen and human lysozyme.* PDB-code means "Brookhaven Protein Data Bank" code (Bernstein et al., 1977). The references are (1) Ramanadham et al. (1989), and (2) Artymiuk and Blake, (1981).

Lysozyme	PDB-code	Resolution	R-factor	# residues	Reference
hen	2LZT	1.97 Å	0.124	95	(1)
human	1LZ1	1.50 Å	0.187	95	(2)

Regarding the program execution performance, we will restrict ourselves to the experiment on hen lysozyme. The experiment on human lysozyme is perfectly comparable. For the 95 hen residues, using the standard rotamer library (without steps), the total number of initially possible global conformations (N) equals 1.3×10^{88} (see Figure 10-3, stage A). After two elimination cycles of template-restricted rotamers, N was reduced to 1.5×10^{56} (stages B1 and B2). Elimination of side-chain incompatible rotamers further reduced N to 1.5×10^{52} (stage C). At this stage 24% of the residues were already uniquely defined. The required computing time was 37 CPU-minutes on an IRIS 4D/310.

The minimum energy conformation of the remaining residues was determined by means of the (E)DEE routines. This resulted in seven steps (D1–D7) in which the DEE and EDEE criteria were alternately applied. Note that this part of the $\log(N)$ curve shows a quasi-exponential decrease. At exhaustion of the (E)DEE routines on single rotamers, N was already

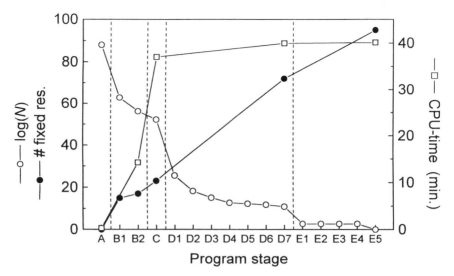

Figure 10-3. *Program performance for the modeling of hen lysozyme.* All rotatable side chains of hen lysozyme are repositioned using the DEE algorithm and the standard rotamer library. In abscissa the consecutive program stages are given as described in Section 3. The stages are separated by vertical lines. The numbers correspond to the iteration cycle. Symbols: open circles, evolution of the logarithm of the total number of possible rotamer combinations ($\log N$). Filled circles, the number of uniquely determined residues at each stage. Open squares, CPU time as a function of the program stage.

as small as 8.3×10^{10}. Then, 76% of the residues were fixed. All DEE calculations together required 3 CPU-minutes.

The remaining 23 residues were then unified into 11 double residues and a rest residue with one element, according to the "maximal average interaction" criterion (equation 21). Iterative application of the generalized (E)DEE theorem for multi-residues resulted in a further reduction of the side-chain conformational space, N, to only 512 possible rotamer combinations (stage E1). Further unification gave no further reduction (stages E2–E4) until the fifth cycle (E5) in which only one super-residue with 23 elements was left. This stage yielded one unique GMEC structure. The steps E1 to E5 required only 10 CPU-seconds.

Analysis of the modeled structure for hen lysozyme using the χ_1, χ_2-criterion gave an overall prediction score of 71%. This result is remarkably similar to that of insulin (72%), haemocyanin (71%; see Desmet et al., 1992) and BPTI (72%). A global prediction score around 70% seems to be a constant for proteins with a crystallographic resolution of ca. 2 Å and a refinement of 15–20%.

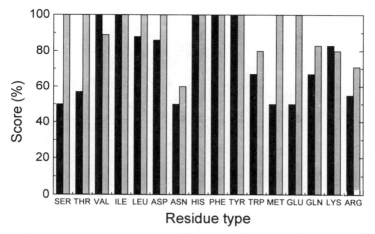

Figure 10-4. *Prediction score for the different residue types of lysozyme.* The modeling of hen and human lysozyme was performed based on the original template structure. Solid bars, hen lysozyme; gray bars, human lysozyme.

We considered this protein sufficiently large to plot the prediction score as a function of the amino acid residue type with statistical relevance (see Figure 10-4, filled bars). It appears that the apolar (Val, Ile, Leu) and the aromatic (His, Phe, Tyr, Trp) residue types are very well predicted, even with this rather coarse rotamer library (overall score = 91%). The other ones, predominantly polar, H-bond-forming residues seem to be more problematic. Especially Ser, Thr, Asn, Gln, Glu, and Arg cause problems (overall score = 52%). Not unexpectedly, this result correlates with a plot of the prediction score versus the solvent accessibility of the residues (Figure 10-5, filled bars). Although a gradual decrease with increasing ASA is now absent, it is still clear that residues with low accessibility are much better positioned than are exposed residues. The overall score for residues with a fractional ASA below 30% is 85%, while the others are predicted with an accuracy of 51%.

A surprisingly good score was obtained for the rebuilding of human lysozyme. This experiment was performed in exactly the same way as for the hen variant. Yet the total prediction score was as high as 87% and *all* Ser, Thr, Ile, Leu, Asp, His, Phe, Tyr, Met, and Glu residues were correctly oriented (see Figure 10-4, gray bars). Figure 10-5 shows again that solvent-directed residues are somewhat more difficult to model than buried ones. A possible explanation for this remarkably good result will be given in Section 5.

4.4.2 Homology modeling of hen and human lysozyme. The ultimate goal of the DEE algorithm is to use it in model-building experiments on pro-

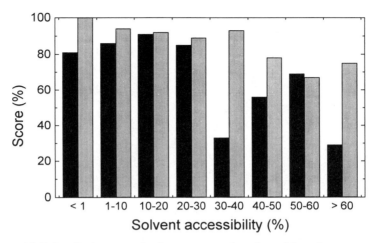

Figure 10-5. *Prediction score for lysozyme as a function of the solvent accessibility.* The modeling of hen and human lysozyme was performed based on the original template structure. Solid bars, hen lysozyme; gray bars, human lysozyme.

teins of unknown structure but for which a homologous protein of known structure exists. In such homology modeling experiments one can use the main-chain structure of the homologous protein as a starting template and model the side chains of the target protein on top of it. This method relies on the experimental finding that most often amino acid substitutions leave the main-chain structure unaffected, especially at surface positions (Greer, 1981) but small shifts of secondary structure elements are sometimes observed (Lesk and Chothia, 1980, 1982). Also, it is usually possible to take into account a certain amount of structural information regarding conserved residues (Lesk and Chothia, 1986; Blundell et al., 1987; Summers et al., 1987; Sutcliffe et al., 1987; Summers and Karplus, 1989). However, in order to illustrate the effect of an improper main-chain template on the results of the method, *all* side chains were rebuilt without considering whether they are conserved or not. Here we will discuss the transformation of human into hen lysozyme and *vice versa*. The results will be evaluated on basis of the x-ray structures and compared with the modeling experiments starting from the original templates.

Hen and human lysozyme have 129 and 130 residues, respectively. At position 48 a Gly residue is inserted in the human variant. Both proteins have a sequence identity of 59% (77 identical residues). The main chains fit rather well. A least-squares fit on the main-chain atoms of the residues [1–46,49–129] (hen numbering) yields an RMS value of 1.07 Å. The C_α-atoms fit even better: RMS = 0.71 Å. The most relevant deviations are observed in the C-terminal lobe, from residue 99 on. This observation correlates with the finding that the B-factors for the same C-terminal part of hen lysozyme are markedly higher than for the rest of the protein (Ramanadham et al., 1989).

4.4.3 Mutation of human into hen lysozyme. The deletion of Gly 48 (in a β-turn) was modeled using an algorithm described by Claessens et al. (1989). This program searches the protein database fragment with the best fit on two anchor regions flanking the deletion site. The anchor regions in human lysozyme were chosen at the residues [43–46] and [50–53]. The best-matching fragment was found in hen lysozyme (RMS on main-chain atoms $= 0.39$ Å), but for evident reasons the next-best fragment was selected, being residues 4–13 of ubiquitin (RMS $= 0.50$ Å).

Next, all 95 rotatable side chains of hen lysozyme were built on the "naked" human template using the DEE algorithm. A comparison of the resulting structure with the x-ray structure of hen lysozyme showed that 62% of the residues were correctly modeled on basis of the χ_1, χ_2-criterion (see Figure 10-6, filled bars). The decrease in prediction score of about 10% relative to previous experiments can be entirely attributed to the nonnative origin of the main-chain structure. For several residues the program signalled complete incompatibility with the template (Lys 13, Leu 25, Lys 33, Ile 98). It is remarkable that all these residues are in helical regions and that they are conserved in both lysozymes. Apparently subtle deviations in template structure can in these cases cause difficulties. The incompatibility problems must be due to a nonideal template rather than a too limited rotamer library, because the rebuilding of hen lysozyme based on the original template showed no template incompatibility at all.

4.4.4 Mutation of hen into human lysozyme. The insertion of Gly 48 was modeled in a way analogous to the previous experiment. The best-matching fragment was found in chymosin b (residues 308–318, RMS $= 0.50$ Å). The transformation of hen into human lysozyme is important especially to study the effect of an imperfect template structure on the modeling results. Note in this respect that the rebuilding of human side chains on the original template yielded a remarkably high prediction score of 87% in all. A comparable result now (say 70–80%) would mean that the prediction accuracy is determined mainly by the properties of the residues while a score of 60–70% would show that the reliability mainly depends on the accuracy of the template.

A conformational analysis of the modeled human lysozyme showed that 24 residues that were correct when using the original human template were wrong in this experiment and that three misoriented residues were

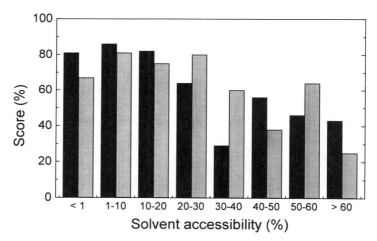

Figure 10-6. *Prediction score for the homology modeling of lysozymes.* The score is expressed as a function of the solvent accessibility of the side chains. Solid bars, hen lysozyme starting from the human template; gray bars, human lysozyme based on the hen template.

now correct. This gives a total score of 62 correctly predicted residues or 65% in all (see Figure 10-6, gray bars). This is a little better than the homology experiment on hen from human lysozyme (62%), only a little worse than what may be expected when using an original template (around 70%) but much lower than the exceptionally high score for the modeling of the human residues starting from the human template (see Table 10-4 for an overview). We can thus conclude that the template accuracy is indeed an important determinant for correct prediction. However, the homology modeling experiments score *only less than 10% worse* than rebuilding experiments. We are convinced that this drop in accuracy can be largely compensated by using a more extended rotamer library or a more detailed energy function (that takes into account solvent interactions, for instance). Also, the information about template and side-chain incompatibilities that is automatically provided by the program in stages B and C could be used to refine the main-chain structure at specific regions.

5. Conclusions

The experiments that are discussed in this work have all come to a good end, i.e., the DEE algorithm has succeeded in cutting back the giant combinatorial tree of initially possible global conformations. Hitherto, a full combinatorial analysis was limited to small protein segments of less than 10

Table 10-4. *(Homology) modeling of lysozymes.* The prediction scores are shown for the repositioning of the side chains of hen and human lysozyme on the original template and for the mutation of both proteins into each other. The scores for buried and solvent exposed residues (using a cutoff criterion of 30% ASA) are given separately.

based on :	Correct residues hen lysozyme (%)		Correct residues human lysozyme (%)	
	Total	< 30% ASA > 30% ASA	Total	< 30% ASA > 30% ASA
hen template	71	85 51	65	75 50
human template	62	79 42	87	94 79

residues (Moult and James, 1986). Moreover, in principle the DEE method does not require arbitrary, user-defined threshold values. (The experiments on insulin have shown that a threshold of 30 kcal/mol for template and side-chain incompatibility is high enough not to eliminate GMEC-compatible rotamers. Still, if computational speed is of no importance, all modeling experiments can be executed by using only DEE routines.) In addition it has mathematically been proven that the (E)DEE criteria cannot exclude any GMEC rotamer. Thus, on condition that the algorithm succeeds to reduce conformational space to such an extent that one or more equienergetic structures are found, one is guaranteed that the absolute global minimum energy conformation is obtained. For this reason, the differences between modeled versus crystal structures must be due solely to one of the following causes: (1) a very much less accurate template, (2) a too approximate potential energy function, (3) a rotamer library that is too coarse or that is not representative for some particular residues *in situ*, (4) inaccuracies and uncertainties in x-ray structures. These factors will hereafter be discussed in the light of the described experiments.

The importance of a correct template structure is inferred from three experimental findings. First, analysis of the rotamer–template interaction energies (results not shown) indicated that the local template structure is the main driving force for the conformation of residue side chains. Second, the structure of human lysozyme on the basis of the human template was much better predicted than that of hen lysozyme on the hen template (see

Table 10-4). The only essential difference between both template structures is the crystallographic resolution and refinement (see Table 10-3). The structure of human lysozyme is determined at very high resolution (1.50 Å) while that of hen lysozyme is merely "high" (1.97 Å). Also, the resolution appears to be more important for prediction accuracy than is the refinement. The R-factor of human lysozyme (0.187) is even higher than that of the hen structure (0.124). Third, we have seen that the very good score for the human-to-human experiment was not maintained in the hen-to-human experiment. The score for the latter experiment (65%) is far from being bad (only 6% less than for the modeling of hen lysozyme on basis of the original hen template), but still 23% lower than for the human to human experiment.

The effect of the exactness of the potential energy function on the reliability of the predictions can be deduced from a comparison between the scores for solvent-exposed and buried residues (see Figures 10-5 and 10-6, and Table 10-4). The side chains of residues that have a solvent accessibility below 30% in the crystal structure are generally very well predicted (score = 80–95%) while the exposed residues are relatively difficult (except in the human-to-human modeling). These problems can probably be attributed to a large extent to the potential energy function that lacks solvent interaction terms. Besides this, the absence of steric constraints among solvent directed residues may also be a plausible explanation. Finally, it has been observed that in many cases the correct H-bonding patterns were not found. This is also suggested by the lower prediction scores for Ser, Thr, Asn, and Glu (see Figure 10-4). A possible improvement to this could be to amplify the energy contributions for H-bonds or to use a more permissive definition for the latter, such as proposed by Kabsch and Sander (1983).

All experiments have been performed using the Ponder and Richards (1987) rotamer library. In the insulin and BPTI experiments additional rotamers were generated by taking discrete steps around the standard χ_1- and χ_2-angles. A comparison between experiments on BPTI performed with the standard rotamer library and with the extended version showed that the latter is to be preferred whenever possible. Yet, we have found that often side-chain conformations as observed in the crystal structure are not at all represented in the rotamer library and are therefore necessarily predicted wrongly (e.g., Lys 69 in human lysozyme: $\chi_1 \approx 0°$). The main problem is encountered for the orientation of the amide and carboxylate plane of Asn/Gln and Asp/Glu, respectively. In the current rotamer library the corresponding dihedral angles are only represented in the 0° (or 180°) conformation, while in reality these angles are rather uniformly distributed over all possible values (Bhat et al., 1979).

As stated in the introduction, crystallographically determined structures are of deceptive rigidity. However, thermal motion is a fact and some regions in a protein are really flexible (e.g., the C-terminal lobe in lysozyme; Ramanadham et al., 1989). Also, we have found that alternative side-chain conformations, leading to equivalent though not identical global energies, do exist. Possible isomers are almost never available from the Brookhaven database. It is clear that these factors might influence the prediction scores in the negative sense.

In conclusion, we believe that the DEE algorithm in its current preliminary version performs in a manner equivalent to alternative methods for side-chain modeling (Summers and Karplus, 1989; Holm and Sander, 1991; Lee and Subbiah, 1991; Tuffery et al., 1991). The algorithm is unique in that it provides a way to obtain the absolute, global minimum energy structure given a particular rotamer library, a potential energy function and a starting template. Other methods are perfectly well capable of generating good, low-energy structures but they cannot guarantee to obtain the "best." Therefore, the DEE method may become a powerful tool to improve its prerequisites, i.e., it may be used to create more representative rotamer libraries and energy functions. Within this respect, we believe we can state that the DEE method is *by no means* a dead-ending method.

Acknowledgments. J.D. would like to thank Professors H. Van Dael and M. Joniau for their scientific cooperation and the "Onderzoeksraad" of the K.U. Leuven for financial support. This research was also financed by a research fellowship of "het Vlaams Instituut voor de bevordering van het wetenschappelijk-technologisch onderzoek in de industrie (IWT)."

REFERENCES

Artymiuk PJ, Blake CCF (1981): Refinement of human lysozyme at 1.5 Å resolution. Analysis of non-bonded and hydrogen-bond interactions. *J Mol Biol* 152:737–762

Benedetti E, Morelli G, Némethy G, Scheraga HA (1983): Statistical and energetic analysis of side-chain conformations in oligopeptides. *Int J Peptide Protein Res* 22:1–15

Bernstein FC, Koetzle TF, Williams GJB, Meyer EF, Brice MD, Rodgers JR, Kennard O, Shimanouchi T, Tasumi M (1977): The Protein Data Bank: A computer-based archival file for macromolecular structures. *J Mol Biol* 112:535–542

Bhat TN, Sasisekharan V, Vijayan M (1979): An analysis of side-chain conformation in proteins. *Int J Pept Protein Res* 13:170–184

Blow D (1983): Molecular structure. Computer cues to combat hypertension. *Nature* 304:213–214

Blundell TL, Sibanda BL, Sternberg MJ, Thornton JM (1987): Knowledge-based prediction of protein structures and the design of novel molecules. *Nature* 326:347–352

Bowie JU, Reidhaar-Olson JF, Lim WA, Sauer RT (1990): Deciphering the message in protein sequences: Tolerance to amino acid substitutions. *Science* 247:1306–1310

Brooks BR, Bruccoleri R, Olafson D, States DJ, Swaminathan S, Karplus M (1983): CHARMM: A program for macromolecular energy, minimization, and dynamics calculations. *J Comput Chem* 4:187–217

Claessens M, Van Cutsem E, Lasters I, Wodak S (1989): Modelling the polypeptide backbone with "spare parts" from known protein structures. *Prot Eng* 2:335–345

Desmet J, De Maeyer M, Hazes B, Lasters I (1992): The dead-end elimination theorem and its use in protein side-chain positioning. *Nature* 356:539–542

Delhaise P, Bardiaux M, Wodak S (1984): Interactive computer animation of macromolecules. *J Mol Graph* 2:103–106

Dill KA (1987): Commentaries. Protein surgery. *Prot Eng* 1:369–372

Greer J (1981): Comparative model-building of the mammalian serine proteases. *J Mol Biol* 153:1027–1042

Holm L, Sander C (1991): Database algorithm for generating protein backbone and side-chain coordinates from a C(alpha) trace. Application to model building and detection of coordinate errors. *J Mol Biol* 193:775–791

IUPAC Commission on Macromolecular Nomenclature (1979): *Pure Appl Chem* 51:1101–1121

James MNG, Sielecki AR (1983): Structure and refinement of penicillopepsin at 1.8 Å resolution. *J Mol Biol* 163:299–361

Janin J, Wodak S, Levitt M, Maigret B (1978): Conformation of amino acid side-chains in proteins. *J Mol Biol* 125:357–386

Kabsch W, Sander C (1983): Dictionary of protein secondary structure: Pattern recognition of hydrogen-bonded and geometrical features. *Biopolymers* 22:2577–2637

Lasters I, Desmet J (1993): The fuzzy-end elimination theorem: Correctly implementing the side chain placement algorithm based on the dead-end elimination theorem. *Prot Eng* 6:717–722

Lee C, Levitt M (1991): Accurate prediction of the stability and activity effects of side-directed mutagenesis on a protein core. *Nature* 352:448–451

Lee C, Subbiah S (1991): Prediction of protein side-chain conformation by packing optimization. *J Mol Biol* 217:373–388

Lesk AM, Chothia C (1980): How different amino acid sequences determine similar protein structures: The structure and evolutionary dynamics of the globins. *J Mol Biol* 136:225–270

Lesk AM, Chothia C (1982): Evolution of proteins formed by beta-sheets. II. The core of the immunoglobulin domains. *J Mol Biol* 160:325–342

Lesk AM, Chothia C (1986): *Phil Trans R Soc London, Ser A*, 317:345–356

Levinthal C (1968): Are there pathways for protein folding? *J Chem Phys* 65:44–45

McGregor MJ, Islam SA, Sternberg MJE (1987): Analysis of the relationship between side-chain conformation and secondary structure in globular proteins. *J Mol Biol* 198:295–310

Miller S, Janin J, Lesk A, Chothia C (1987): Interior and surface of monomeric proteins *J Mol Biol* 196:641–656

Moult J, James MNG (1986): An algorithm for determining the conformation of polypeptide segments in proteins by systematic search. *Proteins* 1:146–163

Ponder JW, Richards FM (1987): Tertiary templates for proteins. Use of packing criteria in the enumeration of allowed sequences for different structural classes. *J Mol Biol* 193:775–791

Privalov PL (1989): Thermodynamic problems of protein structure. *Annu Rev Biophys Chem* 18:47–69

Ramanadham M, Sieker LC, Jensen LH (1989): *Triclinic Lysozyme: Some Features of the Structure at 2 Angstroms Resolution*, Smith-Gill S, Sercarz E, eds. Schenectady, NY: Adenine Press

Reid L, Thornton JM (1989): Rebuilding flavodoxin from C coordinates: A test study. *Proteins* 5:170–182

Reidhaar-Olson JF, Sauer RT (1990): Functionally acceptable substitutions in two alpha-helical regions of lambda repressor. *Proteins* 7:306–316

Schellman JA, Lindorfer M, Hawkes R, Grutter M (1981): Mutations and protein stability. *Biopolymers* 20:1989–1999

Shortle D, Meeker AK, Freire E (1988): Stability mutants of staphylococcal nuclease: Large compensating enthalpy-entropy changes for the reversible denaturation reaction. *Biochemistry* 27:4761–4768

Summers NL, Carlson WD, Karplus M (1987): Analysis of side-chain orientations in homologous proteins. *J Mol Biol* 196:175–198

Summers NL, Karplus M (1989): Construction of side-chains in homology modelling. Application to the C-terminal lobe of rhizopuspepsin. *J Mol Biol* 210:785–811

Sutcliffe MJ, Hayes FRF, Blundell TL (1987): Knowledge based modelling of homologous proteins, part II: Rules for the conformations of substituted sidechains. *Prot Eng* 1:385–392

Taylor W (1992): Protein structure. New paths from dead ends. *Nature* 356:478–480

Tuffery P, Etchebest C, Hazout S, Lavery R (1991): A new approach to the rapid determination of protein side chain conformations. *J Biomol Struct Dyn* 8:1267–1289

Warshel A, Levitt M (1976): Theoretical studies of enzymic reactions: Dielectric, electrostatic and steric stabilization of the carbonium ion in the reaction of lysozyme. *J Mol Biol* 103:227–249

Wodak SJ, De Coen JL, Edelstein SJ, Demarne H, Beuzard Y (1986): Modification of human hemoglobin by glutathione. III. Perturbations of hemoglobin conformation analyzed by computer modeling. *J Biol Chem* 261:14717–14724

11

Short Structural Motifs: Definition, Identification, and Applications

Ron Unger

1. Introduction

Short structural motifs have been studied in the last few years as an important component in our understanding of protein structures. Traditionally, protein structure organization is defined on three levels: the primary level of the amino acids sequence, the secondary level describing local patterns within the structure, and the tertiary structure level that reflects the global conformation of the chain. The classification of secondary structure elements includes a very small number of categories. Thus, it does not seem to be rich enough to describe in full the variety of possible, short structural arrangements of the polypeptide chain. This gives rise to another intermediate level of protein structure classification: the short structural motif. This level of classification is more specific both in structural organization and in its sequence preferences, so it is suggested that this level can serve as a better bridge between the primary level and the tertiary level of protein structure.

In this chapter we will describe how structural motifs are defined and identified, and examine the importance of short structural motifs. Current applications of a structural approach based on this level of representation will be discussed, and we will elaborate on the sequence/structure relationship this representation induces.

2. Background

It was realized very early by Pauling and co-workers (Pauling and Corey,

The Protein Folding Problem and Tertiary Structure Prediction
K. Merz, Jr. and S. Le Grand, Editors
© Birkhäuser Boston 1994

1951; Pauling et al., 1951) (even before the determination of the first x-ray structures of proteins) that certain repetitive patterns of hydrogen bonds can dictate specific arrangements of the polypeptide chain that form secondary structure elements such as α-helices and β-sheets. Some specific kinds of turns were recognized as a third important category (Venkatachalam, 1968). The segments of the chain that are not covered by any of these three categories are usually classified as "random coils." This broad classification does not capture the structural details of short chain segments. On the one hand, there are significant structural differences between some segments with the same secondary structure assignment. On the other hand, there are significant structural similarities between some segments that are classified as random coils. Dealing directly with short segments in terms of three-dimensional coordinates might enable us to better handle the structural analysis of proteins.

Very similar conformations of short protein segments can be found among different proteins. The first work that took advantage of this observation is due to Jones and Thirup (1986), who demonstrated that the C_α skeleton of the retinol binding protein can be reconstructed using fragments of 5 to 10 residues taken from three different proteins. The accuracy of the match of the atomic positions in the fragments ranged between 0.04 Å to 1.28 Å RMS deviation. The small number of proteins that were needed in order to match fragments in the original protein is a clear indication of how often proteins employ the same short structural motifs. Jones and Thirup suggested the use of this observation in crystallography as a tool in the initial stages of constructing a model from the electron density map. Approximated C_α coordinates can be used as templates to fetch similar fragments from solved structures, and thus make the map-fitting process easier. The idea was implemented in later versions of the program FRODO (Jones, 1978).

Additional work in this direction was done by Claessens et al. (1989). In this work, a database of 66 well-refined proteins was used as a source of "spare parts" for reconstruction of a few proteins in an automated procedure. The C_α coordinates of the modeled protein were used to pull out the full backbone atoms of similar fragments from the database. Again, it was found that fragments with an average length of seven or eight residues could be used to replace the original fragments with a resulting backbone deviation of less than 1 Å. It was also shown that similar results can be obtained with a smaller database of only 10 proteins. Furthermore, satisfactory matches could be found even if the original C_α template was perturbed by shifts of up to 0.5 Å.

Recently, Levitt (1992) suggested a significant extension of this idea.

The aim is to start with a C_α representation and to end up with a full atom model that includes side chains.

The algorithm starts by pulling matching segments (including their side chains) from the database on the basis of sequence as well as C_α coordinates. The structures of the best few matches are averaged, and simple energy minimization is used to achieve the final structure. The method works remarkably well, producing structures with averaged RMS deviation of 1.2 Å for all atoms and 0.4 Å for main-chain atoms. The method works well even if the initial C_α skeleton is inaccurate or incomplete.

All of the studies mentioned above employ the observation that almost every short fragment in any protein has at least one close homologous match in the database. They use the main-chain matches as a step towards full atom modeling of a protein. We would like to study this phenomenon from another viewpoint, focusing on backbone conformations. We are interested in a *systematic* study of such homologies, so that we could determine how many distinct conformations there are, how frequent each one of them is, and to what degree they have specific sequence preferences.

3. Distance between Fragments

In order to carry out a systematic study of short structural motifs, one clearly needs an operational definition of these motifs. This definition should be based directly on the data without introducing *a priori* assumptions about the "correct" classification. In our first study (Unger et al., 1989), we measured the RMS distances between all the hexamers (fragments of six residues) of a few proteins. Hexamers are considered in an overlapping manner, which yield $n - 5$ hexamers for a protein of n residues. We decided to use hexamers mainly because we wanted to use short size fragments, and fragments of five residues seemed to be too short to have structural integrity. It has been shown (Kabsch and Sander, 1984) that pentamers with identical sequences can have very different conformations. The comparison of two fragments begins with their C_α representations. The distance between two fragments s and t is measured by superimposing them and then calculating the normalized root-mean-square deviation of their C_α coordinate vectors.

$$\text{RMS} = \left(\frac{\sum_{i=1}^{n} \left(r_i^s - r_i^t \right)^2}{n - 2} \right)^{1/2}, \tag{1}$$

where r^s is the coordinate vector of fragment s, and n is the length of the fragment. We normalize by $n - 2$ to reflect the actual number of degrees of freedom in the system. Note that for $n = 6$, distances calculated in this way are significantly larger than distances calculated by normalizing by n, which has been done in some other works.

Upon calculating the distances between a large number of hexamers, a bimodal distance distribution was found that divided the distances into two groups. In one group (about 10% of the pairs) the distances were less than 1 Å, in the second group most distances were between 1.5 and 5 Å. This observation suggested to us that 1 Å is an appropriate threshold to define similarity for hexamers.

Measuring the RMS distance between two fragments is expensive because of the need to find the best superposition of the structures. The superposition can be done analytically by using eigenvalue decomposition or matrix inversion (Kabsch, 1978; McLachlan, 1979), or numerically by iterative minimization of the rotation angles (Nyburg, 1974; Sippl and Stegbuchner, 1991). All of these methods are computationally expensive. In order to avoid these computations other measures have been suggested. Karpen et al. (1989) suggested to measure the structural deviation in terms of the ϕ and ψ dihedral torsion angles:

$$\text{TOR} = \left(\frac{\sum_{i=1}^{n} \left(\phi_i^s - \phi_i^t\right)^2 - \left(\psi_i^s - \psi_i^t\right)^2}{2n} \right)^{1/2}, \tag{2}$$

where ϕ_i^s and ψ_i^s are the dihedral angles around the $C_{\alpha i}^s$ atom.

Torsional differences are simple to calculate, and this work showed a general correlation between the RMS and the TOR measures. But it also pointed out a few differences. The TOR measure gives the same weight to deviations at every position in the fragments, while the RMS measure is more sensitive to dihedral angle deviations in the center of the fragments, since central deviations have a greater effect on the overall conformation. Another difference is that the TOR measure is not sensitive to *trans–cis* conformational shifts. The study of Karpen et al. (1989) primarily focused on the TOR measure but two other measures that are simple to calculate were also mentioned. One method measures the deviation in the virtual C_α torsion values between the two structures:

$$\text{VIR} = \left(\frac{\sum_{i=1}^{n} \left(\alpha_i^s - \alpha_i^t\right)^2}{n - 3} \right)^{1/2}, \tag{3}$$

where the virtual torsion angle α_i^s measures the angle between the plane defined by the three consecutive C_α atoms $C_{\alpha i-1}^s$, C_α^s, and $C_{\alpha i+1}^s$, and the plane defined by C_α^s, $C_{\alpha i+1}^s$, and $C_{\alpha i+2}^s$. The normalization was done by $n - 3$ to reflect the fact that it takes four consecutive C_α atoms to produce one virtual angle. The other method measures the similarity of the interatomic distances between corresponding C_α in the fragments.

$$\text{DIS} = \left(\frac{\sum_{i<j}^{n} \left(d_{i,j}^s - d_{i,j}^t \right)^2}{0.5n(n-1)} \right)^{1/2}, \tag{4}$$

where $d_{i,j}^s$ is the Cartesian distance between atoms $C_{\alpha i}^s$ and $C_{\alpha j}^s$.

Prestrelski et al. (1992a) combined measures similar to VIR and DIS to estimate the similarity between fragments using a pentamer as a base unit even when comparing longer fragments. For each pair of pentamers, they used a linear combination of the difference between the sum of the distances of the first C_α atom and its four successive C_α atoms and the difference between the corresponding two virtual α-angles in each pentamer. Several coefficients for the linear combination were tried. For comparison of longer fragments Prestrelski et al. (1992a) used a measure based on the maximal distance between corresponding pentamers in the longer fragment.

There are three conditions that define a distance function: self similarity (i.e., $d(A, A) = 0$), symmetry (i.e., $d(A, B) = d(B, A)$), and the triangle inequality (i.e., $d(A, B) + d(B, C) > d(A, C)$). While the first two are clearly met for all the measures discussed above, the third one is not trivial. For the RMS measure that we used, the triangle inequality was empirically tested and was found to hold in more than 99.5% of a large number of triplets of hexamers drawn at random from the database. The small number of exceptions were probably related to small numerical errors in the best fit calculation. It is not clear whether this condition holds for the other suggested measures. For short structural motifs, it seems to us that the RMS distance is the most natural measure, notwithstanding its considerable computational price.

4. Clustering Short Structural Motifs

Having a distance measure and a threshold, the next step is to design a clustering algorithm that can take hexamers from the database and assign them to different classes. Finding an optimal clustering, in the sense that for a given number of clusters, the maximal distance between members assigned to the same cluster will be less than a given threshold, is difficult, as it belongs to the NP-complete class of problems (Garey and Johnson, 1981, p. 281). We designed the following heuristic, two-stage procedure. The distances between all hexamers from four proteins were calculated. In the first stage of the algorithm, clusters were constructed such that every hexamer included in a cluster matches at least one other hexamer in that cluster with a distance less than the threshold of 1 Å. This is done by assigning an initial hexamer to a cluster and then iteratively adding to that cluster all hexamers that are similar (i.e., within the 1 Å distance) to any hexamer already in the cluster. When a cluster is exhausted (i.e., no other hexamers can be added), then one of the remaining hexamers is chosen to serve as the first member of a new cluster, and the process is repeated until all hexamers are assigned. This process is deterministic (i.e., it will produce the same clusters regardless of the order in which the hexamers are used) and guarantees that each hexamer will belong to exactly one cluster. This stage takes time that is of the order of n^2 for n hexamers.

This clustering procedure is highly dependent on single hexamers. Because it takes only one match in order to be included in a cluster, the clusters tend to grow and the distance between a pair of hexamers in a cluster can become quite large. Actually, a single hexamer can act as a "bridge" forcing a merger of really distinct clusters. Thus, in the second stage, we divided each cluster into subclusters. We required that each subcluster have a "central" hexamer, whose distance from any other hexamer in that subcluster will be less then 1 Å. Dividing each cluster into the minimal number of subclusters with the above property is again an NP-complete problem. We used a simple scheme that does not guarantee finding the smallest number of subclusters but in practice produced good results. The hexamer with the most neighbors closer than 1 Å was chosen as the center of a subcluster, and all its neighbors were assigned to that subcluster and "removed" from the list of the cluster. This process was repeated until all hexamers had been assigned to subclusters. Clearly, in this way, the subclusters "overlap" in the sense that a given hexamer might match more than one subcluster. It has been found that in practice most hexamers can fit only to one or two subclusters.

As an additional measure to avoid the situation where a single hexamer can bridge between distinct clusters, and to lower the computational load, the procedure was run for hexamers drawn from a small number of proteins. In order to ensure that the results are not overly dependent on the initial set of proteins, a few different initial sets of proteins were tested and the results were shown to be very similar.

We used the following four proteins: 4HHBb (human deoxyhaemoglobin, β-chain), 5PTI (bovine pancreatic trypsin inhibitor), IBP2 (bovine pancreatic phospholipase A2), and IPCY (oxidized poplar plastocyanin), with a total length of 426 residues (and thus 406 overlapping hexamers). These structures have been very accurately determined, and they represent different classes of proteins. The 82,215 RMS distances between all pairs of hexamers were calculated and served as a basis for the clustering algorithm described above. The first stage of clustering yielded 55 distinct clusters, which indicates that indeed the conformations of short fragments have distinct shapes and they don't form a continuum in the conformational space. Some of the clusters became quite large in diameter (i.e., the distance between the two furthest points), so we used the second stage of clustering and ended up with a total of 103 subclusters. The central hexamer in each subcluster (as defined above) was classified as a short structural motif or a *building block*. Nine of the most common building blocks are shown in Figure 11-1.

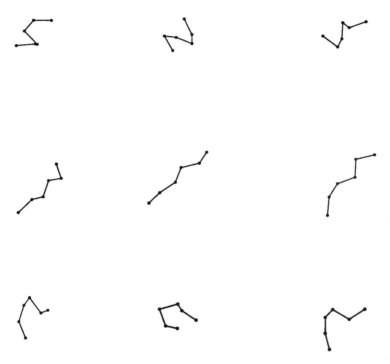

Figure 11-1. Nine building blocks are shown. In the top row the building blocks represent three subclusters of the main helical cluster. In the second row the building blocks represent three subclusters of the main extended cluster. The bottom row shows building blocks that represent three clusters usually associated with turns. The building blocks were aligned such that their N-terminal is in their left lower part. The building blocks are ordered left to right according to the number of hexamers each of them represent.

Next, we checked the level of similarity between hexamers in the database and the set of 103 building blocks. We used 82 well-refined proteins containing 12,973 hexamers. We found that 76% of the hexamers had a distance less than 1 Å to at least one of the building blocks. If we relaxed the similarity threshold to 1.25 Å, then 92% of the hexamers of the database could be represented by one of the building blocks.

Similar results were obtained by Prestrelski et al. (1992a). As mentioned above, they used a combination of the VIR and DIS measures to classify fragments of eight residues (octamers). To facilitate the classification, they calculated "ideal" secondary structures to compare with conformations of actual octamers in the database, although the issue of the clustering method was not emphasized in their paper. They generated 113 representative substructures, which could account for 67% of the octamers in their database of 14 protein (2,378 octamers).

5. The Nature of the Structure of the Building Blocks

The question of the source of the structures of the building blocks was addressed by Unger et al. (1990). We first showed that the conformations adopted by the building blocks are not a mere reflection of the known tendency of the dihedral angles ϕ and ψ to adopt certain values. This kind of study must be based on statistical considerations, since conclusions can't be drawn from a few exemplary hexamers. Thus, we constructed a random population of hexamers that preserved the overall dihedral angle distribution in the database. This population was compared to a population of real hexamers drawn from the database, and it was shown that the two populations were different. The next stage was to construct a population of random hexamers that preserved the frequency of doublet dihedral values of the database. The doublet dihedral values are two consecutive dihedral angle pairs (ϕ_i, ψ_i), (ϕ_{i+1}, ψ_{i+1}). This was accomplished by a Markovian process of biasing the selection of random dihedral values so that the constructed population of hexamers maintained the doublet dihedral distribution. Comparison between this population and populations of real hexamers showed them to be very similar, and thus this process was shown to produce realistic hexamers. The ability to create hexamers with a distribution of conformations similar to that found in actual proteins based solely on doublet dihedral angles preservation is interesting. Further studies will be needed to clarify whether the doublet dihedral distribution is the "driving force" behind the observed conformations of short structural motifs.

6. Applications of the Short Structural Motifs Classification

As mentioned above, one application of the short structural motifs is as templates for electron density fitting, side-chain building, and general modeling. Another kind of application, regarding the relationship between the sequence and the structure associated with short structural motifs, will be discussed in the next section. An interesting application of the building blocks is to use them to assess the quality of protein structure determination. As mentioned above, our building blocks can represent 76% of all hexamers in the database with accuracy better than 1 Å. Single proteins vary in the extent to which their hexamers can be replaced by the standard building blocks. Most of the proteins were in the range of 60%–90%. A lower match would indicate that either the structure is very unusual (i.e., contains a high number of unusual, local three-dimensional conformations) or that the structure is wrong. In our database search we found two outstanding low exceptions, Azotobacter ferredoxin (2FDl) with 14% fit, and alpha bungarotoxin, chain a, (2ABX) with 26% fit. As for 2FDl, a redetermination of the structure (Stout et al., 1988) showed the original structure to be incorrect. The 2ABXa case still has to be checked. Thus, the set of building blocks can provide an automated and objective tool to screen for bad or unusual structures.

Another interesting application was suggested by Prestrelski et al., (1992b). They used their library of building blocks to calibrate measurements of Fourier transform infrared spectroscopy (FTIR). FTIR is a sensitive method for studying protein secondary structure in solution. However, there is not enough data on the exact correlation between a precise local conformation and the actual bands detected in the infrared spectrum. A library of substructures taken from a family of serine proteases enabled Prestrelski et al. (1992b) to detect the spectral differences between similar but not identical fragments, and thus to map the correlations.

7. The Association between the Sequence and Structure of Short Structural Motifs

Classification of protein fragments into short structural motifs is much finer than the traditional classification to a few secondary structure elements. Thus, it is interesting to check whether it is accompanied by a similar refinement in the sequences associated with the structural classification. The sequence of all the hexamers that can be represented by a certain building block can be summarized in a 20 × 6 matrix that reflects the frequency of each amino acid in each position in the hexamers. The sequence matrices associated with building blocks that are part of the standard secondary structure elements show a tendency to the known secondary structure preferences of the amino acids. Furthermore, nonrandom sequence matrices were found for many building blocks that are associated with nonstandard conformations. However, the existence of nonrandom sequence matrices, which can be shown based on statistical tests, does not imply that one can easily identify the structure of a hexamer given its sequence.

An interesting example was presented in our study of extended motifs (Unger and Sussman, 1993). We examined the structures of all 539 hexamers in our database that are classified as extended structures (i.e., their secondary structure assignment using the DSSP program (Kabsch and Sander, 1983) is EEEEEE). We found that the RMS distances among these hexamers can be surprisingly high. The average RMS distance is 1.7 Å, and some pairs have distances greater than 4 Å! We found that these hexamers could be divided into three distinct homogeneous subclasses. We further demonstrated that there is a strong sequence signal associated with each subclassification. It is clear that a window of six residues is too small to be the basis of a real prediction scheme. However, we wanted to check whether there is a predictive advantage in the three specific sequence matrices over using one matrix associated with all 539 hexamers. We were not able to find a way to utilize this signal to predict the subclassification of an extended hexamer, or to predict whether a given hexamer is extended. Our prediction scheme was simple and used only information included in the hexamers. It is possible that a more sophisticated prediction scheme based on a longer window would be able to use the additional sequence specificity induced by the finer structural classification.

The relationship between sequence and structure of short structural motifs was studied extensively by S. Wodak and co-workers. In their earlier studies (Rooman and Wodak, 1988; Rooman et al., 1989; Rooman and Wodak, 1991) the group concentrated on the relationship between short, specific sequence patterns and the secondary structure elements associated with them. It was noted that while there is a significant correlation between many sequence patterns and secondary structure (what was called "high intrinsic predictive power of the associations"), the overall predictive power of a scheme based on those associations was not very high. This was attributed to the limited size of the current database and to long-range interactions that restricted the dependency of local structure on local sequence.

In their recent work, the group shifted its attention from secondary structure elements to studies of sequence/structure relationships in the context of short structural motifs more similar to what we discussed above. The aim is to examine short fragments for which the sequence/structure association is strong (Rooman et al., 1990). They defined short structural motifs in terms of the dihedral angles based on seven structural states that reflect common ϕ, ψ, and ω combinations (Rooman et al., 1991) (ω being the planer amide angle reflecting *trans* or *cis* conformation; six states have a *trans* conformation and one has a *cis* conformation). The sequence preference was expressed not in a simple frequency matrix, but in a more sophisticated way as a potential of mean force (Sippl, 1990). Here, the potential of mean force reflects the ratio of found frequencies of amino acid pairs within a local window versus the expected frequencies based on a random model. The actual "energy" values are calculated based on Boltzmann's law. In Rooman et al. (1991) a structural prediction was performed for peptides whose structure had been experimentally probed, with good results. The group suggested that while in general the sequence/structure relationship is not strong enough to enable prediction, short motifs that have a very high sequence signal and thus are highly predictable may play an important role in folding (Rooman et al., 1992). A strong sequence signal is characterized by a wide gap in the energy values attached to alternative conformations computed for the fragment (i.e., one specific conformation fits the sequence much better than any other). Such fragments can stabilize their structure internally, independent of the state of the rest of the chain, and thus serve as nucleation sites for the folding process (Moult and Unger, 1991). The importance of these fragments was further demonstrated by Rooman and Wodak (1992). They showed that such fragments tend to be preserved in evolution more than other fragments within protein families, suggesting that they have an important role in folding.

Acknowledgments. Joel Sussman, David Harel, Scot Wherland, and John Moult took an important part in our work described in this review. I thank Walter Stevens, Michael Braxenthaler, and Krzysztof Fidelis for critical reading of this manuscript.

REFERENCES

Claessens M, Van Custem E, Lasters I, Wodak S (1989): Modeling the polypeptide backbone with "spare parts" from known protein structures. *Prot Eng* 2:335–345

Garey MR, Johnson DS (1979): *Computers and Intractability: A Guide to the Theory of NP-Completeness*. San Francisco: Freeman Press.

Jones TA (1978): A graphics model building and refinement system for macromolecules. *J Appl Cryst* 11:268–272

Jones TA, Thirup S (1986): Using known substructures in protein model building and crystallography. *Embo J* 5:819–822

Kabsch W (1978): A discussion of the solution for the best rotation to relate two sets of vectors. *Acta Cryst* A34:828–829

Kabsch W, Sander C (1983): Dictionary of protein secondary structure: Pattern recognition of hydrogen-bonded and geometrical features. *Biopolymers* 22: 2577–2637

Kabsch W, Sander C (1984): On the use of sequence homologies to predict protein structure: Identical pentapeptides can have completely different conformations. *Proc Nat Acad Sci* 81:1075–1078

Karpen ME, de Haseth PL, Neet KE (1989): Comparing short protein substructures by a method based on backbone torsion angles. *Proteins* 6:155–167

Levitt M (1992): Accurate modeling of protein conformation by automatic segment matching. *J Mol Biol* 226:507–533

McLachlan AD (1979): Gene duplication in the structural evolution of chymotrypsin. *J Mol Biol* 128:49–79

Moult J, Unger R (1991): An analysis of protein folding pathways. *Biochemistry* 30:3816–3824

Nyburg SC (1974): Some uses of a best molecular fit routine. *Acta Cryst* B30:251–253

Pauling L, Corey RB (1951): Configuration of polypeptide chains with favored orientation around a single bond: Two new pleated sheets. *Proc Nat Acad Sci* 37:729–740

Pauling L, Corey RB, Branson HR (1951): The structure of proteins: Two hydrogen bonded helical configurations of the polypeptide chain. *Proc Nat Acad Sci* 37:205–211.

Prestrelski SJ, Byler DM, Liebman MN (1992a): Generation of a substructure library for the description and classification of protein secondary structure: I. Overview of the method and results. *Proteins* 14:430–439

Prestrelski SJ, Byler DM, Liebman MN (1992b): Generation of a substructure library for the description and classification of protein secondary structure: II. Application to spectra structure correlation in Fourier transform infrared spectroscopy. *Proteins* 14:440–450

Rooman MJ, Kocher JA, Wodak SJ (1991): Prediction of protein backbone conformation based on seven structure assignments: Influence of local interactions. *J Mol Biol* 221:961–979

Rooman MJ, Kocher JA, Wodak SJ (1992): Extracting information on folding from amino acid sequence: Accurate predictions for protein regions with preferred conformation in the absence of tertiary conformation. *Biochemistry* 31:1022–10238

Rooman MJ, Rodriguez J, Wodak SJ (1990): Relations between protein sequence and structure and the significance. *J Mol Biol* 213:337–350

Rooman MJ, Wodak SJ (1988): Identification of predictive sequence motifs limited by protein structure data base. *Nature* 335:45–49

Rooman MJ, Wodak SJ (1991): Weak Correlation between predictive power of individual sequence patterns and overall prediction accuracy in proteins. *Proteins* 9:69–78

Rooman MJ, Wodak SJ (1992): Extracting information on folding from amino acid sequence: Consensus regions with preferred conformation in homologous proteins. *Biochemistry* 31:10239–10249

Rooman MJ, Wodak SJ, Thornton JM (1989): Amino acid sequence template derived from recurrent turn motifs in proteins: Critical evaluation of their predictive power. *Prot Eng* 3:23–27

Sippl MJ (1990): Calculation of conformational ensembles from potential of mean force: An approach to the knowledge based prediction of local structures in globular proteins. *J Mol Biol* 213:859–883

Sippl MJ, Stegbuchner H (1991): Superposition of three dimensional objects: A fast and numerically stable algorithm for the calculation of optimal rotation. *Comput Chem* 15:73–78

Stout GH, Turley S, Sieker LC, Jensen LH (1988): Structure of ferredoxin I from *Azotobacter vinelandii*. *Proc Nat Acad Sci* 85:1020–1022

Unger R, Harel D, Wherland S, Sussman JL (1989): A three-dimensional building block approach to analyzing and predicting structure of proteins. *Proteins* 5: 355–373

Unger R, Harel D, Wherland S, Sussman, JL (1990): Analysis of Dihedral angles distribution: The doublets distribution determines polypeptide conformations. *Biopolymers* 30:499–508

Unger R, Sussman JL (1993): The importance of short structural motifs in protein structure analysis. *J Comp Aid Mol Design.* 7:457–472

Venkatachalam CM (1968): Stereochemical criteria for polypeptide and proteins. V. Conformation of a system of three linked peptide units. *Biopolymers* 6:1425–1436

12

In Search of Protein Folds

Manfred J. Sippl, Sabine Weitckus, and Hannes Flöckner

1. Introduction

The protein folding problem is exciting as well as frustrating. A large body of knowledge on protein folds has accumulated over past decades, but it is still impossible to calculate native structures from amino acid sequences. Our duty is to present the reasoning behind an approach that attempts to derive an energy model of protein solvent systems from the information contained in experimentally determined protein structures. The result of this approach is called *knowledge-based mean field*. The goal is to employ this model in the search for the native fold of amino acid sequences of unknown structure. We present an overview of the principles involved in this approach, assess the current state of the art, and present several applications in protein structure theory.

A common theme is the search for native folds. This is a powerful tool in force-field development, in the judgment of their predictive power, and in fold recognition. We show that the current mean field, although still quite incomplete, is able to recognize the native fold of a given sequence. It fails in only three cases. One of these cases corresponds to a low-resolution structure that has recently been refined and corrected, and in a second case the x-ray structure is in conflict with recent nuclear magnetic resonance studies.

Fold recognition is an active research area with exciting prospects. The goal is to search databases for sequences that are compatible with a given fold known from experiment. We present a case study using the

The Protein Folding Problem and Tertiary Structure Prediction
K. Merz, Jr. and S. Le Grand, Editors
© Birkhäuser Boston 1994

immunoglobulin fold to retrieve sequences from the swissprot database and we investigate the performance of the technique in some detail. We also discuss fundamental limitations.

Protein folding has been an active research area for a long time. The most pressing question is how an amino acid sequence adopts its unique three-dimensional fold, and in spite of an enormous research effort over past decades, no satisfying solution has been found. In fact a literature search on the topic not only results in an endless list of references but it also reveals that there is hardly any new approach that has not yet been tried, be it classical mechanics, quantum mechanics, neural networks, energy minimization, fractals, or chaos. Seemingly the problem is a hard one.

Our contribution deals with yet another approach that is best described by the term "knowledge-based mean fields." Like it or not, this term describes its main concept. The quest is to extract a force field or energy function for protein solvent systems from a database of experimentally determined structures. It is one of our goals to present the reasoning behind this approach. Construction of a force field, knowledge-based or otherwise, is not in itself rewarding. What we need to show then is that the force field has predictive power. This is our second goal. Even predictive power is still not the desired level. What is needed after all is applicability to real problems. Since we cannot ignore this point our third goal will be to demonstrate the applicability of the approach.

The presentation reflects our personal view of protein folding and is not intended as a review of the various approaches in the field. There are many excellent up-to-date reviews available. The material is presented in a way that we hope is digestible for the nonspecialist in the field. There are only a handful of equations (after all it is theoretical work), but they are not essential to follow the main stream of ideas. We hope the text is understandable for the uninitiated but not trivial for the expert. The material has been reviewed recently in a more technical language (Sippl, 1993). We include new and yet unpublished results, notably on fold recognition, which we hope will also attract the professional protein folder.

2. The Classic Protein Folding Problem

When the first structures were revealed by x-ray analysis, biochemists were struck by the beautiful topologies of their backbone folds and soon researchers in the field became eager to collect structures. Much like zoologists and botanists in past centuries they developed systematic schemes and looked for common features among the various families of folds hoping to unravel the underlying theme responsible for their bizarre structures.

A frequent phenomenon in science is that new discoveries and developments provide answers to old questions but at the same time they lead to a collection of new puzzles and problems. The first three-dimensional protein structures determined were myoglobin and hemoglobin (Kendrew et al., 1960; Perutz et al., 1960). They provided the key to a detailed understanding of the molecular processes involved in oxygen binding and transportation, they explained the cause of sickle cell anemia in molecular detail, and the first enzyme structures solved revealed several intriguing mechanisms of biological catalysis and molecular control (see Fersht, 1985, for an introduction to enzyme mechanisms).

But soon a wealth of new problems and questions arose. One of these is the now classic protein folding problem: Given an amino acid sequence, predict its three-dimensional structure (Anfinsen 1973). The commandment is simple and easy to grasp. It is inevitable that the curious student confronted with sequences and structures starts to develop his own hypotheses on the way proteins fold, and some even become badly involved. They fall in love, pushing aside the advice of their wise seniors and suffering from all the ups and downs of such an affair. In spite of tremendous efforts of many research groups and individuals, the problem is still there and largely unsolved.

Presently, many rules are known that are important for the stability and folding of protein chains (the collection by Fasman, 1989, is a bonanza, as are the recent reviews in *Current Opinion of Structural Biology*, e.g., Thornton, 1992; Dill, 1993; Dyson and Wright, 1993; Fersht and Serrano, 1993). In some special cases the structure of proteins has been predicted ahead of experiment (Crawford et al., 1987). Nevertheless correct prediction of structures are singular events (Rost and Sander, 1992) depending largely on a knowledge of homologous proteins whose structures are known (Blundell et al., 1988).

The protein folding problem is one of nature's puzzles that seems to be particularly suited to catch human curiosity. This is but one of the reasons the problem became well known. With the advent of automatic sequencing techniques the number of known protein sequences is exploding but the number of known structures lags far behind. The unknown structures of these sequences prevent a detailed study of their function and biological role. In addition, molecular biology provides all the tools necessary to engineer protein sequences at will, and it is even possible to design new proteins from scratch. This can be done on the sequence level, but it is hard to predict the effects of changes in amino acid sequences on the stability of their three-dimensional structures; in the design of new proteins it is difficult to predict whether the sequence will fold to a unique structure at all (DeGrado et al., 1991). Indeed, the protein folding problem is one of the major bottlenecks of molecular biology. Powerful techniques are available but the properties of their products cannot be predicted. Understanding protein folding and a working solution of the folding puzzle would remove the bottleneck, paving the way for a tremendous number of exciting applications.

3. Approaches to the Folding Problem

There are two main streams of approaches to tackling the protein folding problem: statistics and physics. The statistical approaches rely on the database of known structures solved by experimental means. In physical approaches the common theme is the search for a reasonable description of the energetic features of proteins based on physical principles. Both schools take it for granted that native structures obey a basic set of rules whose knowledge would enable the prediction of structures from sequences, but they start from opposite directions and they talk about these rules in different languages.

The goal of statistical approaches is the extraction of frequencies, propensities, and preferences. These data are combined using probabilistic reasoning in an effort to predict structures or at least some of their properties. Typical questions are: What is the likelihood that valine is at the surface of a protein exposed to the solvent? Or what is the preference of alanine to adopt an α-helical conformation? These preferences and propensities can be evaluated from a database of known structures and once they are compiled they can be recombined and employed in an attempt to predict structures from sequences.

The most popular branch of statistical approaches concerns the prediction of local structures along the protein chain. These methods, called secondary structure prediction, concentrate on the preferences of single amino acid residues to adopt one of three conformational states, α-helix, β-strand, or other (often called random). Preferences are obtained by statistical analyses of protein structures or from experimental data derived from model peptides. These parameters are then used to calculate the probable conformational states of amino acid residues in proteins. A variety of methods and extensions have been investigated in order to maximize the success rate. However, there seems to be a barrier at 60% of overall correct prediction that is impossible to break even if information on neighboring residues is included (for a recent review see Sternberg, 1992). Only if extensive information on homologous proteins is used can the success rate be raised to about 70% (Rost and Sander, 1993a, 1993b). It is now well known that conformational states of single residues strongly depend on the particular environment created by individual protein folds (Kabsch and Sander, 1984). A view of the whole protein is required even if the conformational state of a single amino acid is at stake.

Besides secondary structure prediction there is an enormous variety of statistical studies of protein structure databases. Among the most recent developments are profile methods (Bowie et al., 1991) and the investigation of surface accessibilities (e.g., Holm and Sander, 1992). The common theme is the search for those parameters that catch the most important factors responsible for the shape of native folds, since these are the parameters that necessarily have the strongest predictive power.

The guiding principle in physical approaches is the energy of a protein fold. The basic tenet is that the native structure of a given amino acid sequence has the lowest energy among all alternative conformations. If we are able to calculate the energy of protein folds, then in principle we are able to locate the most favorable structure by minimizing the energy as a function of the variables used to describe the three-dimensional structure. The construction of energy functions from basic physical principles has been the subject of intense studies and presently there are several force fields available that are used in important problems of protein structural research, be it the refinement of structures obtained from electron densities or the calculation of structures from distance constraints obtained in nuclear magnetic resonance studies. For reviews in this field see Karplus and Petsko, 1990; van Gunsteren, 1988; Brooks et al., 1988; or Dill, 1993, and Jernigan, 1992 (simplified potentials and representations).

The total energy of a protein is thought to be a combination of pairwise atomic interactions, in particular covalent bonds and nonbonded terms. The functional forms of these interactions are called potentials. They are derived from first principles like Coulomb's law or from quantum mechanical calculations. The resulting potentials contain parameters that depend on the particular type of interaction, e.g., the partial charges on the interacting atoms. These parameters are derived from experimental data or again from theoretical considerations. The resulting force fields are called semi-empirical, due to the combination of experimental and theoretical results. Once a tool kit of potentials for all interactions that occur in proteins is established, the total energy of a sequence in a particular three-dimensional fold can be evaluated.

The most pressing question is, of course, whether the global minima of the energy functions constructed from the potential-toolbox correspond to the native folds of protein sequences. If not, the energy model derived in this way cannot be used to predict native folds of proteins. Misfolded structures would then have lower energy than the native fold and any search strategy would inevitably go astray. Such an energy function is a misleading and deceptive guide in conformation space.

Unfortunately, it is hard to demonstrate the correspondence between native structures and global minima of model energy functions. An extensive and rigorous demonstration requires the generation of all possible three-dimensional folds of a given amino acid sequence (whose native fold is known from experiment) and the evaluation of the associated energies. This is computationally prohibitive even in the case of small proteins. If it is technically impossible to prove that the global minimum equals the native state property of functions, we can nevertheless try to falsify it. The equivalence of global minimum and native state is falsified, for example, when at least one structure of more favorable energy is found that is considerably different from the known native fold. Using this argument, Novotny et al. (1984) have shown that the energy calculated from a typical semiempirical force field is unable to distinguish a native fold from a grossly misfolded structure, indicating that at the present stage of development, semiempirical force fields do not have the predictive power required to determine the native folds of amino acid sequences (Novotny et al., 1988).

Both approaches to protein folding have their strengths and weaknesses. The strength of the statistical approach is its practical simplicity. The basic operations are collecting and counting data and the comparison of expected values with actual frequencies. The main difficulty is the reduction of complex three-dimensional structures to a simplified representation amenable to statistical analysis. For example, the possible conformational states of amino acid residues are exhaustively described by three categories: α, β, and random. The same is true for the magic number of seven states (Rooman et al., 1991) or any other subdivision of the ϕ–ψ map. But which is best? Often there are numerous possibilities. The optimal choice is hard to find since the question is coupled to the number of data available for statistical analysis. If the description has too much detail we have to face low counts and unreliable frequencies. If the description is too coarse we may save statistics but lose the ability to build unique structures from the reduced description. It is, for example, impossible to calculate the three-dimensional structure of a protein from the single residue states α, β, and random. A success rate of 100% in secondary structure prediction would therefore be insufficient to predict the native fold of a protein.

An additional disadvantage of statistics is the danger of getting lost. Protein folds are complicated, presenting therefore an endless variety of ways to analyze the structures. One gets easily carried away. Alanine, for example, has a strong tendency to adopt α-helical conformations. But not all helix positions are equivalent. Even proline, the helix breaker, is found quite frequently in α-helices, but only at terminal positions. Hence a more reasonable analysis has to take into account positional effects (Richardson and Richardson, 1988). But there is more. There may be a concerted action of several types of residues to start a helix and of course we would like to know the associated rules. But we cannot. At this point we run out of data.

The advantage of the physical approach is its conceptual strength in providing a general framework for the solution of the protein folding problem. All that is needed is an energy function with the property native state equals global minimum. Once such a function is known, finding the native structure will be successful, at least in theory, by exhaustive search. There are no insurmountable theoretical problems along the avenue. The weakness is of course that there are severe practical limitations.

Let us assume for the moment that we have an energy model for proteins which has the property global minimum equals native state. Then, in theory, we have already solved the protein folding problem. But how could we prove that our function has this property? All we have to do is to find the global minimum. We know that in principle we are able to find it. We even have a most simple, robust and exhaustive search strategy. But during our life span we can investigate only a tiny fraction of the approximately 10^{100} possible structures.

Even if we are lucky enough to hit the native structure early, we would only know at the end of our search. There are more sophisticated search strategies available. But none of these techniques is able to locate the global minimum of present-day semiempirical force fields. In fact, there is not a single case in which the global minimum of a semiempirical force field has been located for a protein.

4. Knowledge-based Mean Fields

Let us emphasize once more the advantages of the statistical and physical approaches. A major advantage of statistical approaches is that what you see is what you get. No assumptions about the physics of the protein system are needed when preferences or frequencies are derived from a database of known structures. If we resist the temptations of overinterpretation of data, we can be relatively sure that what we get are the imprints of the rules that govern protein folding.

Physics, on the other hand, provides us with a strong conceptual framework. The underlying concepts are very general in nature and they are very powerful in predicting real-world phenomena. In addition, using physical concepts we have the key to an arsenal of powerful techniques like thermodynamics and statistical mechanics connecting the submicroscopic world with its macroscopic consequences. The drawback is that unless we have really solved a problem (e.g., by finding global minima of functions) we do not see what we get from the basic concepts. Energy models may be good or bad, but due to technical difficulties considerable effort is required to judge their qualities.

Nevertheless, physics provides an avenue to protein folding. The recipe is simple: (1) construct an energy function that reasonably models protein solvent systems (global minimum equals native state), and (2) locate the global minimum, thereby calculating the native structure.

How can we construct a suitable energy function? In principle, there are two ways. We can try to derive an energy function from first principles. Then we have to figure out what the important forces are that form native folds as well as their relative strengths. This is a tough problem, as we have seen. An alternative approach is to derive these forces from the known three-dimensional structures of proteins (Sippl, 1990). The structures reflect the interplay of forces that are at work in protein folding. Looking at structures we see an equilibrium state, where the system is relaxed and large scale movements have ceased. We cannot see the forces directly but we observe their consequences in a relaxed and energetically favorable state. All we need, then, to construct a suitable energy function for protein folding is to extract the forces or energies from the structures. But to achieve this goal we need a powerful tool.

Now the intention is clear. The goal is to combine statistics and physics. In doing so we hope to get the benefits of both approaches by avoiding their drawbacks. In fact statistical physics, the combination of physics and statistics, has a long tradition, and is a most powerful theory. The connection of statistics and physics becomes particularly transparent in Boltzmann's principle, which is at the heart of statistical mechanics. This principle states that probabilities and energies are intimately related physical quantities. Low-energy states are occupied with high frequency but it is unlikely to observe a state of high energy. This is the essence of the principle that is most useful for our purposes. Boltzmann's principle can be written in the form

$$P_{i,j,k,\ldots} = e^{-E_{i,j,k,\ldots}/kT}/Z \tag{1}$$

where the Boltzmann sum

$$Z = \sum_{i,j,k,\ldots} e^{-E_{i,j,k,\ldots}/kT} \tag{2}$$

is used for normalization, so that the quantities $P_{i,j,k,...}$ represent a probability distribution:

$$\sum_{i,j,k,...} P_{i,j,k,...} = 1 \qquad (3)$$

(i.e., the probability of finding the system in any of the possible states is one).

The subscripts i, j, k, ... specify the possible states of the system. What exactly is the system we are looking at and what are the states? The complete system is a collection of native protein structures determined by experiment. The individual proteins are assumed to be in or at least close to equilibrium. We know from physics that atomic pair interactions are important. Native folds are stabilized by a large number of such interactions. Our goal is to derive the energy functions or potentials describing these interactions from the database.

Then our next question is, what are the variables we need to describe pair interactions? Interactions depend on the amino acid types a and b that take part in the interaction, on the particular atom types c and d involved, on the separation k of a and b along the amino acid sequence, and on the spatial separation r of c and d. These are the sub- or superscripts we need in order to apply the formulas to protein structures. To extract a particular potential from the database we have to determine the frequencies f_r^{abcdk} (e.g., how frequently do we find the C^β atoms of valine and alanine, separated by five residues along the sequences at a spatial distance between 5 and 6 Å). Since r is a continuous variable, we have to use distance intervals. After all, counting is straightforward.

What we get from the database are relative frequencies. Relative frequencies approach probability densities in the limit of infinitely many measurements. The current database of proteins is nowhere in the range of infinity. With 200 structures it is very small, in fact almost too small for reliable statistical analyses. We therefore need suitable techniques to deal with sparse data. We shall not dig into this problem here, but such techniques are available (e.g., Sippl, 1990).

The final step is to apply Boltzmann's principle by transforming frequencies to energies:

$$E_r^{abcdk} = -kT \ln(f_r^{abcdk}). \qquad (4)$$

The potentials obtained in this way are potentials of mean force. They describe the combined action of all forces (electrostatic, van der Waals, etc.) between the interacting atoms as well as the influence of the surroundings on this interaction. The potential is obtained by collecting data from a database of diverse proteins. What we get is therefore a potential describing the interaction of particular atoms taking place in a background generated by an average protein including the surrounding average solvent.

In this form, individual potentials contain more or less the same information. What we really need is the specific information contained in a particular potential that distinguishes it from the average. In other words, we need a reference state containing most of the redundant information. After subtraction of this surplus information we are left with the specific features of the interactions (Sippl, 1990). A convenient choice for the reference state is

$$E_r^{cdk} = -kT \ln(f_r^{cdk}),\tag{5}$$

which is an average over all amino acid pairs a and b. The net interaction is then obtained from

$$\Delta E_r^{abcdk} = E_r^{abcdk} - E_r^{cdk} = -kT \ln(f_r^{abcdk}/f_r^{cdk}).\tag{6}$$

This type of potential contains the specific information on the interaction.

5. Quality of Force Fields

Once the potentials have been compiled from a database they can be used to calculate conformational energies of amino acid sequences. The energy function of a molecule of sequence S is constructed from the tool kit of mean force potentials and is thus a recombination of the information extracted from the database of known structures:

$$\Delta E(x_i, y_i, z_i) = \sum_{i,j,c,d} \Delta E_r^{a_i,b_j,c,d,k}.\tag{7}$$

The summation is over all sequence positions i and j and atoms c and d where a_i and b_j denote the amino acids at position i and j in the sequence. For a given amino acid sequence S, the only variables on the right-hand side are the spatial distances r between atoms c and d, which are functions of the atomic coordinates x_i, y_i, and z_i (or any other convenient set of conformational variables). Subscripts and superscripts in the potentials are used to distinguish variables depending on conformation from those that depend on amino acid sequence. If needed, the associated force field can be obtained from the derivatives of $\Delta E(x_i, y_i, z_i)$ with respect to the conformational variables.

This is all rather straightforward. But is the energy model we arrived at reasonable? Does it have predictive power? Does it have the global minimum equal to native state property? For technical reasons we cannot rigorously prove this last property, as discussed above. But we should try, as hard as we can, to falsify it. We can try to confuse the force field by pretending that we have several native-looking conformations but we do not know which is the right one.

This can be done by hiding the known native fold of a particular protein among a large number of alternative conformations. The force field has the sequence to work with and we are curious to see whether it is able to pick its native fold. If one or more alternative conformations has lower energy as compared to the native fold, then the minimum principle does not hold and the force field necessarily is the loser in the game.

The goal is, of course, to make the search as hard as possible. There is no point in hiding the native fold among structures that can be easily rejected by their non-native features (e.g., random chains). We need conformations whose backbones, when deprived of side chains, have all the characteristics of native folds. Such a set of decoys can be prepared from known native folds (Hendlich et al., 1990). For reasons discussed in detail elsewhere (Sippl, 1993; Sippl and Jaritz, 1994) the structures are joined to a polyprotein as shown in Figure 12-1. The native fold of a test protein is hidden somewhere along the polyprotein. Then we start the game. The sequence is placed at the N-terminal position and the total energy of the sequence in the respective conformation is evaluated. When finished, the sequence is displaced by one residue towards the C-terminus of the polyprotein and the energy calculation is repeated. In this way the sequence is shifted through the polyprotein and the conformational energy is recorded at each position. Finally the energies are sorted. In the case that the position corresponding to the native fold has lowest energy, the force field is the winner. In a sense, the conformations encountered by the sequence along the polyprotein represent the accessible conformational space. They represent only a tiny fraction of all possible folds, but many of them are plausible native folds.

Before we present some results, we need to discuss an additional point. The force field employed to obtained the results discussed below is quite incomplete. It contains only the interactions between C^β atoms. This is a reduced description of a protein structure. If it works, i.e., if the force field is able to recognize a large number of native folds by repeating the game for different sequences, so much the better. We could probably refine the model by incorporating more atoms and hence more interactions. The advantage is that if we use only C^β atoms (and/or N, C^α, C', O) we do not have to construct side chains, and at the same time it is impossible to deduce the original sequence from the remaining scaffold. This is important since

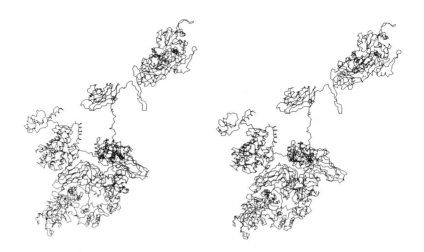

Figure 12-1. Small section of the polyprotein used in the force-field quality assessment. Each fragment of a particular length corresponds to a possible conformation of an amino acid sequence of the same length. The polyprotein represents the conformational space accessible in the hide-and-seek game. The polyprotein is also used as a background to calculate z-scores of folds.

otherwise the argument would be that the native fold could be detected by sequence comparison.

Now we take lysozyme, Brookhaven Protein Data Bank (Bernstein et al., 1997) code 1LZ3, as an example. The native fold of 1LZ3 is hidden somewhere in the polyprotein, the latter consisting of 260 individual protein modules that add up to a total length of 50,000 residues. The number of folds encountered by the lysozyme sequence is $50,000 - 129 \approx 50,000$ and hence of the same order of magnitude. Table 12-1 summarizes the result. The lowest energy of the lysozyme sequence is obtained for its native fold, a quick score for the force field. The next-best choice is a section of the polyprotein corresponding to 1LZ1, a related lysozyme whose structure is similar to 1LZ3. The energy of this fold is considerably higher but there is still a large difference in energy compared to the next structures down the list. The third position is equivalent to 1ALC, α-lactalbumin, which is homologous to lysozymes. The next structures in the sorted list are narrowly spaced in terms of energy. Here we hit the zone of folds that are unrelated to lysozyme.

In terms of energy, the 1ALC conformation does not seem to be a reasonable model for the native fold of the 1LZ3 sequence. Nevertheless, the folds are quite similar. The reason for this discrepancy is that we neglect gaps. The optimal alignment of 1ALC with 1LZ3 requires two gaps, which are not allowed in our game. This is of course unreasonable if our goal is to obtain an optimal fit between a sequence and a structure. We will have more to say on fold recognition below, but here we concentrate on the quality of the mean-field energy in terms of its ability to identify the native fold of a given sequence.

But why not allow for gaps in this game? If so, the game would even be harder. Instead of 50,000 alternatives we would have an enormous number of possible combinations of a sequence with structures along the polyprotein, since every deletion of a residue or insertion of a gap would produce an additional decoy. So why not use this more demanding variant? For a profound reason: If we open gaps and delete residues, we change sequences. It may happen in our game that by deleting a few residues the sequence does not find its native fold. Would this indicate that the force field is an inadequate model of the system? What if the altered sequence indeed adopts a different fold? We simply would not know. In our game we have to investigate different states of a unique system. If we change the sequence, we change the system and we lose control, and our conclusions become invalid (Sippl, 1993).

We have seen that the force field manages to identify the native fold of 1LZ3. It did not confuse the native fold with the related 1LZ1 conformation nor any other alternative fold. This of course does not prove the global minimum equals native fold property, but it indicates that we are on the right track, at least in the case of lysozyme. But what about other proteins? Before we repeat the game for our complete database we have to make sure that we are able to compare the results in a meaningful way.

The guiding principle in the search for the native fold is the total energy calculated from the mean force potentials. The energy can be used to distinguish different folds of one and the same sequence, but the comparison of energies of different sequences is not as straightforward. Proteins of different size, for example, have different energy content. In large proteins there are more interactions resulting in large energies as compared to small proteins. Therefore, the range of energies calculated from the polyprotein depends on the particular sequence chosen. If we want to compare the results obtained for different proteins we have to normalize the energy ranges.

The ability to recognize the native fold is one important feature we want to investigate. But we want more. We want to know what the significance or strength of the force field is in recognizing the native fold of a given sequence. The larger the distance between the energy of a fold and the average energy along the polyprotein, the better. This distance can be normalized by the spread of energies around the mean value, yielding the z-score of a given fold:

$$z = \frac{E - \bar{E}}{\sigma},$$ (8)

where E is the energy of some fold, $\bar{E}$ is the average energy of sequence S with respect to the conformations in the polyprotein, and σ the associated standard deviation.

Table 12-1. Hide-and-seek result for sequence of lysosyme 1LZ3.

Rank	fragment[a]	position[b]	z-score	energy[c]
1	ILZ3	0	−8.15	−174
2	ILZ1	1	−6.26	−117
3	1ALC	−3	−4.12	−53
4	1LFG	−72	−4.08	−52
5	4DFR-A	−48	−3.48	−34

[a] Protein module in polyprotein
[b] Relative position of first residue in sequence to the start of the module in the polyprotein (e.g., 1ALC −3 means that the sequence of 1LZ3 starts three positions upstream from the N-terminus of 1ALC in the polyprotein)
[c] In units of E/kT

Table 12-1 lists the z-scores obtained for the 1LZ3 sequence. The native z-score indicates that the force field is able to recognize the native fold with a significance of more than 8 units of standard deviation. The score for the related 1LZ1 conformation drops by more than two units. The score of 1ALZ is in a range where the results become insignificant.

There is an additional point we want to demonstrate. A quality check of a knowledge-based mean field is only meaningful if it does not contain specific information on the protein under study. In other words, the protein has to be removed from the database and the potentials have to be recompiled before we start the game. This technique is known as the jackknife test. In fact the results assembled in Table 12-1 were obtained using a force field devoid of 1LZ3, 1LZ1, and 1ALC (it did not contain any other related molecule). If information on 1LZ3 is included in the force field, the native z-score rises by almost two units (Table 12-2). Now we are ready to jackknife the database. The mean force potentials are calculated from approximately 200 proteins. We take every protein one at a time, remove it from the database, and recompile the mean force potentials. Then we take the sequence of this protein, shift it through the polyprotein, and record the

Table 12-2. Hide-and-seek result for sequence of lysozyme 1LZ3 when the molecule is not excluded from the mean force potential database.

Rank	fragment	position	z-score	energy
1	ILZ3	0	−10.00	−231
2	ILZ1	1	−7.03	−141
3	1ALC	−3	−4.55	−66
4	1LFG	−72	−4.09	−52
5	4DFA-A	−48	−3.49	−34

energies of the sequence in the various conformations encountered. Finally the energies are transformed to z-scores. This procedure is repeated for all molecules in the database.

The results are summarized in Figure 12-2, where the z-scores of the native folds of the individual proteins are plotted as a function of sequence length. Each diamond corresponds to the result of one hide-and-seek run. With a few exceptions discussed below, in all cases the native fold has lowest energy and score. Hence, for almost all proteins in the database the mean-field energy is able to distinguish the native fold from a large number of alternative conformations. Again, this is no proof of the global minimum equals native fold property, but the result is encouraging.

On the other hand, the game seems to be easy to win. How significant is the result? If we apply a blind test, picking conformations at random, in a single game the odds are one in 50,000 to hit the native fold. Obviously the mean field outperforms the blind test. But we do have a stronger argument. A substantial fraction of folds along the polyprotein do have all the characteristics of compact globular folds. These are indeed genuine decoys. We know from previous studies that it is by no means trivial to distinguish a native fold from a deliberately misfolded structure. To pick the correct fold among many native-like decoys is obviously even more demanding. In fact, our early versions of the mean field still lost out on a substantial fraction of folds (Hendlich et al., 1990). After all, by using genuine protein folds as decoys, we have made the game as hard as we presently can.

An alternative would be to minimize the mean-field energy starting from some arbitrary conformation. If the native folds can be identified, then we would have a general and robust technique for the calculation of native folds from amino acid sequences. This is, in fact, the ultimate goal. Minimization of conformational energies, however, poses difficult technical problems. We have already touched on this issue. The mean-field energy function has some peculiarities of its own, which need to be considered

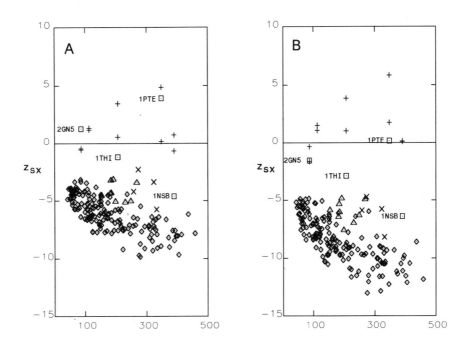

Figure 12-2. Summary of the results obtained for all proteins subjected to the hide-and-seek test. The z-scores of native folds are plotted as a function of sequence length of the respective proteins. (A) z-scores obtained from a force field constructed from cd–cd interactions only; (B) from C?–C? interactions only. The different symbols denote soluble globular proteins (?), chains of the membrane associated photosynthetic reaction center (?), viral coat proteins (A), proteins of unusual z-scores (in), and deliberately misfolded proteins (i), respectively. In B, all proteins except misfolded 2GN5, 1THI, and 1PTE are recognized by the force field as native folds. The rank of 2GN5 with respect to c? interactions is 3326, i.e., there are 3325 fragments in the polyprotein that have lower energy as compared to the 2GN5 fold reported from x-ray analysis. The respective rank obtained for 1THI is 82. In the case of 1PTE the reported fold occupies rank 28857.

in minimization studies. We will come back to this point later, but one of these features is that the force field employed is incomplete.

This is surprising. The native folds of (almost) all proteins can be identified using a knowledge-based mean field that contains C^β–C^β interactions only. If we compare the physical and chemical properties of the C^β atoms of the standard amino acids (except glycine), there is not much difference. If these interactions are modeled from first principles, the functional forms are similar. There may be some difference in the partial charges of the individual C^β atoms, for example, but the differences between the various possible C^β–C^β interactions are marginal. As a consequence two different folds having roughly the same distribution of C^β–C^β distances (e.g., two globular folds) would have indistinguishable energies.

In contrast, the functional forms of mean-force C^β–C^β potentials differ dramatically, depending on the amino acid types involved. This is one of the key features of the mean-field approach. The C^β atoms feel their surroundings. They hang out like probes. When we record their interactions, we capture information on their average environment in the form of mean force potentials. Since the side chains of the respective amino acids are part of this environment, the potentials strongly depend on the amino acid types involved (Sippl, 1990).

Nevertheless, the mean field constructed from the C^β interactions is the most reduced model conceivable. Many important interactions are missing. Interactions of amino acid residues with cofactors and prosthetic groups like the heme in myoglobin are neglected, although these components are important in the stabilization of native folds. In some cases such interactions are essential for the native structure, like the iron-sulfur clusters of small proteins, for example, or the covalently attached heme group in cytochrome. The significance of the results of several such proteins is markedly lowered as indicated by their respective z-scores, and even in these cases the force field is able to recognize the native fold.

The native z-score indicates how well the knowledge-based mean field separates the native fold from the background of misfolded decoys. There is a clear tendency for native z-scores to become more significant with increasing sequence length. At least one reason for this tendency relates to the information content of protein folds. Obviously, a small protein contains less information than a large one, since the number of interactions stabilizing the native fold of a small protein is much smaller as compared to a large protein. The functional form of this dependency of information content on protein size should be somewhere between linear and quadratic. Hence the total energy distinguishes the native fold of a large protein, as compared to a small protein, more precisely. Note, however, that our argument requires that on average the individual interactions in different proteins are of comparable strength and that the folds are globular.

As a control, Figure 12-2 contains the results obtained for deliberately misfolded proteins (+). Misfolded proteins are obtained by combining sequences with unrelated folds. None of the misfolded proteins is identified as a native fold (we would be in trouble otherwise) and their z-scores are far from the bulk of scores obtained for native folds.

The triangles in Figure 12-2 correspond to viral coat proteins. Coat proteins are in some ways unique. Individual coat proteins strongly associate, thereby creating a protective cover for the genetic material. The resulting overall structure is reminiscent of a membrane shielding the nucleic acids from the surroundings. Although many of the individual coat proteins are more or less globular, they often contain large loops protruding from the protein's main body, enabling strong interactions with neighboring molecules. Since the mean field is derived from a set of globular proteins, it is interesting to look at virus coat proteins separately. Their z-scores are less significant but their native folds are recognized.

Membrane-bound proteins are a different class. Depending on the position at or in the membrane, this class of proteins exists in a more or less hydrophobic environment. The four chains of the photosynthetic reaction center associating in varying degrees with membranes are again recognized by the mean field, although the results are less significant and the z-scores of these folds are recognizably separated from those of soluble globular proteins.

Nevertheless, native folds of membrane proteins are recognized by a mean field derived from a set of soluble proteins. This seems surprising at first, since in at least one important aspect soluble and membrane proteins are vastly different. Membrane proteins have hydrophobic surfaces and they are turned inside out as compared to soluble proteins. Why then is it possible to detect these folds by a mean field derived from soluble globular proteins? Perhaps the two classes of proteins are not so different after all. The nature of intramolecular interactions in both types of proteins is likely to be the same. At least it is difficult to imagine that they are fundamentally different. They are, however, modulated by the surrounding environment. Both classes of proteins have much in common while differing in several important aspects. The mean field derived from soluble proteins, therefore, provides a reasonable model for several important aspects of membrane proteins, but for some of the energetic features of membrane associated proteins, the model is insufficient, resulting in a considerably lower significance of native fold detection.

There remain three proteins, represented by squares in Figure 12-2, that are troublemakers. These are soluble and more or less globular proteins whose native folds are not identified. In addition, the native fold of 1NSB, a neuraminidase B subunit, has an unusual z-score, although the native fold is still recognized as the most favorable fold along the polyprotein. But even more striking is that the native fold of 1NSB has a positive mean-field energy, which is unusual for a native fold. Also, the energies calculated from the different folds along the polyprotein are distributed in a comparably narrow range resulting in an unusually small standard deviation. What is the trouble with these proteins? There are two possible answers. The mean field is an incorrect model for the energetic features of these proteins, or the structures do not correspond to the native folds of these proteins. Errors in three dimensions have been detected in several cases, so we have to be aware of this possibility (for reviews see Branden and Jones, 1990; Janin, 1990). In any case neither of the two possibilities is very pleasing, but we have to face it.

With a few exceptions the mean field is able to recognize the native fold in the hide-and-seek game, indicating that the mean-field approach is a possible route to the construction of force fields that do have predictive power. It would, however, be reassuring if the force field successfully passed at least one additional test independent of the hide-and-seek game.

Suppose we have a number of model folds for the native structure of a given amino acid sequence, compatible with a set of rules or parameters, like a set of distance constraints derived from nuclear magnetic resonance studies, for example. Since the structures do not violate these constraints, they are indistinguishable by the parameters used to construct the models. The individual models in the set may nevertheless be quite different from each other. Then we are confronted with the question of which of the structures are the most similar to the native fold.

One approach to the construction of models for native protein folds has been pioneered by Fred Cohen, Michael Sternberg, and William Taylor (Cohen et al., 1982). They address the following questions: Assume that the location of α-helices and β-strands of a particular protein are known, say from a (yet to be developed) reliable secondary structure prediction or from experimental data. Then impose the additional constraint that the polypeptide chain has to adopt a compact more or less globular fold. How many folds are possible? Are there only a few conformations left that are compatible with the prescribed local structure and compactness? Or is it still possible to construct a large number of models using these prescriptions?

To investigate the reduction of possible structures, Cohen et al. (1979, 1982) actually constructed models for myoglobin and ferredoxin using the known helices and strands and the appropriate compactness (which is a function of sequence length). Their combinatorial program produced several folds that match the given constraints, but as expected the number of possible folds is considerably reduced. The folds constructed in this way have many features in common with the native folds of these proteins, but the relative orientation of helices and strands in the models differ from those of the native folds.

An interesting question is whether the native folds can be distinguished from the models by additional criteria. Hence the set of combinatorial models provides an interesting test set of decoys that, as can be inferred from the results reported by Gregoret and Cohen (1990) and Holm and Sander (1992), are not easily distinguished from the native folds.

Figure 12-3 demonstrates that the native folds of myoglobin and flavodoxin have lowest mean force energy as compared to the combinatorial models. Again the native fold is recognized as the most favorable structure in both cases. The Figure also shows that there is a correlation between low energy and spatial similarity of the model structures to the native fold. These results again demonstrate the predictive power of the force field.

6. Mean-Field Analysis of Protein Folds

We now take a look at the energetic architecture of protein folds and the distribution of energies in terms of the mean field. Protein folds can be characterized and even classified in terms of their spatial architecture. Two proteins have similar spatial architecture if their backbones trace out similar paths in three-dimensional space. The three-dimensional scaffold of proteins must be kept in its place. Otherwise the chain would fall apart due to the constant bombardment of thermal agitation. Where is the glue that holds the scaffold together? Are there particularly stable or unstable regions in the native structure? Or is the glue more or less equally distributed along the chain? Looking at two native folds of similar three-dimensional structure, are the stabilizing nuts and bolts at comparable positions in both molecules, or is the energetic architecture of the molecules different in spite of the fact that the folds are similar in three dimensions?

These are but a few questions vital in understanding protein folding, in modeling by homology, or in the rational design of protein folds. To obtain some insight concerning these issues we need glasses that make the energetic architecture of protein molecules transparent.

Right from the start our working hypothesis has been that native folds of protein molecules are stabilized by intramolecular interactions and by interactions of the molecule with the surroundings. From the results obtained in the hide-and-seek game we may assume with some confidence that the mean field obtained from the database of known structures is a reasonable albeit approximate energy model of proteins in an (aqueous) solvent system. But now instead of calculating total energies of structures we look at the components that add up to the total energy.

Consider a particular residue in the native fold of a protein. Does it feel comfortable in its surroundings? Does it contribute to the stability of

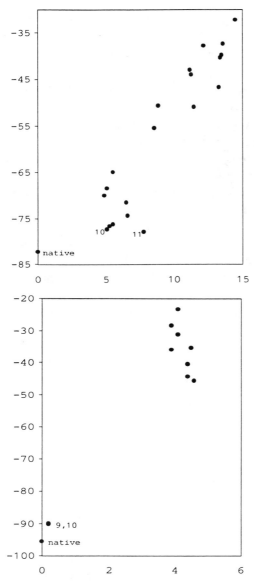

Figure 12-3. Energy of combinatorial models of myoglobin (top) and flavodoxin (bottom) plotted as a function of the root mean square error of optimal superimposition of the respective models to the native folds. In both cases the native fold has lowest energy.

the native fold by favorable interactions? Or is it a source of instability in the protein? We can address such questions by calculating the interaction energy of a residue with its surroundings from the mean field. When these energies are plotted as a function of sequence position we obtain a snap shot of the energy distribution in the structure from a certain viewpoint. Since in native folds the energetic interactions have come to equilibrium, high energies reveal strained sections along the chain, which have to be stabilized by regions of low energy.

What is the shape of such energy graphs calculated from native folds? If the graph is plotted with highest resolution on a per residue basis, the graph fluctuates strongly. Stabilizing residues alternate with destabilizing residues in a seemingly erratic manner. However, if the graph is averaged using a window size of, say, 10 residues, patterns become discernible. Figure 12-4 assembles energy graphs of several native protein structures of diverse spatial architecture, ranging from 1ADK, adenylate kinase, consisting mainly of α-helices to 1FB4-L, the light chain of an immunoglobulin, whose two domains are assembled from β-strands.

Our main interest here is not the fine structure of energy graphs, but rather their overall shape. A large window size of 50 residues is appropriate for this purpose. The main feature common to native folds is that the resulting graphs stay well below the zero baseline, indicating the balanced energy distribution in the three-dimensional folds. At this low resolution some features are still visible, like the two-domain structure of 1FB4-L.

What is the shape of the energy graph of a deliberately misfolded structure? In Figure 12-5 we compare the native folds of 1HMQ, hemerythrin, a four-helix bundle, and the variable domain of the immunoglobulin 2MCP-L, a β-strand sandwich, with two misfolded variants generated by exchanging the sequences of the two folds. Graphs of the misfolded variants stay well above the zero baseline, pointing to extremly strained conformations that clearly disqualify these structures as candidates for the native folds of the respective sequences. We have to emphasize that this strain does not arise from steric clashes. The energy is calculated from the C^β interactions only.

Figure 12-4. Energy graphs of dihydrofolate reductase (3DFR), leghemoglobin (1LH1), adenylate kinase (3ADK), and immunoglobulin light chain (1FB4-L). The graphs are typical for native sequence structure pairs. Graphs obtained from a small window size (10 residues) have only a few small positive peaks. Graphs obtained from large window sizes (50 residues) stay below zero. The examples cover a wide range of protein architectures. 1LH1 belongs to the all-α class and 1FB4-L is an all-β protein. In this and the following figures thin lines correspond to an averaging window size of 10 residues, bold lines to 50 residues, respectively. *See figure on following page.*

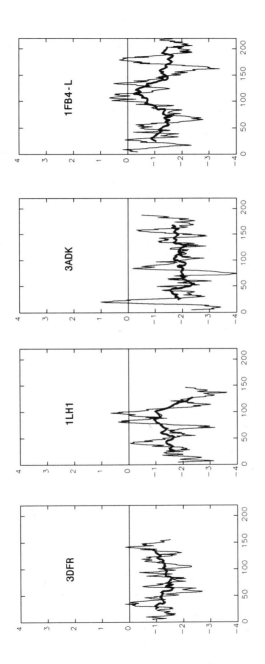

The high energies are a consequence of the incorrect backbone traces of the misfolded structures.

In Figure 12-6 we compare another two folds determined by x-ray analysis with their two misfolded variants obtained by crossbreeding. One of these folds, 2GN5, caused troubles in the hide-and-seek game. The mean field was unable to recognize this structure as the native fold of the 2GN5 sequence. Now the energy graph calculated from the x-ray structure of 2GN5 points to a strained conformation, comparable to the two misfolded variants as shown in the same figure.

Let's look at the energy graphs of the remaining troublemakers, the one obtained for 1THI, thaumatin, for example (Figure 12-7). In this case the graph reveals a highly imbalanced energy distribution. High energies on the N-terminal side point to a strained conformation in this section of the protein chain. Then the energy drops steeply, staying below the zero baseline for the remainder of the chain. The fold of 1PTE, D-alanin-D-alanyl carboxypeptidase, produces an energy graph of overall high energies and finally the graph of 1NSB, neuraminidase B, fluctuates around the zero baseline with occasional positive and negative peaks (Figure 12-7).

We are still faced with two possible causes for the abnormal behavior of these proteins: (1) The force field is inadequate for these proteins, or (2) something is wrong with these folds. Some experimentally determined protein folds have previously been proved to be wrong by repeated structure determination. At low resolution it is often hard to correctly trace the polypeptide backbone or to identify amino acid side chains. Frequent errors in early phases of structure determination are mistraced backbones or frame-shift errors, i.e., a correct backbone trace where the sequence is out of register by one or more residues.

There are some recent additional experimental data available for two of the structures causing troubles. The structure of thaumatin has been recently refined from 3.1 to 1.65 Å resolution (Ogata et al., 1992). The original structure (1THI) has several frame-shift errors (i.e., at several positions the amino acid sequence is out of register by one or several residues) which

---

Figure 12-5. Comparison of the observed (x-ray) hemerythrin (lHMQ) and the variable domain of mouse myeloma immunoglobulin (2MCP) structures and two deliberately misfolded pairs. The misfolded pairs are obtained by exchanging the sequences of 2MCP and 1HMQ. 1HMQ on 2MCP corresponds to the 1HMQ sequence folded in the 2MCP conformation. 2MCP on 1HMQ corresponds to the 2MCP sequence folded in the hemerythrin conformation. Graphs of misfolded proteins (high energies) are in marked contrast with graphs obtained for observed sequence structure pairs (low energies). *See figure on following page.*

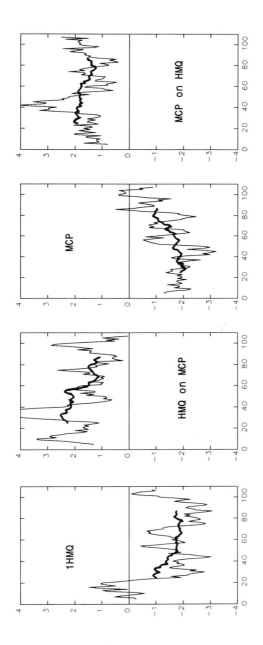

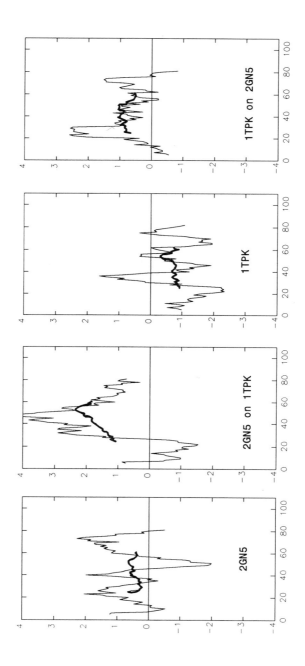

have been corrected in the refinement process. In addition loop regions had to be re-modeled. The errors in 1THI exactly match the high energy regions in its energy graph.

The second case is 2GN5, which has been determined to a resolution of 2.3 Å (Brayer and McPherson, 1983). Recent nuclear magnetic resonance studies on a variant containing histidine at sequence position 41 instead of tyrosine in the wild type are in conflict with the structure obtained from x-ray analysis. Although, in the latter study the molecule has been shown to consist predominantly of β-strands the location of these strands is in marked contrast to the x-ray structure (Folkers et al., 1991). No additional experimental data are available on 1PTE. 1PTE (like 1THI) is a low-resolution structure (2.8A) and the Brookhaven file contains only the C^{α}-trace of the molecule (Kelly et al., 1985).

1NSB has been determined to 2.2 Å resolution. In addition to the protein monomers, 446 water molecules have been included in the refinement process, yielding a very low R-factor of 14.8% with good bond length and valence angle geometry in the final model (Burmeister et al., 1992). Low R-factors are indicative of high-quality structures, but inclusion of water molecules in the refinement process may minimize R-factors without improving the model of the protein chain. The structure of the related neuraminidase A is structurally similar to 1NSB and yields similar results in our analysis.

Let us finally resolve the troubles as far as we can. The results obtained on 1THI are in agreement with the recent structure refinement. 2GN5 is in conflict with nuclear magnetic resonance studies and it is a strained structure in terms of the mean field. In the case of 1PTE we have to await further refinement. There remains only one (medium resolution) structure, 1NSB, for which the mean field seems to be an inappropriate model. It will be interesting to isolate the features that seem to confuse the mean field.

7. Prediction of Protein Folds

From the results obtained so far we have gained some confidence in the knowledge-based mean field. Of course, the model is still incomplete and

Figure 12-6. Comparison of the 2GN5 and 1TPK folds obtained from x-ray analysis with deliberately misfolded pairs. 2GN5 on 1TPK corresponds to the 2GN5 sequence folded in the 1TPK conformation and 1TPK on 2GN5 corresponds to the 1TPK sequence folded in the 2GN5 conformation. The graph of 1TPK is typical for native-like sequence structure pairs. The 2GN5 graph resembles misfolded energy graphs. *See figure on preceding page.*

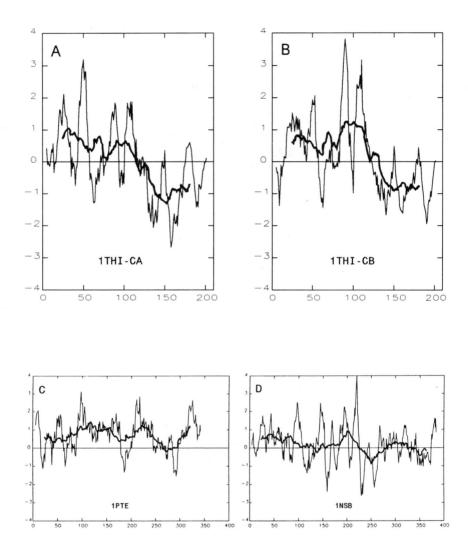

Figure 12-7. Energy graphs of thaumatin (1THI) are calculated from C^α–C^α interactions (A) and C^β–C^β interactions (B). The energy graphs of D-alanyl carboxypeptidase (1PTE) (C) and neuraminidase B (1NSB) (D) are calculated from C^β–C^β interactions. Both 1THI graphs resemble those of misfolded chains in the N-terminal section but they are native-like at the C-terminus. Comparison of the C^α and C^β graphs reveals only minor differences. Both graphs are reminiscent of partially misfolded proteins. It is noteworthy that energy graphs can be calculated from the C^α trace of a conformation, a feature that is of particular value in early interpretations of electron densities.

there is ample room for improvement. Present main goals in our research concern the explicit treatment of protein–solvent interactions, addition of side-chain atoms, and the approximation of probability density functions from sparse data, to mention a few. The predictive power is considerably raised if explicit solvent terms are added to the force field. In spite of the still incomplete force field, temptations are strong to explore its ability to predict structures and to apply the current model in protein structural research.

In order to actually predict structures we have to follow the second advice in our physical recipe. We must find techniques that enable the calculation of low-energy structures. The most general method is the global minimizer. If the energy function used to describe the protein solvent system has the global minimum equals native state property, and if we have a global minimizer that is fast enough to locate the global minimum using present-day computer technology, then we have a general technique at hand that solves the protein folding problem. Unfortunately, there is no global minimizer at hand.

But do we really need a global optimizer? It depends on the complexity of the energy function. The dream of any protein folder is a function having a single minimum corresponding to the native fold. Then protein folding would be a very enjoyable indeed. The native fold would be easy to find by rolling down the energy surface until we hit the ground. Such a function has a built-in global minimizer. At the other extreme there is the nightmare of a function shaped like an egg box with an enormous number of local minima, all of roughly the same depth as the global minimum. Imagine stumbling around blindly on such a rough surface, with no idea where to go. One might even stand on the global minimum without recognizing it, continuing an aimless search.

What are the features of the real energy function of the protein–solvent system? We do not want to speculate too much about this issue, but what we can state is that the real-world energy function cannot be too far from the ideal case. If the real function has many minima separated by high-energy barriers, then thermal agitation would be insufficient to drive the folding of proteins. The individual molecules would hang around randomly, unable to find their common native fold.

What is the shape of the mean-field energy surface? We do not know yet, but the function seems to be well behaved. It is of course impossible to visualize a complex, high-dimensional energy surface, but nevertheless we can get several clues on its shape. Consider Figure 12-3, for example, where the energy of folds is plotted as a function of spatial similarity to the native structure. The more the model resembles the native state, the lower the energy of the model. If the function would be comparable to an egg-box, the correlation between spatial similarity and energy would be hard to explain.

There are more diagnostic results. Figure 12-8 shows the energy graphs of two conformations of α-cobratoxin. 1CTX denotes a model obtained from an early interpretation of electron density, 2CTX a refined version. Both graphs show unusually high energy peaks in the region of residues 30 to 40, but we want to emphasize that the refinement resulted in an overall improvement of the energy (note that the mean field was not used in the refinement). The example demonstrates that the energy drops when a refined structure is approached from an unrefined model.

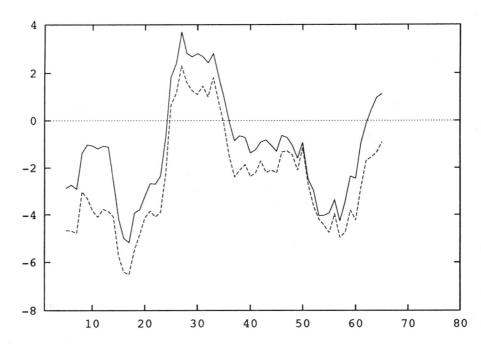

Figure 12-8. Energy graphs of two α-cobratoxin structures using a window size of 10 residues. The solid line is calculated from 1CTX corresponding to 2.8 Å resolution and the dashed line from 2CTX, which is at a resolution of 2.4 Å.

What happens if the mean-field energy is minimized starting from an arbitrary conformation? This is a tough question for several reasons. First of all, the present energy model is incomplete. There are no side chains, for example. If the energy is minimized only as a function of C^β interactions we have to watch out for artifacts enabled by the reduced geometric constraints. There are additional problems. Most minimizers rely on the derivatives of functions that they require to find the most promising direction in which to go. The mean-force energy is a discrete function, however. It is possible to calculate derivatives of discrete functions numerically, but one has to figure out the best way to do it. A similar question concerns the choice of a suitable minimizer, since its performance strongly depends on the type of functions involved. These are technical problems for which a working solution has to be found before general conclusions can be drawn.

One of the molecules we use to study these questions is the homeobox domain of the antennapedia peptide whose structure has been solved by Kurt Wuethrich's group at the ETH in Zurich (Billeter et al., 1990) using nuclear magnetic resonance. The native fold consists of three α-helices that fold to a compact core. Both ends of the chain are flexible, having little discernible structure. The question is whether the native fold can be found when we pretend that we do not know the structure.

After varying degrees of excitement and disappointment we found a recipe. First, the conformational preferences of all possible hexapeptide fragments are calculated as described by Sippl (1990). Then the overlapping fragments are assembled to a complete model as reported by Sippl et al. (1992). As a byproduct the calculations yield variabilities that identify stable parts of the chain. At this stage the three helices of the antennapedia peptide are formed. The resulting structure is minimized using a Monte Carlo technique that preferably changes those parts of the chain that have high variability as determined in the assembly process. The chain folds quickly to a compact fold. In our initial studies, minimization usually terminated at a compact globular fold built from three helices consistent with the experimental data. But the topology of the folds was usually wrong, and even worse, they had lower energy as compared to the experimentally determined folds.

In parallel, however, we were constantly improving the force field. The first published version still missed a lot of native folds in the hide-and-seek game. But when the quality of the force field approached the current state, minimization of the antennapedia peptide terminated in folds similar to those obtained from experiment (Hendlich and Sippl, unpublished). These results cannot be generalized. There are still many open technical issues concerning the possibility of *ab initio* folding using the mean field. But again we see from these studies that the function is well behaved. Minimization does not get stuck in local minima and the region of lowest energy is in the region of the native fold in spite of the incompleteness of the energy function and the neglect of side chains.

8. Recognition of Folds

There is still some work to do before tertiary structures can be calculated from scratch. On the other hand, the incomplete version of the mean field seems to be competent in recognizing the native backbone fold of a given amino acid sequence. Is there a possibility to exploit this feature for the prediction of protein structures by circumventing the *de novo* construction of folds? In fact there is.

The field of protein structure theory has seen exciting new developments over the last few years. Among these are inverse folding and fold recognition. The impetus for these approaches comes from several directions. It is now well known that three-dimensional structures of proteins are far more conserved in the course of evolution than amino acid sequences. Initially it came as a surprise that the structures of vertebrate globins are similar to a globin isolated from legume root nodules, although the similarity on the sequence level is at best sparse.

With the determination of more and more three-dimensional protein folds, it became clear that this is a frequent phenomenon. Proteins that are unrelated on the sequence level frequently adopt similar folds in three dimensions. There are now several examples of proteins having very similar three-dimensional topology but diverse function (for reviews see Murzin and Chothia, 1992; Overington, 1992). Nowadays it becomes increasingly likely that the architecture of a newly determined protein structure is similar to some protein fold already known, although this could not have been inferred from the amino acid sequences.

But the number of different fundamental protein architectures may be small, so that sooner or later x-ray crystallographers will have determined all the folds adopted by biological proteins. This number has been estimated recently by Cyrus Chothia (1992) to be in the order of 1000 distinct folds. Compare this number to the estimated 100,000 genes in the human genome, for example; then on average 100 proteins in the human genome adopt a related fold.

A second impetus is exerted by the growing database of known DNA and protein sequences. The number of sequenced genes is exploding, and in spite of the accelerated pace of structure determinations facilitated by new technical advances in x-ray analysis and nuclear magnetic resonance, the number of determined structures lags far behind. The coming years will see an additional acceleration of genome sequencing and by the turn of the millennium it is expected that a large fraction of the human genome and the genomes of other organisms will be deciphered. The ultimate interest is in the structure and function of the encoded proteins, and as long as their folds are unknown, little can be inferred from their sequences unless homologous proteins of known structures can be identified.

Here we arrive at the playground of inverse folding and fold recognition. The quest is to identify sequences that fit into a given fold as outlined in Figure 12-9. The hunting district has enormous dimensions. Presently, the number of different folds is on the order of 300 as opposed to several thousands of sequences whose structure is unknown, not counting those whose structure can be inferred from sequence comparison techniques. Using the current databases the sequence structure combinations that can be formed are on the order of millions. And the hunting ground is rapidly expanding. There is no doubt that techniques that are able to recognize native-like sequence structure pairs will have an enormous impact on protein structure theory and molecular biology. Consequently, the development of techniques for protein fold recognition and inverse folding is an active research area.

Inverse folding has been pioneered by Eisenberg and co-workers at UCLA and already their first algorithms produced promising results (Bowie et al., 1991). The rules of the game are as follows: Take a protein fold and translate the three-dimensional structure into a linear array so that each sequence position is described by a set of parameters, like surface accessibility or secondary structure type. In a similar way, translate an amino acid sequence into the same language. The codebook for translation is a set of amino-acid-type dependent structural parameters obtained from a statistical survey of known structures. In this way, the preferences of the individual amino acid types to be exposed on the surface of protein structures, to be in a helix, and so on, can be evaluated.

With this codebook, sequences and structures can be translated into the same language (called profiles), and most importantly they can be compared. The objective is clear. If the preferences derived from the sequence at (almost) every position closely resemble those obtained from the structure, then the sequence is thought to fit into that fold. Using this technique, the results obtained by Eisenberg and co-workers (Bowie et al., 1991) are encouraging. When a globin fold is used as a structural template, their program is able to retrieve most of the known globin sequences from the sequence database. Results obtained for other folds were less clear, but these also confirmed the potential of the method.

There are several groups actively engaged in the search for workable solutions for the fold recognition problem, like Chris Sander's group at the EMBL in Heidelberg, or David Jones, Janet Thornton, and William Taylor in London. Important advances have been published recently (Jones et al., 1992; Luethy et al., 1992; Ouzounis et al., 1993; Wilmanns and Eisenberg

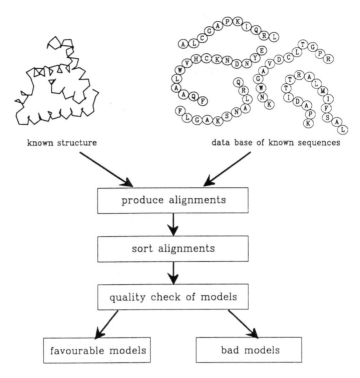

Figure 12-9. Outline of fold recognition and inverse folding strategies. A known structure is combined with sequences from a database. Following alignment production, structure sequence alignments are sorted according to their score or energy. Finally the quality of the models has to be judged.

1993), and still more are on the way. Below we concentrate on our own developments in fold recognition.

When the prototype mean field was published (Hendlich et al., 1990; Sippl, 1990) it was clear that the technique was able to recognize a substantial number of native folds. More detailed studies of the globin family revealed that the resolving power of the mean field was high enough to detect distantly related globin folds and to rank the folds in the correct order (Sippl and Weitckus, 1992), and we continued to develop techniques that could be used to align sequences with known structures using the mean field as the driving force.

Obviously fold recognition and the hide-and-seek game employed to check the quality of the mean field have a common core: In both cases the goal is to identify the native fold of a given sequence. But there is also a fundamental difference, which has to do with insertions and deletions.

As long as sequence homologies are strong, the three-dimensional folds of two proteins are very similar. But with diminishing homology the folds become increasingly dissimilar. Loops are deleted and inserted, elements of secondary structure are shortened and enlarged, and the relative orientations of structural building blocks change. As with sequences (Sander and Schneider, 1991) there is a zone of structural homology where the common theme becomes blurred.

The realm of fold recognition begins beyond the point where homologies can be detected by sequence comparison techniques. The method must be able to align a sequence whose unknown fold resembles the query structure. For distant relationships, this is bound to be a tough problem. The minimal requirement is that the method is able to delete pieces of sequence and/or structure so that gaps can be opened. This is not in itself a problem, but just a technical issue. The real trouble is that if we delete or insert residues in a given amino acid sequence, then we change the system and lose control, as noted above.

Let us illustrate the point once more. Assume we know the native fold of a given sequence. Now take the sequence and delete one or more residues. Will it still adopt the same fold as before? To be honest, we don't know! From multiple sequence comparisons we do know that a particular protein family can tolerate a seemingly large number of diverse substitutions, insertions, and deletions, but we only see the survivors. We are still unable to predict the effect of small changes in amino acid sequences on the native fold. We know that changes in loop regions exposed to the surface are less likely to disrupt the structure as compared to changes made in the protein interior and that single substitutions often do not change the features of the molecule. But at the same time small variations may completely change or destroy the native state.

Examples of dramatic structural consequences caused by minute changes in sequence have been documented. A particularly striking case is α_1 antitrypsin, an inhibitor of neutrophil elastase. The inhibitor becomes deactivated when the active site loop is cleaved (a gap is opened). Cleavage has a dramatic structural consequence. The exposed active site loop inserts into a β-strand and the two residues at the cleavage site become separated by 70 Å (Bode and Huber, 1991, 1992; Perutz, 1992). The molecule relaxes into a more favorable state, which is accessible after cleavage but inaccessible to the intact molecule. When gaps are allowed in fold recognition, we have to be aware of similar consequences.

In terms of fold recognition, the situation is even more dramatic. Here we have to deal with distantly related sequences and folds and we get into trouble. The amino acid sequence of the query fold is at best distantly related to the sequences we have to align. Hence there are a lot of substitutions. Furthermore, the alignment technique employed has to allow gaps, since the chance that two distantly related proteins can be aligned without gaps is very small. But by removing a few residues here and inserting a few residues there, the sequence becomes transformed. And here is the real danger. By introducing these changes we may end up with a sequence that nicely fits into the given structure, which is fine, but the original sequence may adopt a different conformation.

The reason for this seemingly paradoxical situation is that sequence structure alignment does not correspond to a real physical process (Sippl, 1993). When amino acid sequences fold they cannot open gaps and throw away clothes they don't like. And nobody is there to tell them that they should try to fold into a prescribed framework. In contrast, the goal of sequence structure alignment is to graft sequences on a given structure as optimally as possible by allowing to throw away residues that do not fit. But we have to pay the price. Even if the altered sequence adopts this fold, the original sequence may not. It is a fundamental problem. There is no escape.

One lesson is that right from the beginning we have to be aware of artifacts and false positives. Even if we have the most reliable technique conceivable, the most we can say is that the changed sequence fits into the structure and that the alignment provides a possible first model for the unknown fold. To proceed, one needs to build a complete model that accommodates the continuous chain and all its residues, where the alignment can be used as a starting point. During refinement one may encounter insurmountable difficulties identifying the model as a false positive.

Hence another lesson is that we should not expect to develop a comfortable algorithm in the sense that there will be a clear distinction between correct models and incorrect ones. This is a severe drawback concerning the development of fold recognition algorithms. There is no clear criterion

---

Figure 12-10. Each residue in the sequence S is plugged into the structure C and its energy is evaluated in the field generated by the sequence of fold C. The possible combinations of sequence and structure positions yield a matrix. The optimal path through this matrix is obtained by dynamic programming techniques. The path is equivalent to an alignment of sequence S and fold C. The final model, corresponding to the bold lines of the fold shown at the bottom, is obtained by removing all unpaired residues in the sequence and/or structure. Alignments can be produced for arbitrary sequence–structure combinations. A crucial final step in fold recognition is therefore an extensive quality check of the model followed by a refinement of the model. *See figure on following page.*

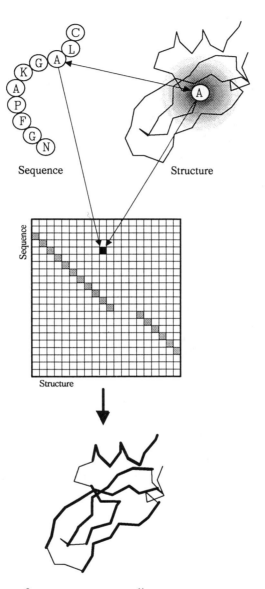

Figure 12-10. Strategy of structure sequence alignment.

for optimality. We can search for a technique that produces many correct alignments with as few false positives as possible, but this requires hard work.

There is again a straightforward recipe to tackle the fold recognition problem. (1) Produce the best alignments you can, and (2) get rid of wrong alignments as far as you can. Our current implementation of alignment production is outlined in Figure 12-10. The core of the technique is a dynamic programming algorithm developed by Needleman and Wunsch (1970) for the optimal alignment of two amino acid sequences. The algorithm requires a matrix whose elements m_{ij} describe the similarity of the amino acid at position i of sequence A to the amino acid at position j of sequence B. The output of the Needleman and Wunsch algorithm is an optimal path through the matrix equivalent to an optimal alignment of the sequences A and B in terms of the criteria used to define similarity between residues.

In fold recognition the goal is to optimally align a sequence and a structure. Hence one sequence, say B, is replaced by conformation C and we need a suitable redefinition of the matrix element m_{ij}. This time we are not interested in the similarity of two residues. What we need to know is how well residue i of sequence A fits into position j of conformation C. To evaluate this fitness we employ the mean field. The current amino acid at position j of C is replaced by the amino acid at position i of A and the interaction energy of this residue is evaluated. If the energy is low, then the amino acid fits nicely into this position, but if it is high the situation is unfortunate. In any case, the energy obtained becomes element m_{ij} of the matrix. The process is repeated for all combinations of i and j resulting in the complete matrix array. Applying dynamic programming techniques like the algorithm reported by Needleman and Wunsch (1970), we get a low-energy path through this matrix equivalent to a low-energy thread of sequence A in fold C. The final model is obtained by removing all unpaired residues contained in gaps from sequence A and fold C and the total mean-field energy of the model is calculated. This completes alignment production.

But is the model just obtained a good one? We can look at the total energy. In general the total energy is a diagnostic but unreliable tool for the quality of the alignment. A more reliable parameter is the z-score of the model obtained from the hide-and-seek game. If this parameter is native-like, we have a much stronger point. We may still have an artifact as discussed above, but we are also quite confident that the trimmed sequence fits the model fold.

Alignment production is efficient. Alignment of one structure with all sequences in the swissprot database is completed within a few workstation-hours. But hide-and-seek takes some computer time. To save time, following alignment production, the models are sorted according to their energy and the hide-and-seek test is performed for a certain range of low-energy alignments only. This completes the general outline of the method.

9. A Case Study of Fold Recognition

As a case study we present the results obtained for the constant domain of the immunoglobulin 1FB4 light chain shown in Figure 12-11 (Marquart et al., 1980). The immunoglobulin family is an interesting object for several reasons. It has many distantly related members (see Kuma et al., 1991, for example) whose sequences are stored in the swissprot database (Bairoch and Boeckmann, 1991), and this is what we need in order to judge the quality of fold recognition techniques. In addition there are a large number of proteins that are thought to contain one or more immunoglobulin-like domains, although no structure is actually known for most of these sequences and the similarities on the sequence level are sparse. So there are a lot of hard test cases and the question is whether or not the method will pick at least some of these sequences. A more interesting question is whether the method picks up sequences that are not known to be related to immunoglobulins.

Our main interest lies in distant relationships. All obvious cases having more than 35% sequence homology to the sequence of 1FB4-L are therefore removed from the database prior to our search. The remaining sequences (still more than 20,000) are threaded through the constant domain of 1FB4-L. After alignment production the 200 models of lowest energy are subjected to the hide-and-seek test and re-sorted according to their native z-score. The result is assembled in Table 12-3.

The z-score obtained for the alignment indicates the quality of the model. The original sequence in its native fold yields a z-score of -8.39. But the scores are rapidly decreasing in absolute value if we go down the list. Where should we stop? Where does the zone of insignificant z-scores begin, telling us that the models produced are probably wrong? This is a complex question, but for the moment let's cut the list at z-scores of -5.0. We will come back to that point later.

The column labeled IG contains an asterisk if the swissprot entry of the respective sequence is known or suspected to have an immunoglobulin-like fold. There are many asterisks to begin with. The method picks up a variety of immunoglobulins (remember we removed the obvious cases), sequences of the histocompatibility complex (class I as well as class II), and distantly related proteins like β_2-microglobulin, tyrosine kinase, CD33 (myeloid cell surface antigen), CD1 (T-cell surface glycoprotein), Zn-α_2-glycoprotein, and more.

The column labeled h% shows percentage of sequence homology between the sequence of the query structure 1FB4-L and the respective sequence retrieved. The numbers calculated from the sequence–structure alignment are obtained, as opposed to sequence–sequence alignment, by

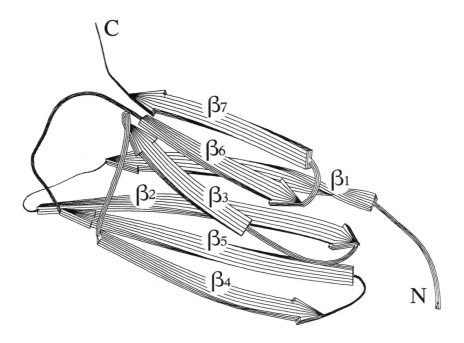

Figure 12-11. The constant domain of the light chain of immunoglobulin 1FB4 (human antibody molecule XOL, class IG) used in the fold recognition case study.

comparing the original amino acid at position j in the structure of 1FB4-L with the amino acid that resides at this position after a sequence has been aligned to the 1FB4-L scaffold. From the homologies we get an idea of the qualities of the alignments produced. Many entries having an asterisk also have sequence homologies of 20% or more. But this is only possible if the sequence is correctly aligned with the structure.

In these cases sequence homology is significant and could also be detected by sequence comparison techniques. Maybe, by using the protein fold recognition technique, we have just done conventional sequence alignment in a complicated way. There is some strength in this argument, since we used the mean field generated by the original 1FB4-L sequence to construct the matrix needed for dynamic programming and we should abandon the fancy technique and return to standard sequence comparison if fold recognition cannot do more.

After all, the goal of fold recognition is to identify sequences having no discernible sequence homology to the sequence of a given fold in spite of the

Table 12-3. *Part 1 of 3*: Fold recognition search of swissprot database using the constant domain of 1FB4-L.

Sequence	$g_a{}^a$	$g_b{}^a$	$h\%{}^c$	Z-score^d	IG^e	Molecule	Source
1FB4_LC	0	0	100	−8.39	*	IG FAB C	human
GC1_RAT	1	2	22.4	−8.01	*	IG GAMMA-I CHAIN C	rat
MUC_MESAU	′	1	28.0	−7.57	*	IG MU CHAIN C	hamster
HB2L_CHICK			24.3	−7.41	*	CLASS II HC AG	chicken
HA15_MOUSE			20.6	−7.33	*	H2-CLASS I HC AG	mouse
GC3_MOUS′			′3	−7.30	*	IG GAMMA-3 CHAIN C	mouse
MUCM_				−7.14	*	IG MU CHAIN C	rabbit
MUC				7.14	*	IG MU CHAIN C	rabbit
M′				′	*	IG MU CHAIN C	mouse
					*	IG MU CHAIN C	mouse
					*	HLA CLASS I HC AG	human
						RT1 CLASS I HC AG	rat
						RT1 CLASS I HC AG	rat
						′ CLASS I HC AG	mouse
						CLASS I HC AG	mouse
						_A CLASS II HC AG	human
B2′						BETA-2-MICROGLOBULIN	chicken
VP4_NCDV						VP4 HEMAGGLUTININ	virus
HA1Z_MOUSE	′					H-2 CLASS I HC AG	mouse
HA11_MOUSE	1	∪			*	H-2 CLASS I HC AG	mouse
HA1L_MOUSE	1	0	15.′	′.68	*	H-2 CLASS I HC AG	mouse
HA11_RAT	2	0	19.6	−6.85	*	CLASS I HC AG	rat
HB23_MOUSE	1	1	25.2	−6.85	*	H-2 CLASS II HC AG	mouse
HA1B_MOUSE	1	0	15.9	−6.84	*	H-2 CLASS I HC AG	mouse
HB22_MOUSE	1	1	24.3	−6.84	*	H-2 CLASS II HC AG	mouse
HB21_MOUSE	1	1	25.2	−6.83	*	H-2 CLASS II HC AG	mouse
VIO6_VACCV	1	2	9.3	−6.80		PROTEIN I6	virus
B2MG_MOUSE	3	0	18.7	−6.77	*	BETA-2-MICROGLOBULIN	mouse
GCAB_MOUSE	1	0	25.2	−6.76	*	IG GAMMA-2A CHAIN C	mouse
HB2I_MOUSE	1	1	25.2	−6.75	*	H-2 CLASS II HC AG	mouse
HB2J_MOUSE	1	1	25.2	−6.75	*	H-2 CLASS II HC AG	mouse
VP4_ROTH6	1	2	10.3	−6.75		VP4 HEMAGGLUTININ	virus
HB24_MOUSE	1	1	25.2	−6.72	*	H-2 CLASS II HC AG	mouse
HA1K_MOUSE	1	0	15.9	−6.67	*	H-2 CLASS I HC AG	mouse
HA13_MOUSE	1	0	15.9	−6.65	*	H-2 CLASS I HC AG	mouse
HA12_MOUSE	1	0	15.9	−6.63	*	H-2 CLASS I HC AG	mouse
HA1Y_MOUSE	1	0	15.9	−6.63	*	H-2 CLASS I HC AG	mouse
HA14_MOUSE	1	0	17.8	−6.56	*	H-2 CLASS I HC AG	mouse
HA10_MOUSE	1	0	15.9	−6.52	*	H-2 CLASS I HC AG	mouse
HA18_MOUSE	1	0	15.9	−6.52	*	H-2 CLASS I HC AG	mouse
HA1F_CHICK	2	0	15.9	−6.51	*	CLASS I HC AG	chicken
HA17_MOUSE	1	0	15.9	−6.49	*	H-2 CLASS I HC AG	mouse
MUC_SUNMU	0	1	26.2	−6.47	*	IG MU CHAIN C	shrew
GC3M_MOUSE	1	0	25.2	−6.45	*	IG GAMMA-3 CHAIN	mouse

(a) Number of gaps opened in sequence. (b) Number of gaps opened in structure. (c) Number of identical residues in sequence structure alignment. The residues in the aligned sequence are compared to the sequence of the constant domain of 1FB4-L at corresponding positions. (d) Z-score of model calculated using the conformations in the polyprotein as a background. (e) If the row contains an asterisk the respective swissprot sequence file has an IG_MHC prosite entry.

Table 12-3. *Part 2 of 3*: Fold recognition search of swissprot database using the constant domain of 1FB4-L.

Sequence	ga^a	gb^a	$h\%^c$	Z-score^d	IG^e	Molecule	Source
HA1E_PANTR	2	0	19.6	−6.43	*	CHLA CLASS I HC AG	chimpanzee
EPC_HUMAN	1	0	23.4	−6.42	*	IG EPSILON CHAIN C	human
HA2R_HUMAN	1	1	19.6	−6.33	*	HLA CLASS II HC AG	human
HA19_CANFA	1	1	18.7	−6.27	*	DLA CLASS I HC AG	dog
MUC_CHICK	0	1	20.6	−6.26	*	IG MU CHAIN C	chicken
HPRT_SCHMA	2	0	7.5	−6.23		PH-RIBOSYLTRANSFERASE	fluke
ALC1_HUMAN	0	2	29.0	−6.20	*	IG ALPHA-I CHAIN C	human
ALC2_HUMAN	0	2	18.7	−6.20	*	IG ALPHA-2 CHAIN C	human
MUC_HUMAN	0	1	26.2	−6.20	*	IG MU CHAIN C	human
HA16_HUMAN	2	0	19.6	−6.18	*	HLA CLASS I HC AG	human
ALC1_GORGO	0	2	29.0	−6.16	*	IG ALPHA-I CHAIN C	gorilla
DTCM_MOUSE	0	1	20.6	−6.16	*	IG DELTA CHAIN C	mouse
DTC_MOUSE	0	1	20.6	−6.16	*	IG DELTA CHAIN C	mouse
DPP4_RAT	1	2	6.5	−6.15		DIPEPTIDYL PEPTIDASE IV	rat
B2MG_CAVPO	1	1	13.1	−6.14	*	BETA-2-MICROGLOBULIN	pig
HA1U_MOUSE	1	0	15.9	−6.10	*	H-2 CLASS I HC AG	mouse
HA12_PONPY	1	0	18.7	−6.04	*	HLA CLASS I HC AG	orangutan
KPC1_DROME	1	0	12.1	−5.96		PROTEIN KINASE C	fly
VG10_BPT4	0	2	9.3	−5.93		BASEPLATE STRUCT. GP10	phage
KFLT_HUMAN	0	1	15.9	−5.91	*	REC-RELATED TYR KINASE	human
HA1Z_HUMAN	1	0	18.7	−5.90	*	HLA CLASS I HC AG	human
HA13_HUMAN	1	0	19.6	−5.89	*	HLA CLASS I HC AG	human
RENS_MOUSE	0	2	9.3	−5.88		RENIN PRECURSOR	mouse
CD33_HUMAN	0	3	10.3	−5.85	*	MYELOID CELL SURFACE AG	human
DTC_HUMAN	0	1	19.6	−5.85	*	IG DELTA CHAIN C	human
BPHC_PSES1	0	1	6.5	−5.84		DIOXYGENASE	bacteria
HA1K_HUMAN	1	0	19.6	−5.84	*	HLA CLASS I HC AG	human
ZA2G_HUMAN	2	0	20.6	−5.84	*	ZINC-ALPHA-2- GLYCOPROTEIN	human
RENI_MOUSE	0	2	10.3	−5.81		RENIN PRECURSOR	mouse
HA11_HUMAN	1	0	19.6	−5.76	*	HLA CLASS I HC AG	human
CD1D_HUMAN	2	0	16.8	−5.74	*	T-CELL SURFACE GP	human
HA1A_HUMAN	1	0	19.6	−5.73	*	HLA CLASS I HC AG	human
DUT_HSVI1	0	2	10.3	−5.69		DUTP PYROPHOSPHATASE	virus
HA1D_HUMAN	1	0	18.7	−5.68	*	HLA CLASS I HC AG	human
ALC_MOUSE	0	1	13.1	−5.64	*	IG ALPHA CHAIN C	mouse
HA1A_BOVIN	1	0	16.8	−5.64	*	BOLA CLASS I HC AG	bovine
HA1H_HUMAN	1	0	18.7	−5.61	*	HLA CLASS I HC AG	human
FGR3_HUMAN	1	1	7.5	−5.60	*	FIBROBL. GROWTH FACT. REC.	human
ENV_MPMV	1	2	11.2	−5.56		ENV POLYPROTEIN	virus
HA1B_BOVIN	1	0	15.9	−5.56	*	BOLA CLASS I HC AG	bovine
CG2A_XENLA	0	1	8.4	−5.52		G2/MITOTIC-SPECIFIC CYCLIN	frog
HA1F_HUMAN	1	0	18.7	−5.52	*	HLA CLASS I HC AG	human
HA1G_HUMAN	1	0	18.7	−5.52	*	NLA CLASS I HC AG	human
DNLI_VACCV	0	2	12.1	−5.50		DNA LIGASE	virus
V1A_CMVQ	0	3	8.4	−5.47		1A PROTEIN	virus

(a) Number of gaps opened in sequence. (b) Number of gaps opened in structure. (c) Number of identical residues in sequence structure alignment. The residues in the aligned sequence are compared to the sequence of the constant domain of 1FB4-L at corresponding positions. (d) Z-score of model calculated using the conformations in the polyprotein as a background. (e) If the row contains an asterisk the respective swissprot sequence file has an IG_MHC prosite entry.

Table 12-3. *Part 3 of 3*: Fold recognition search of swissprot database using the constant domain of 1FB4-L.

Sequence	$g_a{}^a$	$g_b{}^a$	$h\%{}^c$	Z-score^d	IG^e	Molecule	Source
HA1F_HUMAN	1	0	18.7	−5.52	*	HLA CLASS I HC AG	human
HA1G_HUMAN	1	0	18.7	−5.52	*	HLA CLASS I HC AG	human
DNLI_VACCV	0	2	12.1	−5.50		DNA LIGASE	virus
V1A_CMVQ	0	3	8.4	−5.47		1A PROTEIN	virus
HA1C_PANTR	1	0	18.7	−5.46	*	CHLA CLASS I HC AG	chimpanzee
HA1A_PANTR	1	0	18.7	−5.44	*	CHLA CLASS I HC AG	chimpanzee
HA1D_PANTR	1	0	18.7	−5.44	*	CHLA CLASS I HC AG	chimpanzee
HA1L_HUMAN	1	0	18.7	−5.44	*	HLA CLASS I HC AG	human
HA1N_HUMAN	1	0	18.7	−5.44	*	HLA CLASS I HC AG	human
HA1N_PANTR	1	0	18.7	−5.44	*	CHLA CLASS I HC AG	chimpanzee
HA1O_HUMAN	1	0	18.7	−5.44	*	HLA CLASS I HC AG	human
HA1P_HUMAN	1	0	18.7	−5.44	*	HLA CLASS I HC AG	human
HA1U_HUMAN	1	0	18.7	−5.44	*	HLA CLASS I HC AG	human
HA1V_HUMAN	1	0	18.7	−5.44	*	HLA CLASS I HC AG	human
HA1W_HUMAN	1	0	18.7	−5.44	*	HLA CLASS I HC AG	human
DNLI_VACCC	0	2	12.1	−5.43		DNA LIGASE	virus
HA1B_PANTR	1	0	18.7	−5.43	*	CHLA CLASS I HC AG	chimpanzee
HA1A_RABIT	1	1	17.8	−5.42	*	RLA CLASS I HC AG	rabbit
HA1B_RABIT	1	1	17.8	−5.42	*	RLA CLASS I HC AG	rabbit
HA1S_HUMAN	1	0	18.7	−5.42	*	HLA CLASS I HC AG,	human
HA1Q_HUMAN	1	0	18.7	−5.40	*	HLA CLASS I HC AG	human
HA1T_HUMAN	1	0	18.7	−5.40	*	HLA CLASS I HC AG	human
HUCM_ICTPU	0	0	20.6	−5.39	*	IG MU CHAIN C	catfish
HA1X_HUMAN	1	0	19.6	−5.37	*	HLA CLASS I HC AG	human
HA1M_HUMAN	1	0	18.7	−5.34	*	HLA CLASS I HC AG	human
HA1M_PANTR	1	0	18.7	−5.33	*	CHLA CLASS I HC AG	chimpanzee
NRAM_IARI5	0	2	10.3	−5.24		NEURAMINIDASE	virus
P3P_LACLC	1	1	7.5	−5.22		PIII-TYPE PROTEINASE	bacteria
FLAA_CAMCO	0	3	5.6	−5.18		FLAGELLIN A	bacteria
VAT1_NEUCR	0	2	11.2	−5.15		VACUOLAR ATP SYNTHASE	fungi
SY65_DROME	1	1	8.4	−5.12		SYNAPTOTAGMIN	fly
VR2B_BPT4	0	3	4.7	−5.09		RIIB PROTEIN	phage
VG05_BPT4	1	1	11.2	−5.08		TAIL-ASSOCIATED LYSOZYME	phage
HEMO_HYACE	0	1	10.3	−5.07	*	HEMOLIN PRECURSOR	insect
CARP_POLTU	1	1	8.4	−5.06		POLYPOROPEPSIN	fungi
CG12_YEAST	1	2	7.5	−5.04		G1/S-SPECIFIC CYCLIN	yeast
POLG_HRV1B	0	1	9.3	−5.02		GENOME POLYPROTEIN	virus

(a) Number of gaps opened in sequence. (b) Number of gaps opened in structure. (c) Number of identical residues in sequence structure alignment. The residues in the aligned sequence are compared to the sequence of the constant domain of 1FB4-L at corresponding positions. (d) Z-score of model calculated using the conformations in the polyprotein as a background. (e) If the row contains an asterisk the respective swissprot sequence file has an IG_MHC prosite entry.

fact that their native structures have a similar topology. So the interesting cases in Table 12-3 are those sequences that are not marked by an asterisk. The first such sequence we encounter in Table 12-3 is VP4_NCDV (call it VP4), a viral hemagglutinin (Nebraska calf diarrhea virus, as the swissprot entry tells us, determined by Nishikawa et al., 1988), with a z-score of −6.95. VP4 is a large capsid glycoprotein of 775 amino acid residues. The

Structure	Alignment	RMS a	RMS b	z-Value

l-l10-216 — RMS a 0,0 · RMS b 0,0 · z-Value -8,38

```
LGQPKANPTVTLFPPSSEELQANKATLVCLISDFYPGAVTVAWKADGSPVKAGVETTKPS
||||||||||||||||||||||||||||||||||||||||||||||||||||||||||||
LGQPKANPTVTLFPPSSEELQANKATLVCLISDFYPGAVTVAWKADGSPVKAGVETTKPS

KQSNNKYAASSYLSLTPEQWKSHRSYSCQVTHEGSTVEKTVAPTECS
|||||||||||||||||||||||||||||||||||||||||||||||
KQSNNKYAASSYLSLTPEQWKSHRSYSCQVTHEGSTVEKTVAPTECS
```

h-l20-239 — RMS a 2,45 · RMS b 3,44 · z-Value -7,45

```
.....LGQPKANPTVTLFPPSSEELQANKATLVCLISDFYPGAVTVAWKADGSPVKAGVE
       ||                    ||
TPVTVSSASTKGPSVFPLAFSSKSTSGGTAALGCLVKDYFPQPVTVSWN--SGALTSGVH

TTKPSKQSNNKYAASSYLSLTPEQWKSHRSYSCQVTHEGSTVEKTVAPTECS..
     ||||                              ||
TFPAVLQSSGLIYSLSSVVTVPSSSLGT-QTYICNVNHKPSNTKVDK-RVEPKSC
```

l-1-109 — RMS a 6,45 · RMS b 11,15 · z-Value -4,76

```
LGQPKANPTVTLFPPSSEELQANKATLVCLISDFYPGAVTVAWKADGS-----PVKAGVE
                                          ||
......ESVLTQPPSASGTPGQRVTISCTGTSSNIGSITVNWYQQLPGMAPKLLIYRDAM

TTKPSKQSNNKYAASSYLSLTPEQWKSHRSYSCQVTHEGSTVEKTVAPTECS....
                              ||
RPSGVPTRFSGSKSGTSASLAISGLEAEDESDYYCASWNSSDNSVFGTGTKVTV
```

h-1-l19 — RMS a 6,43 · RMS b 15,22 · z-Value -3,52

```
LGQPKANPTVTLFPPSSEELQANKATLVCLIS--DFYPGAVTVAWKADGSPVKAGVETTK
                                          ||
......EVQLVQSGGGVVQPGRSLRLSCSSSGFIFSSYAMYWVRQAPGKGLEWVAIIWD

PSKQSNNKYAASSYLSLTPEQWKSHRSYSCQVTHEGSTVEKTVAPTECS...........
                              ||
DGSDQHYADSVKGRFTISRNDSKNTLFLQMDSLRPEDTGVYFCARDGHGFCSSASCFGPDYWGQG
```

region aligned to the constant domain of 1FB4-L is between residues 270 and 370.

The significance of the model is more than one standard deviation less than that of the native 1FB4-L fold, but it is surrounded by immunoglobulin-like sequences, having a similar score. So have we just predicted that the aligned part of VP4 adopts the immunoglobulin fold? A nasty question. Wrong predictions are not welcome. Let us try as hard as we can to find an answer.

Obviously, the most important number we get is the z-score of the model. According to probability theory a z-score of more than 5.0 units of standard deviation (in absolute values) is quite unlikely to arise by chance, and even more so for 8.4, which we obtained for the native protein. But what do these numbers really mean in terms of structural similarity of the model obtained from fold recognition and the real native structure? Distantly related proteins may share a common architecture, but we have to expect that there are substantial differences. In other words, the sequence may not quite fit into the query fold. It would do still better in its genuine native fold. Can we get some idea of these structural differences between the model obtained from fold recognition and the actual native fold from the z-score?

To address this point we can study the z-scores obtained when the sequence of the constant domain of 1FB4-L is aligned with several immunoglobulin domains of known structure of varying degrees of structural homology. This should give some indication of the real significance of the z-score in terms of the quality of the models obtained for the 1FB4-L sequence as well as of the quality of the alignment technique itself. Figure 12-12 assembles the results for four domains of 1FB4: the light chain constant domain (l 110-216) (native fold), the constant domain of the heavy

---

Figure 12-12. Structure sequence alignments of the light-chain constant domain sequence of 1FB4 with several immunoglobulin folds. The structures used are the four domains of 1FB4: light chain constant (l 110-216), heavy chain constant (h 120-239), light chain variable (l 1-109), and heavy chain variable (h 1-119). The alignments produced by the algorithm are shown. In the alignments the upper sequence corresponds to the sequence of 1FB4 (11-109) and the lower sequence to the sequences of the respective domains. RMSa is the root mean square error of optimal superimposition of the various domains on the constant light-chain domain. RMSb is the same error when the equivalences implied by the structure sequence alignments are used in the superimposition. The z-scores indicate that the alignments with variable domains correspond to poor models as compared to the alignment obtained for the heavy chain constant domain. *See figure on preceding page.*

chain (h 120-239), the light chain variable domain (l 1-109), and the heavy chain variable domain (h 1-119).

All four domains have the canonical immunoglobulin fold, but there are substantial differences between constant and variable domains, indicated by the large root mean square deviation after superimposition of the various folds on the constant domain of the light chain. The models obtained for the 1FB4-L sequence from the alignment technique are of distinct quality. Combination with the constant domain of the heavy chain yields a model whose z-score of -7.45 is close to the result obtained for the native fold. But the scores calculated for the models derived from the variable domains are poor: -4.74 for the light chain and -3.52 for the heavy chain. The alignments are not consistent with those obtained from the superimposition of structures. This is not very satisfying, but at least the result indicates that the score discriminates bad models from good ones.

Why did the fold recognition technique fail to properly align the constant domain sequence with the variable domain structures? Part of the answer is that a proper alignment of variable and constant domains requires a number of gaps. The two types of domains can be superimposed with a much smaller error if we introduce many gaps as shown in Figure 12-13. This means, however, that many amino acids do not have a partner in the alignment and that the common core of structural similarity is confined to a small fraction of residues.

The alignment technique has a few important parameters that we have neglected up to this point. One of these parameters is a penalty for gap opening. This parameter is necessary to avoid excessive fragmentation of the aligned chains. If its value is high, the optimal path through the matrix of Figure 12-10 can afford only a very limited number of gaps. If it is very low, many gaps can form. For distantly related proteins a low gap penalty is required to allow all the gaps necessary for an optimal alignment. The results assembled in Table 12-3 and Figure 12-12 were obtained with

---

Figure 12-13. Improved alignments of variable domain structures and light chain constant domain sequence obtained by using a low gap penalty in alignment production. Structure alignments are obtained from an algorithm that optimizes the superimposition of two structures. The algorithm (Flöckner and Sippl, unpublished) was allowed to open as many gaps as needed in order to make the root mean square error as small as possible. This results in a rather small core of equivalent residues. (The larger rms-deviations shown in Figure 12-12 have been obtained using a higher gap penalty). The structure–sequence alignments obtained with a comparably small gap penalty point to a considerably more significant fit as compared to the alignments in Figure 12-12. *See figure on following page.*

Region **Structure-Structure Alignment**

1-1-109

```
LGQPKANFTVT-LFPPSSEELQANKATLVCLISDF-YPG---AVTVAWKAD--GSP-V---------
      |     | |  |                     |                |
.....ESVLTQPPSASGT---P---GQRVTISCTGTSSNIGSITVNWYQQLPGMAPKLLIYRDAMRPS

K-A-GVETTKPSKQSNNKYAASSYLSLTPEQWKSHRSYSCQVTHE------GSTV-EKTVAPTECS
    | |                                                 |
GVPTRFSGGS-KSG------TSASLAISGL--EA--EDESDYCASWNSSDNSYVFGTGTKVTV....
```

RMS = 3,21

h-1-119

```
LGQPKANFTVTLFPPSSEELQANKATLVCLISDFYPGAVTVAWKAD-----GSPVK---------A----
      | |  | | |                                       |
......EVQLVQSGGGV-VQPGRSLRLSCSSSGFIFSSYAMYWVRQAPGKGLEWVAIIWDDGSDQHYADSV

---GVETTKPSKQSNNKYAASSYSL-TPEQWKSHRSYSCQVTHEGST--------V-EKTVAYPTECS
                                                         |
KGRFTISRNDS------KNTLFLQMDSLRPEDT--GVYFCARDGGHGFCSSASCFGPDYWGQG.......
```

RMS = 3,98

Sequence-Structure Alignment

```
LGQPKANFTVTLFPPSSEELQANKATLVCLISDFYPGA--VTVAWKADGSPVKAGVET--
      |        | |  |                                 |
......ESVLTQPPSASGT-PGQRVTISCTGTSSNIGSITVNWYQQLPGMAPKLLIYRDA

TKPSKQSNNKY--AASSYLSLTPEQWKSHRS--YSCQVTHEGSTVEKTVAPTECS...
|  |                              | |                 |
MRPSGVPTRFSGSKSGTSASLAISGLEAEDESDYYCASWNSSD--NSYVFGTGTKVTV
```

RMS = 11,14   z-Value = -6,92

```
LGQPKANFTVTLFPPSSEELQANKATLVCLIS--DFYPGAVTVAWKADGSPVKAGV-------
      |        | |                                    |
......EVQLVQSGGGVVQ-PGRSLRLSCSSSGFIFSSYAMYWVRQAPGKGLEWVAIIWDDGSD

ETTKPSKQSNNKYAASS-----YLSLTPEQWKSHRSYSCQV----THEGSTVEKTVAPTECS.
|  |                                                       |
QHYADSVKGRFTISRNDSKNTLFLQMDSLRPEDTGVYFCARDGGHGFCSSA--SCFGPDYWGQG
```

RMS = 11,28   z-Value = -6,23

a high penalty, however. If the penalty is lowered, then the alignments of a constant domain sequence and a variable domain structure improve dramatically. In this case the models obtained from the fold recognition technique have z-scores of -6.92 (light chain) and -6.23 (heavy chain), as shown in Figure 12-13.

In terms of the z-scores these are good models. But the error of super-imposition between the models and their native folds is still high, in marked contrast to the structural alignments. This seems to contradict the conclusion that the models resemble the native folds. Well, yes and no. The reason for this discrepancy is that in the structural alignment only those residue pairs are retained that yield a low error of superimposition (i.e., it is the goal of the algorithm to minimize this error). This can only be achieved by a very large number of gaps, leaving many residues unpaired, resulting in a small common core.

The optimization that takes place in sequence structure alignment is different. There the goal is to combine a sequence with a given fold in order to minimize the energy of that sequence. If the framework used is only distantly related to the native fold of the sequence, then the fine geometric details do not matter (in contrast to superimposition of structures). If the sequence fits into that framework then there is no need to get rid of loops protruding in different directions relative to the common core. Although different from the native loops, their energy contribution may be favorable and the alignment would be suboptimal if these loops are discarded. The native conformations of such regions will do even better, but the algorithm only has the given scaffold to work with.

In summary, the large structural deviation of native structures and models obtained from fold recognition as compared to the geometric similarity have at least two sources. One is that sequence structure alignment does not discard regions that have large structural deviations unless they are energetically unfavorable. The second reason is that in the case of distant relationships the resolving power of the mean field is inferior to a detailed geometric comparison.

We now come back to our original question: What is the significance of the z-score in terms of structural similarity between models and native folds? What we have learned from the few examples is that if a model is wrong (like those corresponding to incorrect alignments in Figure 12-12), then the z-score is low and that significant z-scores are indicative of similar architecture—at least we have no example to the contrary. But in terms of errors of superimposition, the deviations can be large.

Obviously, the alignments are much better if gap penalties are low. A high gap penalty prevented the production of useful alignments between variable and constant domains. In our case study using the constant domain of 1FB4-L to retrieve immunoglobulin-like sequences from the swissprot database, we used a high gap penalty, missing all the variable domains in our search. Why not use a small gap penalty then?

Because small gap penalties allow the production of a large number of alignments, which have interesting z-scores. This may seem exciting, but we have to keep in mind that there is still this fundamental uncertainty. By removing residues we may succeed in fitting a sequence into a given framework. The result may be that these manipulations completely change the molecule. The answer we get may be correct in one sense, i.e., this sequence may in fact adopt the framework fold. But the conclusion that the original unchanged sequence also adopts this fold can be incorrect.

This is why we want to be restrictive in fold recognition. Then we are bound to miss a lot of sequences that actually would fit the scaffold but at the same time we hope to reduce the odds of producing artifacts. So let us return again to the VP4 case. Is the result an artifact or is it likely that VP4 contains an immunoglobulin-like domain?

The structure of VP4 is unknown, but the structure of the influenza virus hemagglutinin (2HMG) has been solved (Weis et al., 1988). Unfortunately, VP4 and 2HMG are only distantly (but seemingly functionally) related and it is therefore not clear to what extent their structures are similar. Moreover, to our knowledge no structural similarity between 2HMG and immunoglobulins has been reported. But let us try to fit the 2HMG sequence onto the constant domain of 1FB4-L.

If the 2HMG sequence is aligned with the constant domain of 1FB4-L, a model is obtained whose z-score of -4.66 is low. This points to a poor fit of the aligned 2HMG sequence with the constant domain of 1FB4-L. But let's see if there are any similarities. The aligned region is part of the headpiece of 2HMG as shown in Figure 12-14. The region is composed of β strands and in this respect this part of 2HMG is similar to the immunoglobulin fold as shown in Figure 12-15. Both contain seven strands. When the strands, starting at the N-terminus, are labeled from β_1 to β_7, it is clear that equivalent strands point in the same direction. However, the relative arrangement is different. β_1 and β_2 are adjacent in 1FB4-L, but they are separated in 2HMG. On the other hand there are pairs that have similar orientation like β_4 and β_5, for example.

In short, the 2HMG region is in many respects similar to the constant domain of 1FB4-L but it is not quite an immunoglobulin-like fold. We could go on now and investigate whether 2HMG fits better into the variable domain, for example, but we would rather stop here. Our goal was to get some idea whether the high z-score of VP4 indicates an immunoglobulin-like domain

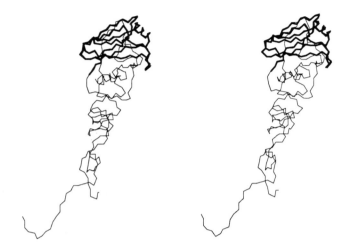

Figure 12-14. Stereo pair drawing of the influenza virus hemagglutinin a chain (2HMG). The sequence of this molecule was aligned with the structure of the constant domain of the 1FB4 light chain. The region drawn in bold face corresponds to that part of the sequence that was picked up by the immunoglobulin fold. The z-score obtained from the combination of the 2HMG sequence with the immunoglobulin fold is −4.66.

in this protein. We have seen that 2HMG has some immunoglobulin-like features and chances are that the aligned region of VP4 is even more similar to the immunoglobulin architecture. In fact, there is one report of a computer-assisted immunoglobulin sequence database search that picked up yet another hemagglutinin variant (vaccinia virus).

The fold recognition case study picked up a number of additional sequences not known to be related to immunoglobulins. This indicates that these proteins may have immunoglobulin-like domains or that their structures at least resemble the immunoglobulin architecture in several respects. The case of 2HMG showed that we may expect structural similarities even down to a significance of −5.0 and beyond. But we should be cautious. There may also be artifacts.

10. Conclusion

Here we reach the end of our survey of the mean-field approach to protein folding. We tried to demonstrate the predictive power of the approach and its applicability to problems in structural biology. We discussed the capabilities of fold recognition techniques based on the mean field using

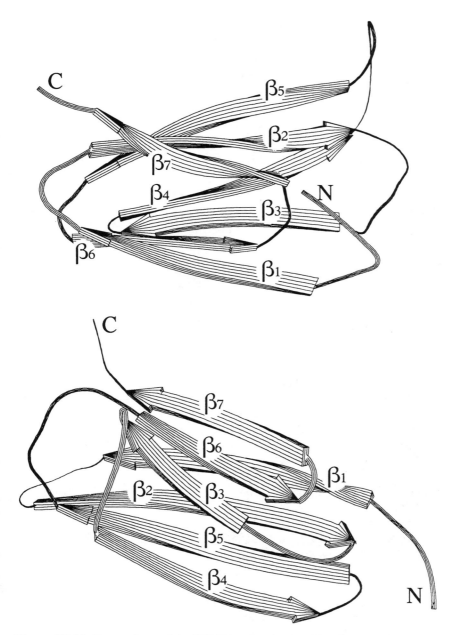

Figure 12-15. Comparison of the 2HMG region (top), which fits into the constant domain fold of 1FB4 (bottom) with a *z*-score of −4.66. Both structures have many features in common but at the same time there are significant differences.

the immunoglobulin fold as an example and touched several problems concerning the *de novo* prediction of protein folds. There would be more to say about the physical content of mean force potentials (e.g., Casari and Sippl, 1992), for example, the optimization and completion of the current mean field (Sippl and Jaritz, 1994) or additional results on fold recognition (Sippl et al., in preparation). This is a wide area and the prospects are promising. There is still much work ahead and ample room for improvement.

Acknowledgments. We are indebted to our colleagues Peter Lackner, Manfred Hendlich, Georg Casari, Markus Jaritz, and Maria Ortner. They contributed substantially to the results presented in the text. This work was supported by the Fonds zur Forderung der Wissenschaftlichen Forschung (Austria), project number 8361-CHE, and the Jubilaeumsfonds der Oesterreichischen Nationalbank, project number 4356.

REFERENCES

Anfinsen CB (1973): Principles that govern the folding of protein chains. *Science* 181:223–230

Bairoch A, Boeckmann B (1991): The SWISS-PROT protein sequence data bank. *Nucl Ac Res* 19:2247–2248

Bernstein FC, KoetzleTF, Williams GJB, Meyer EF Jr, Brice MD, Rodgers JR, Kennard O, Shimanouchi T, Tasumi M (1977): The protein data bank: A computer based archival file of macromolecular structures. *J Mol Biol* 112:535–542

Billeter M, Qian Y, Otting G, Mueller M, Gehring WJ, Wuethrich KJ (1990): Determination of the three-dimensional structure of the antennapedia homeodomain from *Drosophila* in solution by 1H nuclear magnetic resonance spectroscopy. *J Mol Biol* 214:183–197

Blundell T, Carney D, Gardner SP, Hayes F, Howlin B, Hubbard T, Overington J, Singh DA, Sibanda BL, Sutcliffe M (1988): Knowledge based protein modelling and design. *Eur J Biochem* 172:513–520

Bode W, Huber R (1991): Ligand-binding: Proteinase and proteinase inhibitor interactions. *Cur Opin Struct Biol* 1:45–52

Bode W, Huber R (1992): Natural protein proteinase inhibitors and their interaction with proteinases. *Eur J Biochem* 204:433–451

Bowie JU, Clarke ND, Pabo CO, Sauer RT (1990): Identification of protein folds: Matching hydrophobicity patterns of sequence sets with solvent accessibility patterns of known structures. *Proteins* 7:257–264

Bowie JU, Luethy R, Eisenberg D (1991): A method to identify protein sequences that fold into a known three-dimensional structure. *Science* 253:164–170

Braenden CI, Jones A (1990): Between objectivity and subjectivity. *Nature* 343:687–689

Brayer GD, McPherson A (1983): Refined structure of the gene 5 DNA binding protein from bacteriophage fd. *J Mol Biol* 108:565–596

Brooks CL, Karplus M, Pettitt BM (1988): *Proteins: A Theoretical Perspective of Dynamics, Structure, and Thermodynamics.* New York: John Wiley and Sons

Burmeister WP, Ruigrok RWH, Cusack S (1992): The 2.2 Å resolution crystal structure of influenza B neuraminidase and its complex with sialic acid. *EMBO J* 11:49–56

Casari G, Sippl MJ (1992): Structure-derived hydrophobic potential. Hydrophobic potential derived from X-ray structures of globular proteins is able to identify native folds. *J Mol Biol* 224: 725–732

Chothia C (1992): One thousand families for the molecular biologist. *Nature* 357: 543–544

Cohen FE, Richmond TJ, Richards FM (1979): Protein folding: Evaluation of some simple rules for the assembly of helices into tertiary structures with myoglobin as an example. *J Mol Biol* 132:275–288

Cohen FE, Sternberg MJ, Taylor WR (1982): Analysis and prediction of the packing of α-helices against a β-sheet in the tertiary structure of globular proteins. *J Mol Biol* 156:821–862

Crawford IR, Niemann T, Kirschner K (1987): Prediction of secondary structure by evolutionary comparison: Application to the alpha subunit of tryptophan synthase. *Proteins* 2:118–129

DeGrado WF, Raleigh DP, Handel T (1991): De novo protein design: What are we learning? *Cur Opin Struct Bio* 1:984–993

Dill KA (1993): Folding proteins: Finding a needle in a haystack. *Cur Opin Struct Biol* 3:99–103

Dyson HJ, Wright PE (1993): Peptide conformation and protein folding. *Cur Opin Struct Biol* 3:60–65

Fasman GD, ed. (1989): *Prediction of Protein Structure and the Principles of Protein Conformation.* New York and London: Plenum Press

Fersht A (1985): *Enzyme Structure and Mechanism.* New York: WH Freeman

Fersht AR, Serrano L (1993): Principles of protein stability derived from protein engineering experiments. *Cur Opin Struct Biol* 3:75–83

Folkers PJ, van Duynhoven PM, Jonker AJ, Harmsen BJ, Konings RN, Hilbers CW (1991): Sequence-specific 1H-NMR assignment and secondary structure of the Tyr41-His mutant of the single-stranded DNA binding protein, gene V protein, encoded by the filamentous bacteriophage M13. *Eur J Biochem* 202:349–360

Gregoret LM, Cohen FE (1990): Novel method for the rapid evaluation of packing in protein structures. *J Mol Biol* 211:959–974

Hendlich M, Lackner P, Weitckus S, Floeckner H, Froschauer R, Gottsbacher K, Casari G, Sippl MJ (1990): Identification of native protein folds amongst a large number of incorrect models. *J Mol Biol* 216:167–180

Holm L, Sander C (1992): Evaluation of protein models by atomic solvation preference. *J Mol Biol* 225:93–105

Janin J (1990): Errors in three dimensions. *Biochimie* 72:705–709

Jernigan RL (1992): Protein folds. *Cur Opin Struct Biol* 2:248–256

Jones DT, Taylor WR, Thornton JM (1992): A new approach to protein fold recognition. *Nature* 358:86–89

Kabsch W, Sander C (1984): On the use of sequence homologies to predict protein structure: Identical pentapeptides can have completely different conformations. *Proc Natl Acad Sci* 81:1075–1078

Karplus M, Petsko GA (1990): Molecular dynamics simulations in biology. *Nature* 347:631–639

Kelly JA, Knox JR, Moews PC, Hite GJ, Bartolone JB, Zhao H (1985): 2.8 Å structure of penicillin-sensitive D-alanyl carboxypeptidase-transpeptidase from Streptomyces R61 and complexes with β-lactams. *J Biol Chem* 260:6449–6458

Kendrew JC, Dickerson RE, Strandberg BE, Hart RJ, Davies DR, Phillips DC, Shore VC (1960): Structure of myoglobin: A three-dimensional Fourier synthesis at 2 Å resolution. *Nature* 185:422–427

Kuma K, Iwabe N, Miyata T (1991): The immunoglobulin family. *Cur Opin Struct Biol* 1:384–393

Luethy R, Bowie JU, Eisenberg D (1992): Assessment of protein models with 3D profiles. *Nature* 356:83–85

Marquart M, Deisenhofer J, Huber R, Palm W (1980): Crystallographic refinement and atomic models of the intact immunoglobulin molecule KOL and its antigen-binding fragment at 3.0 Å and 1.9 Å resolution. *J Mol Biol* 141:369

Murzin AG, Chothia C (1992): Protein architecture: New superfamilies. *Cur Opin Struct Biol* 2:895–903

Needleman SB, Wunsch CD (1970): A general method applicable to the search for similarities in the amino acid sequences of two proteins. *J Mol Biol* 48:443–453

Nishikawa K, Taniguchi K, Torres A, Hoshino Y, Green K, Kapikian AZ, Chanock RM, Gorziglia M (1988): Comparative analysis of the VP3 gene of divergent strains of the rotaviruses simian SA 11 and bovine Nebraska calf diarrhea virus. *J Virol* 62:4022–4026

Novotny J, Brucceroli RE, Karplus M (1984): An analysis of incorrectly folded protein models. Implications for structure predictions. *J Mol Biol* 177:787–818

Novotny J, Rashin AA, Bruccoleri RE (1988): Criteria that discriminate between native proteins and incorrectly folded models. *Proteins* 4:19–30

Ogata CM, Gordon PF, de Vos AM, Kim SH (1992): Crystal structure of a sweet tasting protein Thaumatin I, at 1.65 Å resolution. *J Mol Biol* 228:893–908

Overington JP (1992): Comparison of the three-dimensional structures of homologous proteins. *Cur Opin Struct Biol* 2:394–401

Perutz MF, Rossmann MG, Cullis AF, Muirhead G, Will G, North AT (1960): Structure of haemoglobin: A three-dimensional fourier synthesis at 5.5 Å resolution, obtained by X-ray analysis. *Nature* 185:416–422

Perutz MF (1992): *Protein Structure. New Approaches to Disease and Therapy.* New York: W.H. Freeman

Ouzounis C, Sander C, Scharf M, Schneider R (1993): Prediction of protein structure by evaluation of sequence-structure fitness. Aligning sequences to contact profiles derived from 3D structures. *J Mol Biol* 232:805–825

Richardson JS, Richardson DC (1988): Amino acid preferences for specific locations at the ends of alpha-helices. *Science* 240:1648–1652

Rooman MJ, Kocher J-P, Wodak SJ (1991): Prediction of protein backbone conformation based on seven structure assignments: Influence of local interactions. *J Mol Biol* 221:961–979

Rost B, Sander C (1992): Jury returns on structure prediction. *Nature* 360:540

Rost B, Sander C (1993a): Prediction of protein secondary structure at better than 70% accuracy. *J Mol Biol* 232:584–599

Rost B, Sander C (1993b): Improved prediction of protein secondary structure by use of sequence profiles and neural networks. *Proc Nat Acad Sci* 90:7558–7562

Sander C, Schneider R (1991): Database of homology-derived protein structures and the structural meaning of sequence alignment. *Proteins* 9:56–68

Sippl MJ (1990): Calculation of conformational ensembles from potentials of mean force. An approach to the knowledge-based prediction of local structures in globular proteins. *J Mol Biol* 213:859–883

Sippl MJ (1993): Boltzmann's principle, knowledge based mean fields and protein folding. *J Comp Aid Mol Des* 7:473–501

Sippl MJ, Hendlich M, Lackner P (1992): Assembly of polypeptide and protein backbone conformations from low energy ensembles of short fragments: Development of strategies and construction of models for myoglobin, lysozyme, and thymosin β_4. *Protein Science* 1:625–640

Sippl MJ, Jaritz M (1994): Predictive power of mean force pair potentials. In *Protein Structure by Distance Analysis*. Bohr H, Brunak S, eds. Amsterdam: IOS Press

Sippl MJ, Weitckus S (1992): Detection of native-like models for amino acid sequences of unknown three-dimensional structure in a data base of known protein conformations. *Proteins* 13:258–271

Sternberg MJE (1992): Secondary structure prediction. *Cur Opin Struct Biol* 2:237–241

Thornton JM (1992): Lessons from analyzing protein structures. *Cur Opin Struct Biol* 2:888–894

van Gunsteren WF (1988): The role of computer simulation techniques in protein engineering. *Prot Eng* 2:5–13

Weis W, Brown JH, Cusack S, Paulson JC, Skehel JJ, Wiley DC (1988): Structure of the influenza virus haemagglutinin complexed with its receptor, sialic acid. *Nature* 333:426–431

Wilmanns M, Eisenberg D (1993): 3D profiles from residue-pair preferences: Identification of sequences with β/α-barrel fold. *Proc Natl Acad Sci* 90:1379–1383

13

An Adaptive Branch-and-Bound Minimization Method Based on Dynamic Programming

Sandor Vajda and Charles Delisi

1. Introduction

Computational studies are potentially useful for analyzing various putative mechanisms of protein folding. The general question that can be addressed is whether or not it is possible to obtain native-like conformations of a protein on the basis of a well-defined set of physical and chemical principles. In particular, computation is the only tool available for studying the relations between native protein structures and the conformations that correspond to various free energy minima (Skolnick and Kolinski, 1989; Chan and Dill, 1993). If the "thermodynamic hypothesis" is valid and the free-energy evaluation is of acceptable accuracy, then the native structure can be identified in principle by systematic evaluation of the energy of every possible conformation. However, using a detailed model of protein geometry and a force field of near-atomic resolution, we are at present unable to explore the conformational space even for the smallest proteins.

In view of the above limitation, a number of recent studies are based on the use of simplified proteins models that involve a discrete approximation of the conformational space (Moult and James, 1986; Bruccoleri and Karplus, 1987; Lipton and Still, 1988; Vajda and DeLisi, 1990; Shortle et al., 1992) or a lattice representation of the protein geometry (Go and Taketomi, 1978; Taketomi et al., 1988; Lau and Dill, 1989; Sikorski and Skolnick, 1989; Skolnick and Kolinski, 1989, 1990; Shaknovich et al., 1991; Brower et al., 1993). These studies have been successful in reproducing some essential qualitative features of protein folding, such as the

The Protein Folding Problem and Tertiary Structure Prediction
K. Merz, Jr. and S. Le Grand, Editors
© Birkhäuser Boston 1994

uniqueness of the of the folded state, and demonstrate that useful results can be obtained using simple interaction potentials. However, some important questions remains unanswered or even controversial. For example, Dill and co-workers (Lau and Dill, 1989; Chan and Dill, 1991, 1993; Shortle et al., 1992) introduced and studied a simple two-dimensional lattice model in which a protein is represented by two types of residues, H (hydrophobic) and P (polar), on a square lattice subject to excluded volume and an HH attraction energy between any two nonbonded H-type elements that occupy neighboring positions. On this lattice an exhaustive exploration of the full conformational space is possible, at least when the chain is short ($N < 17$). Molecules with certain sequences were found to fold to conformations that have low free energy, high compactness, and a core of H residues (Lau and Dill, 1989). The most interesting conclusion is that, even with only the H and P discrimination among residues, such a folding sequence is most likely to have only a single conformation with the lowest energy.

By contrast, Skolnick and co-workers (Sikorski and Skolnick, 1989; Skolnick and Kolinski, 1990) have shown by a series of Monte Carlo simulations that on three-dimensional lattices similar nonbonded interactions yield large sets of folded conformations, and uniqueness is attained only by introducing local conformational preferences among nearest-neighbor and second-nearest-neighbor residues. Since these simulations were restricted to particular sequences that may be nonfolding sequences in the terminology used by Lau and Dill (1989), it is not clear whether the above results show a genuine difference between 2D and 3D representations. Furthermore, Monte Carlo simulations tend to be trapped in local minima for some sequences, without reaching the global minimum (Shaknovich et al., 1991; Dill et al., 1993). Thus, further studies are required using 3D lattice models, or even more realistic discrete representations of protein geometry. The development of efficient search methods is essential for making such calculations possible.

In this paper we describe an efficient combinatorial search algorithm to determine the global minimum of a function over the configuration space of discrete protein conformations. The method is based on dynamic programming. Although dynamic programming is one of the most general approaches to combinatorial optimization, in its original form it is applicable only to problems that satisfy the condition called "the principle of optimality" (Bellman, 1975). Here we discuss a branch-and-bound type generalization of the classical algorithm to a much broader class of problems that satisfy this condition in some approximate sense. The use of the extended procedure is demonstrated by solving a 2D lattice polymer folding problem.

The selection of a cutoff parameter is critical to the use of extended dynamic programming. The selected value must be large enough to find the global minimum, but not so large as to substantially reduce the computational efficiency of the search. Although the cutoff parameter has a well-defined physical meaning, one can relatively easily obtain a meaningful estimate only in some simple cases such as in the lattice polymer problem to be discussed here. Earlier we used the extended dynamic programming in more realistic conformational studies, but with an entirely heuristic selection of the parameters (Vajda and DeLisi, 1990; Vajda et al., 1993). In this work we introduce an additional iteration that enables us to identify the lowest cutoff value necessary in such more general situations. Under rather mild assumptions that are satisfied in a large variety of problems, the resulting procedure always finds the global minimum while attaining reasonable numerical efficiency. The method is demonstrated by determining finite-state, minimum RMSD approximations of backbone conformations in 11 proteins. Although this problem is simpler than locating the minima of a potential function, it exhibits some of the difficulties associated with optimization of a function defined on the space of discrete protein configurations.

2. The Problem of Discrete State Assignment

We assume that each residue of a protein can adopt one of the m conformational states. Such discrete states can represent different directions from a lattice point (Lau and Dill, 1989; Skolnick and Kolinski, 1989) or points on the $(\phi, \psi,)$ plane (Moult and James, 1986; Rooman et al., 1991). The number of different conformational or structural states may depend on the type of the amino acid residue, but without loss of generality we assume that m represents the maximum number of states for all residues. For a protein of N residues, a particular conformation in this space is described by the sequence $(k_1, k_2, \ldots, k_N)$ of integers, where k_i is the index of the conformational state assigned to the i-th amino acid residue.

Let Q denote a function defined on this finite dimensional conformational space. The problem of state assignment considered in this paper is to minimize Q, i.e., to find the state assignment $(k_1^*, k_2^*, \ldots, k_N^*)$ such that $Q(k_1^*, k_2^*, \ldots, k_N^*) \leq Q(k_1, k_2, \ldots, k_N)$ for all $(k_1, k_2, \ldots, k_N)$. Combinatorial optimization, i.e., locating the minimum of a function over a finite configuration space, differs substantially from the problem of minimizing a function in the space of a number of continuously variable parameters. Local continuous minimization is relatively simple: The traditional methods try to move "downhill" in a favorable direction. In the most general case, combinatorial optimization does not have this principle, and the concept of direction may not have any meaning (Press et al., 1986).

Since the objective function Q is defined over a discrete configuration space, in principle the optimal solution(s) can be found in a finite number of steps. However, the number of state assignments increases exponentially with N, the number of residues, and hence apart from very short peptides the number of configurations cannot be explored exhaustively. Similar combinatorial optimization problems are encountered in a variety of contexts. A famous example is the traveling salesman problem of finding the shortest cyclical itinerary for a traveling salesman who must visit each of the N cities in turn (Press et al., 1986). The elements of the configuration space in this case are the possible orders on N cities. Both the traveling salesman and the classical unconstrained protein folding problem belong to the large class of NP-complete problems. Solving such problems by any exact method requires computing time that grows faster than any polynomial in N (the number of cities, residues, etc.). However, a number of methods have been developed for obtaining approximate solutions to particular NP-complete problems with computational requirements proportional to small powers of N (Hu, 1982).

3. State Assignment by Dynamic Programming

The problem of discrete state assignment can be formulated in terms of finding the path in a graph from node N to node C, selecting exactly one node from each intermediate stage (Figure 13-1). The nodes represent the structural or conformational states available to the particular amino acid residue of the sequence, whereas nodes N and C represent the amino and carboxyl termini of the chain, respectively. The goal of the search is to locate the path on which the cost function Q attains its minimum (Figure 13-1).

The complete state assignment for a sequence of N amino acid residues can be viewed as the result of a sequence of N individual decisions. For some of such sequential or stagewise problems an optimal sequence of decisions may be found by making the decisions *independently* one at a time and never making an erroneous decision. Such local *greedy* approach works only for a narrow class of problems (Horowitz and Sahni, 1979). For most problems, a sequence of locally optimal decisions does not yield an optimal solution of the entire problem.

A safe but very inefficient way of solving stagewise problems is enumerating all possible decision sequences and then picking out the best.

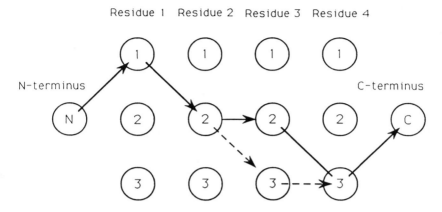

Figure 13-1. Graph representation of the state assignment problem.

Dynamic programming is a relatively general algorithm for solving stage-wise optimization problems. It drastically reduces computational efforts by avoiding the enumeration of some decision sequences that cannot possibly be optimal (Horowitz and Sahni, 1979). In dynamic programming an optimal sequence of decisions is arrived at by making explicit appeal to the "principle of optimality" (Bellman, 1975). This principle states that an optimal sequence of decisions has the property that whatever the initial state and decision are, the remaining decisions must constitute an optimal decision sequence with regard to the state resulting from the earlier decisions.

Applied to the multistage graph problem shown in Figure 13-1, the principle of optimality is equivalent to the assumption that any subpath of the optimal path must be an optimal subpath. Figure 13-1 illustrates this condition: If $(1, 2, 2, 3)$ is the optimal path, then $(1, 2, 2)$ must be the optimal path from state 1 of the first residue to state 2 of the third residue. Thus, paths containing suboptimal subpaths cannot be optimal if the optimality principle holds, and so will not (as far as possible) be generated (Horowitz and Sahni, 1979). The elimination of such suboptimal paths leads to the efficiency of the search in dynamic programming.

A classical application of dynamic programming is to the shortest path problem in which the objective function Q is some measure of the path length. Since any such measure is necessarily additive, the principle of optimality applies. Figure 13-2 shows a simple way of implementing dynamic programming. For the moment we leave aside the physical interpretation of stages, nodes, and paths, but retain the stagewise character of the problem, and also retain the single nodes in the two terminal stages. We first seek the shortest path to node 1 of stage 2. The possible paths are (1, 1), (2, 1), and (3, 1), where the two integers in parenthesis are the indices of nodes in the particular path. Let (1, 1) be the shortest among these competing paths to node 1 of stage 2. We similarly compare the paths (1, 2), (2, 2), and (3, 2) leading to node 2 of stage 2, and select the shortest, say, (1, 2). The decision process within stage 2 is completed by selecting the shortest path to node 3, say, (1, 3). Consider now node 1 within stage 3. Due to the decisions already made, to find the shortest path leading to node 1 of stage 3 one has to compare only the paths (1, 1, 1), (1, 2, 1), and (1, 3, 1). The algorithm continues until the single node in the last stage is reached. Dynamic programming requires $(N-1)m^2$ function evaluations for solving a problem of N stages, with m nodes in each stage. Even for a moderate N this is much less than the m^N function evaluations in an exhaustive search.

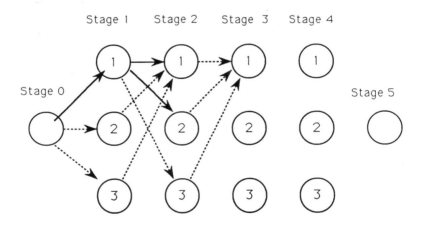

Figure 13-2. Stagewise decisions in dynamic programming.

In the state assignment problems considered in the present work the stages correspond to monomers of a chain or to amino acid residues of a protein. As mentioned, the nodes represent the different conformational states that can be assigned to each monomer. In order to show the various levels of difficulty one can encounter in assignment problems, we consider a polymer chain of N monomers, represented by filled circles in Figure 13-3. The conformational states for each monomer are a number of nearby points in space (empty circles in Figure 13-3). We define three different assignment problems as follows:

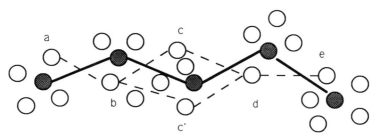

Figure 13-3. A simple state-assignment problem. The goal is to approximate a polymer chain (represented by filled circles) using given points (empty circles) in space as conformational states.

(a) The goal is to select N points in order to attain the minimum root mean square deviation (RMSD) between the chain defined by these points and the original polymer, i.e., $Q =$ RMSD. The problem can be trivially solved by the greedy algorithm: For each monomer select the point closest in space, and consider the chain defined by these points.

(b) The problem is similar to the one just described, but the objective function is given by $Q = $ RMSD $+ \sum a_i (l_i - l_i^0)^2$, where the second term represents deviations from standard bond lengths The greedy algorithm, based on selecting the optimal point in each stage separately is no longer applicable because it is possible that these optimal points can be connected only by distorting the bond lengths substantially. It can be easily verified that the principle of optimality holds. Let the path (a, b, c, d, e) represents the optimal solution (Figure 13-3), then each subpath must be an optimal path itself, e.g., (b, c, d) is the best path from point b to point d. Indeed, if this were not the best, then we could replace it by a better path, say (b, c', d), and form the path (a, b, c', d, e) such that $Q(a, b, c', d, e) < Q(a, b, c, d, e)$ which contradicts the assumed optimality of (a, b, c, d, e). Thus, the problem can be solved by dynamic programming.

(c) Assume now that the objective function is given by $Q = $ RMSD $+ \sum a_i (l_i - l_i^0)^2 + \sum b_j (t_j - t_j^0)^2$, where the third term represents deviations from the standard bond angles. To show that the principle of optimality does not necessarily apply, assume again that path (a, b, c, d, e) represents the optimal solution (Figure 13-3), but the subpath (b, c', d) is better than segment (b, c, d) of the optimal path. In contrast to the problem discussed in (b), replacing (b, c, d) by (b, c', d) now does not necessarily make $Q(a, b, c', d, e) < Q(a, b, c, d, e)$. In fact, possible changes in the bond angles at points b and d may actually increase Q.

Although the principle of optimality does not apply to problem (c), it still has some relevance. While a subpath of the optimal path is not necessarily optimal, it must have a function value that is reasonably close to the lowest value attained on any path connecting the same two nodes, or else the solution for the entire problem would not be optimal.

We have a similar situation when the goal is to minimize an energy function that includes nonbonded interactions among monomers. The principle of optimality does not hold, because a fragment of the lowest energy conformation does not necessarily have the lowest energy among all the conformations of that length and given endpoints. However, the optimal solution cannot include segments that have very high internal energies. In the next section we show that the large class of such problems that do not satisfy the classical principle of optimality can still be efficiently solved by an appropriate extension of the dynamic programming algorithm.

4. Extended Dynamic Programming

Consider a polymer divided into two segments A and B, where A denotes the chain of the first q monomers, with the q-th monomer in a fixed conformational state. The second fragment B is in an arbitrary fixed conformation. We want to assign states to the $q - 1$ free monomers in A in order to minimize the energy of the entire molecule AB. If A and B did not interact, the energy of the full chain AB would be $E = E_A + E_B$, where E_A and E_B represent the internal energies of A and B, respectively. Let A' denote another conformation of the first segment (i.e., state assignment to monomers $1, 2, \ldots, q - 1$) with internal energy $E_{A'}$. The energy of the complete chain, again in the absence of interaction, would then be $E' = E_{A'} + E_B$. Then $E < E'$ if and only if $E_A < E_{A'}$. Thus, in the absence of interaction, for any particular conformation of B the minimum energy conformation of the chain involving a particular state assigned to monomer q would be obtained by always retaining the minimum energy assignment to the first segment A. This is the standard dynamic programming situation discussed in problem (b) of the previous section.

When interactions are introduced, the optimal state assignment for the monomers in A will of course depend on the state assigned to the monomers in B. The central question concerns the sensitivity of this dependence (Vajda and DeLisi, 1990). Now the energy of the entire chain is given by $E = E_A + E_B + E_{AB}$, where E_{AB} represents the interaction energy between A and B. Similarly, for the conformation A' of the first segment we have the total energy $E' = E_{A'} + E_B + E_{A'B}$. Since $E - E' = E_A - E_{A'} + E_{AB} - E_{A'B}$, we have $E < E'$ if and only if $E_A < E_{A'} - E_{AB} - E_{A'B}$. Introduce the notation $C = \max |E_{AB} - E_{A'B}|$, where the maximum is taken over all the possible state assignments to A, A', and B. Then the worst case analysis shows that $E_{A'} > E_A + C$ always implies $E' > E$. Therefore, in order to minimize the energy E, in each stage we can disregard any assignment A' to the current segment if $E_{A'} > E_A + C$, where E_A is the minimum energy found for the segment of q chain elements, with the last element in a fixed conformational state

The above considerations suggest a straightforward extension of the dynamic programming. In contrast to the classical algorithm that proceeds from one stage to the next by retaining only the best path leading to each node, we will retain every path if it is not much worse than the optimum path leading to the same node, i.e., the two values differ by less than a cutoff C. We used the extended dynamic programming for conformational analysis to study the minimum energy conformations of polypeptides (Vajda and DeLisi, 1990). Precomputed single-residue energy minima were used as conformational states. Apart from an additional local energy minimization when evaluating the energy of a conformation, the procedure retained its combinatorial character as described here. The efficiency of the extended dynamic programming enabled us to study the low-energy conformations of medium-size polypeptides such as melittin and avian pancreatic polypeptide (Vajda et al., 1993). Since peptides in this size range have conformations that do not depend on the environment as heavily as the conformations of short linear peptides (with less than 10 residues), the calculations provided interesting information on the global behavior of semi-empirical potential functions.

The main difference between the classical and extended dynamic programming algorithms is due to the cutoff parameter C. With $C = 0$ the procedure is reduced to dynamic programming, whereas by increasing C we retain more and more conformations and proceed toward an exhaustive search. Thus, the choice of C is critical both for reliability and numerical efficiency. In spite of its importance, we had only heuristic rules for applying the extended dynamic programming to polypeptides (Vajda and DeLisi, 1990; Vajda et al., 1993). In the remainder of this chapter we discuss the meaning of C by considering a simple assignment problem, and describe a very general adaptive procedure for its selection in more general cases.

5. A Lattice Model of Polymer Folding

In principle, the value of the parameter C required in a step of the extended dynamic programming applied to the polymer problem just described can

be obtained from the values of the interaction energies between segments
A and B. Since C is the maximum deviation between the energies that
correspond to the weakest and strongest interactions in any conformation
of the two segments, a relatively tight lower bound on C can be found only
in simple problems. In this section we describe such a simple application
that will also demonstrate the basic idea of extended dynamic programming
and its potential advantages.

The problem involves a restricted self-avoiding walk on a two-dimensional square lattice. The restriction means that the direction changes either
by $+90°$ or by $-90°$ at every lattice point. Thus we allow only two conformational states for each residue, denoted by 0 and 1, representing turns by
$-90°$ and by $+90°$, respectively. Restricting considerations to two states
per monomer does not change the essential properties of the problem, but
it is convenient. In particular, the set of all configurations for a polymer
of N elements is equivalent to the set of all binary integers of N digits.
Let $r_i = (x_i, y_i)$ denote the coordinates of the i-th monomer on the lattice. Then a particular conformation can also be described by the vector
$(r_0, r_1, \ldots, r_N)$, where $r_0 = (0, 0)$, $r_1 = (1, 0)$, and $r_2 = (1, 1)$ are fixed
in order to define the coordinate system. We introduce an energy function
by assuming an attractive interaction $-\varepsilon$ between two monomers i and j
if they occupy neighboring positions on the lattice but are not connected
by a bond. The energy of a particular conformation is the sum of all such
pairwise interaction terms. For example, the dashed lines in Figure 13-4
indicate 11 favorable interactions, and the total energy is $E = -11\varepsilon$. The
energy raises beyond bound if the chain is not self-avoiding.

The problem is to find all configurations for a chain of N monomers
which have minimum energy E. In this case the principle of optimality does
not hold, and we have to use extended dynamic programming (EDP). As
shown in the previous section, the cutoff parameter C in the EDP algorithm
is determined by the maximum deviation between the largest and the smallest values of the interaction term for any two non-overlapping segments of
the chain. Simple considerations show that the maximum interaction is
attained when the two segments are of nearly equal length and form a
hairpin-type configuration. It can be readily verified that the interaction
energy E_{AB} between two segments in such a configuration never exceeds
$[\text{int}(N/4) - 1]\varepsilon$. Since $C = \max |E_{AB} - E_{A'B}|$, this value also provides
an upper bound on the cutoff parameter C if we assume that there exists
an alternative conformation A' for the first segment such that $E_{A'B} = 0$.
While such a conformation can be constructed, its overall energy always
substantially exceeds the lowest energy among all conformations that have
the same state assigned to the last monomer (which is the within-path min-

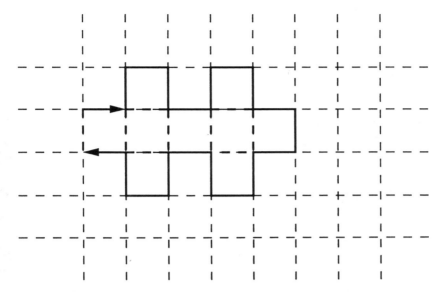

Figure 13-4. Definition of allowed moves and the interaction potential on the 2D lattice model of a polymer chain for $N = 19$. The energy of the configuration shown is $E = -11\varepsilon$.

imum in terms of dynamic programming). Thus, in all applications the required cutoff is less than the bound given by $[\text{int}\,(N/4) - 1]\,\varepsilon$.

We solve the problem for $N = 30$. While $\text{int}\,(N/4) - 1 = 8$, the same set of lowest-energy conformations is obtained for any $C > 6\varepsilon$. According to the previous paragraph, this shows that in all lowest-energy conformations the magnitude of the interaction energy between two segments must be at least 2ε. Figure 13-5 shows the set of 14 equivalent minima with $E = -19\varepsilon$. The 14 solutions form 7 symmetry-related pairs indexed by i and i', respectively, where $i = 1, \ldots, 7$. The occurrence of such pairs follows from the symmetry of the problem, i.e., we do not distinguish between the two ends of the polymer chain. With $C = 6\varepsilon$, the number of conformations retained during the dynamic programming increases from 2 up to 2512, attained in stage 25, and decreases slightly afterward. Although the total number of function evaluations is over 5.5×10^4, it is 5000-fold smaller than the 2^{28} function evaluations in an exhaustive search.

We also solved the lattice problem by simulated annealing (Metropolis et al., 1953; Kirkpatrick et al., 1983), performing a series of simple Monte Carlo simulations. A monomer (i.e., one node on the chain) was selected randomly, using the same probability for all nodes, and at the selected

Figure 13-5. Optimal solutions of the 2D lattice polymer problem for $N = 30$ ($E = -19\varepsilon$).

lattice point the angle of the chain was changed by 180° (e.g., from +90° to −90°). The resulting conformation is either accepted or rejected according to the Metropolis algorithm, on the basis of the energy change ΔE. For $\Delta E < 0$ the conformation is always accepted, whereas for $\Delta E > 0$ it is accepted with probability $p = \exp(-\Delta E / kT)$, where k is the Boltzmann constant and T is the generalized temperature (Kirkpatrick et al., 1983).

Since in this problem both ε and T are arbitrary parameters and only their ratio matters, for simplicity we select $\varepsilon = -k$. Then the smallest unfavorable energy change is $\Delta E = k$, and at $T = 3$ the probability of accepting a conformation with this energy increase is about 0.7. To perform simulated annealing, this initial temperature is decreased in multiplicative steps, each amounting to a 10% decrease in T. Each value of T is held constant for $100N$ configurations, or for $10N$ successful configurations, whichever comes first, where N is the length of the chain. When efforts to reduce E further become sufficiently discouraging, we stop (Press et al., 1986).

After some experimentation with the above parameters, we attained the minimum energy $E = -19$ in about 35,000 function evaluations. Convergence was achieved only from self-avoiding initial conformations. Since only a small fraction of all binary numbers of N digits represent self-avoiding chains, we had to use a simple buildup procedure to find initial self-avoiding conformations. Many of the conformations generated during the Monte Carlo search are not self-avoiding, and hence the acceptance ratio did not exceed 1% even at high T. We recognize that this value can likely be increased through the use of concerted local moves for generating new conformations on the lattice (see, e.g., Skolnick and Kolinski, 1990). However, the simple Monte Carlo method discussed here has been used for comparison in a recent study of genetic algorithms (Unger and Moult, 1993).

While local moves will certainly increase the efficiency of simulated annealing, the qualitative conclusions do not seem to change. In the simple lattice problem, dynamic programming may require somewhat more function evaluations than simulated annealing. However, the latter finds only one minimum, whereas several minima can be found simultaneously by EDP. It is interesting to note that the above lattice problem is not particularly suitable to demonstrate the power of the EDP algorithm. In fact, in the lattice polymer problem, the bound on the cutoff parameter C (6ε for $N = 30$) is relatively high compared to the magnitude of the minimum energy. In other terms, the interaction energy between two segments is comparable to the internal energies of the segments. As we will discuss, in a variety of problems the interaction terms are substantially smaller than the within-segment values of the objective function. Although in such problems a subpath of the optimal path is not necessarily optimal, the objective function on the segment of the optimal path does not substantially exceed the lowest function value attained on any subpath connecting the same two states. While the EDP algorithm is particularly efficient in problems with the above property, it has advantages over simulated annealing even in the 2D lattice problem considered here: Finding all of the 14 minima takes less than twice the CPU time of finding a single minimum by simulated annealing.

In addition to representing the EDP algorithm on a graph (Figure 13-1), it is useful to describe it as a tree. At the i-th level the nodes represent the conformations generated in the i-th stage of dynamic programming, and for each of these conformations the tree shows the branches or "children conformations" obtained from it at the next level. This formalism reveals that EDP can be considered as a branch-and-bound type process. In fact, we do not branch from a node (i.e., conformation) if the corresponding function value exceeds by more than the cutoff C the lowest function value attained so far on any path connecting the same two states.

6. Adaptive Selection of the Cutoff Parameter C

Return now to the more general problem of assigning discrete conformational states to the amino acid residues of a sequence in order to minimize an objective function Q. The choice of the cutoff parameter C determines the success and numerical efficiency of the EDP algorithm in a particular application. Since a different value of C can be used in each stage of the buildup, we face the problem of selecting the parameters $C_1, C_2, \ldots, C_N$, where N is the number of stages (i.e., the number of residues). In spite of its importance, the selection of the cutoff values involved some heuristics even in the simple problem described in the previous section, and was completely heuristic when we applied the algorithm to polypeptides (Vajda and DeLisi, 1990). In this section we introduce a procedure that iteratively updates the values of the cutoff parameters, and under very mild assumptions guarantees that the global minimum is found while preserving the numerical efficiency of EDP as much as possible. Thus, with the additional iteration the optimization method is equivalent to an exhaustive search, although the function Q is evaluated only for a small fraction of all configurations.

Let $(k_1^*, k_2^*, \ldots, k_N^*)$ denote the optimal state assignment for a sequence of N residues, and let $(k_1', k_2', \ldots, k_q')$ denote the optimal state assignment for the segment of $q < N$ residues such that $k_q' = k_q^*$, i.e., the last (q-th) residue has the same state assignment in both conformations. Since the principle of optimality does not necessarily hold,

$$Q(k_1', k_2', \ldots, k_q^*) \leq Q(k_1^*, k_2^*, \ldots, k_q^*), \tag{1}$$

i.e., a q-residue-long segment of the optimal solution is not necessarily optimal among segments of the same length. Let C_q denote the cutoff parameter in the q-th stage of EDP; then for each conformational state k_q' of the last residue, we retain all conformations $(k_1, k_2, \ldots, k_q')$ that satisfy the condition

$$Q(k_1, k_2, \ldots, k'_q) \le Q(k'_1, k'_2, \ldots, k'_q) + C_q. \tag{2}$$

As we discussed, this procedure finds the global minimum of Q if C_q is sufficiently large for all q. The exact cutoff values required to attain the minimum in a particular problem can easily be established *a posteriori*, when the solution is already known. The lower bound C_q^{lb} on C_q is given by

$$C_q^{lb} = Q(k_1^*, k_2^*, \ldots, k_q^*) - Q(k'_1, k'_2, \ldots, k'_q). \tag{3}$$

In the q-th stage of dynamic programming the path $(k_1^*, k_2^*, \ldots, k_q^*)$ leading to the global minimum is retained if and only if C_q satisfies the condition

$$C_q^{lb} \le C_q. \tag{4}$$

The cutoff value C_q is said to be critical if $C_q < C_q^{lb} + \Delta C$, where ΔC is a small positive number. The selection of ΔC will be further discussed in this section. Here we only note that decreasing a critical cutoff parameter by more than ΔC leads to the loss of the global minimum.

Since the bounds C_q^{lb} are not *a priori* known during the conformational search, we develop an iterative procedure for determining reasonable estimates C_q^e for C_q^{lb}. Let $C_1, C_2, \ldots, C_N$ denote an initial selection of the cutoff parameters, and assume that with these values the lowest value of Q is attained at the state assignment $(k_1^e, k_2^e, \ldots, k_N^e)$. Replacing the q-residue-long segment $(k_1^*, k_2^*, \ldots, k_q^*)$ of the globally best solution by the corresponding segment $(k_1^e, k_2^e, \ldots, k_q^e)$ of the current best solution $(k_1^e, k_2^e, \ldots, k_N^e)$ in equation (3) yields the estimates

$$C_q^e = Q(k_1^e, k_2^e, \ldots, k_q^e) - Q(k'_1, k'_2, \ldots, k'_q), \tag{5}$$

where $(k'_1, k'_2, \ldots, k'_q)$ denotes the assignment that gives the minimum for the segment under the condition $k'_q = k_q^e$. This assignment is necessarily calculated during the dynamic programming. In the particular application to proteins, each path of the dynamic programming is identified by the conformational state assigned to the last residue in the current stage, and the conformation $(k'_1, k'_2, \ldots, k'_q)$ is the minimum within the path that leads to the state k_q^c, the conformational state assigned to the same residue in the currently best solution for the entire chain. Thus, condition (4) becomes

$$C_q^e \le C_q, \tag{6}$$

and the following adaptive procedure can be used to identify the sequence $C_1, C_2, \ldots, C_N$ of cutoff values that are large enough to attain the global minimum, but exceed the corresponding lower bounds as little as possible:

(i) Select a uniform starting cutoff $C_q = C$.

(ii) Apply the EDP algorithm with the current cutoff values to obtain a conformation of the entire chain.

(iii) Calculate estimates C_q^e of the lower bounds C_q^{lb} by equation (5), and check conditions (6).

(iv) Terminate the search if no cutoff value is critical.

(v) If the cutoff C_q in the q-th step is critical, increase the parameter to $C_q^e + \Delta C$, and return to step (ii).

The above procedure is based on the assumption that the configuration space of the state assignment problem is densely populated. To define this latter notion consider a solution $(k_1, k_2, \ldots, k_N)$ that is not the global minimum, and let $Q_i = Q(k_1, k_2, \ldots, k_i)$ denote the objective function for the first i residues at this particular assignment. The configuration space is densely populated if, for any non-optimal solution, there exists a positive constant ΔC, an index i, with $0 < i < N$, and another configuration $(k_1^a, k_2^a, \ldots, k_N^a)$ such that $Q(k_1, k_2, \ldots, k_i) - Q(k_1^a, k_2^a, \ldots, k_i^a) < \Delta C$. In other words, this means that any ΔC-neighborhood of a non-optimal solution contains another solution, where the neighborhood is defined in terms of the norm $\|Q\| = \max_i |Q(k_1, k_2, \ldots, k_i)|$.

The search tree in EDP is constrained by the choice of the cutoff parameters $C_1, C_2, \ldots, C_N$. If any of these values is not large enough, the global minimum is outside of the feasible set of nodes on the tree, and is not found. However, if the configuration space is dense, then there exists a solution very close to the boundary of this set. Thus, using the lower bound C_q^e calculated from equation (5), the cutoff value C_q is critical, and $C_q \leq C_q^e + \Delta C$. By increasing C_q we can eliminate criticality and decrease the minimum found. Although a cutoff value may become critical at a different stage of the algorithm, eliminating all critical values guarantees that all global minima are found.

7. Example: Finite-State Backbone Approximation

In this section we assign conformational states, defined as points on the (ϕ, ψ, ω) map, to residues of a protein so as to obtain backbones that are most similar to the native structure. The bond lengths and bond angles are assumed to be standard. Backbone similarity is measured in terms of root mean square deviations (RMSD) between superimposed N, C_α, and C atoms. Although minimizing the RMSD value from a known conformation is easier than locating the global minimum of a potential, the problem is far from trivial and requires the use of combinatorial search techniques.

As shown by Rooman et al. (1991), a relatively good approximation of observed backbone conformations can be attained by appropriate assignment of seven conformational states. The procedure they used starts at the N-terminus and proceeds along the chain, assigning conformational states to residues one-by-one, each time evaluating the RMSD between the segment composed of all residues up to the current one and the equivalent native fragment. The fragment is lengthened if the RMSD is lower than a threshold. Otherwise the program backtracks and generates alternative conformational states for each residue until a backbone segment with RMSD lower than the threshold is found. The entire procedure is repeated until all solutions complying with the imposed threshold are determined. Rooman and co-workers retain 100 structures with the lowest RMSD in which the native disulfide bridges are identifiable. The algorithm does not guarantee to find the optimal solution, because the best-fitting chain may contain segments with RMSD higher than the cutoff value, which are then rejected.

The iteratively updated EDP algorithm applied to this problem not only guarantees that the global minimum is found, but also reduces the required computational efforts by at least an order of magnitude. The objective function Q is the RMSD from the native backbone. Rooman et al. (1991) defined six *trans* ($\omega = 180°$) conformational states (Figure 13-6A) and one additional *cis* state ($\omega = 0°$). To see whether a distinction between *cis* and *trans* states follows from the requirement of minimum RMSD, we allowed the *cis* state to be assigned not only to Pro but also to Gly residues. Values for standard bond lengths and bond angles are the ones given by Momany et al. (1972).

Figure 13-7 shows the lower bounds C_q^e on the cutoff values, derived by the iterative update procedure, for the 11 small proteins also considered by Rooman et al. (1991). The proteins are denoted by their Brookhaven Data Bank codes (Bernstein et al., 1977). The C_q^e values are calculated by equation (5) and expressed as functions of the percentage segment length $100q/N$. While the required cutoff values vary from problem to problem, there exists an envelope value, denoted by the thick solid line in Figure 13-7, that is large enough for all proteins considered here. As we discussed, the largest cutoff values are expected in the stages of dynamic programming around $N/2$. These maximum values weakly correlate with the total chain length (Figure 13-8), but the C_q^e values do not scale well with these maxima or the chain length N.

Table 13-1 shows finite-state approximation of the backbone conformations in the 11 proteins. Amino acid sequences are given in the first row, using single letter codes. The second row shows the regions of the (ϕ, ψ, ω)

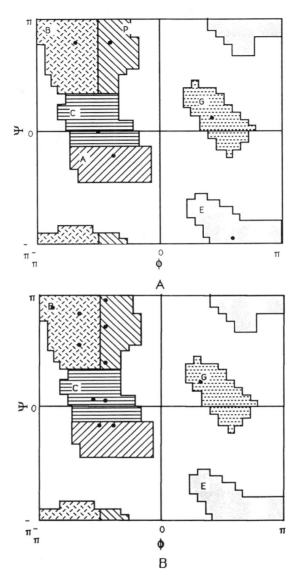

Figure 13-6. Conformational states in the finite-state backbone approximation problem. (A) Points and regions defined by Rooman et al. (1991); (B) points defined by Moult and James (1986) for residues except for Pro and Gly.

map in which the dihedral angles observed in the x-ray structure occur. The regions are defined as shown in Figure 13-6A. 'Y' indicates that the dihedral

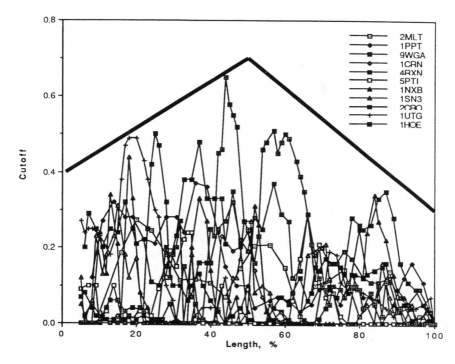

Figure 13-7. Lower bounds on cutoff values (in Å^2) required for attaining the globally optimum finite-state backbone approximation in 11 test proteins. The thick solid line shows the envelope values that yield global minima for any of these proteins.

angles of the corresponding residue occur outside the permitted domains. 'X' indicates that at least one of the angles is undefined, which is always the case for the first and last residues in each sequence. In the same row we show the minimum RMSD that can be obtained using a backbone model with standard bond lengths and bond angles, but without any constraints on dihedral angles. These values were calculated by a standardization procedure, i.e., by fitting a standard model to the given backbone (Vajda et al., 1993), and are shown here to represent the best approximation attainable by a standard model. The 3rd and 4th rows, respectively, give the backbone assignments obtained by Rooman et al. (1991), and in this work by EDP. Accordingly, the RMSD value in row 3 is the one reported by Rooman et al. (1991), whereas the RMSD in row 4 is the outcome of dynamic programming. Notice that the two numbers cannot be directly compared, since we used a slightly different polypeptide geometry. Nevertheless, the minimum RMSD attained is almost the same for all proteins, although the state

Table 13-1. Finite-state backbone approximation for 11 test proteins.

Protein	Amino acid sequence and structure assignments based on best global fit	r.m.s.
2MLT	GIGAVLKVLTTGLPALISWIKRKRQQ	
	XAAAAAAAAABYAAAAAAAAAAAAAP	0.19
	PAAAAAAAAAACAAAAAACACAAAAX	1.18
	ECAAAAAAAABGACAAAAAAAAGBGB	1.22
	BAAAAAAAABYAAACAAAAAAAACC	0.54
1PPT	GPSQPTYPGDDAPVEDLIRFYDNLQQYLNVVTRHRY	
	XPPPPPPCEACPPAAACAAAAAAAAAAAAAAACGBCA	0.29
	APBPBPPAECCEoAACAAAAAAAAAAAACACACGBCX	1.18
	PCEBBBPCECCAEAAAAAAAAAAAAAACACACBAPCA	1.14
	EPPPPPPCEAPBPAAAAAAAAAAAAAACACAACGPCB	0.67
9WGA	XRCGEQGSNMECPNNLCCSQYGYCGMGGDYCGKGCQDGACWTSK	
	XBBEACCGGPBPPGGPPBPCCGBPBBECCACPPGPABECPABPP	0.44
	PBBEACCGGPBPPGGPPAEACGAGAEEAACBEPGCGEEAPEPPX	1.54
	GBBEAACECPBPPECPBAEACEECGPEACABGCGCGPoBBCBPC	1.50
	GBPECCYGGPPPPCCGPPPPACGPPYPYGACBYBGPCBEPPCAGA	1.08
1CRN	TTCCPSIVARSNFNVCRLPGTPEAICATYTGCIIIPGATCPGDYAN	
	XBBBCBAAAAAAAAAAACACGPPAAAAAACGPBBBCPCBPPCCBCC	0.62
	BBBBCBACAAAAACACACACGAEAAACAACBCBBPBAPAPBPPGCEX	1.22
	PBBBABAAAAAAAAAAAAAGEEBAAAAAAACGBBPBAPAPBPAAAEB	1.03
	PBBBCBAAAAAAAAAACCACGPPAAACAAAAGBPBPAPCPBPAACPB	0.60
4RXN	MKKYTCTVCGYIYDPEDGDPDDGVNPGTDFKDIPDDWVCPLCGVGKDEFEEVEE	
	XBPBBPCACGPBPBACAEBACCGBPPGPPACCPPCCPBPAACGPPCCCPBPPBA	0.58
	CPBBBPACBCBBPBACAEAGGCGBPPGAEACCBPAAPBBEAPCAECACBBPPCX	1.33
	PPPBBPACBCBPBBACAEAGGCGBPPGAEACCBPCCBAGBACEEBCAAPBPPCG	1.34
	BPBBBPAABCPBPPAAPPPAACGPBPGPBAACPPACPBPCABCBPAAPBBPPBB	0.75
5PTI	RPDFCLEPPYTGPCKARIIRYFYNAKAGLCQTFVYGGCRAKRNNFKSAEDCMRTCGGA	
	XPACCCPPPBAECPCPBBPBBBPBCACGBPBPBBBCGBGPBAPBBCBAAAAAAACPYA	0.61
	CBCACCBPPBAEAPCPPBCPBBPPABEBEBBPBPPBPEPPCEGBBABAACACAABPBX	1.42
	EBACACPPPBAECBCPPBCABBPPABEBEBBPBPPBPEPPCEGBBABAACACAABGCC	1.40
	BPCAAPBPPBAECPCPPBBBPBBPAABAPBBPBPBCEGBCPCAPBCBAAAAAAAACPG	0.82
1NXB	RICFNQHSSQPQTTKTCSPGESSCYHKQWSDFRGTIIERGCGCPTVKPGIKLSCCESEVCNN	
	XBBYCBPYYYPBBBYPPPYPCPBBBBBBBBEAYEBBBBBBBEYYPYYYGYGBBBBYBPGACY	1.20
	ABBPAEGGGAAGBBBBCECAGAEBPBEEEEEPGEBBBPCPAPAEAEEEABPBBBBCPAAECX	1.67
	PBBAEPPECAoEBBPPPBECBBBAGPBAEEEEPGEPBBBPoGPGAEBBoCPBBBBBCPAAEAA	1.66
	PBBPCPBGGPPBBBPBCPPYBCPBBBBBPPBAAYBBBBBYBEBAGPCAGBBBBBBPAPPGCAB	1.08
1SN3	KEGYLVKKSDGCKYGCLKLGENEGCDTECKAKNQGGSYGYCYAFACWCEGLPESTPTYPLPNKSC	
	XBEPPBPAACGBPBPBABPEPBCAAAAAACPCAAGPABBBBBGGBPBBBGPPACPPBPoPCABPA	0.55
	AEEPPBBAABCBPPBBAPBEPEGCACAAACBEABBCGEEEEEACEBBPBGAECABPCEGPAAPBX	1.45
	AEEPPBBAABCPBBPPAPBEPEGGACAACAAGACGPCEEEEEPGEBBBBEEAGCBPBGEEGGAGA	1.42
	BBEPPBPAAAGBPBPBPBAYBBAAAAAAACPAAAGBPBBBBBGGBPPBBGPPCAPPPCAGCABPA	0.96
2CRO	MQTLSERLKKRRIALKMTQTELATKAGVKQQSIQLIEAGVTKRPRFLFEIAMALNCDPVWLQYGT	
	XCBAAAAAAAAAAAACGBBAAAAAAAACGBPCCAAAAAACGAPABPACAAAAAAACGBPAAAAACGX	0.30
	ACBAACAAAAACACBCPPAACACAACGPPAAAAAAAACGABAGABGCAAAAACCGEGAACAABAX	1.38
	AAPAAAAAAAAAAAAAGBPAAACAACCGEGAACACCAAAGAAGGAPPCAACAAABCPPAAAABGEC	1.32
	CAPAAAAAAAAACACGBPAACAAAAAGPPAAAAACACAGABBGPPGAAAAAAACGBPCAAAAAGB	0.70
1UTG	GICPRFAHVIENLLLGTPSSYETSLKEFEPDDTMKDAGMQMKKVLDSLPQTTRENIMKLTEKIVKSPLCM	
	XBPAAAAAAAAAAACBPAAAAAAAACACGPPAAAAAAAAAAAAACCPPAAAAAAAAAAAAAACCPACCP	0.43
	CEPAAAAAAAAAABECGACAAAACACAGAEAAACACAAAAAAACACBEPACACAACACAAAAACBBAACX	1.32
	CEPAAAAACACAAPGoGAAACACACACAGAEAAACACAAAAAAAAEPACACACAAAAAAAACCPACCB	1.17
	ABPCAAAAAAAAAAABEPCAACAAACACCGPPAAAAAAAAAAAAACACAPPAAAAAAAAAAAAAACPAACA	0.60
1HOE	DTTVSEPAPSCVTLYQSWRYSQADNGCAETVTVKVVYEDDTEGLCYAVAPGQITTVGDGYIGSHGHARYLARCL	
	XBCABPPPPACBBBBBBACBBBBBPCBCBPBBBBPBPACYPBPPBBPBPGBBBPCBPPCCPCCEPPABBBPBB	0.43
	CAGAPPPAEPGAEBAEBEAPBBBBACGBABBAGPBBBACEEBBPBBPBCEGEPCEEEBPAABGEAGCPPPPAEX	1.49
	PBCPEBAEPCABBPPCPCCEEEGPBCEGEPAGBBPAEPGGPBBPPCGPBoPBAEEGBAECAPAAoEGABBBAEE	1.56
	BBPCBPPPPCABBBPBBAABBPAGPBBCAGPPBBPBPCCGBPPPPBPBBBPGBBBPPEPPCBEAAEPBCPPPPPB	1.12

assignments frequently differ. The different assignment corresponding to almost identical RMSD values clearly reflect that the configuration space is densely populated, as discussed in the previous section.

It is emphasized by Rooman et al. (1991) and confirmed in this section that backbone conformations can be well described by using as few as seven discrete conformational states. However, there are large errors in dihedral angles. In fact, using only a few points representing entire regions

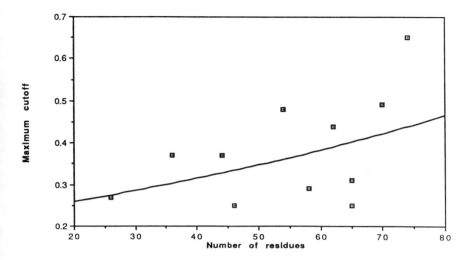

Figure 13-8. Dependence of the cutoff value required on the length of the protein.

of the (ϕ, ψ, ω) map, small RMSD can be attained only by assigning a conformational state to some residues from a region that does not include the observed point of the (ϕ, ψ, ω) map (see Table 13-1). Furthermore, minimization of the RMSD from the x-ray structure does not automatically lead to physically meaningful structures. In particular, we assigned *cis* conformations not only to Pro but also to Gly residues. Another problem is that in Pro the dihedral ϕ must be between $-85°$ and $-50°$ due to the geometry of the ring. States B, G, and E of Rooman et al. (1991) do not satisfy these conditions, although they are frequently assigned to Pro (Table 13-1). The RMSD increases considerably if these states are not allowed for prolines.

The approximation improves considerably if the number of states is increased and specific conformational states are used for Gly and Pro. We repeated the calculations using the 11 conformational states for non-Gly and non-Pro residues as proposed by Moult and James (1986) and shown in Figure 13-6B. Notice that regions A, B, C, and P are now represented by two or three points, whereas region E is forbidden for these residues. For Gly we used 23 conformational states described earlier (Vajda and DeLisi, 1990), with three points in region G, one point in E, and eight points in the region forbidden for all the other residues. Another set of 14 points were used for Pro, with dihedral angle ϕ restricted at $-75°$. The resulting state assignments are shown in row 5 of Table 13-1. The RMSD deviations are decreased, in some cases by a factor of two.

The EDP algorithm is much more efficient for solving the backbone approximation problem than the backtracking used by Rooman et al. (1991). The efficiencies can be compared in terms of CPU times, since the same Silicon Graphics 340 processors were used in both studies. For the 26-residue melittin (2MLT) the backtracking requires of the order of 100 CPU-minutes. For the a-amylase inhibitor (1HOE), a protein three times that size, the time required is multiplied by a factor of about 200 (about 300 hours). Using the envelope cutoff values shown in Figure 13-7, the EDP calculation for 2MLT is completed in 14 minutes. For the same molecule, the adaptive updating procedure requires two iterations, and less than four minutes. For 1HOE the procedure was completed in four iterations, together requiring somewhat less than 19 hours of CPU time. Thus, on the average, we attain a 15-fold increase in efficiency. It is interesting to note that increasing the number of conformational states from 6 to 11, i.e., using the conformational states proposed by Moult and James (1986) and shown in Figure 13-6B, there is less than twofold increase in CPU time required by dynamic programming, because the cutoff values identified by the iterative algorithms were considerably decreased.

8. Conclusions

Exploring the configuration space of a protein for the minima of a function, such as potential energy, is an NP-complete optimization task even in the case of simplified representations of the protein geometry. Although the principle of optimality does not hold in most of these problems, it is approximately satisfied in a large variety of cases. In terms of protein conformations this means that while segments within the lowest-energy conformation of a protein do not attain their lowest energies, the energy differences are moderate. Under these conditions an extended version of the dynamic programming is a numerically efficient optimization tool. The use of extended dynamic programming (EDP) requires the selection of a cutoff parameter that must be large enough in order to find the global minimum, but cannot be too large, in order to retain numerical efficiency. The iterative procedure for selecting the cutoff parameters overcomes these difficulties, and yields an algorithm that guarantees finding the global minimum with reasonable computational efforts in a number of problems.

Acknowledgment. This work was supported in part by Grant AI30535 from the National Institute of Allergy and Infectious Diseases.

References

Bellman R (1975): *Dynamic Programming*. Princeton: Princeton University Press

Bernstein F, Koetzle T, Williams G, Meyer E, Brice M, Rodgers J, Kennard O, Shimanoushi T, Tasumi M (1977): The protein data bank: A computer-based archival file for macromolecular structures. *J Mol Biol* 112:535–542

Brower R, Vasmatzis G, Silverman M, DeLisi C (1993): Exhaustive conformational search and simulated annealing for models of lattice peptides. *Biopolymers* 33:329–334

Bruccoleri RE, Karplus M (1987): Prediction of the folding of short polypeptide segments by uniform conformational sampling. *Biopolymers* 26:137–168

Chan HS, Dill KA (1991): Polymer principles in protein structure and stability. *Ann Rev Biophys Biophys Chem* 20:447–490

Chan HS, Dill KA (1993): The protein folding problem. *Phys Today* 2:24–32

Covell DG, Jernigan R L (1990): Conformations in folded proteins in restricted spaces. *Biochemistry* 29:3287–3294

Crippen GM (1991): Prediction of protein folding from amino acid sequence over discrete conformational spaces. *Biochemistry* 30:4232–4237.

Dill KA, Fiebig K, Chan SH (1993): Cooperativity in protein-folding kinetics. *Proc Natl Acad Sci USA* 90:1942–1946

Go N, Taketomi H (1978): Respective roles of short- and long-range interactions in protein folding. *Proc Natl Acad Sci USA* 75:559–563

Horowitz E, Sahni S (1979): *Fundamentals of Computer Algorithms*. Rockville: Science Press.

Hu, TC (1982): *Combinatorial Algorithms*. Reading, MA: Addison-Wesley.

Kirkpatrick S, Gellatt CD, Vecchi MP (1983): Optimization by simulated annealing. *Science* 220:671–680

Lau KF, Dill KA (1989): A lattice statistical mechanics model of the conformational and sequence space in proteins. *Macromolecules* 13:3986–3997

Lipton M, Still WC (1988): The multiple minima problem in molecular modeling. Tree searching internal coordinate conformational space. *J Comp Chem* 9:343–355

Metropolis N, Rosenbluth AW, Rosenbluth MN, Teller AH, Teller E (1953): Equations of state calculations by fast computing machines. *J Chem Phys* 21:1087–1092

Moult J, James MNG (1986): An algorithm for determining the conformation of polypeptide segments in proteins by systematic search. *Proteins* 1:146–163.

Press WH, Flannery BP, Teukolsky SA, Vetterling WT (1986): *Numerical Recipes*. Cambridge: Cambridge University Press.

Rooman MJ, Kocher J-PA, Wodak SJ (1991): Prediction of protein backbone conformation based on seven structure assignments. *J Mol Biol* 221:961–979

Shaknovich E, Farztdinov G, Gutin AM, Karplus M (1991): Protein folding bottlenecks: a lattice Monte-Carlo simulation. *Phys Rev Lett* 67:1665–1668

Shortle D, Chan HS, Dill K (1992): Modeling the effects of mutations on the denatured states of proteins. *Protein Science* 1:201–215

Sikorski A, Skolnick J (1989): Monte Carlo studies of equilibrium globular protein folding. α-helical bundles with long loops. *Proc Natl Acad Sci USA* 86:2668–2672

Skolnick J, Kolinski A (1989): Computer simulations of globular protein folding and tertiary structure. *Annu Rev Phys Chem* 40:207–205

Skolnick J, Kolinski A (1990): Simulations of the folding of a globular protein. *Science* 250:1121–1125

Taketomi H, Kano F, Go N (1988): The effect of amino acid substitution on protein-folding and -unfolding transition studied by computer simulation. *Biopolymers* 27:527–559

Unger R, Moult J (1993): Genetic algorithms for protein folding simulations. *J Mol Biol* 231:75–81

Vajda S, DeLisi C (1990): Determining minimum energy conformations of polypeptides by dynamic programming. *Biopolymers* 29:1755–1772

Vajda S, Jafri MS, Sezerman U, DeLisi C (1993): Necessary conditions for avoiding incorrect polypeptide folds in conformational search by energy minimization. *Biopolymers* 33:173–192

14

Computational Complexity, Protein Structure Prediction, and the Levinthal Paradox

J. Thomas Ngo, Joe Marks, and Martin Karplus

1. Perspectives and Overview

A protein molecule is a covalent chain of amino acid residues. Although it is topologically linear, in physiological conditions it folds into a unique (though flexible) three-dimensional structure. This structure, which has been determined by x-ray crystallography and nuclear magnetic resonance for many proteins (Bernstein et al., 1977; Abola et al., 1987), is referred to as the native structure. As demonstrated by the experiments of Anfinsen and co-workers (Anfinsen et al., 1961; Anfinsen, 1973), at least some protein molecules, when denatured (unfolded) by disrupting conditions in their environment (such as acidity or high temperature) can spontaneously refold to their native structures when proper physiological conditions are restored. Thus, all of the information necessary to determine the native structure can be contained in the amino acid sequence.

From this observation, it is reasonable to suppose that the native fold of a protein can be predicted *computationally* using information only about its chemical composition. In particular, it should be possible to write down a mathematical problem that, when solved, gives the native conformation of the protein. This procedure would be self-contained, in the sense that no additional information about the biology of protein synthesis would be required. Further, it is reasonable to hope that this procedure could be accomplished without requiring an astronomical amount of computer resources, given the observation that polypeptide chains do fold to their

The Protein Folding Problem and Tertiary Structure Prediction
K. Merz, Jr. and S. Le Grand, Editors
© Birkhäuser Boston 1994

Table 14-1. Glossary of problem-related terms.

---

problem instance   The task of finding the global minimum[a] of a particular function $f(x_1, x_2, \ldots, x_n)$, e.g., the task of finding the value of x that will give the smallest possible value of $f(x) = 18x^7 + x^3 - 3x^2$. Note that the problem instance must be self-contained—everything necessary to specify the problem fully must appear.

problem   A well-defined class of problem instances, e.g., the task of globally minimizing $f(x_1, x_2, \ldots, x_n)$, where f is any polynomial in the variables $x_1, x_2, \ldots, x_n$.

restriction   Some stipulation about what constitutes an instance of the problem. In the example we have been using so far, it was stated that f is restricted to be a polynomial in one or more real variables.

restricted problem   One problem, X, is said to be a restricted version of Y if X comprises a subset of the instances of Y. For example, the problem of globally minimizing quadratic functions is a restricted version of the problem of globally minimizing polynomial functions, because all quadratic functions are polynomials.

instance description   The information needed to specify a particular instance of a given problem. In the example we have been using so far, the instance description might contain the following information: (a) the number of variables x_i is 1; (b) the polynomial is of degree 7; (c) the coefficients of the polynomial terms are, in descending order of degree, $\{18, 0, 0, 0, 1, -3, 0, 0\}$. Note that it is possible to construct a precise instance description like this only when the restrictions on a problem are well defined.

instance size   The amount of memory needed to store a particular instance description.

---

[a] For simplicity in this discussion, we assume that the only computational tasks of interest to the reader are those that involve global optimization of a function.

---

native structures in an amount of time far less than required for an exhaustive search.

Based partly on these suppositions, computational chemists have developed and are refining expressions for the potential energy of a protein as a function of its conformation (Momany et al., 1975; Brooks et al., 1983; Weiner et al., 1986, for example). According to the Thermodynamic Hypothesis (Epstein et al., 1963), finding the global minimum of a protein's potential-energy function should be tantamount to identifying the protein's native fold.[1] (All notes are at the end of the chapter.) Unfortunately, the task of finding the global minimum of one of these functions is not easy, in

Table 14-2. Glossary of algorithm-related terms.

---

algorithm A computer program that takes an instance description as input and supplies an answer to the given problem instance as output. Note that an algorithm is always associated with a particular problem, including any associated restrictions; otherwise, the instance description may not make any sense to the program.

correctness An algorithm is said to be correct for a given problem *instance* if it gives the right answer when fed with the description of that instance. It is said to be correct for a *problem* if it is correct for all possible instances of the problem. When an algorithm is simply said to be *correct*, it is understood to be correct for the problem for which it was designed.

efficiency An algorithm is efficient for a given problem if is guaranteed to return *some* answer, right or wrong, to every possible instance of the problem, within an amount of time that is a polynomially bounded function of the instance size. Note that there is no notion of efficiency for a single problem instance.

It must be emphasized that this definition of the word "efficiency," which we employ throughout this review, does not necessarily correspond to the practical notion of efficiency. An algorithm whose running time is proportional to n^{100}, where n is the size of a problem instance, is far from practical; nevertheless, it is considered efficient by this definition. Thus, while it is usually reasonable to assume that an inefficient (exponential-time) algorithm is too slow to be of any practical use except for small problem instances, it is not always reasonable to assume that an efficient algorithm is fast enough to be of practical use.

guarantee A statement about the behavior of an algorithm that can be proved rigorously. For example, an algorithm is not considered to be efficient for a problem unless polynomial time bounds can be proved; it is not enough to be able to say that the algorithm has met the given time bounds for all instances of practical interest with which it has been tested.

polynomially bounded function A function $f(n)$ is polynomially bounded if there exists some polynomial function $g(n)$ such that $f(n) \leq g(n)$ for every positive value of n. In the interest of brevity, computer scientists call a function *polynomial* if it is polynomially bounded, even though some functions that are not normally considered polynomial (such as $\log n$) are included in this definition. Similarly, a function is called *exponential* if it is not polynomially bounded, even though some functions that fit this description (such as $n^{\log n}$) are not normally thought of as exponential (Garey and Johnson, 1979).

exhaustiveness An algorithm is said to search its solution space exhaustively if it tests every possible candidate solution. An algorithm that is exhaustive will always be correct; but to be correct, an algorithm need not be exhaustive.

---

particular because the potential-energy surface of a protein (as represented by an empirical potential) contains many local minima (Elber and Karplus, 1987, for example). Many clever and extremely creative techniques have been employed to try to escape from these local minima (Piela et al., 1989; Gordon and Somorjai, 1992; Head-Gordon and Stillinger, 1993), but so far no practical method for globally optimizing the potential energy of a protein has been produced.

The central question addressed in this review is this: Is there some clever algorithm, yet to be invented, that can find the global minimum of a protein's potential-energy function reliably and reasonably quickly? Or is there something intrinsic to the problem that prevents such a solution from existing?

Many measures of a problem's difficulty are possible. They range from the informal to the formal, and they focus on various sources of difficulty. Each can be useful in its own right. For example, proteins are certainly "hard to model with quantitative accuracy," the more realistic energy functions are "complicated to code as computer subroutines," and the algorithms that one uses to try to find the minima of these functions can consume seemingly unlimited amounts of supercomputer time.

Instead of such qualitative statements, the results reviewed here focus on a formal measure of difficulty called *intractability*. A problem is said to be *tractable* if there exists an algorithm for it that is *guaranteed* to be *correct* and *efficient*. It is said to be *intractable* if no such algorithm exists. (Several terms such as "correct" and "efficient" are used in a precise technical sense in this chapter. Tables 14-1 and 14-2 list their definitions.) When a problem is intractable, there is generally a rather unforgiving limit to the size of the problem instances that can be solved correctly before running times become astronomically large, and this limit is relatively uninfluenced by implementation details such as cleverly designed program codes and improvements in computer hardware.

It is widely believed that the problem of locating the global minimum of a protein's potential energy function is intractable by this definition. However, the conventional reasoning underlying this belief is fallacious. The conventional argument proceeds as follows. Although the bond lengths and angles in a protein can be predicted easily since they cannot vary much, torsion angles (rotations about bonds) are not easily predicted. Rotatable torsions tend to have three preferred values, none of which can be ruled out *a priori*. If there are N rotatable torsions in the protein molecule, then there are 3^N possible combinations of those torsions. To be sure to find the best combination, a computer program would have to try them all. But 3^N is a huge number for typical values of N because it is exponential in N. For typical values of N (~ 100 or larger) and the speed of current hardware, the expected running time of this exhaustive algorithm is astronomically long.

This reasoning is fallacious. For some variants of the problem, it is wrong.[2] Every problem in combinatorial optimization[3] has an exponential number of candidate solutions. It will therefore require an exponential amount of time to solve *any* such problem by an exhaustive search of its candidate solutions. However, it is rarely necessary to proceed by such brute-force tactics (see Figure 14-6). With some thought, it is nearly always possible to do many orders of magnitude better than exhaustive search. Moreover, with many combinatorial optimization problems, algorithms can be found that are both efficient and correct (have polynomial time bounds and always give a right answer—see Table 14-2).

Is global potential-energy minimization *inherently impossible* to accomplish efficiently without sacrificing correctness, or is an efficient, correct algorithm waiting to be found? Recently, efforts have been made to answer this question using the formal tools of the theory of NP-completeness (Ngo and Marks, 1992; Unger and Moult, 1993). Introduced in the 1970's, NP-completeness theory (Lewis and Papadimitriou, 1978; Garey and Johnson, 1979; Lewis and Papadimitriou, 1981) was developed to help discover why certain problems in combinatorial optimization seem to be intractable, whereas others are not. A problem that is found to be NP-complete or NP-hard by an analysis of this type is intractable if P$\neq$NP. (The meaning of the proposition "P$\neq$NP" is summarized in Figure 14-1.)

One motivation for undertaking this line of inquiry pertains to the development of structure-prediction algorithms. Practical solutions to NP-complete problems do exist. They are compromises that entail well-understood tradeoffs between guarantees of efficiency, correctness, and generality (Papadimitriou and Steiglitz, 1982), and the forms of these compromises have been studied extensively. Thus, the mere fact that a problem is known to be NP-complete can guide algorithm developers to existing classes of heuristic solutions. Moreover, the *details* of an NP-completeness proof can expose the sources of a particular problem's complexity. With this knowledge, the algorithm developer can know in advance that certain algorithms are bound to fail, and might identify restricted forms of the problem that can be solved efficiently. These issues are discussed in Section 4.

The other goal is to obtain an improved understanding of the Levinthal paradox (Levinthal, 1968, 1969). The Levinthal paradox refers to the observation that although a protein is expected to require exponential time to achieve its native state from an arbitrary starting configuration, the process of folding is not observed to require exponential time. But why is a protein expected to require exponential time to fold? The conventional justification for this premise requires the use of a model of protein behavior that leads to incorrect physical consequences. A reformulation of the Levinthal paradox with a more rigorous reason to expect exponential-time folding is discussed in Section 5.

"P≠NP"—What does it mean?

The proposition P≠NP, whose truth (or falsehood) has not been proved, is a pivotal conjecture in the theory of NP-completeness (Lewis and Papadimitriou, 1978). P is the class of problems for which correct algorithms with polynomial time bounds exist; NP is the class of problems for which a correct answer can be verified in polynomial time. The underlying theory and the precise meaning of the designation "NP" are too subtle to treat properly in this review, but the consequences of the proposition are easily summarized.

- Computer scientists have identified several classes of problems whose intractability is likely, but not yet proved. Two of the most important such classes are called NP-complete and NP-hard. Membership in these classes is well-defined whether or not P=NP. Hundreds of NP-complete and NP-hard problems have been identified; many of these problems are of great practical significance.
- If P=NP, then all NP-complete problems are efficiently solvable. If P≠NP, then all NP-complete problems are intractable.
- The class of NP-hard problems includes all NP-complete problems and others. An NP-hard problem can be thought of as being "at least as hard as" an NP-complete problem. If the NP-complete problems are intractable, then all of the NP-hard problems are intractable. But if the NP-complete problems are efficiently solvable, some NP-hard problems will still be intractable.
- (Almost) nobody believes that P=NP.

Figure 14-1. Meaning of the proposition "P≠NP."

The conclusions that may be drawn from the results described here are rigorous but qualified. In addition to reviewing the results themselves, much of the space in this chapter is devoted to identifying the caveats associated with each possible inference. There are some fundamental limitations of the scope of this approach that we state in advance.

First, the theory of NP-completeness can be used to address only certain aspects of the protein-folding problem. The protein-folding problem can be defined as encompassing, but not being limited to, the following questions:

1. Why does a protein have a unique native structure, i.e., why is such a small portion of a protein's conformational space significantly populated under physiological conditions?
2. What can be said about the pathway(s) in conformation space by which proteins reach their native states?
3. What accounts for the observed rate at which proteins fold?
4. Can a protein's three-dimensional structure be predicted from its amino acid sequence, and if so, how?

The results that we review here are related directly to questions about structure prediction (4) and indirectly to the consideration of folding rates (3), but they have little to do with the existence of unique native structures (1) and the pathway(s) by which a protein folds (2). (This is not to say that the four questions are unrelated to each other. All aspects of the protein-folding problem are determined by the potential-energy function that describes the polypeptide chain. Thus, the population of a unique native structure is clearly related to the prediction of that structure.) Many aspects of the protein-folding problem are addressed in a recent volume edited by Creighton (1992).

Second, it is incorrect to state that "the protein-structure prediction problem has been shown to be NP-hard." There are numerous approaches (Fasman, 1988) to protein-structure prediction that either do not employ global potential-energy minimization at all, or include stipulations on the nature of the solution in addition to the energy function itself. Secondary-structure prediction, when based on statistical rules derived from known structures, is one approach in which no potential-energy function is used. While NP-completeness theory certainly could be used to analyze secondary-structure prediction, the analysis might be irrelevant because models for secondary-structure prediction are usually designed with efficient (usually linear-time) algorithms in mind. The difficulty of predicting secondary structure arises from the inadequacy of the underlying models in predicting what occurs in reality, not from the time required to solve the computational tasks that arise in the context of those models. The results reviewed here have a formal bearing only on algorithms that operate by attempting to solve a self-contained[4] global minimization problem.

It is also incorrect to state that "global potential-energy minimization for a protein has been shown to be NP-hard," even though this statement is much closer to the truth than the previous one. An NP-completeness proof can, at best, address a form of the problem that is more general than protein-structure prediction by energy minimization—and therefore, possibly more difficult. Put another way, protein-structure prediction might be an easy restricted form (special case) of the problems (Sections 3.3, 3.4, and 3.5) that are known to be NP-hard. This subtle point, which we discuss in Sections 4 and 6, turns out to be central to understanding the limitations of the results.

Third, for some purposes, intractability itself is not as bad as it sounds. The intractability of a problem means that no algorithm for it can be efficient and correct. These qualities would make a protein-structure prediction algorithm, perhaps literally, "too good to be true." Most developers of protein-structure prediction algorithms gave up on such high standards long ago; they focus efforts on developing algorithms that fall short of the ideal in some way. Thus, the NP-hardness of a problem is a somewhat weak statement, even given the very likely assumption that $P \neq NP$. An objective of this review is to explore what may be inferred from this weak statement, given that it is one of very few statements about the difficulty of protein-structure prediction that are known rigorously to be true.

In writing this review, we have tried to focus on objectives that are appropriate for this young line of inquiry. First, we believe that the question most important to the reader, given that he accepts that the proofs being reviewed are mathematically correct, is what they mean. Accordingly, we include an exposition of the theory of NP-completeness (Section 2) that is intended to give an accurate picture of the form of an NP-completeness proof without becoming mired in technical details. Similarly, in explaining the proofs themselves (Sections 3.3, 3.4, and 3.5), we describe the essential steps—it is impossible to understand exactly what is proved otherwise—but do not attempt to persuade the reader of their mathematical correctness. We invite the reader who is interested in the technical details to refer to the original papers (Ngo and Marks, 1992; Fraenkel, 1993; Unger and Moult, 1993) and to existing references on the theory of NP-completeness (Lewis and Papadimitriou, 1978; Garey and Johnson, 1979; Lewis and Papadimitriou, 1981).

Second, we believe that one of the functions of a review is to reinterpret and evaluate existing results. Therefore, interspersed with a straightforward recitation of the facts, the reader will find our opinions and speculations about the implications of this line of reasoning, both for the development of algorithms (Section 4) and for the behavior of real proteins (Section 5).

Third, from the limitations of the existing results, it is clear that the use of computational complexity theory for tasks in protein-structure prediction is by no means a closed book. In Section 6 we point out areas in which continued analysis might be of value, particularly given the results that have already been established.

2. Introduction to NP-Completeness Theory

Before the advent of computational complexity theory, computer scientists were in the quandary of not being able to describe or characterize accurately the difficulty of several important computational problems. One well-known example was the Traveling Salesman Problem (TSP), described in Figure 14-2. Enumerating all possible routes (or *tours*) for the salesman is the basis for a simple algorithm for the TSP, but for all but trivial cases such an algorithm would take far too long. So the central question was this: Were these problems in some way inherently difficult, or was it merely that nobody had been clever enough to discover efficient algorithms for them?

Problem instance: A list of cities and a table of the distances between them. Also, the maximum distance M that the salesman is allowed to travel.

Question: Can the salesman visit each city once and return to the original city from which he departed without traveling further than the maximum distance, M?

Figure 14-2. The Traveling Salesman Problem, stated as a decision problem. Readers more familiar with the optimization form of the problem—that of finding the minimum-length tour—will find the relationship between optimization problems and their corresponding decision problems explained in the discussion of NP-hardness.

Computational complexity theory (Garey and Johnson, 1979; Lewis and Papadimitriou, 1981) provides a rigorous foundation for considering the intrinsic difficulty of computational problems. For our purposes there are three key aspects to the theory:

1. measures of complexity,
2. the principle of reducibility, and
3. the theory of NP-completeness.

2.1 Measuring Complexity

When computer scientists speak of computational complexity, they usually mean one of two things. The *complexity of an algorithm* is a measure of how much time it requires to execute; the *complexity of a problem* is a measure of the time required for the theoretically best possible algorithm for that problem to execute. This distinction is explained in more detail below.

2.1.1 Algorithmic complexity. The most relevant way to measure the complexity of an algorithm is in terms of how much time it takes to run. (Space, i.e., memory, can also be used as the basis for a complexity measure, but time is usually the more useful measure.) For a complexity measure to reflect the intrinsic nature of the algorithm, it should summarize performance on all problem instances, and it should not reflect inconsequential implementation details. These requirements suggest the following three strategies:

1. express complexity as a function of problem-instance size,
2. consider only what can be guaranteed about the running time of the algorithm across all possible problem instances, and
3. focus on the functional dependence of the running time on problem-instance size, not on some absolute measure of execution time.

The combination of these ideas gives us the notion of *worst-case asymptotic time complexity*, which is the complexity measure most often used by computer scientists. A useful way of expressing this complexity measure is by means of "Big Oh" notation: A function $f(n)$ is said to be $O\big(g(n)\big)$ (have an upper bound of "order $g(n)$") if there exist positive constants c and n_0 such that $|f(n)| \leq c \cdot |g(n)|$ for any $n > n_0$. Using this notation, we can say that Bubblesort, a simple sorting algorithm, has $O(n^2)$ worst-case time complexity (Aho et al., 1982), which means that the time required to solve a sorting problem of size $n > n_0$ is bounded above by the value cn^2 for some values of c and n_0. Similarly, we can say that the naive algorithm for the TSP that requires enumerating each possible tour of the cities is not $O\big(f(n)\big)$, where $f(n)$ is any polynomial in n. By convention, an *efficient* algorithm is one that runs in $O\big(f(n)\big)$ time, where $f(n)$ is a polynomial in n. Thus exponential-time algorithms are not considered to be efficient. (Recall that this formal definition of the term "efficient" does not always correspond to the practical meaning of the word—see glossary in Table 14-2.)

2.1.2 Problem complexity. Complexity measures for algorithms allow us to compare different ways of solving a computational problem. But we can also think about the complexity of a problem, which is defined to be the complexity of the best possible algorithm for that problem. Thus problem complexity is a measure that extends across the space of all possible algorithms for a particular problem. For example, one can show that sorting has complexity $O(n \log n)$ (Aho et al., 1974), which means that no algorithm for sorting[5] can have asymptotic worst-case time complexity better than $O(n \log n)$.

An *intractable* problem is one for which no efficient, correct algorithm exists, i.e., for which no correct algorithm exists whose time complexity can be bounded by a polynomial function of problem-instance size. The complexity of the TSP is not known, although TSP is thought to be intractable since no known algorithm for it is both correct and efficient. However, nobody yet has been able to *prove* that the TSP has exponential time complexity.

2.2 Reducibility

Determining the complexity of a problem is not always an easy task. In fact, there are many interesting and practical problems for which worst-case asymptotic time complexity has not been determined. The TSP is one such problem. However, there are other ways in which computational complexity can be measured. The concept of *reducibility* allows us to make useful statements about the complexity of the TSP and many other problems.

Problem instance: A list of cities and a table showing which pairs of cities are connected by air routes.

Question: Can the salesman visit each city exactly once and return to the original city from which he departed, if he travels solely by air?

Figure 14-3. The Hamiltonian Circuit Problem.

Reducibility is best understood by means of an example. Consider the Hamiltonian Circuit Problem (HCP), which is described in Figure 14-3. HCP is *reducible* to TSP, which means that any HCP instance can be transformed efficiently (in polynomial time) into a related TSP instance with the same yes-or-no answer. So if an efficient, correct algorithm for solving TSP were to exist, we could use it to solve any instance of HCP by doing the following:

1. Copy the set of cities from the HCP instance to the TSP instance

2. For each pair of cities that are connected in the HCP instance, set the distance between the corresponding TSP cities to be 1

3. Set all other intercity distances[6] to 2

4. Pose this question for the TSP algorithm: Is there a tour of length n or less, where n is the number of cities in the HCP instance?

By inspecting this construction, it is easy to see that there can be no TSP tour of length less than n. More important, however, is the fact that a tour of length n for the TSP instance will exist if and only if the original HCP instance can be answered in the affirmative. To see this, consider a TSP tour of length n; the TSP tour would require all intercity distances traveled to be of length 1 exactly. But the pairs of cities in the TSP instance that are unit distance apart correspond exactly to those that are directly connected in the HCP instance, so the TSP tour would be a valid HCP solution. Thus the HCP instance can be answered in the affirmative if and only if the derived TSP instance can be answered in the affirmative.

From the existence of this transformation, we can state that *the TSP is at least as hard as the HCP*; if an efficient algorithm for the TSP were to exist, then an efficient algorithm for the HCP would exist. The concept of polynomial-time reducibility thus gives us a way of relating the inherent difficulty of different problems, which is the key to deriving more rigorous arguments about intractability.

Problem instance: A Boolean expression, such as

(a or b or c) and (a or not b or not c or d) and (not a or not d).

Question: Is there a set of assignments to the logical variables that makes the expression true? (In the example given above, one possible set of satisfying assignments is [c *true*], [d *true*], and [a *false*].)

Figure 14-4. Satisfiability, the first known NP-complete problem.

2.3 NP-Completeness

It turns out that a reduction in the reverse direction also exists: the TSP is polynomial-time reducible to the HCP. The reduction is more complicated than the one given above, but similar in principle. Thus, the HCP is also at least as hard as the TSP. In fact, both problems are members of a class of problems called *NP-complete*.[7] A defining characteristic of the class of NP-complete problems is the property of mutual reducibility: Each NP-complete problem is polynomial-time reducible to every other NP-complete problem.[8] A direct consequence of this property is that an efficient, correct algorithm for any NP-complete problem would lead, via reducibility transformations, to efficient, correct algorithms for all NP-complete problems.

Does any NP-complete problem admit efficient solution? Or equivalently, can all NP-complete problems be solved efficiently? The answer to this question is unknown. However, the fact that nobody has found a polynomial-time algorithm for any of these problems—there are several hundred known NP-complete problems, many of great practical significance—is taken as a strong indication that NP-completeness implies intractability. Showing that a problem is NP-complete is a very strong argument that the problem is intractable.

2.4 NP-Hardness

Sometimes a problem can be shown, via the principle of reducibility, to be at least as hard as some existing NP-complete problem, but not vice versa. (Showing reducibility in the opposite direction might be either impossible or too difficult to be worth the effort.) Such a problem is said to be NP-hard.[9]

Many NP-hard problems are optimization problems. The theory of NP-completeness applies to *decision problems*, in which the task is to answer a yes-or-no question. For example, the TSP decision problem is typical: Is there a tour whose length does not exceed M? The corresponding Traveling Salesman *optimization problem* is: What is the shortest tour? NP-complete decision problems often have related optimization problems.

What can be said about the computational complexity of these optimization problems? The most important observation is that such optimization problems are at least as hard as their corresponding NP-complete decision problems, and are therefore NP-hard. Consider the TSP and its optimization variant. If we had an efficient algorithm for the optimization problem, we could use it to solve the decision problem efficiently—we could simply compute the cost of the best tour, and then compare it to the tour length M in the decision problem. Thus, the theory of NP-completeness can be extended from decision problems, via the concept of NP-hardness, to optimization problems.

3. Arguments That Protein-Structure Prediction Is Intractable

Complexity arguments regarding structure prediction have evolved in terms of the assumptions that they require to be rigorous. In this section we review this evolution from traditional arguments to recent formal results that use the tools of computational complexity theory.

3.1 Traditional Counting Arguments

The simplest arguments about the complexity of structure prediction are based on counting. This line of reasoning is summarized by Reeke (1988) as follows:

[T]he folding problem is combinatorially formidable. Each residue may lie in one of three or more stable regions in ϕ, ψ space. Therefore, even if it were possible to refine the angles to an optimal structure from a trial structure with each residue assigned only to the correct region, there would still be on the order of 3^N starting structures to check for an N-residue protein. Since the time required for refinement increases as a power of N, the complexity of the whole problem increases exponentially with a power of N.

While such counting arguments rule out an exhaustive search of a protein's conformational space, they do not rule out the possible existence of a more selective technique for finding the conformation that corresponds to the global energy minimum. An analogous argument would "prove" the intractability of sorting: The number of ways in which a list of numbers can be permuted is exponential in the length of the list,[10] therefore sorting numbers is a very difficult problem. Of course sorting is known to be easy (Aho et al., 1982), whereas structure prediction is considered very difficult. Nevertheless, crude counting arguments that attempt to *prove* the intrinsic difficulty of either problem are equally fallacious. An informal discussion that is closely related to this point may be found in Figure 14-6.

3.2. Crippen's Analysis: A More Sophisticated Counting Argument

A more refined argument in this vein is due to Crippen (1975). Like the argument outlined in Section 3.1, it does not involve the theory of NP-completeness—at the time, the theory of NP-completeness was still in its infancy and not yet widely known. He considers the task of finding the global minimum (or minima) of a function $f(\mathbf{x})$, i.e., the vector(s) $\mathbf{x}_{\min}$ such that $f(\mathbf{x}_{\min}) \leq f(\mathbf{x})$ for all $\mathbf{x}$. It is assumed that $f(\mathbf{x})$ is single-valued and real, and obeys a Lipschitz condition[11]

$$|f(\mathbf{x}_a) - f(\mathbf{x}_b)| \leq L \cdot \|\mathbf{x}_a - \mathbf{x}_b\|$$

for some known Lipschitz constant L. Nothing else is assumed about the function f.

Crippen analyzes how a search algorithm might exploit this Lipschitz condition if the globally minimal value $f_{\min}$, but not the value(s) of $\mathbf{x}_{\min}$, is already known. First, the search space can be discretized by the introduction of a tolerance parameter Δf. Each dimension in $\mathbf{x}$ is discretized with resolution $\Delta x \equiv \Delta f / L$, and the task is restated as identifying the set D of points $\mathbf{x}_D$ for which $f(\mathbf{x}_D) - f_{\min} \leq \Delta f$.[12] The true global minima are guaranteed not to be missed by this criterion; i.e., it is guaranteed that every true global minimum $\mathbf{x}_{\min}$ is within a distance Δx of one of the elements of D.

Second and more importantly, whenever f is evaluated for some point x_{probe}, any other point within a distance $(f(x_{probe}) - f_{min})/L$ may be eliminated from consideration since, by the Lipschitz condition, the function cannot vary fast enough for x_{min} to be within that region. Thus, every time a function evaluation is performed, a local region[13] of solutions can be eliminated;[14] and higher values of $f(x_{probe})$ eliminate larger regions (this can be seen by inspecting the construction in Figure 14-5). An *essential point* is defined to be one that cannot be eliminated by such a criterion for *any* value of x_{probe}. A search algorithm that uses no information about f other than the Lipschitz condition must at least try every essential point to be sure to find the global minimum.

Crippen identifies the minimum and maximum number of evaluations that such a search algorithm might perform as it seeks the global optimum. The best case can occur when f contains a high value at some point x_{high} that is further away from x_{min} than the distance between any two other points in the domain, and is such that $f(x_{high}) - f_{min} = L \cdot \|x_{high} - x_{min}\|$. If x_{high} happens to be the first value of x tried, this evaluation eliminates all of the space except for x_{min}, and the global optimum x_{min} is found by the second evaluation of f.

The worst case identified by Crippen is one in which all values of f lie in the range $f_{min} \leq f(x) < f_{min} + \Delta f$. In that case, all points are essential, so an exhaustive search is required. Note that this case is very pathological indeed—all points in the discretized space are in D, the solution set! Nonetheless, it is this worst-case limit that Crippen applies to the problem of predicting protein structures. He cites an algorithm of Shubert (1972a, 1972b) as optimal in the class of search techniques that exploit only the Lipschitz condition, then shows how it will require time (and space!) that is exponential in the number of degrees of freedom (usually torsions) that are permitted to vary in the molecule. (Crippen does go on to consider average-case performance from information-theoretic considerations, but he does not employ these results in drawing his final conclusion about the exponential-time nature of protein-structure prediction.)

Crippen points out the main limitation of his argument, which is that it applies only to an algorithm that employs no information about the function being optimized, other than a Lipschitz condition. If more information is used, then the proof does not apply. For example, consider the problem of finding the minimum of the Gaussian function $-e^{-x \cdot x}$. This function has a maximum absolute slope of $\sqrt{2/e}$, so it obeys a Lipschitz condition. An algorithm that uses only the Lipschitz condition requires exponential time to be sure to find the global optimum of this function, since the value of the function never gets large enough to eliminate an area of the search

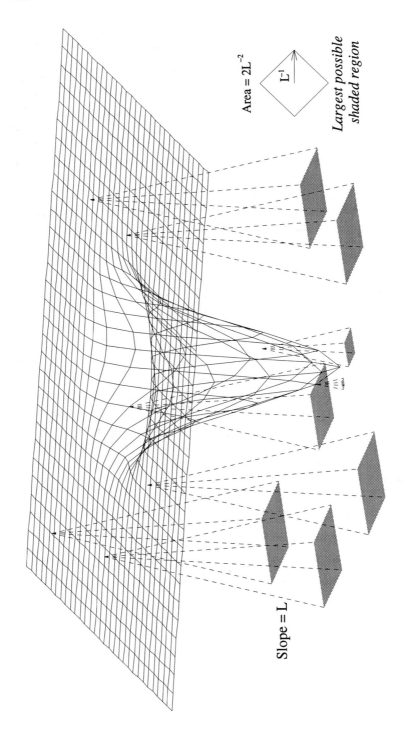

Area = $2L^{-2}$

L^{-1}

*Largest possible
shaded region*

Slope = L

space larger than $2L^{-2} = e$ (Figure 14-5; see also the informal reasoning in Figure 14-6). However, if the algorithm is designed specifically for Gaussian functions, it can employ a technique appropriate for that class of functions—in this trivial example, a standard gradient-descent procedure will do, because a Gaussian function is known not to contain any local minima. Thus, Crippen's arguments are rigorous, but they apply to a limited class of algorithms, namely those that use no information about the function other than a Lipschitz constant.

Crippen closes his paper with a key insight: "Only by establishing additional conditions on the energy function, on an either mathematical or physical basis, and then by employing algorithms which make use of these additional features, [might the global minimization problem become feasible]." We are left with two questions. First, what additional features might help in solving the problem, and can we know in advance what other features will not be sufficient to solve the problem efficiently? Second, can the first question be answered in a manner that is independent of the type of algorithm employed?

3.3 Ngo & Marks Reduction: PSP Is NP-Hard

The principle of reducibility, described in our discussion of NP-completeness (Section 2), might be a way to discover what features of the potential-energy function could be used by an algorithm for global minimization to achieve efficiency. Ngo and Marks (1992) analyze the case in which the algorithm is permitted to exploit the functional form of the potential-energy function describing the protein. Specifically, they define a problem called Polymer Structure Prediction (PSP), described in Figure 14-7, that contains a generalized form of the potential-energy function used by existing molecular-mechanics programs such as CHARMM (Brooks et al., 1983) and AMBER (Weiner et al., 1986).

Figure 14-5. How to be sure to find all of the global minima of a function using a search algorithm that is armed only with a Lipschitz constant for the function and the value (but not the location) of its global minimum. This Gaussian function has, of course, only one minimum, but the algorithm must operate on the assumption that undulations in the surface can appear anywhere because only a Lipschitz condition is known. It must therefore operate by a process of elimination, and at each step only a local region of the search space can be eliminated. The shaded regions delineate points eliminated by a few selected function evaluations. Note that with the Gaussian function, which has a finite dynamic range, the shaded regions cannot exceed a certain maximum size. *See figure on preceding page.*

When is an exhaustive search truly necessary?

A source of worry to some readers who are not already familiar with computational complexity theory is the notion that, to be sure to find the global optimum of a complicated function, one must search its domain exhaustively. The purpose of this panel is to provide some insight, by informal reasoning independent of NP-completeness theory, as to the origin of this belief and why it is generally incorrect.

Function optimization is often thought about in terms of a topographical metaphor. The domain (e.g., x) is taken to represent geographic location, and the range (e.g., $f(x)$) denotes terrain height. This intuitive device is very appropriate for understanding how local minimization works (by "rolling downhill") and for seeing why such methods fail to find the global minimum when local minima are present.

However, the topographical analogy can be pushed too far when it is applied to global optimization. In the physical surfaces to which we are accustomed, points far away from each other have heights that are essentially independent; a ditch can be dug in Florida without causing topographical changes in Maine. If one were to ask the U.S. Geological Survey for a topographical map of a square kilometer of the Rocky Mountains at 10-meter resolution, one would get back a table of 10^4 independent numbers, irreducible to simpler form. To find the highest point in the area one would genuinely have to consider all 10^4 numbers.

By contrast, a surface generated by a mathematical function, even a rugged one, is usually controlled by a relatively small number of parameters. Such a function need not contain as many independent parameters as there are, say, local minima in the surface. In other words, the locations and depths of the local minima are related to each other. When are these relationships simple enough to permit the development of an efficient global-optimization procedure? This type of question can sometimes be answered by applying the theory of NP-completeness.

Figure 14-6. Some informal reasoning about the need for exhaustive searches.

Arguments based on the use of NP-completeness theory are essentially independent of the nature of the algorithm; they test hypotheses about the algorithm's generality and performance without using any assumptions about the algorithm's inner workings. For the conclusions of the analysis to hold true for a particular algorithm, it is required only that the formal description of the problem being analyzed contain an accurate picture of what information *about the problem instance* is available to the algorithm. (In PSP, the algorithm is given the connectivity of the atoms and their chemical types, as well as all of the coefficients of the potential-energy function.)

Problem instance: A list of the atoms in a molecule, their chemical identities, and their connectivity; also, parameters[a] for the empirical potential function: equilibrium values for every bond length and angle, equilibrium values and periodicities for every torsion (possibly many per torsion), equilibrium distances for every possible pair of nonbonded atoms, and force constants for every one of these terms.

Task:[b] Find the globally minimal energy $U^{\min}$ for the molecule, where the energy function is

$$U = \sum_b K_b^{\text{bond}}(l_b - l_b^0)^2$$
$$+ \sum_a K_a^{\text{angle}}(\theta_a - \theta_a^0)^2$$
$$+ \sum_t K_t^{\text{torsion}}\left(1 - \cos[n_t(\phi_t - \phi_t^0)]\right)$$
$$+ \sum_{i>j} K_{ij}^{\text{nonlocal}} f(r_{ij}/r_{ij}^0).$$

The dimensionless function $f(x)$ can be any that has a unique minimum at $x = 1$; for example, it might be the Lennard–Jones potential, $f(x) \propto x^{-12} - 2x^{-6}$. All variables have their usual meanings (Momany et al., 1975; Brooks et al., 1983; Weiner et al., 1986).

---

[a] Although no restrictions on the values of the parameters are named explicitly in this problem description, the proof construction holds for a restricted form of the problem in which all energy coefficients and equilibrium values for the bond-length, angle, and dihedral terms are restricted to lie in their usual ranges. The notion of proving NP-hardness for a restricted form of PSP, and therefore the relevance of this comment, is discussed in Section 4.6.

[b] For clarity, we present the optimization problem, as opposed to its related decision problem.

Figure 14-7. The problem "PSP."

To orient the reader with respect to the use of this proof methodology, it is worth noting a few consequences that accrue automatically from the use of NP-completeness theory:

- The proof applies even if the algorithm employs some "mathematical trick" that involves modifying the potential as the run proceeds (Crippen and Scheraga, 1969, 1971; Purisima and Scheraga, 1986, for example), as long as the final objective is to find the global optimum of a function of the form that appears in the description of PSP. (Algorithms of this type are discussed in greater detail in Section 4.7.)

- The proof applies even if the algorithm has access to information about proteins *in general*, in addition to the potential-energy function—e.g., information from a rotamer library (Vásquez and Scheraga, 1985) or other database (Robson et al., 1987; Summers and Karplus, 1990). (As required by the proof, the algorithm cannot be correct and efficient for *all possible instances* of PSP. This requirement leaves open the possibility of an algorithm that is correct and efficient for proteins only [Section 4.6].)

- However, the proof does not apply if, with every problem instance, the algorithm is given extra information *about the instance* that does not appear in the instance description in Figure 14-7—e.g., experimental data about the molecule whose structure is being predicted, such as interatomic distance constraints derived from nuclear Overhauser effect (NOE) data (Brünger et al., 1986).

The strategy used in the proof is very similar to what is employed in the sample reduction from the Hamiltonian Circuit Problem to the Traveling Salesman Problem (Section 2), and in all other proofs of NP-hardness. Suppose that there were to exist a correct, efficient algorithm for global minimization of potential-energy functions, as described in Figure 14-7. Ngo and Marks (1992) show how such an algorithm, if one were to exist, could be used to solve efficiently any instance of a known (Garey and Johnson, 1979) NP-complete problem called Partition (Figure 14-8),[15] which would imply that all NP-complete problems could be solved efficiently. They do this by demonstrating how a given instance of Partition can be transformed to an instance of PSP that is equivalent in the sense that answering the PSP instance is answering the Partition instance. The PSP instances thus generated may not be what the developers of the hypothetical PSP algorithm had in mind—for example, they are not constructed from amino acids. Nonetheless, they conform to the formal definition of PSP given in Figure 14-7. That is all that is required by the supposition, given earlier, that the hypothetical algorithm is correct and efficient for all possible instances of PSP.

The Partition instance consists of a bag of integers; and one would like to know whether the bag can be split into two parts of equal sum. To find out whether this can be done, one describes a molecule (admittedly a strange-looking one) that has these integers encoded into its chemical structure. One then asks the postulated correct, efficient PSP program to compute the molecular configuration of globally minimal energy. By inspecting

---

Problem instance: A bag of integers $A \equiv \{a_1, a_2, \ldots, a_{|A|}\}$.
Question: Can the bag of integers be split into two parts of equal sum?

Figure 14-8. The Partition Problem

this energy, one can tell whether the original instance of Partition could be answered in the affirmative.

Specifically, one designs the molecule and the potential-energy function so that every term in the function can assume its minimal value simultaneously if and only if the Partition instance can be answered in the affirmative. The steps involved in describing the molecule are described here informally; a formal presentation may be found in the original paper (Ngo and Marks, 1992):

1. Set up the molecule as an unbranched alkane, $CH_3(CH_2)_n CH_3$. Hydrogen atoms may be omitted. Use equilibrium bond lengths that are normal for alkanes, and equilibrium bond angles that are tetrahedral. Set all force constants to be any positive numbers, except where specified otherwise.

2. Set the force constants of all but 12 nonbonded terms to zero. The 12 nonzero terms and their equilibrium distances are chosen so that they cannot all be satisfied unless three atoms at one end (the variable end; see below) and four atoms at the other end (the scaffolding) adopt certain relative positions.

3. In one part (approximately half) of the chain, called the *variable* region, set each torsion to be either "one-fold" or "three-fold." A one-fold torsion prefers anti (180°) over any other value. A three-fold torsion has three equally preferred values, anti (180°) or ±gauche(±60°). *Here is where the integers a_i in the Partition instance enter into the specification of the molecule.* Along the torsions in the variable region, place one three-fold torsion, then $2a_1 - 1$ one-fold torsions, then one three-fold torsion, then $2a_2 - 1$ one-fold torsions, then one three-fold torsion, then $2a_3 - 1$ one-fold torsions, and so on (Figure 14-9).

   The three-fold torsions are the only terms in the potential function of this hypothetical molecule that can achieve their globally minimal values in more than one way.

4. In the other part of the chain, called the *scaffolding* (Figure 14-10), set torsional periodicities to 1, and choose the equilibrium torsional values so that when all potential-energy terms in the scaffolding are at

their globally minimal values, the ends of the scaffolding have relative position and orientation determined by the sum of the numbers in the Partition instance.

All of these parameters are chosen such that the molecule is frustrated (i.e., not all of the terms in its potential-energy function can adopt their globally minimal values) if and only if the original instance of Partition cannot be answered in the affirmative. From the existence of this construction, it follows that PSP is NP-hard.

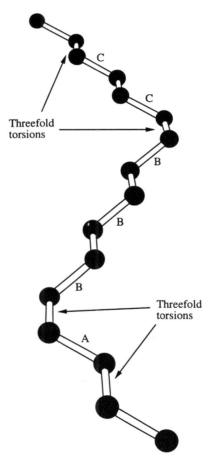

Figure 14-9. Sample variable region. The integers in the PARTITION instance are 1, 3, and 2, and they map onto the subsequences of bonds labeled A, BBB, and CC.

A few points about this transformation are worth noting. First, in contrast to Crippen's analysis, which applies only to any "weak" algorithm that

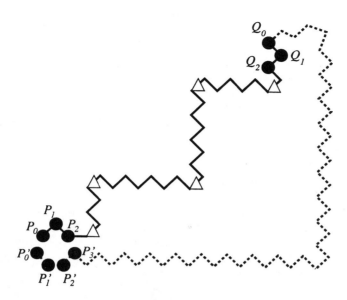

Figure 14-10. Schematic representation of the scaffolding (broken lines), which is flexible but has a unique zero-energy conformation. The variable region is drawn with solid lines; threefold torsions are marked with triangles. The twelve non-bonded interactions specified in the reduction are between three variable-region atoms $\{P_0, P_1, P_2\}$ and four scaffolding atoms $\{P'_0, P'_1, P'_2, P'_3\}$.

is permitted to assume nothing about the function being optimized other than a Lipschitz condition, the foregoing analysis of PSP applies to any algorithm that is "stronger," in the sense that it is permitted to assume that the function being optimized is of the form given in Figure 14-7. Second, nothing at all is assumed about *how* the postulated efficient PSP algorithm goes about finding the globally minimal energy of this molecule; it is assumed merely that the algorithm is capable of solving the problem efficiently and correctly. Third, nothing is required with respect to the stiffness of the molecule. All of its geometric parameters are free to vary continuously, and the force constants need not be set to large values—just any positive values. (A similar point about the need for endpoint constraints is made in Figure 14-11.) Fourth, the molecule used in the construction is quite different in some respects from a real protein. A key objective of Section 4 will be to discuss what limitations of the proof arise as a result of this fact.

3.4 Unger & Moult Reduction: LPE Is NP-hard

The NP-completeness proof of Unger and Moult (1993) is complementary to the one just outlined. The Ngo–Marks reduction relies on the difficulty of

---

PSP and endpoint constraints

To the reader who is accustomed to reasoning from physical grounds, it may appear as if the proof that PSP is NP-hard depends on the existence of hard geometric constraints on the endpoints of a segment of the chain, represented in this case by the variable region. Based on this reasoning, it would be essential to ask whether a segment of the polypeptide chain is ever subject to endpoint constraints that are as restrictive as those that arise in the PSP reduction. In fact, that question is not of concern, because (as the reduction shows), the form of the potential given in PSP is sufficient to create decisions for the prediction algorithm that are as difficult as those produced by a set of hard endpoint constraints.

In a proof based on physics, one usually begins with simplifying assumptions. The assumption that endpoints of a segment are perfectly constrained is one that could plausibly appear in a chain of physical reasoning. The conclusions of such a physical argument would inevitably depend on the validity of the assumptions on which they are based.

A proof based on computational complexity theory is different. No assumption is made about the physical system. Instead, the argument begins with a hypothesis about the existence of an algorithm; it is assumed that there exists an algorithm that solves all PSP instances efficiently and correctly. From this hypothesis it is inferred that all NP-complete problems can be solved efficiently and correctly, thus showing that the hypothesis is unlikely to hold. The reduction involves setting the energy coefficients so as to mimic endpoint constraints. By the hypothesis, this is entirely legal. The possibility of mimicking endpoint contraints arises automatically from the form of the energy function; that is what the reduction shows formally.

Figure 14-11. Computational and physical reasoning: A fundamental difference, explained in the context of endpoint constraints.

finding a geometric configuration that simultaneously satisfies local (bond, angle, and torsion) and nonlocal (nonbonded) potential-energy terms. The Unger–Moult reduction focuses on the difficulty of arranging objects with varying propensities to be adjacent to each other in a manner that satisfies those propensities optimally.

In the problem addressed by Unger and Moult, the protein is represented as a self-avoiding polymer embedded in a cubic lattice. The geometric portion of the model is as follows:

- The protein consists of a chain of beads.
- Space is a finite cubic lattice, i.e., a finite grid of points at integer

Cartesian coordinates. The lattice is defined to be just large enough to accommodate any possible configuration of the chain.

- Every bead must occupy a lattice site.
- The chain must be connected, i.e., any two beads that are neighbors on the chain must occupy adjacent lattice sites.
- The chain must be self-avoiding, i.e., no two beads can occupy the same lattice site.

Lattice-polymer models of this type are typically employed in Monte Carlo simulations designed to help elucidate the dynamics of protein folding (Chan and Dill, 1991). They are used in place of full protein models because they are much less expensive to simulate computationally—in many cases, the folding of a lattice polymer can be simulated from start to finish. Such models, it is argued, have some of the essential properties of proteins. Two important protein-like properties that arise from the features listed above, connectivity and self-avoidance, lead to phenomena involving restrictions on the motions of the polymer chain that may be important in the folding of real proteins.

Further protein-like properties are added by the introduction of an energetic component to the model. There are two broad approaches to designing the energy function for a lattice polymer that is to be used in a folding simulation. In one approach (Sikorski and Skolnick, 1990, for example), parts of the chain are designed carefully with particular protein secondary structures in mind. These secondary structural elements are typically lattice analogues of α helices and β sheets. In the other approach (Shakhnovich et al., 1991, for example), the beads are given "colors"[16] analogous to amino acid identities, and energy coefficients for nonbonded interactions for the possible pairs of colors are assigned either at random, or by rules that are not intrinsically predisposed to the formation of protein-like secondary structures. Protein-like folding of the chain is an emergent property of such a model.

Unger and Moult prove the NP-hardness of a problem whose underlying model would fall into the second category. The problem, which we call Lattice Polymer Embedding (LPE),[17] is described in Figure 14-12. The strategy used by Unger and Moult is to demonstrate the existence of a reduction to LPE from a known (Garey and Johnson, 1979) NP-complete problem called Optimal Linear Arrangement (OLA, Figure 14-13).

The OLA instance consists of N people, some pairs of whom are friends; and one would like to know whether they can be seated in a row of N chairs such that the sum of distances between pairs of friends does not exceed a certain value. To find out whether such a seating arrangement exists,

---

Problem instance: A chain of beads $S \equiv \{s_1, s_2, \ldots, s_{|S|}\}$ to be embedded in a three-dimensional cubic lattice $L \equiv \{1, 2, \ldots, |S|\}^3$. An affinity coefficient $c(s_i, s_j)$ for every pair of beads $s_i, s_j \in S$. A nonbonded-function value $f(|\Delta x|, |\Delta y|, |\Delta z|)$ for every $0 \leq |\Delta x| \leq |S|$, $0 \leq |\Delta y| \leq |S|$, and $0 \leq |\Delta z| \leq |S|$.

Task: Find the globally minimal energy $U^{\min}$ of the chain. The chain must be embedded in the lattice such that neighboring beads along the chain occupy neighboring lattice sites (diagonal chain links are permitted), and no two beads occupy the same lattice site. The expression for U is

$$U = \sum_{i<j} c(s_i, s_j) f(|x(s_i) - x(s_j)|, |y(s_i) - y(s_j)|, |z(s_i) - z(s_j)|),$$

where $(x(s_i), y(s_i), z(s_i)) \in L$ is the lattice point at which the bead s_i is positioned.

Figure 14-12. The Lattice Polymer Embedding Problem.

Problem instance: A list of N people, and a chart showing which pairs of people are friends. Also, a maximum distance $D_{\max}$.

Question: Can the people be seated in a row of N chairs in such a way that the sum D of distances between friends does not exceed $D_{\max}$? The distance between two friends is taken to be 1 if they are seated next to each other, 2 if they are separated by one chair, and so forth; and the sum is done over all pairs of friends. No two people can occupy the same chair.

Figure 14-13. The Optimal Linear Arrangement Problem.

one describes a bead sequence, interaction coefficients, and nonbonded-function values for an LPE polymer such that the globally optimal energy of the polymer is guaranteed to be equal to the globally optimal value of the distance function in the OLA instance. Then one asks the postulated correct, efficient LPE program to compute the energy of the optimal embedding of the polymer. By comparing the optimal energy with the distance bound $D_{\max}$ from the original instance of OLA, one can know whether the OLA instance could be answered in the affirmative. The steps involved in determining the bead sequence are described here informally:

1. Build the chain from two types of beads: beads labeled with the names of people from the OLA instance ("real beads"), and extra beads added to give the chain flexibility ("fake beads"). Real beads are placed in

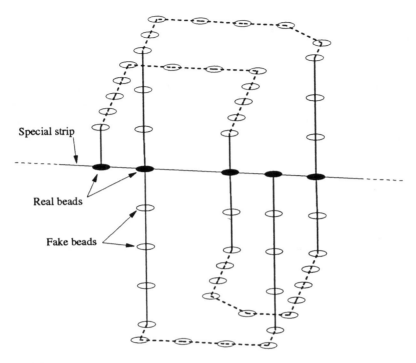

Figure 14-14. Schematic diagram of the Unger–Moult construction.

the chain in any order, and sequences of k fake beads are interposed between the real beads. The choice of the integer k is explained below.

2. *This is where the friendships from the original OLA instance enter into the specification of the chain.* Set the affinity coefficient $c(s_i, s_j)$ to one for every pair of real beads that are labeled with the names of friends from the OLA instance. Set all other affinity coefficients to zero. With this choice of affinity coefficients, the energy of the chain will be a function only of the locations of pairs of friends.

3. Set the nonbonded-function values f as follows:

$$f(|\Delta x|, |\Delta y|, |\Delta z|) = \begin{cases} |\Delta x| & \text{if } |\Delta y| = |\Delta z| = 0, \\ N^3/6 & \text{otherwise.} \end{cases}$$

The purpose of choosing nonbonded-function values in this way is to guarantee that in the configuration of globally optimal energy, all of the real beads involved in friendships lie in a row in the x direction.[18] We refer to this row as the "special strip."

4. Choose the integer k to be large enough so that the sequences of fake beads are flexible enough not to disallow any of the possible permutations of the real beads in the special strip. Unger and Moult demonstrated rigorously that this could be done. (See Figure 14-14.)

It is guaranteed that in the optimal embedding, all real beads involved in friendships lie in the special strip, and that the energy of this embedding is equal to the smallest distance sum that can be obtained in the original instance of OLA. From the existence of this construction, it follows that LPE is NP-hard.

Problem instance: A collection of beads S, to be embedded in a two-dimensional[a] square lattice L. An electrostatic charge $C(s_i) \in \{-1, 0, 1\}$ for each bead $s_i \in S$. A table listing the distance constraint $d^0(s_i, s_j)$ for every pair of beads that are connected by a bond. Finally, a distance cutoff $d_{\max}$, beyond which the electrostatic interaction between two nonbonded atoms is taken to be zero.

Task: Find the globally minimal energy $U^{\min}$ of the chain. The chain must be embedded in the lattice such that $d(s_i, s_j) = d^0(s_i, s_j)$ for every bonded atom pair (s_i, s_j). The expression for U is

$$U = \sum U(s_i, s_j),$$

summed over all nonbonded atom pairs, where

$$U(s_i, s_j) = \begin{cases} C(s_i)C(s_j)/d(s_i, s_j) & \text{if } d(s_i, s_j) \le d_{\max}, \\ 0 & \text{otherwise.} \end{cases}$$

The quantity $d(s_i, s_j)$ is the discretized Euclidean distance between s_i and s_j, i.e., $\left\lceil \sqrt{[x(s_i) - x(s_j)]^2 + [y(s_i) - y(s_j)]^2} \right\rceil$. Bonds may cross each other and may touch atoms, but at most one atom may occupy any given lattice site.

---

[a] Fraenkel also shows that a three-dimensional version of CGE is NP-hard.

Figure 14-15. The Charged Graph Embedding Problem

3.5 Fraenkel Reduction: CGE Is NP-Hard

Fraenkel (1993) has also addressed the computational complexity of protein-structure prediction. Like Unger and Moult (1993), Fraenkel analyzes a computational task that is based on a lattice model of folding. The problem, which we call Charged Graph Embedding (CGE),[19] is described in Figure 14-15. One potential source of concern over the model employed in CGE is the use of a distance cutoff for the electrostatic interaction; in fact, in the reduction to be described, the distance cutoff is set equal to the grid spacing so that only neighbor-neighbor contacts (in the four cardinal directions) can produce energy contributions. A distance cutoff for the Van der Waals interaction, which decays very rapidly (usually as r^{-6}) with distance, is commonly used in simulations of continuous protein models (Brooks et al., 1983). The use of such a cutoff for the electrostatic interaction is considered less acceptable because the r^{-1} potential remains significant over a range comparable to the diameter of a protein. (Instead, multipole expansions (Greengard, 1987, for example) are used to reduce computational requirements without sacrificing accuracy.) Nonetheless, the presence of an electrostatic distance cutoff in CGE should not be considered to be a weakness of the proof; in lattice simulations, interactions that depend on neighbor-neighbor contacts are the norm (Chan and Dill, 1991).

CGE is shown to be NP-hard by reduction from a known (Garey and Johnson, 1979) NP-complete problem called Three-Dimensional Matching (3DM, Figure 14-16). Because of the complicated nature of the reduction, what follows is not a self-contained explanation of the proof. Rather, we give a broad overview that may also serve as a guide for the reader who wishes to consult the original paper (Fraenkel, 1993). Taken out of context, many of the terms used in the original presentation (e.g., pillow, handle, bloc, cache) may seem unintuitive. We have chosen to use them here for the sake of consistency.

Problem instance: A set of N shirts, N trousers, and N hats. Also, a chart showing which combinations of shirts, trousers, and hats constitute matching outfits. The number of matching combinations is M, where $0 \leq M \leq N^3$.

Question: Can the clothes be arranged into N matching outfits? Every article of clothing must be used exactly once.

Figure 14-16. The Three-Dimensional Matching Problem. Two-Dimensional Matching, in which the instance description contains only shirts and trousers (no hats), is tractable.

A 3DM instance consists of N shirts, N trousers, and N hats, not all combinations of which match; and one would like to know whether these articles of clothing can be arranged into N matching outfits. To find out whether this can be done, one describes a collection of beads, their charges, and distance constraints such that the molecule consists of a collection of rigid domains. These domains are constructed such that the molecule as a whole cannot have energy lower than $U_{\text{match}} = -8(2M + 3N)$, and it can have energy equal to U_{match} only if a solution to the original 3DM instance exists. The construction is drawn schematically in Figure 14-17 and summarized here:

1. The basic building block of the molecule is a *ladder*, a collection of atoms with distance constraints that, when satisfied, force the atoms to lie in a pair of adjacent rows (or columns). The smallest possible ladder, called a *square,* consists of four atoms connected by four bonds of unit length. In general, a ladder may contain bonds with lengths greater than unity. Also, using additional bonds, ladders may be abutted at right angles to each other.

2. Set the distance cutoff d_{max} to unity so that charges interact only when they are at neighboring lattice sites.

3. Encode the constraints of the original 3DM instance into the domains by distributing charges as follows. It is instructive to think in terms of two fundamental types of charge distribution, a charged square and a charged hole. A hole occupies 32 lattice sites in a 6×6 region; four points in the middle of the region are left unoccupied. The unoccupied lattice sites in the middle of a hole are intended to accommodate a square with a charge distribution complementary to that of the hole. There are two kinds of holes and two kinds of squares. One type of hole, a *chest element*, is complementary to a (square) *handle element* and will be discussed below. The other type of hole, a *pillow*, is complementary to a (square) *bloc element*.

4. All pillows are identical to each other and all bloc elements are identical to each other. The construction centers around a horizontal row of $3N$ pillows, bound together into a rigid domain. Each pillow's position in the row represents an article of clothing. The energy bound U_{match} will be attained if and only if every pillow can be filled with exactly one bloc element, corresponding to the observation that in an instance of 3DM that can be answered in the affirmative, each article of clothing must be used exactly once.

5. Encode the information in the chart of M permissible matches into M rigid domains called *blocs*. Each bloc is a single ladder of length $3N$ that consists of three bloc elements. The locations of these bloc elements along the bloc represent particular articles of clothing; and the bloc represents a particular allowable match of these clothes. The

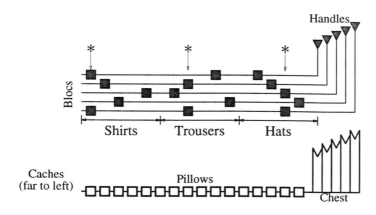

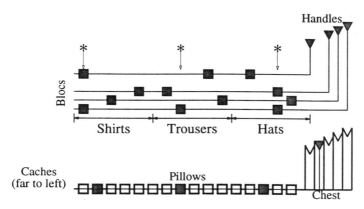

Figure 14-17. Schematic overview of the construction used in Fraenkel's reduction from 3DM to CGE, with no bloc embedded in the pillows (upper diagram), and with one bloc embedded (lower diagram). The matches permitted by the instance of 3DM are encoded into the blocs. The instance of 3DM is answerable in the affirmative if and only if N blocs can be embedded simultaneously in the pillows. The handles and chest ensure that if a bloc is embedded in the pillows, its horizontal alignment is correct. Each asterisk (*) indicates a pair of matches that are mutually exclusive because they require using the same article of clothing. Unused blocs dock with caches (to left of diagram). The cache, pillows, and chest form one rigid domain. Each bloc-and-handle assembly constitutes another rigid domain. The rigid domains are connected to each other by infinitely stretchable bonds.

energy bound U_{match} will be attained if and only if N of the M blocs can be embedded simultaneously in the pillows (see Figure 14-18).

6. The rest of the construction serves two purposes. First, the *cache* is

a rigid domain composed of holes that are designed to accommodate the $M - N$ unused blocs. The cache is attached to the pillows by long bonds. Second, the handle-and-chest assembly ensures proper horizontal alignment of the blocs. The horizontal alignment of a given bloc is enforced by a handle, to which it is bound at right angles. A handle is a ladder that contains a distribution of handle elements to ensure a unique position in the chest. The chest is a rigid domain that is composed of chest elements and is bound to the bed.

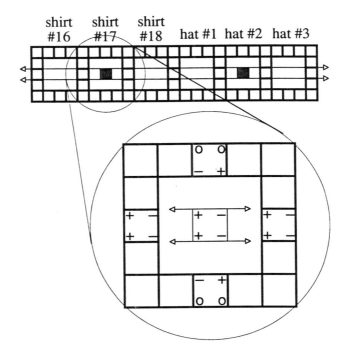

Figure 14-18. Schematic diagram of one bloc and the row of pillows in the Fraenkel construction. One bloc (fine lines) is shown embedded among the pillows (bold lines). The bloc encodes the information that shirt #17 is to be matched with hat #2; a distant bloc element (not shown) encodes a pair of trousers to complete the permissible combination. The magnified view shows the complementary charge distributions of one bloc element and one pillow.

This construction is submitted to the postulated correct, efficient CGE program, which computes its minimum-energy configuration. If the configuration has minimum energy equal to the energy bound U_{match}, then it is known that the original instance of 3DM could be answered in the affirmative. However, 3DM is known to be NP-complete (Garey and Johnson, 1979). It follows that CGE is NP-hard.

3.6 Comparison of Available NP-Hardness Results

There are now three proofs based on the theory of NP-completeness that are concerned with the difficulty of the protein-structure prediction problem. Throughout the remaining sections in this review, we discuss ramifications of these results for protein-structure prediction. We focus exclusively on the implications of the NP-hardness of PSP, for the following reasons:

1. While the proofs that LPE and CGE are NP-hard are rigorously correct, we believe that PSP is more relevant to protein-structure prediction, because PSP employs a continuous model and an energy function that has the form found in existing molecular-mechanics programs.[20] By contrast, LPE and CGE are based on lattice models, so the relationships between them and protein-structure prediction are ones of analogy (Section 4.2).

   It is common to suppose that continuous problems are always at least as difficult as their discrete counterparts (Unger and Moult, 1993). While this has been found to be generally true (from experience) in terms of the actual time required to run a "protein-folding" simulation, it is not necessarily true in terms of the possibility of developing efficient algorithms (i.e., algorithms with polynomial time bounds).[21] It is possible for a discrete problem to be intractable, but for its continuous counterpart to be efficiently solvable! The computational complexity of a continuous problem and a discrete version of it must be determined independently, unless a formal reduction between the two can be demonstrated. A classic example is discussed in Section 4.6.

2. LPE differs from existing lattice models in two fundamental ways. The first of these is that the nonbonded function f is present in the instance description. This implies that the developer of an algorithm for LPE is not permitted to know the functional form of the nonbonded potential in advance. Thus, an unreasonable level of difficulty may be built into LPE by virtue of the way it is defined. In existing lattice models, in PSP, and in protein-structure prediction algorithms based on potential-energy minimization, the form of the nonbonded potential is known in advance,[22] and might be exploited by the algorithm to simplify the problem. For example, one might imagine a hypothetical efficient algorithm for PSP that relies for its efficiency on the form of

the Lennard–Jones potential,[23] $f(x) \propto x^{-12} - 2x^{-6}$. The reduction from Partition to PSP rules out the existence of such an algorithm unless all NP-complete problems can be solved efficiently. A similar statement cannot be made for LPE.

In addition, the reduction from OLA to LPE requires that a particular direction in the lattice be made special—the same polymer configuration can have three different energies, depending on its orientation. This anisotropy could, in principle, be a source of complexity not present either with more conventional lattice models, or with real proteins.

3. CGE differs significantly from LPE and from conventional lattice models (Sikorski and Skolnick, 1990; Shakhnovich et al., 1991; Chan and Dill, 1991), in terms of the bonds permitted by the model. A conventional lattice polymer is a linear sequence of beads with adjacent beads constrained to be neighbors in space. By contrast, a lattice molecule of the variety employed in CGE can contain a bond between any pair of beads, and a bond can be arbitrarily long. Moreover, not all bonds have associated length constraints; in fact, such bonds are superfluous because they do not influence the set of configurations available to the molecule in any way. The lattice molecules generated in the Fraenkel reduction consist of rigid domains connected to each other by these infinitely flexible bonds.

4. Implications for Protein-Structure Prediction

This section addresses two key questions that pertain to the relationship between the NP-hardness of PSP and protein-structure prediction. First, what is really proved, given that the hypothetical molecules used in the proof constructions are more like alkanes than like proteins? Second, how might the result influence the development of algorithms in a positive manner?

The answers to these two questions are best discussed together. In answer to the first question, we explain the unambiguous consequences for protein-structure prediction. In answer to the second, we enumerate the "loopholes" that remain—each such loophole corresponds to a class of algorithms that might in some way avoid the limitations set by the existing results. A systematic analysis of these loopholes (either the development of algorithms that exploit them or the identification of further complexity results that eliminate them) may constitute a rigorous basis for guiding the otherwise intuitive process of developing useful prediction algorithms.

4.1 The Generality of PSP Is a Weakness of the Proof

PSP is more general than conventional statements of empirical potential-energy minimization (Momany et al., 1975; Brooks et al., 1983; Weiner et al., 1986); i.e., there exist computational tasks that qualify as instances of PSP but not as instances of protein-structure prediction. In terms of what can be concluded from the NP-hardness of PSP, this degree of generality is at once a strength and a weakness. It is a strength because it permits a number of rigorous negative statements that otherwise would be based on informal arguments by analogy. We will elaborate on this point in Section 4.2. However, in the sometimes counterintuitive world of negative proofs about worst-case behavior, it is a weakness.

The weakness of the existing results is fundamental to the present discussion. Algorithms for solving optimization problems lie on a spectrum from the general to the domain-specific. One end of the spectrum comprises algorithms that attempt to solve global optimization problems in a universal manner—they do not rely on specific information about the function being optimized, other than what can be determined by evaluating the function (and possibly its derivatives) at particular points in the search space. Of these, perhaps the best known are based on Monte Carlo simulation (Metropolis et al., 1953). In particular, simulated annealing (Kirkpatrick et al., 1983) has enjoyed significant attention for its practical value in solving a number of large-scale problems. Although approaches at this end of the spectrum have sometimes met with considerable success, in many cases algorithms that are blind to domain-specific information are too slow to be of any practical use. At the other end of the spectrum are fast solutions to certain restricted problems. For example, one would not seriously consider using simulated annealing to find the roots of a polynomial in one variable; problems of this type can be solved quickly and reliably with algorithms that are designed specifically for this purpose; they achieve their speed by exploiting mathematical properties of polynomials (Press et al., 1988).

In keeping with the relationship between the efficiency of an *algorithm* and the extent to which it utilizes information specific to the problem being solved, the intrinsic difficulty of a *problem* is closely tied to its degree of generality; special cases are often easier to solve than more general ones. Thus it remains possible, at least formally, that protein-structure prediction is some "easy special case" of PSP. What the details of the PSP NP-hardness proof help determine is just how restricted a special case of PSP must be before it is worth considering as a candidate for efficient solvability.

To clarify this point, let us recapitulate the central piece of reasoning in the interpretation of the NP-hardness result for PSP. What would happen if, in the future, one were to invent an efficient, correct algorithm for PSP— the algorithm described in the introduction as "too good to be true"? It would, of course, be used to predict protein structures. In addition, it could be used to solve efficiently any instance of any NP-complete problem from any one of a number of domains. For the sake of discussion, let us pick one particular domain, networking. Let us say that the telephone company would like to know whether switching sites can be placed so as to serve a given set of communities, but using at most a given length of wire. This is an instance of the Geometric Steiner-Tree Construction Problem, which is NP-complete (Garey and Johnson, 1979). It would be converted to a problem in Satisfiability, the first known NP-complete problem (Figure 14-4), by means of the transformation in Cook's Theorem (Cook, 1971). Then by a sequence of other well-known transformations it would be converted to an instance of Three-Satisfiability, then to Three-Dimensional Matching, then to Partition (Garey and Johnson, 1979). By the transformations described by Ngo and Marks (1992) the instance of Partition would be converted to an instance of PSP. *The efficient algorithm for PSP would then be used to generate the correct yes-or-no answer to the original instance of the Steiner-Tree problem.* All of this would take place in polynomial time, because each step requires only polynomial time. Because the possibility of this plan is so essential to the interpretation of the NP-hardness proof for PSP, we refer to it in this chapter as the *Central Scheme*, summarized here:

Geometric Steiner-Tree Construction $\longrightarrow$ Satisfiability $\longrightarrow$
$\longrightarrow$ 3-Satisfiability $\longrightarrow$ 3D Matching $\longrightarrow$ Partition $\longrightarrow$ PSP.

Thus, an algorithm for PSP that is powerful enough to make the Central Scheme work cannot exist unless P=NP. But a PSP algorithm that is not correct and efficient in all cases can exist without violating that pivotal assumption of the theory. Herein lies the weakness of the result that PSP is NP-hard: *an algorithm need not be powerful enough to make the Central Scheme work to be useful for predicting protein structures.* For example, an algorithm that relies upon information about proteins (such as secondary-structure propensities or homology with some protein whose structure is already known) may not work correctly and efficiently on the alkane-like PSP instances generated by transformation from Partition, but may still be useful for protein-structure prediction. The purpose of this section is to elaborate on the types of algorithms that have not been ruled out by the NP-hardness of PSP, because PSP is so general.

4.2 The Generality of PSP Is Also a Strength of the Proof

The fact that PSP is a formal generalization of the global-minimization problem that many existing structure-prediction techniques attempt to solve, cited in Section 4.1 as a weakness of the proof, is also a strength because it implies that the relationship between PSP and protein-structure prediction is not merely one of analogy.

The use of highly regular geometric constructions in inductive lines of reasoning—most notably, simulations (Sikorski and Skolnick, 1990) and thermodynamic analyses (Shakhnovich and Gutin, 1989) of lattice models—is necessitated by the difficulty of manipulating more realistic protein models computationally or analytically. Typically the use of such a simplified model is justified by analogy—it is asserted that the model in question captures enough of the essential properties of a protein to exhibit the behaviors being studied in a qualitatively correct manner. The proof of Unger and Moult (1993) appears to fall into this category because the problem that is finally shown to be NP-complete is a cubic lattice model.[24] In such work, the observations pertain to particular artificial systems, and any extrapolation to the nature of more realistic protein models is by inductive reasoning.

By contrast, the logic that relates the NP-completeness of Partition and the NP-hardness of PSP to the running time of protein-structure prediction algorithms is deductive because of the generality of PSP. In particular, the existence of the transformation from Partition to PSP implies that a global potential-energy minimization algorithm for PSP cannot be simultaneously efficient, correct, and general enough to accommodate the instances of Partition that arise in the reduction given in Section 3.3. A PSP algorithm that is practical for proteins may exist, but it *must lack either guaranteed efficiency, guaranteed correctness, or generality*—unless P=NP. The remaining subsections in Section 4 explore what types of algorithm might be expected to "sidestep" the NP-hardness of PSP by sacrificing one of these properties while remaining practical.

4.3 "Improved" Exponential-Time Systematic Searches

It is worth noting that exponential-time systematic algorithms are sometimes of significant practical value. Such techniques may be classified into two categories.

In the first category, efficiency is compromised but correctness and generality are not. Algorithms in this category are guaranteed to find globally optimal solutions (if permitted to run to completion) and lack polynomial time bounds, but are much faster than exhaustive searches. In one common technique, *branch-and-bound* (Papadimitriou and Steiglitz, 1982), the improvement is accomplished by pruning a search tree whose interior nodes represent partial solutions. A branch is eliminated when it can be determined with certainty that the global optimum is not lost—the pruning criterion must not deliver any "false positives." Caflisch et al. have employed such a pruning criterion in a discrete search problem that arises in the design of ligands for protein binding sites (Caflisch et al., 1993). Their pruning criterion is guaranteed not to eliminate the globally optimal solution because it is based on evaluation of a heuristic cost function that is always less than or equal to the true cost function—it consists of a subset of the terms in the true cost function, all of which are positive definite. (One branch-and-bound technique, A*, can be shown to be optimally fast when such a heuristic cost function is available [Nilsson, 1980].) Unfortunately, only a discrete combinatorial problem can be solved by a "true" branch-and-bound technique (one that is guaranteed to deliver the globally optimal solution); a continuous problem generally cannot be solved this way unless discretization can be shown to leave the global optimum unaffected.

In the second category, the guarantee of correctness is sacrificed in addition to efficiency. Numerous torsional tree-search techniques for molecular-structure prediction (Moult and James, 1986; Bruccoleri and Karplus, 1987; Lipton and Still, 1988; Chang et al., 1989) fall into this category. Unlike "true" branch-and-bound techniques, these techniques are not guaranteed to find the global optimum because continuous coordinates are discretized, and consequently the underlying search tree is not exhaustive. Moreover, in many cases the pruning criteria cannot be guaranteed not to eliminate the best solution in the discretized space. (For example, in torsional tree searches of cycloheptadecane [Saunders et al., 1990], every partial structure containing a +gauche/−gauche pentane fragment was pruned. The purported global optimum that was identified independently by other methods contains no such fragments, but this fortunate fact could not have been determined without resorting to the other methods.) These techniques can often, in practice, generate many near-optimal solutions; also, their utility can be greatly enhanced if the search is ordered so that better solutions tend to be found sooner rather than later, for example, by examining more promising branches of the search tree earlier than others.

It is rare that an exponential-time algorithm in either of the two categories runs in a practical amount of time when presented with large problem instances. A classic example of an algorithm that lacks provable polynomial time bounds but runs fast in practice is Dantzig's Simplex algorithm (Dantzig, 1963) for Linear Programming. While the Simplex method has exponential worst-case asymptotic time complexity, the technique can in practice be used to solve problem instances involving hundreds or thousands of variables. However, this example may not be representative of what can be expected for an NP-hard problem since after years as an open problem, Linear Programming was found to be efficiently solvable (Khachiyan, 1979). (The experience with Linear Programming is a specific example of how theoretical results can help guide algorithm development. Part of the impetus for finding this efficient algorithm came from a result [Garey and Johnson, 1979] that suggested that Linear Programming was unlikely to be NP-complete.[25] A discussion of Linear Programming aimed at a different issue, the effect of discretization on NP-hardness, appears in Section 4.6.)

4.4 Relaxed Guarantees of Efficiency, Correctness, and Generality

An NP-completeness proof is a worst-case analysis. It does not preclude the existence of an algorithm that works reasonably fast and with reasonable accuracy "on average" or "for typical problem instances." Such an algorithm would carry no guarantee of efficiency, correctness, or generality, but might do reasonably well by all three criteria. Many of the most promising techniques for protein-structure prediction are possible candidates for this category. Thorough reviews of the many techniques that have been proposed are available elsewhere (Fasman, 1988; Reeke, 1988). To place our comments about performance guarantees in the context of protein-structure prediction, we consider a few examples.

4.4.1 Build-up procedures.

The "build-up" procedure of Scheraga and co-workers (Simon et al., 1991) involves a very reasonable set of tradeoffs: it is specialized for proteins and is not *guaranteed* to be correct, but it generates reasonable structures and the time requirements of the algorithm are kept under control. The method begins with the construction of peptide fragments containing three amino acids in standard geometries, then combines these to form larger fragments.[26] At every step, high-energy structures are discarded. Because of its aggressive use of the hierarchical nature of protein structures and of existing libraries of standard amino acid structures, the algorithm does not attempt to achieve the level of generality and correctness that would make it a candidate for participation in the Central Scheme. The build-up procedure is one of many techniques (Fasman, 1988) that might fall into this category; we use it here only as an example.

The principle on which the build-up procedure is founded is similar in origin to an element of the diffusion-collision model (Karplus and Weaver, 1976), in which the protein molecule is considered to be divided into microdomains, "each short enough for all conformational alternatives to be searched through rapidly." Translating this principle into algorithmic terms, one might be tempted to ask whether an NP-hard problem such as PSP can be solved in less than exponential time by dividing it into small pieces, and solving each piece individually. Unfortunately, the presence of such "simplyfing structure" in a problem is precisely what the theory of NP-completeness tests. If PSP could be solved in divide-and-conquer fashion, it would not be NP-hard. However, it remains possible that a restricted form of PSP (Section 4.6) is tractable.

4.4.2 Pseudoatom representations. One approach to reducing the computational cost of structure prediction is to replace groups of atoms, or even entire residues, by "pseudoatoms" whose properties are intended to mimic the collective behavior of the constituent atoms. The pseudoatom model is not as computationally intensive to manipulate as the all-atom model because of the reduced number of variables. Doing this "sacrifices correctness," because the globally optimal state of the pseudoatom system does not necessarily match that of the full-atom system. Are algorithms that employ pseudoatoms subject to the known NP-hardness results? This question can be interpreted in two ways. First, would an efficient, correct algorithm for the problem employing pseudoatoms be an efficient, correct algorithm for the corresponding all-atom problem? That question pertains to the accuracy of the pseudoatom model, not its computational complexity. Second, would the existence of an efficient, correct algorithm for the pseudoatom problem imply that P=NP, given the result that PSP is NP-hard? The answer depends on the form of the potential function used with the pseudoatoms. If the function is analogous to the full-atom potential, i.e., it contains bond-length, bond-angle, torsional, and nonbonded terms, then it may be possible to use the reduction described in Section 3.3 to show that the pseudoatom problem is NP-hard. Without demonstrating such a reduction, however, it is incorrect to presume that the pseudoatom problem is automatically NP-hard.

4.4.3 Stochastic algorithms. When used as function optimizers, randomized methods such as Monte Carlo searches (Metropolis et al., 1953) and genetic algorithms (Dandekar and Argos, 1992; LeGrand and Merz, 1993; Sun, 1993) are very general techniques for trying to overcome local-minimum problems. Because of their probabilistic nature, they carry no guarantees of correctness or efficiency. Thus, a stochastic algorithm for PSP automatically satisfies the conclusions that follow from PSP's NP-hardness, regardless of its average-case behavior. It is not clear whether there is a correlation between the difficulty of a problem, as determined by proofs of NP-completeness, and the ability of stochastic algorithms to solve them. In other words, there is no formal reason to expect that NP-hard problems are more difficult to solve by stochastic means (either exactly or approximately) than those that are not NP-hard. One thing, however, is clear (from experience, not proof). In keeping with the inverse relationship between the generality of a problem and its potential for tractability (Section 4.6), stochastic algorithms that attempt to be universal by being "blind" (not exploiting domain-specific information) are successful only rarely, and usually only with smaller problem instances. For a stochastic algorithm to work well for a particular problem, it usually must make use of information about the problem's origin and structure (Davis, 1991).

Many other approaches to protein-structure prediction that, at first sight, may seem appropriate for this list are not because they are not attempts at solutions to PSP—they use different mathematical statements of the prediction problem. Secondary-structure prediction (when based on statistical rules derived from known structures), template matching, pattern recognition, distance geometry, and methods based on packing are all examples of such approaches (Fasman, 1988).

A word of warning about algorithms with relaxed guarantees of efficiency is in order. Quotes of average-case performance must be interpreted with care, as they depend strongly on the assumed distribution of inputs. Experience with Satisfiability, the earliest known NP-complete problem, illustrates this point (Mitchell et al., 1992). Because Satisfiability problems arise ubiquitously in artificial-intelligence applications, significant effort has been devoted to identifying algorithms that solve typical instances quickly. In one well-cited result (Goldberg, 1979; Goldberg et al., 1982), for example, it was claimed that Satisfiability is readily solved "on average" in $O(n^2)$ time. However, in subsequent analysis (Franco and Paull, 1983), it was shown that the distribution used in those tests consisted predominantly of easy instances.

Recently (Cheeseman et al., 1991; Mitchell et al., 1992), *a priori* criteria for distinguishing hard and easy distributions of instances of particular NP-complete problems have been proposed. These criteria are usually very specific to the NP-complete problem at hand, and therefore may not be applicable to protein-structure prediction. They are based on ideas that are analogous to thermodynamic concepts; an order parameter is defined, and a phase transition is identified. Far from the phase transition, the decision problem is either easily answered in the affirmative or easily answered in the negative. Near the phase boundary, extensive computation is required to get the answer right. It is not clear whether a similar analysis could be applied to the protein-structure prediction problem.

4.5 Efficient Heuristics

One particular approach to solving NP-complete problems, that of developing efficient heuristics, involves settling for suboptimal solutions without sacrificing either generality or efficiency. This approach has been applied successfully to certain optimization problems.

If the solutions produced by an efficient heuristic have provably bounded error, the heuristic is said to be an approximation algorithm. An *approximation algorithm*, therefore, is one that handles all instances of the given problem and is guaranteed to terminate in polynomial time, but instead of being guaranteed to locate the globally optimal solution, is only guaranteed to locate one with bounded error. (Note that an approximation algorithm is not just an "approximate" algorithm, i.e., one that is not very good at finding correct answers.) The Minimum Matching algorithm for solving the Traveling Salesman problem is an approximation algorithm; it runs in polynomial time and is guaranteed to return a tour that is at most 50% longer than the optimal tour (Christofides, 1976). An approximation algorithm for protein-structure prediction might be one that is guaranteed to produce a structure whose energy is within, say, 1 kcal/mol of the global optimum.[27]

It has been observed that some NP-complete problems have approximation algorithms that guarantee solutions close to optimal, whereas others have resisted similar solution despite continued efforts. Is there an approximation algorithm for global potential-energy minimization? For some problems it is possible to prove that an approximation version of an optimization problem (with a particular performance guarantee) is NP-hard, and therefore that the existence of an efficient approximation algorithm with that performance guarantee would imply that P=NP. For example, it is known that the error bound guaranteed by the Minimum Matching algorithm for Traveling Salesman cannot be improved upon unless P=NP (Garey and Johnson, 1979). Moreover, some problems can be shown by similar *ad hoc* methods not to have *any* approximation algorithm unless P=NP. Many of these relatively isolated results have been unified by the recent, much-publicized, characterization of a new complexity class called MAX SNP-hard (Papadimitriou and Yannakakis, 1991). If a problem is MAX SNP-hard, then no polynomial-time approximation scheme exists for it unless P=NP (Arora et al., 1992).[28] To our knowledge, the possible existence of an approximation algorithm for protein-structure prediction has not been addressed either on a specific, *ad hoc* basis, or using the more general techniques introduced by Papadimitriou and Yannakakis (1991), and by Arora et al. (1992).

An approximation algorithm might be of significant practical use in protein-structure prediction, because exactness is not an absolute requirement. If the guaranteed error bound were sufficiently small, an approximation algorithm might be useful for generating crude structures that, though not necessarily folded correctly, are close enough to the native structure to be useful. If not, merely knowing that the energy of the optimal structure is below a certain threshold (by running the approximation algorithm) could still be of use as part of a larger scheme. For example, a recently proposed stochastic algorithm based on the multicanonical ensemble (Berg and Neuhaus, 1992) (as opposed to the canonical ensemble employed in simulated annealing) appears to be most practical when the energy of the global optimum can be estimated accurately. Also, for certain problems, an approximation algorithm can be joined with an existing stochastic algorithm to form a hybrid that is better than either of its components alone (Davis, 1991). The design of such hybrid algorithms is presently more art than science; few general guidelines are known.

4.6 Algorithms for Restricted Forms of PSP

The only techniques for sidestepping NP-completeness that we have not yet discussed are those that lack generality but retain guaranteed efficiency and correctness. For the sake of this discussion, it is useful to recall some definitions in Table 14-1. A *problem* consists of a set of possible *instances*, and to be general for that problem, an algorithm must produce the correct outputs for all possible instances. If it solves only some well-defined *subset* of the possible problem instances correctly, it is said to handle a *restricted form* of the problem (Table 14-1). In this subsection we discuss the possible existence of efficient algorithms for restricted forms of PSP. The study of restricted forms of a problem is an essential part of characterizing the computational complexity of a problem. It may also be relevant for more physical aspects of the protein-folding problem because it is not yet clear what aspects of PSP, in its general form, are not appropriate for real proteins.

One must be careful about the use of the word "restricted." In particular, a common technique for speeding a search is to discretize coordinates. This "restricts" the search space enormously, but it will not necessarily make the problem easier according to the notion of intractability. In fact, Figure 14-19 contains an example of a tractable continuous problem that becomes NP-complete when it is discretized. Discretizing coordinates does restrict the set of possible inputs, but it also imposes a new stipulation on the form of an acceptable solution. A restriction can help if it is an additional condition on the "Problem instance," but not if it is an additional condition on the

---

Linear Programming and the effect of discretization

Linear-programming (LP) problems arise frequently in operations research. Every LP problem can be stated in standard form as follows: Given the rational-valued matrix A and vectors b and c, with dimensions $m \times n$, m, and n, respectively ($m < n$), find a rational-valued vector x that minimizes the cost function $c \cdot x$ and satisfies the constraints $Ax = b$ and $x \geq 0$.

Integer linear programming (ILP) is very closely related to linear programming; integer linear programs have the same standard form as linear programs, except that the elements of x are required to be integers. Zero-one linear programming (ZOLP) is even more strict: The elements of x must all be either 0 or 1.

LP, a continuous problem, can be solved exactly by an algorithm (Khachiyan, 1979) that is efficient and correct. ILP and ZOLP appear to be more restricted versions of the problem because their search spaces are discrete. It is therefore tempting to conclude that ILP and ZOLP must be easier than LP—but this conclusion is erroneous, both in theory and in practice. ILP and ZOLP are both NP-complete, and only very small ILP and ZOLP problems can be solved exactly using current techniques (Papadimitriou and Steiglitz, 1982).

Figure 14-19. An illustration that discretizing a problem does not necessarily make it more likely to be solvable in polynomial time.

"Question." (These are the headings used to label the two parts of a formal problem description, as in Figure 14-2.)

A trivial example of a restricted form of PSP that can be solved efficiently is the subset of instances in which the coefficients of all nonlocal interactions are zero. This restricted form of PSP can be solved in linear time, simply by setting each geometric parameter to its equilibrium value. Unfortunately, an algorithm for this restricted form of PSP would be of little practical value because it does not include any realistic instances of protein-structure prediction. Such restricted forms of PSP that are tractable but fail to accommodate proteins may be of little practical use, but they can be of theoretical interest because they help to identify the sources of a problem's complexity. For example, the tractability of the restricted form of PSP without nonlocal interactions turns out to be useful in understanding why a bead model proposed by Zwanzig et al. (1992) does not completely resolve the Levinthal paradox (end of Section 5.3).

What would a more useful algorithm for a restricted form of PSP look like? To be able to guarantee correctness, the algorithm would be unlikely to be stochastic. To guarantee efficiency, the algorithm would very likely exploit protein-specific information in some way that prohibits it from being a general solution to PSP. We do not know, at present, precisely what protein-specific information could lead to the development of such an algorithm. Some possible examples are given later in this subsection.

Barahona's complexity analysis (Barahona, 1982) of another class of physical models, that of Ising spin glasses, is an example of how restricted forms can be analyzed, via both NP-completeness proofs and the development of efficient algorithms. Barahona proves the following results:

- For any finite 3D spin-glass lattice in the presence of a magnetic field, both finding the ground states and computing the magnetic partition function are NP-hard.

- For the special case in which the magnetic field is zero, both tasks are still NP-hard.

- If the problem is 2D but a magnetic field is present, the two tasks are NP-hard.

- However, in the special case of the 2D problem in which the magnetic field is zero, both problems can be solved efficiently.

The general problem that Barahona considers includes all finite 3D spin-glass lattices in the presence of magnetic fields. He examines the complexity of two restricted forms: the subset of instances in which the magnetic field is zero, and the subset of instances in which the lattice is two-dimensional. Both are NP-hard, but a restricted form that comprises the intersection of the two subsets is tractable. It is therefore reasonable to say that the presence of a nonzero magnetic field and the presence of spins in a third dimension are both sources of complexity for finite Ising spin-glass models.

Knowledge of the details of an NP-completeness proof can help in this line of reasoning. It is known that any restricted form of PSP remains NP-hard if it includes all problem instances that can be generated in the reduction from Partition. For example, since the total absence of nonlocal interactions[29] makes PSP tractable, one might wonder whether there is an efficient algorithm that handles all instances of PSP in which there are very few nonlocal interactions—say, fewer than 20. From the form of the proof one can know in advance that such an algorithm does not exist unless P=NP, because the instances of PSP generated in the reduction from Partition contain fewer than 20 nonlocal interactions. (In fact, they contain 12 nonlocal interactions. However, this does not mean that a form of PSP in which the number of nonlocal interactions is restricted to be smaller than 12 is tractable.) Similarly, an earlier proof (Ngo and Marks,

1991) employed a reduction from a cubic lattice model to the continuous case, and it was hypothesized that molecules with 90° bond angles might represent pathological cases that could be avoided by a conformational search algorithm in which all bond angles are required to be obtuse. This possibility, too, is eliminated by the reduction described in Section 3.3, in which the bond angles (and all other equilibrium geometric parameters) fall well within the naturally observed ranges of values.

As we have emphasized, the generality of PSP makes the result that it is NP-hard a relatively weak statement. The reason is that proofs about the NP-hardness of restricted problems are always more powerful than proofs about the NP-hardness of more general problems. It is not difficult to specify a restricted form of PSP that admits proteins, but is not known to be NP-hard because the reduction from Partition given in Section 3.3 fails. Consider, for example, a restricted form of PSP in which it is stipulated that the backbone is a polypeptide. This restricted form (i.e., this subset of PSP instances) clearly excludes the PSP instances generated in the reduction from Partition, in which the backbone contains a sequence of one-fold and three-fold dihedrals determined by the integers in the instance of Partition. A future result showing the NP-hardness of this restricted form of PSP would have more sweeping consequences than the currently available result. The technique of examining the complexity of restricted forms of an NP-hard problem is of great potential value, but it may be difficult to use in the case of protein-structure prediction. In Section 6, we discuss this issue at greater length.

A restricted form of PSP whose complexity is of interest is one in which the coefficients of all nonlocal interactions are *non*zero—recall that the reduction from Partition to PSP called for a PSP algorithm that is able to handle a situation in which all but 12 of the possible pairwise nonlocal terms are set to zero, so that the globally optimal solution contains very few nonbonded contacts. Could there exist an efficient and correct algorithm that relies for its efficiency on having nonzero terms for all possible pairwise nonbonded interactions? This proposal is the *opposite* of eliminating the nonbonded interactions, which (as already stated) does permit the design of an efficient algorithm. (The dimensionality-reduction algorithm of Purisima and Scheraga [1986, 1987], in which pairwise distances for *all* atom pairs are initially satisfied in a space of high dimensionality, might have been a candidate had it carried a guarantee of correctness.) The answer to this question is not known because the reduction from Partition to PSP relies on the ability of the algorithm to handle problem instances in which a subset of the nonlocal interactions are set to zero.

In a related observation, real globular proteins are compact because of van der Waals and other attractive nonbonded interactions, and because of the hydrophobic effect (Creighton, 1992). The requirement of compactness drastically reduces the size of the search space because it is known that the number of compact, self-avoiding configurations available to a long chain is much, much smaller than the total number of configurations (Chan and Dill, 1991). Moreover, in lattice-polymer simulations (Šali et al., 1994), folding was observed to proceed via a molten-globule intermediate (Ptitsyn, 1987); after initial collapse to a semi-compact state and a rate-limiting transition to a native-like compact state, subsequent adoption of the native structure occurred rapidly. In light of these physical observations, it is useful to consider how the computational complexity of structure-prediction algorithms might be influenced by incorporating the knowledge that the native state is compact.

A compactness condition might appear in one of two places in a modified statement of PSP. First, it might appear as an additional stipulation on the "Task" (see the definition of PSP in Figure 14-7). That is, the "Task" might be to "Find the configuration of globally minimal energy . . . whose volume is no greater than vN, where N is the number of atoms in the molecule, and v is the highest permissible volume per atom." This modified problem would not be a restricted form of PSP; rather, it would be a different problem without any formal relationship to PSP in terms of reducibility. Nevertheless, because of its possible relation to protein folding, as described above, its computational complexity would be of interest. Although the search space generated by an instance of this problem would be much smaller than that generated by the corresponding instance of PSP, there is no reason to expect that the modified problem should have reduced computational complexity. In fact, adding stipulations to a tractable problem can sometimes render it NP-hard. (Figure 14-19 describes an example of this phenomenon, as observed with linear programming.)

Alternatively, the requirement of compactness might appear in the instance description (i.e., under the "Problem instance" heading in the formal definition of PSP). The instance description would contain a stipulation to the effect that only instances whose global solutions are expected to be compact are to be submitted to the algorithm. If such a stipulation could be specified in terms of the variables that appear in the instance description, the modified problem would be a restricted form of PSP, and a suitable candidate for continued complexity analysis.

4.7 Implications for "Smoothed-Potential" Methods

The NP-hardness of PSP may have a direct influence on the future development of algorithms that involve deforming the potential surface to make it smoother (Piela et al., 1989; Gordon and Somorjai, 1992; Shalloway, 1992; Head-Gordon and Stillinger, 1993; Amara et al., 1993). We consider one of these techniques, the diffusion-equation method (DEM), which has been described briefly thus:

> Another method to surmount the multiple-minima problem is based on the possibility of deforming the complex hypersurface in successive stages so that higher-energy minima disappear and only a descendant of the global minimum remains. A reversal of the deformation procedure then recovers the global minimum of the original potential function. (Scheraga, 1992)

The problem for which this technique is designed is even more general than PSP, and therefore NP-hard; the technique can be applied to any function $f(\mathbf{x})$ whose Laplacian $\nabla^2 f$ exists. It is efficient; its running time has been quoted as being polynomial— $O(N^3)$, or at most $O(N^4)$ (Kostrowicki and Scheraga, 1992). Most promisingly, it has turned out to be correct for a number of small but nontrivial[30] test cases (Piela et al., 1989; Kostrowicki and Piela, 1991; Kostrowicki et al., 1991), including a terminally-blocked alanine (Kostrowicki and Scheraga, 1992). Thus, it would appear that the DEM has the potential to become an algorithm for global minimization that is "too good to be true"—efficient, correct, and general.

From the NP-hardness of PSP we may predict that the DEM is very likely to require significant modification before it can be used as a practical method for protein-structure prediction. Here are some of the possibilities:

- The most optimistic possibility (aside from P=NP) is that, after some minor modification (e.g., the use of an exponential transformation [Kostrowicki and Scheraga, 1992]), the method will be efficient and correct for some subset of instances of PSP, and that this subset will include some or all naturally occurring proteins. This possibility would raise very interesting questions as to why naturally occurring proteins should correspond to easier instances of the global minimization problem, especially given the lack of *a priori* mathematical reasoning as to why this should be so.

- Another, perhaps more likely, possibility is that the algorithm will have to be modified to pursue multiple decision paths.[31] As the deformation of the surface is reversed, the single potential well splits into many sub-wells—an exponential number of sub-wells (Section 3.1) by the time the potential surface returns to its original form. The success of the DEM method, as currently formulated (Piela et al., 1989), is contingent on the ability of the algorithm to decide correctly which of the sub-wells will lead to the global optimum, each time the potential well under consideration bifurcates. The detection and handling of these bifurcations has been cited as a significant obstacle that needs to be overcome if the DEM technique is be a reliable method for finding the global optimal structure of a polypeptide (Wawak et al., 1992).

- Even if no method is developed for guaranteeing that the global minimum of the function is found, the DEM could form the basis for an approximation algorithm (Section 4.5); in other words, it may be possible to guarantee that the local minimum found is worse than the global minimum by no more than some predictable energy difference. The DEM has the ingredients of a typical approximation algorithm— it is a polynomial-time procedure in which decisions are made by a deterministic, greedy[32] heuristic.

Comments of this type are less applicable to the so-called *antlion* method (Head-Gordon et al., 1991), in which full generality is not sought. In particular, the antlion method is intended for use as the second stage in a two-stage process. In the first stage, anticipated propensities for particular dihedral angles and nonbonded contacts, predicted by some other method (e.g., a neural network [Head-Gordon and Stillinger, 1993]) are incorporated into the potential function by the addition of penalty functions and the modification of existing terms. This deformation of the potential surface is designed to leave a single basin of attraction. In the second stage, the surface is deformed gradually to its original shape, just as with the diffusion-equation method.

The important difference between this approach and the DEM is that it is intrinsically specialized for proteins; this provides an *a priori* reason to believe that it will perform efficiently for proteins but not for the hypothetical molecules produced in the Central Scheme. More formally, the two-stage algorithm (including the use of the neural network) is designed for a restricted form of PSP. Taken alone, the latter half of the algorithm is not subject to the NP-hardness of PSP because it solves a different problem, one whose instance description includes anticipated dihedral angles and nonbonded contacts.

4.8 Summary

The result that PSP is NP-hard goes a step further than previous arguments as to the intractability of protein-structure prediction. Traditional counting arguments (Section 3.1) show that an exhaustive search of the conformational space of a protein is out of the question. Crippen's analysis (Section 3.2) demonstrates that an algorithm must be expected to exhibit exponential worst-case time complexity unless it utilizes information about the potential-energy function in addition to (or other than) a Lipschitz constant. The proof that PSP is NP-hard suggests that exploiting the functional form of the potential energy is not enough to make the problem efficiently solvable.

Useful algorithms for NP-hard problems do exist. These include exponential-time searches (Section 4.3) and other heuristic algorithms (Section 4.4) that are "fast enough" and "correct enough" to be practical; approximation algorithms (Section 4.5); and algorithms for restricted forms of the problem (Section 4.6). The latter two types of algorithm might be eliminated from consideration by future theoretical results (Section 6).

The result that PSP is NP-hard has little bearing on the computational complexity of a problem that is neither a restricted form nor a generalization of PSP, i.e., a problem whose formal description contains a "Task" that is different from the one found in PSP (cf. Figure 14-7).

The task of determining the minimum-energy conformation of a protein is a special case, or restricted form, of PSP. Is it an "easy special case" for which an efficient algorithm exists? The answer to this question is not yet known.

5. Relevance to the Rate of Protein Folding

We now turn our attention to the implications of this line of inquiry for the physical behavior of proteins. Does the proof that PSP is NP-hard (Section 3.3) in some way contradict the observation that some proteins fold in seconds? It is tempting to surmise that there is a universal relationship between the physical behavior of a system and the computational tractability or intractability of a model of that system—provided, of course, that the model is an accurate representation of the physical system. No universal relationship is known. In the first part of this section, we discuss why this is so. However, subject to certain stipulations, the NP-hardness of PSP can be used in a rigorous reformulation of the Levinthal paradox. The second part of this section is devoted to that issue. Other rigorous reformulations of the Levinthal paradox that are not based on computational complexity theory may be possible; the last part of this section touches on one of these.

5.1 Computational Complexity and Physical Systems in General

The behavior of a physical system is governed by its potential-energy surface. When a system is in thermal equilibrium, it is expected to adopt the state of minimum free energy. The relative free energies of its states are determined, in turn, by the form of the potential surface. In fact, it is usually possible to assume that when a system is at its global free-energy minimum, its global potential-energy minimum is significantly populated.[33] Thus, the physical process of reaching thermal equilibrium and the computational task of global potential-energy minimization have similar elements, and it is reasonable to consider whether proofs about the computational task are relevant to the corresponding physical phenomenon.

If computationally identifying the globally optimal state of a given physical system is intractable, will the physical system itself spend an astronomical amount of time in suboptimal states before achieving its global optimum given an arbitrary starting configuration? Is the reverse true? Let us examine three intuitively attractive propositions along these lines.

Proposition A ("Computational tractability implies rapid attainment of thermal equilibrium"): *If global potential-energy minimization for a particular physical system can be performed efficiently, then that physical system should not have long-lived metastable states.*

This proposition is easy to refute by counterexample. Consider a cubic box at room temperature, filled with a stack of N square slabs of varying density. In the presence of gravity, the potential energy of this system is minimized when the slabs are arranged in order of their density, with the heaviest on the bottom. The optimal configuration for this system can be identified *computationally*, simply by sorting a list of the slabs' densities. This requires only $O(N \log N)$ time (as we have pointed out in Section 2.1); yet the slabs in the *physical* system cannot be expected to sort themselves in order of their density in any reasonable amount of time, because they cannot pass through each other. The trouble with Proposition A is that thermal accessibility is an issue for the physical process but not for the computational solution.

Note that Proposition A also "compares apples and oranges": the statement about long-lived metastable states pertains to absolute times, whereas computational complexity theory addresses only the functional form of algorithmic running times, ignoring the coefficients.[34]

Proposition B ("Computational intractability prohibits guaranteed polynomial-time attainment of thermal equilibrium"): *If global potential-energy minimization for a particular physical system is intractable, then that physical system cannot be guaranteed to reach its global potential-energy minimum from an arbitrary starting configuration before spending exponential time in other states.*

With certain stipulations regarding the computational complexity of dynamical simulation (see below), Proposition B is rigorously correct. Consider its contrapositive, which is logically equivalent to it. If a physical system is guaranteed to reach its globally stable state in polynomial time from an arbitrary starting configuration, then the corresponding computational task of global potential-energy minimization can be guaranteed to be solved in polynomial time; dynamical simulation of the system is that efficient algorithm for solving the problem, if two provisos are met. (The importance of these issues has been noted by Unger and Moult [1993].) The provisos are as follows:

First, simulating a system's dynamics on a digital computer requires discretizing time into small intervals. If these time intervals are too large, i.e., the frequency F at which velocities and accelerations in the system are recalculated is too low, the simulation will become physically incorrect. For a dynamical simulation to behave in a physically reasonable manner, F must exceed a certain minimum value. This value may depend on the force-integration algorithm used in the simulation, and it may depend on the size of the system, N; i.e., $F = F(N)$. But for Proposition B to be valid, this minimum value of $F(N)$ must not depend exponentially on the system size. (This appears to be true for proteins; the time resolution required for accurate simulation of an all-atom model is generally on the femtosecond timescale for a protein of any size.)

Second, the CPU time $S(N)$ required to simulate a single time step must not be exponential in the system size. (For some physical systems, this condition is not presently satisfied; for example, because of singular cases, correctly simulating one time step of a system of nonpenetrating rigid bodies by the most physically accurate known algorithm [Baraff, 1991] requires solving a problem in Quadratic Assignment, which is known to be NP-hard. However, the proviso does appear to hold for protein dynamics, in which a single time step may be executed in $O(N^2)$ time or better.[35])

Given the original premise that the physical system is guaranteed to reach its globally optimal state in polynomial real time $T(N)$, it follows that the dynamical simulation must do the same in CPU time $S(N) F(N) T(N)$, which is polynomial in N because a product of polynomials is itself a polynomial.[36] Thus, Proposition B is formally true for proteins.

However, there is a danger in taking Proposition B at face value because it is prone to misinterpretation. In particular, the following apparent corollary, which is interesting because it seems to preclude equating the native state of a protein with its global potential-energy minimum, is fallacious.

Proposition C *Because it has been shown that PSP is NP-hard, a real protein cannot reach its global potential-energy minimum in less than exponential time.*

Proposition C is based on fallacious reasoning. Nevertheless, a form of this proposition has led Unger and Moult (1993) to the inference that "as there is no general efficient feasible way to find the lowest free energy conformation of a protein we cannot assume that the native state is at the free energy global minimum." In addition to the relatively insignificant possibility that P=NP (in which case, NP-hardness would not imply intractability) and the possibility that the "correct" potential function of a protein is not of the form given in PSP, some serious objections arise because intractability is a worst-case measure. In the previous section we discussed this issue in the context of predictive algorithms; here we focus on the physical process of protein folding.

First, the premise of Proposition C is not just that of Proposition B, reworded for the context of protein-structure prediction. The NP-hardness of PSP does not imply the intractability of protein-structure prediction; even if P$\neq$NP, protein-structure prediction may be a tractable restricted form of PSP (Section 4.6).

To understand better what this means physically, let us retrace the chain of logical steps that lead from Proposition C to the inviolate assumption, P$\neq$NP. A given instance of an NP-complete problem (say, the telephone-company project described earlier) would be fed through the Central Scheme, producing some instance of PSP whose yes-or-no answer is correct for the original question about situating telephone switchboards. Proposition C states, in effect, that the PSP instance can be answered efficiently if a real protein can reach its global potential-energy minimum in polynomially bounded time; dynamical simulation is the algorithm for PSP. The argument breaks down because the dynamical simulation might achieve the global minimum in polynomially bounded time for naturally occurring proteins, but not for the instances of PSP that can be generated by reduction from Partition.

Second, Propositions C and B do not match in terms of what they state about the guarantee involved. Proposition B states that the physical system "cannot be *guaranteed* to reach" the globally optimal state in polynomially bounded time; Proposition C, on the other hand, contains the much stronger statement that the protein "*cannot* reach" it in polynomial time.

Current models of how a protein folds (Karplus and Shakhnovich, 1992) do not focus directly on the issue of polynomial time bounds; instead, they address the ability of a protein to fold in a "reasonable" amount of time (milliseconds to tens of seconds), which for practical purposes can be considered equivalent to polynomial time. The results reviewed here do not help distinguish among these various models of folding. Because they are based on random processes, these models would predict that a protein is neither guaranteed, nor (at the other extreme) unable, to fold in a reasonable amount of time. Random fluctuations are essential to the diffusion-collision model (Karplus and Weaver, 1976, 1979), which calls for portions of the protein chain (microdomains) to "flicker" in and out of their native conformation, becoming stable only when they diffuse, collide, and coalesce into larger structures. In the similar framework model (Baldwin, 1989; Kim and Baldwin, 1990) the microdomains are explicitly identified as secondary structural elements, which form only transiently until stabilized by packing interactions. Some nucleation models (Levinthal, 1966; Tsong et al., 1972; Wetlaufer, 1973) differ from diffusion-collision in that a protein is expected to contain just one special site, possibly equivalent to a microdomain, whose chance adoption of the native structure triggers a "domino effect" in which the remainder of the protein folds; others (Moult and Unger, 1991) assume multiple nucleation sites and are essentially identical to the diffusion-collision model. In the jigsaw-puzzle model (Harrison and Durbin, 1985), the single, well-defined, sequence of events is replaced by a number of different folding pathways, any of which might be followed. In any of these cases, the stochastic nature of the folding models would prevent either a guarantee of polynomially bounded folding time, or a guarantee to the contrary.

What if, however, the guarantees are replaced by probabilities very close to unity? This and other questions are addressed in the ensuing discussion of the Levinthal paradox.

5.2 Computational Complexity and the Levinthal Paradox

The Levinthal paradox (Levinthal, 1968, 1969) raises a question about the rate at which a protein can be expected to fold. The essence of the paradox is that in theory a protein is expected to require exponential time to fold, given an arbitrary starting configuration, whereas in practice proteins are observed to fold within seconds to minutes, independent of size (Jaenicke, 1987). In the usual form of the paradox, exponential-time folding is expected because of the exponential size of the protein's conformational space:

> In a standard illustration of the Levinthal paradox, each bond connecting amino acids can have several (e.g., three) possible states, so that a protein of, say, 101 amino acids could exist in $3^{100} = 5 \times 10^{47}$ configurations. Even if the protein is able to sample new configurations at the rate of 10^{13} per second, or 3×10^{20} per year, it will take 10^{27} years to try them all. Levinthal concluded that random searches are not an effective way of finding the correct state of a folded protein. Nevertheless, proteins do fold, and in a time scale of seconds or less. This is the paradox. (Zwanzig et al., 1992)

We argue that this basis for exponential-time folding is unjustified both physically and computationally, but that other, more rigorous, statements can serve as the basic premise. One such restatement will be based on the result (Section 3.3) that PSP is NP-hard.

The assumption underlying the usual form of the paradox is that a protein samples configurations in a *completely random* fashion until it encounters its native state. If this random-search assumption is to be quantitatively correct, the protein must not get any "clues" as to the location of its native state in conformational space until it chances upon the right configuration. This extreme picture of the folding process, sometimes referred to as a "golf-course" model (Bryngelson and Wolynes, 1989) because a golf ball cannot know where the hole is until it is actually there, has incorrect physical consequences.

One incorrect physical consequence pertains to the apparent existence of hierarchy in observed protein structures. Brown and Klee (1971), for example, found that a 13-residue helical fragment of a protein had significant helical character even when excised from the rest of the protein. Baldwin and co-workers (Bierzynski et al., 1982; Shoemaker et al., 1985, 1987; Marqusee and Baldwin, 1987) conducted extensive experiments that indicate that such behavior is not unusual. NMR experiments by Dyson et al. (1988a, 1988b) also provided evidence for residual structure in very short protein fragments. These and other experiments suggest that a protein *does* have clues as to the nature of its native state even when it is quite far away from being folded. These clues come from the propensities of parts of the chain for native-like structure.

Recent developments in experimental technique have permitted the observation of full proteins (not fragments) during the folding process. These experiments, first conducted with two proteins quite different in structure (one predominantly α-helical [Roder et al., 1988] and the other predominantly β-sheet [Udgaonkar and Baldwin, 1988]) detected significant formation of native-like secondary structure in specific parts of the protein before full folding had taken place. Thus, it is clear that the protein's pathway(s) to the native state is (are) not completely random.

The computational version of the random-search argument fails as well because the required reasoning is tautological—random search is, by nature, an exponential-time algorithm. Any problem in combinatorial optimization with a unique solution is expected to take exponential time to solve by random search, regardless of the problem's computational complexity.

While these observations (in particular, those regarding the hierarchical nature of protein structures) do not automatically resolve the Levinthal paradox, our opinion is that they make its conventional statement, as given above, less compelling than a true paradox—which should be, by one definition, "an argument that apparently derives self-contradictory conclusions by valid deduction from acceptable premises" (Webster, 1991). Given the foundation provided by Proposition B above and the currently available NP-hardness results, we are able to propose one possible rigorous reformulation of the Levinthal paradox, and enumerate its possible resolutions. The restatement of the paradox is:

> Proteins do not require exponential time to fold even though PSP has been shown to be NP-hard. (Ngo and Marks, 1992)

For this paradox to be resolved, *at least one of the following statements must be true*:

1. **P=NP.**

   This possibility is considered highly unlikely because its truth would imply that all NP-complete problems can be solved correctly and efficiently. It is included here for the sake of completeness.

2. **PSP is intractable because P≠NP, but every protein that can spontaneously recover its native state *in vitro* corresponds to an instance of PSP that can be solved exactly and efficiently. The set of such instances would constitute a restricted form of PSP** (Section 4.6) **that is tractable.**

   This statement includes the possibility that the potential function in Figure 14-7 lacks energy terms that may be essential for proper folding—for example, a hydrophobic penalty function. (Merely adding energy terms to the potential function leaves the problem NP-hard; the only necessary modification to the reduction from Partition is that the coefficients of the new energy terms be set to zero. To generate a variant of PSP that is not known to be NP-hard, one must restrict the problem by imposing stipulations on the ranges of values permissible for the newly introduced energy terms.)

   If spontaneously folding proteins do constitute a tractable restricted form of PSP, then some property of naturally occuring proteins that is not common to all instances of PSP (for example, the presence of a hydrophobic effect of a certain magnitude, as discussed above) may be responsible. However, if it is not true, i.e., not every spontaneously refolding protein corresponds to an instance of PSP that can be solved exactly and efficiently, then by Proposition B we are forced to conclude one of the remaining possibilities.

3. **A protein is not guaranteed to encounter its native state in polynomial time.**

The mathematical world of exact computation and the world of physics may be somewhat at odds in the interpretation of this statement. Is a physical system ever *guaranteed* to do anything, given that quantum-mechanical first principles are probabilistic? Consideration of the classical limit, where essentially deterministic results are possible, is adequate for the present discussion.

In that sense, it is worth pointing out that *near guarantees* may be of considerable value, even in the context of the theory of computation, in which exact guarantees are the norm. Let us consider two examples of near guarantees, one physical and one computational:

- When probabilistic events occur involving large numbers (such as Avogadro's number, 6×10^{23}), their collective effect can have a probability distribution that is so sharply peaked that the variance of the distribution is zero for all practical purposes. For instance, even though the particles in an ideal gas have a distribution of velocities, the pressure exerted by the gas on the boundaries of its container is essentially constant. Perhaps more to the point, a digital computer is a physical system; yet we feel confident in making exact predictions about its behavior.

- There exist randomized algorithms that come very close to guaranteeing correctness. An oft-cited example is that of testing whether a given number K is prime. This problem (which has not been shown either to be NP-complete or to be efficiently solvable [Garey and Johnson, 1979]) can be solved to any desired degree of certainty ϵ using $\lceil - \log_4 \epsilon \rceil$ iterations of an algorithm due to Rabin (1976, 1980).[37]

If a particular protein is guaranteed to encounter its native state in polynomial time (or "nearly guaranteed" to do so, in the sense just described), but a similar guarantee cannot be made regarding global minimization of the potential energy of that same protein, then (by Proposition B) the remaining possibility is implied:

4. **A protein's native fold can be predicted from its potential-energy function by some computational procedure other than global minimization, albeit one whose computational complexity is not yet known.**

This possibility has been raised by Levinthal (1968), who pointed out that the folding process may lead to a "uniquely selected metastable state, in which the configurational energy is at a local minimum but not necessarily at an absolute minimum." (The hypothesis that this "uniquely selected metastable state" is not the global minimum was found not to hold for a lattice-polymer model; in Monte Carlo folding simulations, it was observed that lattice polymers either folded consistently to the global minimum, or did not fold consistently to any unique structure [Šali et al., 1994].) Unger and Moult (1993) suggested the related possibility that many polypeptide chains fold to a uniquely (perhaps kinetically) determined metastable state, but that only those proteins for which this state coincides with the global minimum have survived natural selection.

Traditional counting arguments (Section 3.1) suggest that *any* uniquely selected state should be as hard for the protein to find as the global optimum of its potential energy. However, as we have pointed out, such counting arguments are fallacious. (For amplification, recall the reasoning in Figure 14-6.) If native protein structures are uniquely defined by some process other than global minimization, then it may be possible to formulate a variant of PSP based on this process. The computational complexity of that problem, not that of PSP, would then be of concern.

5.3 Other Possible Reformulations of the Levinthal Paradox

Analysis of computational complexity may not be the only way to make the Levinthal paradox more precise. If it is one day found that the structure of every naturally occuring protein can be predicted exactly and efficiently by a computer, will the Levinthal paradox have been resolved? In other words, would the computational tractability of global potential-energy minimization for a particular sytem imply that the system should be able to find its globally optimal state quickly? From our refutation of Proposition A (p. 483), this is not so.

It is therefore worth asking what specific postulates about the *physical behavior* of proteins would lead to expected exponential-time folding. For reference, let us reconsider the conventional form of the paradox. If each amino acid residue in a protein could find its native conformation independently of the others, there would be no search problem. At the opposite extreme, if folding were fully cooperative, so that each residue were to have native-like bias only in the context of the otherwise fully folded protein (the "golf-course" model), the process of folding would be similar to random search, and the conventional form of the Levinthal paradox would be justified.

The "golf-course" argument is fundamentally about the *lack* of information available to the protein about its native state, when it is not in its native state. As we have pointed out, existing experimental evidence more or less rules out arguments of this form; a protein does have clues as to the location of its native state. The arguments that we have proposed, those based on the NP-hardness of PSP, are about the *computational complexity* of exploiting that information to find the configuration of globally minimal energy—in other words, the difficulty of deciding which nativelike propensities to satisfy, given that they cannot all be satisfied at once.

A third avenue for justifying the premise of the Levinthal paradox might be the *physical difficulty* of getting from the unfolded state to the folded state. If, to fold from an arbitrary starting configuration, a protein must surmount an energy barrier whose height $\Delta G^{\dagger}$ is proportional to the number of atoms N in the protein, then the rate at which folding occurs is expected to be proportional to $\exp(-CN/RT)$ for some energy constant C, giving an expected folding time that is exponential in N. It seems unlikely that such a claim about energy-barrier heights is true, though we cannot prove the contrary. However, it is known experimentally that for naturally occurring single-domain proteins, folding times do not increase with size.

A thorough exploration of how the Levinthal paradox might be restated is beyond the scope of this review. Nonetheless, we wish to emphasize the importance of distinguishing the various possible formulations of the Levinthal paradox from each other. To illustrate the point, examine one model from which Zwanzig et al. concluded that "Levinthal's paradox becomes irrelevant to protein folding when some of the interactions between amino acids are taken into account" (Zwanzig et al., 1992). The model was described as follows:

- The protein contains N [rotatable] bonds.
- Each bond has two states: "correct" (c) and "incorrect" (i).
- The probability that a given bond will change from i to c is k_1; the probability that it will change from c to i is k_0.
- The c state is energetically favorable, so that $k_0 = k_1 n \exp(-U/kT)$, where U is the energy gap for a given bond and n is the degeneracy of the i state.
- Changes in different bonds occur independently.

The mean first-passage time from a typical state to the all-c state was found to be, asymptotically,

$$t = \frac{1}{Nk_0} \left[1 + n \exp(-U/kT)\right]^N ,$$

which depends steeply on the energy gap U. Given the assumptions that $N = 100$ and $n = 2$, it was found that in the limit $U \to 0$, the first-passage time is nearly 10^{30} years. However, a modest change to the value of U, say $U = 2kT$, lowers the first-passage time to under one second. (The base of the exponential, $1 + n \exp(-U/kT)$, is equal to 3 when $U = 0$, but 1.27 when $U = 2kT$.)

The analysis of Zwanzig et al. resolves a form of the Levinthal paradox in which the absence of clues about the form of the native state is the sole basis for expecting exponential-time folding. However, it does not resolve the form of the paradox based on computational complexity, since the optimization problem implied by the underlying model can be solved trivially in linear time. The reason for the tractability of the underlying model is the lack of long-range interactions, which are critical to rendering PSP NP-hard (Ngo and Marks, 1992), and essential for cooperativity (Karplus and Shakhnovich, 1992).

6. Future Work

It is not known whether there exists an efficient algorithm for predicting the structure of a given protein from its amino acid sequence alone. Decades of research have failed to produce such an algorithm, yet Nature seems to solve the problem. Proteins do fold! The "direct" approach to structure prediction, that of directly simulating the folding process, is not yet possible because contemporary hardware falls eight to nine orders of magnitude short of the task. However, while this difference is large, it is not astronomical. Would this "direct" approach constitute an efficient and correct algorithm for protein-structure prediction? Too little is known about protein folding, and about the future of computing technology, to be able to answer this question at this time.

The results reviewed here (Section 3) do not completely rule out the existence of a protein-structure prediction algorithm that is both efficient and correct, in the precise senses of those words used throughout this chapter. In particular, it remains formally possible that there is a restricted form of PSP that is efficiently solvable, but subsumes protein-structure prediction. How can this possibility be investigated?

A standard strategy in the analysis of any NP-hard problem is to examine restricted forms of the problem systematically, classifying each as tractable or NP-hard, and thereby exposing the sources of the complexity. Barahona's results with Ising spin-glass models, which were described briefly in Section 4, are exemplary of this approach. While the particular restrictions chosen by Barahona for spin glasses (reduction of dimensionality and removal of the magnetic field) are not suitable for protein-structure prediction, the overall strategy of examining restricted forms is appropriate. Some restricted form of PSP in which compactness plays a critical role is a candidate for this type of analysis (Section 4.6).

The approach of considering restricted forms has worked well for dozens of important problems that are relatively "clean" and abstract (Garey and Johnson, 1979), but it may be difficult to pursue in the case of protein-structure prediction. In the former case, the problem shown to be NP-hard is usually as general as would actually be required in practice. In the latter case, what is desired is not an algorithm that can handle all possible instances of PSP (Section 3), but merely one that works for proteins. Thus, the fact that PSP is a generalization of protein-structure prediction makes the result that PSP is NP-hard less limiting than it could be.

Ideally, one would like to demonstrate the NP-hardness of a problem that is more *specific*, not more general, than protein-structure prediction, because that would automatically prove the NP-hardness of protein-structure prediction itself. This would entail finding an efficient transformation from some existing NP-complete problem that generates instances of PSP that are proteins by every conceivable criterion.[38] It is difficult to see how such a transformation might proceed.[39]

An alternative approach that may be nearly as instructive is to use the currently available result regarding PSP as a baseline in a continuing comparative analysis—to find restricted forms of PSP that are NP-hard but as specialized as possible, and to find others that are tractable but as general as possible. The motivations for pursuing this methodology are both practical and theoretical:

- Every NP-hardness result permits us to know in advance that a certain group of algorithms is likely to fail, and is therefore not worth pursuing (Section 4).

- Conversely, every NP-hardness result helps identify a source of complexity in protein-structure prediction, and therefore what must be stripped away from the problem before it is reasonable to attempt efficient solution.

The work of Finkelstein and Reva (1992) is a good example; an approach to structure prediction with a guaranteed polynomial time bound was developed. The critical assumption behind the algorithm is that only nonbonded interactions between nearest neighbors along the chain are significant. Because of this assumption, the algorithm cannot solve all instances of PSP, but instead is restricted to instances in which only nonbonded interactions between nearest neighbors along the chain are nonzero.[40] This violates the requirements of the reduction from Partition to PSP, in which nonbonded interactions between sites distant from each other along the chain are essential. Thus, the problem is similar in character to that examined by Zwanzig et al. (Section 5.3). While the Finkelstein–Reva algorithm was not inspired by an NP-hardness result, the underlying strategy is similar to how NP-hardness results might be used; they removed from the problem what they observed to be a source of complexity. However, in this case, removing the source of complexity led to a problem different from that posed by protein folding, in which long-range interactions play an essential role.

- The NP-hardness of PSP serves as the premise for a reformulation of the Levinthal paradox (Section 5), whose conventional form is based on a model of folding that is in conflict with known experimental results. A motivation for pursuing an analysis of the computational complexity of protein-structure prediction is to assist in the constructive role of the Levinthal paradox—to help focus attention on the key questions in protein folding.

A small number of reasonably well-defined potential resolutions to the computational-complexity form of the Levinthal paradox were listed in Section 5. One of the possible resolutions is that protein-structure prediction is tractable. NP-hardness results with restricted forms of PSP would make that possible resolution less likely, thus lending credence to the alternatives.

Attempts to resolve the Levinthal paradox, which play a valid and useful role in helping to understand how proteins fold, can lead to confusion because the premises of the original form of the paradox are not well formulated. In particular, one such proposed resolution (Zwanzig et al., 1992) can be shown unequivocally not to resolve the computational complexity form of the paradox, and in related arguments (Karplus and Shakhnovich, 1992) has been shown to lead to physically incorrect consequences (Section 5.3). For the paradox to be meaningful, it must be "falsifiable"—it must be possible to know when the paradox has been resolved.

In addition to restricted forms of PSP, it would be useful to know the computational complexity of other tasks in structure prediction that appear easier than the general problem, but whose complexities are none the less uncertain.

The task of computing side-chain conformations given full knowledge of a protein's backbone conformation is one such problem. Case studies using simulated annealing (Lee and Subbiah, 1991) have suggested that packing effects may suffice to determine, in part, the side-chain conformations in a protein's core. The computational complexity of this packing problem is unknown. Because only short-range effects are present, the graph of possible side-chain–side-chain interactions can be known in advance, is sparse, and consists of vertices of low degree. Previous experience—for instance, with Ising spin-glass models (Barahona, 1982), graph colorability (Garey and Johnson, 1979, p. 191) and cartographic labeling (Formann and Wagner, 1991; Marks and Shieber, 1991)—illustrates that such neighborhood interactions can, on their own, give rise to NP-hardness. On the other hand, many problems that contain such neighborhood interactions are tractable if restrictions can be placed on the nature of the graph (Garey and Johnson, 1979), suggesting that the problem of finding a mutually acceptable set of side-chain conformations for a protein could be tractable. (One currently known algorithm for predicting side-chain conformations based on backbone positions achieves 70% to 80% accuracy for χ_1 and χ_2 angles [Dunbrack and Karplus, 1993].) Not knowing the computational complexity of side-chain structure prediction leaves the algorithm developer in the quandary of not knowing whether inexact methods are truly necessary, given the possible existence of a superior exact algorithm.

Acknowledgments. We thank Ron Unger for answering detailed questions and providing a preprint (Unger and Moult, 1993). Aviezri Fraenkel also kindly provided a preprint (Fraenkel, 1993). We thank Harry Lewis, Eugene Shakhnovich, and Jim Clark for reading and commenting on the manuscript. JTN is grateful for a Graduate Fellowship from the Fannie and John Hertz Foundation. This research was supported in part by grants from the National Science Foundation and the National Institutes of Health.

NOTES

[1] The Thermodynamic Hypothesis states that a protein's native fold is the configuration of globally minimal free energy. However, it is generally assumed that a protein's states of lowest free energy are similar enough in entropy to justify the use of potential energies instead of free energies as a computational convenience; potential energies are much faster and more straightforward to compute.

[2] For example, if only nonbonded interactions between nearest neighbors along the chain are significant, the global minimum structure can be predicted efficiently (Finkelstein and Reva, 1992).

[3] The term *combinatorial optimization* is normally reserved for problems in which the solution space is discrete. Throughout this chapter we use the term to refer

to continuous problems as well, on the assumption that such problems can be discretized (perhaps very finely, if necessary). However, this should not be taken to indicate that the NP-completeness result reviewed in Section 3.3 applies only to a discretized form of the problem; the problem whose complexity is addressed, and to which the result applies, is a continuous one.

[4] More precisely, what is meant in this sentence by "self-contained" is that all of the data available to the algorithm about the problem instance being solved must appear in the instance description. This point is amplified in Section 3.3.

[5] The $O(n \log n)$ result applies only to sorting algorithms that are permitted only to do pairwise comparisons and swaps. Several such algorithms that achieve this $O(n \log n)$ asymptotic worst-case time complexity are known.

[6] This distance can also be any number greater than 2 and the proof remains unaffected.

[7] The term "NP-complete" can be used both as a noun (referring to the class of problems) or as an adjective (describing a member of the class).

[8] The first known NP-complete problem, a problem in Boolean logic called Satisfiability (Figure 14-4), obviously could not have been proved NP-complete by means of mutual reducibility. It was shown to be NP-complete by Cook's Theorem (Cook, 1971), which states that any problem in NP (see Figure 14-1) is polynomial-time reducible to Satisfiability. Thus, Satisfiability is the "hardest" problem in NP. From the property of mutual reducibility, all other NP-complete problems are "equally hard."

[9] All NP-complete problems are NP-hard, but not all NP-hard problems are NP-complete. The only practical reason to make the additional effort to show that a problem is NP-complete, and not merely NP-hard, is to help build up the arsenal of currently known NP-completeness results for the sake of future reductions.

[10] A list of length n can be arranged in $n!$ different ways, and $n! \approx \sqrt{2\pi n}\left(\frac{n}{e}\right)^n$ by Stirling's formula.

[11] The distance between two points, $\|\mathbf{a} - \mathbf{b}\|$, is here defined as $\sum_{i=1}^{n}|a_i - b_i|$, where n is the dimensionality of the vectors, and subscripts denote the taking of components.

[12] The choice of Δf is arbitrary. A smaller value of Δf produces a finer search grid; a larger value causes the set D to include a larger fraction of the search space.

[13] The locus of solutions eliminated is not a sphere, because the metric employed in the Lipschitz condition is not Euclidean. The arguments can be modified to work with a Euclidean metric, but the analysis is more complicated.

[14] How one might record and retrieve efficiently the eliminated candidate solutions is not considered in Crippen's analysis, but this obviously would be a serious concern for any practical algorithm that seeks to exploit this strategy.

[15] Ngo and Marks (1992) present the reduction in two stages. The reduction described here, which is logically equivalent, is done in one stage, directly from Partition.

[16] In some models, the beads are not grouped into chemical types; i.e., every bead has a different color.

[17] Unger and Moult use the name "Protein Folding" (PF). As with PSP, we use a different name to distinguish this problem from the "true" protein-folding problem. For the sake of clarity and accuracy, our presentation differs from that of Unger and Moult in several details. (For example, we use a finite lattice in place of an infinite lattice; otherwise, the instance description is infinitely large.) However, the overall structure of the proof is unchanged.

[18] Consider a configuration in which some real beads s_i and s_j involved in a friendship are not aligned in the x direction, i.e., $y(s_i) \neq y(s_j)$ or $z(s_i) \neq z(s_j)$. From step (2.), $c(s_i, s_j) = 1$; and from step (3.), $f(|x(s_i) - x(s_j)|, |y(s_i) - y(s_j)|, |z(s_i) - z(s_j)|) = N^3/6$. Since all of the energy terms that are generated by the construction are nonnegative, it follows that the total energy of this configuration is at least $N^3/6$. This cannot be the configuration of globally minimal energy; a configuration with lower energy can be constructed trivially by placing all of the real beads in a row of contiguous lattice sites aligned in the x direction. The energy of such a configuration is at most $(N^3 - N)/6$.

[19] Fraenkel uses the name "Minimum Free Energy Conformation of Protein" (MEP). As with PSP and LPE, we use a different name to distinguish this problem from the "true" protein-folding problem.

[20] While it is true that the details of the reduction from Partition to PSP involve considering the properties of a diamond lattice (Ngo and Marks, 1992), PSP itself is a continuous problem, and the reduction to it has been demonstrated formally (Ngo and Marks, 1992). It is not clear whether LPE could be recast as a continuous problem without requiring a new, different proof.

[21] It is true that lattice folding simulations generally require less computer time than simulations based on continuous models. Reasons for the reduced computational requirements include the use of integer coordinates, and a reduction in the size of the search space. While these factors affect the magnitude of the running time, they do not necessarily change its asymptotic functional form in a favorable manner.

[22] In PSP, the form of the nonbonded potential is known in advance—only the effective pairwise interaction coefficients K_{ij}^{nonlocal}, which are analogous to the affinity coefficients $c(s_i, s_j)$ in LPE, are not known in advance. By contrast, in LPE, the nonbonded function f can also differ from one problem instance to another, so an algorithm for LPE cannot exploit any assumed properties of f (such as short-range nature or the existence of a unique minimum).

[23] Many existing algorithms do exploit the short-range nature of the repulsive portion of the Lennard–Jones potential. For example, some conformational search programs (for example, Bruccoleri and Karplus, 1987) test for a high nonbonded energy between two atoms only after an initial distance check is performed. Of course, in no case has this property of the Lennard–Jones potential been the key to achieving efficient (polynomial-time) global potential-energy minimization, or the protein-structure prediction problem would not be considered intractable.

[24] The problem shown by Unger and Moult to be NP-hard, which is referred to in this review as LPE, is neither a special case nor a generalization of PSP. Therefore, there is no formal complexity-theoretic relationship between LPE and PSP.

[25] Linear Programming was shown to lie both in the computational class NP (see Figure 14-1) and in another called co-NP. Membership in both of these classes simultaneously is considered strong evidence that a problem is *not* NP-complete. This situation is very rare.

[26] An earlier version of the algorithm (Vásquez and Scheraga, 1985; Gibson and Scheraga, 1986) employed additional information about interatomic distances in the correct structure, which was known in advance. The earlier version was therefore not subject to the proof that PSP is NP-hard. However, the current version of the algorithm (Simon et al., 1991) is subject to the proof.

[27] Most known approximation algorithms guarantee solutions with error bounds multiplicatively related to the optimal cost, e.g., "at most 50% worse than optimal." Because absolute energies have little meaning, a useful guarantee for molecular-structure prediction would be additive, e.g., "at most 1 kcal/mol worse than optimal." There is no obvious reason that additive guarantees should be more difficult to obtain than multiplicative ones, since an optimization problem does not change if its cost function f is replaced by $\log f$.

[28] The standard technique for proving that a problem is MAX SNP-hard is to demonstrate the existence of an L-reduction from an existing MAX SNP-complete problem. An L-reduction (linear reduction) from one optimization problem X to another optimization problem Y specifies a constant λ such that the existence of an approximation algorithm for Y with worst-case error ϵ would imply the existence of one for X with worst-case error $\lambda\epsilon$.

[29] In the context of this chapter, a "nonlocal" interaction is defined as one that can exist between atoms separated by *arbitrarily* many covalent bonds. For example, "1,4" interactions are local by this definition.

[30] One of the test cases involved structure prediction for a cluster of argon atoms (Kostrowicki et al., 1991). This problem instance falls within a restricted form of PSP in which all local coefficients are required to be zero. The reduction from Partition to PSP fails for this restricted form of the problem. Thus, the restricted form is not known to be NP-hard; on the other hand, it is not known to be tractable.

[31] The number of decision paths to be followed will rise exponentially if all possible paths are considered. One way to help control this combinatorial explosion might be to employ a branch-and-bound technique, described in Section 4.3. This would involve being able to rule out, with certainty, some fraction of the sub-wells at every stage during the reversal of deformation.

[32] The term *greedy* is used by computer scientists to describe any algorithm in which decisions are made on the assumption that always taking the best short-term choice will lead to satisfactory results in the long run.

[33] This assumption is supported in the case of a protein by the fact that the native structure of a protein appears to correspond to a unique state of relatively low entropy.

[34] Nevertheless, exponential-time *algorithms* are usually assumed to be slow in absolute terms for problems of practical size (this is usually a reasonable assumption), while polynomial-time algorithms are assumed to be fast (this is not always a reasonable assumption).

[35] Straightforward simulation requires computing $O(N^2)$ pairwise interactions. Grid-based neighbor lists (van Gunsteren et al., 1984) and multipole expansions (Greengard, 1987; Greengard and Rokhlin, 1989; Schmidt and Lee, 1991) may be employed to achieve near linear-time performance, which of course would also satisfy the proviso.

[36] If the guaranteed physical folding time $T(N)$ is known, and the computational speed of the simulation $S(N)F(N)$ can be determined, then the method for finding the globally optimal state by dynamical simulation is to record the state of lowest energy attained by the simulation within time $S(N)F(N)T(N)$. The dynamical simulation must, of course, be conducted under the same conditions that make the protein fold.

[37] The algorithm is based on a theorem that states the following. For a given large integer K that is composite (not prime), at least $\frac{3}{4}(K-1)$ of the numbers in the interval $1 < i < K$ are "witnesses to the compositeness" of K, i.e., either they are factors of K (modulo K), or they violate Fermat's relation, $i^{K-1} \neq 1 \bmod K$. To test for primality of K, one simply generates random integers smaller than K. If a witness to the compositeness of K is found, then K is not prime. Otherwise, with every test, the probability that one is correct in stating "K is not prime" decreases by 75%. After $\lceil -\log_4 \epsilon \rceil$ iterations, this probability is ϵ. For example, if a number survives 170 tests without being found to be composite, the certainty that it is prime is unity, to 99 digits of precision.

[38] One problem is that for a reduction from some existing NP-complete problem to be valid, it must work for all possible instances of that problem, including arbitrarily large ones. This would mean constructing arbitrarily large proteins, which beyond a certain size are unlikely to look like existing proteins.

[39] The inability to prove that a problem is NP-hard does not constitute evidence that the problem is tractable. In fact, there exists a class of problems that are neither tractable nor NP-hard. This class, called NPI, has been proved to be non-empty if P$\neq$NP (Ladner, 1975).

[40] The problem solved by the Finkelstein–Reva algorithm is not formally a restricted form of PSP. One reason is the way space is discretized. More importantly, their problem description provides for the presence of a "molecular field" produced by a pre-existing molecule or fragment. Because of the presence of this molecular field, the computational complexity of this problem might be analyzed by means of the proof technique employed by Unger and Moult (Section 3.4).

REFERENCES

Abola EE, Bernstein FC, Bryant SH, Koetzle TF, and Weng J (1987): Protein data bank. In Allen FH, Bergerhoff G, Sievers R, eds. *Crystallographic Databases— Information Content, Software Systems, Scientific Applications*, pp. 107–132. Data Commission of the International Union of Crystallography, Bonn/Cambridge/Chester

Aho AV, Hopcroft JE, Ullman JD (1974): *The Design and Analysis of Computer Algorithms*. Reading, MA: Addison-Wesley

Aho AV, Hopcroft JE, Ullman JD (1982): *Data Structures and Algorithms*. Reading, MA: Addison-Wesley

Amara P, Hsu D, Straub JE (1993): Global energy minimum searches using an approximate solution of the imaginary time Schrödinger equation. *Journal of Physical Chemistry* 97:6715–6721

Anfinsen CB (1973): Principles that govern the folding of protein chains. *Science* 181(4096):223–230

Anfinsen CB, Haber E, Sela M, White FH (1961): The kinetics of formation of native ribonuclease during oxidation of the reduced polypeptide chain. *Proceedings of the National Academy of Sciences, USA* 47:1309–1314

Arora S, Lund C, Motwani R, Sudan M, Szegedy M (1992): Proof verification and hardness of approximation problems. In *Thirty-Third Annual Symposium on Foundations of Computer Science (FOCS)*

Baldwin RL (1989): How does protein folding get started? *Trends Biochem Sci* 14:291–294

Baraff D (1991): Coping with friction for non-penetrating rigid body simulation. *Computer Graphics* 25(4):31–40

Barahona F (1982): On the computational complexity of Ising spin glass models. *Journal of Physics A: Mathematics and General* 15:3241–3253

Berg BA, Neuhaus T (1992): Multicanonical ensemble: A new approach to simulate first-order phase transitions. *Physical Review Letters* 68(1):9–12

Bernstein FC, Koetzle TF, Williams GJB, Meyer EF Jr., Brice MD, Rodgers JR, Kennard O, Shimanouchi T, Tasumi M (1977): The protein data bank: A computer-based archival file for macromolecular structures. *Journal of Molecular Biology* 112:535–542

Bierzynski A, Kim PS, Baldwin RL (1982): A salt bridge stabilizes the helix formed by isolated C-peptide of RNAse A. *Proceedings of the National Academy of Sciences, USA* 79:2470–2474

Brooks BR, Bruccoleri RE, Olafson BD, States DJ, Swaminathan S, Karplus M (1983): CHARMM: A program for macromolecular energy, minimization, and dynamics calculations. *Journal of Computational Chemistry* 4(2):187–217

Brown JE, Klee WA (1971): Helix-coil transition of the isolated amino terminus. *Biochemistry* 10(3):470–476

Bruccoleri RE, Karplus M (1987): Prediction of the folding of short polypeptide segments by uniform conformational sampling. *Biopolymers* 26:137–168

Brünger AT, Clore GM, Gronenborn AM, Karplus M (1986): Three-dimensional structure of proteins determined by molecular dynamics with interproton distance restraints: Application to crambin. *Proceedings of the National Academy of Sciences, USA* 83:3801–3805

Bryngelson JD, Wolynes PG (1989): Intermediates and barrier crossing in a random energy model (with applications to protein folding). *Journal of Physical Chemistry* 93:6902–6915

Caflisch A, Miranker A, Karplus M (1993): Multiple copy simultaneous search and construction of ligands in binding sites: Application to inhibitors of HIV-1 aspartic proteinase. *Journal of Medicinal Chemistry* 36:2142–2167

Chan HS, Dill KA (1991): Polymer principles in protein structure and stability. *Annual Reviews of Biophysics and Biophysical Chemistry* 20:447–490

Chang G, Guida WC, Still WC (1989): An internal coordinate Monte Carlo method for searching conformational space. *Journal of the American Chemical Society* 11:4379

Cheeseman P, Kanefsky B, Taylor WM (1991): Where the really hard problems are. In *Proceedings of IJCAI '91*, pp. 163–169

Christofides N (1976): Worst-case analysis of a new heuristic for the travelling salesman problem. Technical report, Graduate School of Industrial Administration, Carnegie-Mellon University, Pittsburgh, PA

Cook SA (1971): The complexity of theorem-proving procedures. In *Proceedings of the Third Annual ACM Symposium on the Theory of Computing*, pp. 151–158

Creighton TE, ed. (1992): *Protein Folding*. New York: WH Freeman

Crippen GM (1975): Global optimization and polypeptide conformation. *Journal of Computational Physics* 18:224–231

Crippen GM, Scheraga HA (1969): Minimization of polypeptide energy. VIII. Application of the deflation technique to a dipeptide. *Proceedings of the National Academy of Sciences, USA* 64:42–49

Crippen GM, Scheraga HA (1971): Minimization of polypeptide energy. XI. Method of gentlest ascent. *Archives of Biochemistry and Biophysics* 144:462–466

Dandekar T, Argos P (1992): Potential of genetic algorithms in protein folding and protein engineering simulations. *Protein Engineering* 5(7):637–645

Dantzig GB (1963): *Linear Programming and Extensions*. Princeton, NJ: Princeton University Press

Davis L (1991): *Handbook of Genetic Algorithms*. New York: Van Nostrand Reinhold

Dunbrack RL Jr., Karplus M (1993): Backbone-dependent rotamer library for proteins: Application to side-chain prediction. *Journal of Molecular Biology* 230:543–574

Dyson HJ, Rance M, Houghten RA, Lerner RA, Wright PE (1988a): Folding of immunogenic peptide fragments of proteins in water solution. I. Sequence requirements for the formation of a reverse turn. *Journal of Molecular Biology* 201(1):161–200

Dyson HJ, Rance M, Houghten RA, Wright PE, Lerner RA (1988b): Folding of immunogenic peptide fragments of proteins in water solution. II. The nascent helix. *Journal of Molecular Biology* 201(1):201–217

Elber R, Karplus M (1987): Multiple conformational states of proteins: A molecular dynamics analysis of myoglobin. *Science* 235:318–321

Epstein CJ, Goldberger RF, Anfinsen CB (1963): The genetic control of tertiary protein structure: Studies with model systems. *Cold Spring Harbor Symposium on Quantitative Biology* 28:439–449

Fasman GD, ed. (1988): *Prediction of Protein Structure and The Principles of Protein Conformation*. New York: Plenum Press

Finkelstein AV, Reva BA (1992): Search for the stable state of a short chain in a molecular field. *Protein Engineering* 5(7):617–624

Formann M, Wagner F (1991): A packing problem with applications to lettering of maps. In *Proceedings of the Seventh Annual Symposium on Computational Geometry*, pp. 281–288, North Conway, NH: ACM

Fraenkel AS (1993): Complexity of protein folding. *Bulletin of Mathematical Biology* 55(6):1199–1210

Franco J, Paull M (1983): Probabilistic analysis of the Davis-Putnam procedure for solving the satisfiability problem. *Discrete Applied Mathematics* 5:77–87

Garey MR, Johnson DS (1979): *Computers and Intractability: A Guide to the Theory of NP-Completeness*. San Francisco: WH Freeman and Company

Gibson KD, Scheraga HA (1986): Predicted conformations for the immunodominant region of the circumsporozoite protein of the human malaria parasite. *Proceedings of the National Academy of Sciences, USA* 83:5649–5653

Goldberg A (1979): On the complexity of the satisfiability problem. Courant Computer Science Report 16, New York University

Goldberg A, Purdom PW Jr., Brown CA (1982): Average time analysis of simplified Davis-Putnam procedures. *Information Processing Letters* 15:72–75. See also "Errata," vol. 16, 1983, p. 213

Gordon HL, Somorjai RL (1992): Applicability of the method of smoothed functionals as a global minimizer for model polypeptides. *Journal of Physical Chemistry* 96:7116–7121

Greengard L (1987): *The Rapid Evaluation of Potential Fields in Particle Systems.* ACM Distinguished Dissertations. Cambridge, MA: MIT Press

Greengard L, Rokhlin V (1989): On the evaluation of electrostatic interactions in molecular modeling. *Chemica Scripta* 29A:139–144

Harrison SC, Durbin R (1985): Is there a single pathway for the folding of a polypeptide chain? *Proceedings of the National Academy of Sciences, USA* 82:4028–4030

Head-Gordon T, Stillinger FH (1993): Predicting polypeptide and protein structures from amino acid sequence: Antlion method applied to melittin. *Biopolymers* 33(2):293–303

Head-Gordon T, Stillinger FH, Arrecis J (1991): A strategy for finding classes of minima on a hypersurface: Implications for approaches to the protein folding problem. *Proceedings of the National Academy of Sciences, USA* 88:11076–11080

Jaenicke R (1987): Protein folding and protein association. *Progress in Biophysics and Molecular Biology* 49:117–237

Karplus M, Shakhnovich E (1992): Protein folding: Theoretical studies of thermodynamics and dynamics. In Creighton TE, ed., *Protein Folding*, chapter 4, pp. 127–196. New York: WH Freeman

Karplus M, Weaver DL (1976): Protein-folding dynamics. *Nature* 260:404–406

Karplus M, Weaver DL (1979): Diffusion-collision model for protein folding. *Biopolymers* 18:1421–1437

Khachiyan LG (1979): A polynomial time algorithm in linear programming. *Soviet Math Dokl* 20:191–194

Kim PS, Baldwin RL (1990): Intermediates in the folding reactions of small proteins. *Annual Reviews of Biochemistry* 59:631–660

Kirkpatrick S, Gelatt CD Jr., Vecchi MP (1983): Optimization by simulated annealing. *Science* 220:671–680

Kostrowicki J, Piela L (1991): Diffusion equation method of global minimization: Performance for standard test functions. *Journal of Optimization Theory and Applications* 69(2):269–284

Kostrowicki J, Piela L, Cherayil BJ, Scheraga HA (1991): Performance of the diffusion equation method in searches for optimum structures of clusters of Lennard-Jones atoms. *Journal of Physical Chemistry* 95(10):4113–4119

Kostrowicki J, Scheraga HA (1992): Application of the diffusion equation method for global optimization to oligopeptides. *Journal of Physical Chemistry* 96: 7442–7449

Ladner RE (1975): On the structure of polynomial time reducibility. *Journal of the Association of Computing Machinery* 22:155–171

Lee C, Subbiah S (1991): Prediction of protein side-chain conformation by packing optimization. *Journal of Molecular Biology* 217:373–388

LeGrand S, Merz K Jr. (1993): The application of the genetic algorithm to the minimization of potential energy functions. *Journal of Global Optimization* 3:49–66

Levinthal C (1966): Molecular model-building by computer. *Scientific American* 214

Levinthal C (1968): Are there pathways for protein folding? *Journal de Chimie Physique* 65(1):44–45

Levinthal C (1969): In *Mössbauer Spectroscopy in Biological Systems*, pp. 22–24. Urbana, IL: University of Illinois Press. Proceedings of a meeting held at Allerton House, Monticello, IL

Lewis HR, Papadimitriou CH (1978): The efficiency of algorithms. *Scientific American* 238(1):96–109

Lewis HR, Papadimitriou CH (1981): *Elements of the Theory of Computation.* Englewood Cliffs, NJ: Prentice-Hall

Lipton M, Still WC (1988): The multiple minimum problem in molecular modeling. Tree searching internal coordinate conformational space. *Journal of Computational Chemistry* 9(4):343–355

Marks J, Shieber S (1991): The computational complexity of cartographic label placement. Technical Report TR-05-91, Harvard University, Cambridge, MA

Marqusee S, Baldwin RL (1987): Helix stablization by Gly-Lys salt bridges in short peptides of *de novo* design. *Proceedings of the National Academy of Sciences, USA* 84:8898–8902

Metropolis N, Rosenbluth AW, Teller AH, Teller E (1953): Equation of state calculations by fast computing machines. *Journal of Chemical Physics* 21:1087–1092

Mitchell D, Selman B, Levesque H (1992): Hard and easy distributions of SAT problems. In *Proceedings of the Tenth National Conference on Artificial Intelligence (AAAI '92)* pp. 459–465. San Jose, CA: AAAI Press/MIT Press

Momany FA, McGuire RF, Burgess AW, Scheraga HA (1975): Energy parameters in polypeptides. VII. Geometric parameters, partial atomic charges, nonbonded interactions, hydrogen bond interactions, and intrinsic torsional potentials for the naturally occurring amino acids. *Journal of Physical Chemistry* 79(22):2361–2381

Moult J, James MNG (1986): An algorithm for determining the conformation of polypeptide segments in proteins by systematic search. *PROTEINS: Structure, Function, and Genetics* 1:146–163

Moult J, Unger R (1991): An analysis of protein folding pathways. *Biochemistry* 30:3816–3824

Ngo JT, Marks J (1991): Computational complexity of a problem in molecular structure prediction. Technical Report TR-17-91, Harvard University, Cambridge, MA. Older version in which 90° angles were employed

Ngo JT, Marks J (1992): Computational complexity of a problem in molecular-structure prediction. *Protein Engineering* 5(4):313–321

Nilsson NJ (1980): *Principles of Artificial Intelligence.* Palo Alto, CA: Tioga Publishing Company

Papadimitriou CH, Steiglitz K (1982): *Combinatorial Optimization: Algorithms and Complexity.* Englewood Cliffs, NJ: Prentice-Hall

Papadimitriou CH, Yannakakis M (1991): Optimization, approximation, and complexity classes. *Journal of Computer and System Sciences* 43:425–440

Piela L, Kostrowicki J, Scheraga HA (1989): The multiple-minima problem in the conformational analysis of molecules. Deformation of the potential energy hypersurface by the diffusion equation method. *Journal of Physical Chemistry* 93(8):3339–3346

Press WH, Flannery BP, Teukolsky SA, Vetterling WT (1988): *Numerical Recipes in C. The Art of Scientific Computing.* Cambridge, UK: Cambridge University Press

Ptitsyn OB (1987): Protein folding: Hypotheses and experiments. *Journal of Protein Chemistry* 6:273–293

Purisima EO, Scheraga HA (1986): An approach to the multiple-minima problem by relaxing dimensionality. *Proceedings of the National Academy of Sciences, USA* 83:2782–2786

Purisima EO, Scheraga HA (1987): An approach to the multiple-minima problem in protein folding by relaxing dimensionality. Tests on enkephalin. *Journal of Molecular Biology* 196:697–709

Rabin MO (1976): Probabilistic algorithms. In Traub JF, ed., *Algorithms and Complexity: New Directions and Recent Results*, pp. 21–39. New York: Academic Press

Rabin MO (1980): Probabilistic algorithms for testing primality. *Journal of Number Theory* 12(1):128–138

Reeke GN Jr. (1988): Protein folding: Computational approaches to an exponential-time problem. In *Annual Reviews of Computer Science* 3:59–84; Annual Reviews, Inc.

Robson B, Platt E, Fishleigh RV, Marsden A, Millard P (1987): Expert system for protein engineering: Its application in the study of chloroamphenicol acetyltransferase and avian pancreatic polypeptide. *Journal of Molecular Graphics* 5(1):8–17

Roder H, Elöve GA, Englander SW (1988): Structural characterization of folding intermediates in cytochrome *c* by H-exchange labelling and proton NMR. *Nature* 335:700– 704

Šali A, Shakhnovich EI, Karplus M (1994): Kinetics of protein folding: A lattice model study of the requirements for folding to the native state. *Journal of Molecular Biology* 235:1614–1636

Saunders M, Houk KN, Wu Y-D, Still WC, Lipton M, Chang G, Guida WC (1990): Comformations of cycloheptadecane: A comparison of methods for conformational searching. *Journal of the American Chemical Society* 112:1419–1427

Scheraga HA (1992): Some approaches to the multiple-minima problem in the calculation of polypeptide and protein structures. *International Journal of Quantum Chemistry* 42:1529–1536

Schmidt KE, Lee MA (1991): Implementing the fast multipole method in three dimensions. *Journal of Statistical Physics* 63(5/6):1223–1235

Shakhnovich EI, Farztdinov GM, Gutin GM, Karplus M (1991): Protein folding bottlenecks: A lattice Monte-Carlo simulation. *Physical Review Letters* 67(12):1665–1668

Shakhnovich EI, Gutin AM (1989): Frozen states of a disordered globular heteropolymer. *Journal of Physics* A22(10):1647–1659

Shalloway D (1992): Application of the renormalization group to deterministic global minimization of molecular conformation energy functions. *Journal of Global Optimization* 2:281–311

Shoemaker KR, Kim PS, Brems DN, Marqusee S, York EJ, Chaiken, IM, Stewart JM, Baldwin RL (1985): Nature of the charged group effect on the stability of the C-peptide helix. *Proceedings of the National Academy of Sciences, USA* 82:2349–2353

Shoemaker KR, Kim PS, York EJ, Stewart JM, Baldwin RL (1987): Tests of the helix dipole model for stabilization of α-helices. *Nature* 326:563–567

Shubert BO (1972a): A sequential method seeking the global maximum of a function. *SIAM Journal on Numerical Analysis* 9(3):379–388

Shubert BO (1972b): Sequential optimization of multimodal discrete function with bounded rate of change. *Management Science* 18(11):687–693

Sikorski A, Skolnick J (1990): Dynamic Monte Carlo simulations of globular protein folding/unfolding pathways. II. α-helical motifs. *Journal of Molecular Biology* 212:819–836

Simon I, Glasser L, Scheraga HA (1991): Calculation of protein conformation as an assembly of stable overlapping segments: Application to bovine pancreatic trypsin inhibitor. *Proceedings of the National Academy of Sciences, USA* 88:3661–3665

Summers NL, Karplus M (1990): Modeling of globular proteins: A distance-based data search procedure for the construction of insertion-deletion regions and pro reversible non-pro mutations. *Journal of Molecular Biology* 216(4):991–1016

Sun S (1993): Reduced representation model of protein structure prediction: Statistical potential and genetic algorithms. *Protein Science* 2(5):762–785

Tsong TY, Baldwin RL, McPhie P (1972): A sequential model of nucleation-dependent protein folding: Kinetic studies of ribonuclease A. *Journal of Molecular Biology* 63(3):453–475

Udgaonkar JB, Baldwin RL (1988): NMR evidence for an early framework intermediate on the folding pathway of ribonuclease A. *Nature* 335:694–699

Unger R, Moult J (1993): Finding the lowest free energy conformation of a protein is a NP-hard problem: Proof and implications. *Bulletin of Mathematical Biology* 55(6):1183–1198

van Gunsteren WF, Berendsen HJC, Colonna F, Perahia D, Hollenberg JP, Lellouch D (1984): On searching neighbors in computer simulations of macromolecular systems. *Journal of Computational Chemistry* 5(3):272–279

Vásquez M, Scheraga HA (1985): Use of buildup and energy-minimization procedures to compute low-energy structures of the backbone of enkephalin. *Biopolymers* 24:1437–1447

Wawak RJ, Wimmer MM, Scheraga HA (1992): An application of the diffusion equation method of global optimization to water clusters. *Journal of Physical Chemistry* 96:5138–5145

Webster (1991): *Webster's Ninth Collegiate Dictionary*. Springfield, MA: Merriam-Webster

Weiner SJ, Kollman P, Nguyen D, Case DA (1986): An all-atom force field for simulations of proteins and nucleic acids. *Journal of Computational Chemistry* 7(2):230–252

Wetlaufer DB (1973): Nucleation, rapid folding, and globular intrachain regions in proteins. *Proceedings of the National Academy of Sciences, USA* 70:697–701

Zwanzig R, Szabo A, Bagchi B (1992): Levinthal's paradox. *Proceedings of the National Academy of Sciences, USA* 89:20–22

15

Toward Quantitative Protein Structure Prediction

Teresa Head-Gordon

1. Introduction

We review a constrained optimization strategy known as the *antlion* method for the purpose of protein structure prediction. This method involves the use of neural network predictions of secondary and tertiary structure to systematically deform a protein energy hypersurface to retain only a single minimum near to the native structure. Successful constrained optimization as applied to protein folding relies on (1) an understanding of the chemistry that distinguishes the native minimum from other metastable structures, (2) the incorporation of such information as robust constraints on the energy function to isolate the native structure minimum, and (3) progress toward providing a quantitative representation of the potential or free energy function. We provide a discussion of completed work by us that begins to affect these three problem areas as we move toward our goal of quantitative protein structure prediction.

The realization of a solution to the protein folding problem would include such benefits as the reengineering of defective proteins indicted in disease, the design of synthetic proteins relevant for biotechnical applications, and a general understanding of how molecules self-assemble. While the benefits are clear, the criteria for success in the protein folding area is more complicated. A detailed understanding of the kinetic mechanism of the entire folding event would comprise a complete solution, but the com-

The Protein Folding Problem and Tertiary Structure Prediction
K. Merz, Jr. and S. Le Grand, Editors
© Birkhäuser Boston 1994

plexity of the problem ensures that such a solution will remain elusive in the near future. A robust mapping algorithm between amino acid sequence and tertiary structure would be highly desirable, even without full physical insight, especially in light of the rapid progress of the human genome project. Alternatively, tremendous satisfaction could be gained from physical models based on first principles that provide general understanding of the folding process, without necessarily predicting explicit structures to compare with experiment.

The intellectual division of the protein folding problem into empirical protein folding algorithms and *ab initio* models is evident in the literature. The former approach seeks a predictive mapping between amino acid sequence and tertiary structure using rules derived empirically from a structural database or single example. These algorithms would encompass such examples as statistical analysis of sequence-structure relationships (Chou and Fasman, 1974; Lim, 1974; Garnier et al.,1978; Levin et al., 1986; Gibrat et al., 1987; Ptitsyn and Finkelstein, 1989; Wilcox et al., 1990), virtual representations of folding (Levitt and Warshel, 1975; Levitt, 1976), conformational searching (Scheraga, 1992), inverse folding approaches (Eisenberg et al., 1992; Godzik and Skolnick, 1992; Wilmanns and Eisenberg, 1993), and the recent use of neural network schemes (Qian and Sejnowski, 1988; Rooman and Wodak, 1988; Holley and Karplus, 1989; McGregor et al., 1989; Bengio and Pouliot, 1990; Bohr et al., 1990; Kneller et al., 1990; Friedrichs et al., 1991; Hirst and Sternberg, 1991; Vieth and Kolinski, 1991; Ferran and Ferrara, 1992; Goldstein et al., 1992; Hayward and Collins, 1992; Hirst and Sternberg, 1992; Muskal and Kim, 1992; Stolorz et al., 1992; Head-Gordon and Stillinger, 1993b). In general the success of these algorithms can be directly assessed by comparison of a predicted structure with a known experimental structure. On the other hand, simple *ab initio* models of protein folding (Bryngelson and Wolynes, 1987; Kolinski et al., 1988; Chan and Dill, 1991; Shakhnovich et al., 1991; Zwanzig et al., 1992; Head-Gordon and Stillinger, 1993a) try to understand the important underlying physics that explains the general folding features, regardless of details of sequence and structure. Lattice models (Kolinski et al., 1988; Chan and Dill, 1991), spin glass theories (Bryngelson and Wolynes, 1987), and perfectly performing neural network models (Head-Gordon and Stillinger, 1993a) described below provide examples of such approaches. Comparison to experiment is much less direct in these cases, and therefore their impact in protein folding remains unclear. While individual research projects in protein folding generally fall cleanly into one category or the other, researchers themselves recognize the importance of both intellectual approaches: Analysis of a successful mapping algorithm may provide deeper insight into how proteins fold, while successful simple models may suggest further improvements that can lead to more detailed predictions. Our research in the protein folding area, the subject of this review, sets forth our philosophy that the success of predictive folding algorithms also depends on a basic understanding of the underlying physics of the folding process.

Because we seek protein folding algorithms with a high degree of predictive accuracy and physical insight, our criteria for success are two-fold. First, we must demonstrate that the folding algorithms can genuinely predict the structure of proteins with rich tertiary features by providing models with full atomic detail to compare directly with experiment. While this level of detail may seem foolhardy, it simply indicates that the protein folding problem should encompass research in the area of quantitative structure prediction. Whether that implies improved empirical protein and solvent force fields, or broadening the application of electronic structure methods like density functional theory to complex condensed phase systems, remains an open research question. Second, we require that the folding algorithms not strictly be black boxes, but instead reveal both a protein structure and some chemical knowledge as to what determines that structure. We believe that it is the latter point that gives credence to simple model studies. While simple models must not be oversold as providing a solution to the protein folding problem, they are highly useful in dissecting the problem for our understanding of the underlying physics. The seemingly slow progress that such modeling may exhibit in the present will pay off in the longer term. However, we appreciate that the insight gained by simple models must ultimately converge with the goal of quantitative protein structure prediction.

Therefore, a tandem approach of using simple models for insight in order to develop folding algorithms that provide quantitative protein structure prediction is our research aim. We begin in Section 2 with a review of the constrained optimization algorithm—the antlion method (Head-Gordon et al., 1990; Head-Gordon and Stillinger, 1993b)—and illustrate it using blocked alanine dipeptide (Head-Gordon et al., 1990) and the 26-residue polypeptide, melittin (Terwilliger and Eisenberg, 1982; Head-Gordon and Stillinger, 1993b). While the melittin pilot study demonstrates that constrained optimization provides the framework for a predictive atomic structure algorithm, it also reveals the algorithmic components that are weak: (1) an understanding of the chemistry that distinguishes the native minimum from other metastable structures, (2) incorporating such information as robust constraints on the energy function to isolate the native structure minimum, and (3) the need for progress toward providing a quantitative representation of the potential (free energy) function. We begin to address these problems by first discussing some insight gained in a simple model study that advances the poly-L-alanine hypothesis (Head-Gordon et al., 1992) in Section 3. The main conclusions reached in this section are that only a subset of the entire amino acid sequence actually indicates the final fold outcome, and that the problem of protein folding should be refined as one of weak and strong sequence-structure relationships. Section 4 discusses our research in the area of neural network design (Head-Gordon and Stillinger, 1993a; Stillinger et al., 1993) where we demonstrate the existence of network topologies that predict tertiary structure perfectly for simple, two-dimensional model chemical systems using only the sequence as input. The model networks reveal that optimal network architectures for real proteins should display the anticipated weak and strong sequence-structure mappings, and that thoughtful design can genuinely impact on the ability of the network to learn these desired mappings. Section 5 provides a discussion of electronic structure calculations on glycine and alanine oligomers (ubiquitously used as structural models for force field design of larger proteins of interest) in order to unravel the relative role of sequence and environment on the conformational preferences of these small peptides (Head-Gordon et al., 1989, 1991; Shang and Head-Gordon, 1994). The final section summarizes our current position, and discusses our plans for research in the near future.

2. Constrained Optimization—The Antlion Method

The protein folding problem from an optimization point of view can be stated as the relation between the physical chemistry encoded in the amino acid sequence and the stationary point on the potential energy surface corresponding to the protein native structure. There are two broad categories in mathematical optimization research for obtaining relevant solutions to large nonlinear systems with numerous local minima. Constrained optimization methods rely on the availability of sufficiently well-defined constraints so that the desired solution is the only available minimum in the optimization phase of the algorithm. Global optimization techniques avoid the problem of determining such constraints by systematically searching the potential energy surface to find all low-lying minima including the global energy minimum.

From the above stated viewpoint of optimization, the predictive capacity problem in protein folding is made obvious. It is generally believed that the number, W, of distinct local minima rises approximately exponentially with N, the number of residues

$$\ln \Omega \approx \alpha N. \tag{1}$$

A rough range of N for naturally occurring proteins is 100 to 1000, while a probably lies in the range of 1 to 100. Current successful global optimization algorithms, however, can only address those problems with very few degrees of freedom, and protein folding clearly involves a hypersurface complexity that is well beyond that manageable by these methods. While constrained optimization is capable of handling the size regime of proteins, the problem of protein folding is currently unsolved because we cannot provide constraints with sufficient predictive accuracy to isolate the relevant minimum corresponding to the native state. Nonetheless, greater progress has been made in the constrained optimization approach since protein folding research is so intimately tied to understanding the chemistry (constraints) that uniquely differentiates the native state (relevant minimum) from other metastable structures.

The application of constrained optimization to the protein folding problem has been investigated by us using the so-called "antlion" method (Head-Gordon et al., 1990), replacing the complicated protein hypersurface by one for which $a = 0$ in equation (1). The constrained optimization algorithm requires a protein hypersurface description, the objective function, where we have used a "gas phase" empirical protein force field (Momany et al., 1990) of the form (Hagler et al., 1974; Momany et al., 1974; Brooks et al., 1983; Weiner et al., 1986; Jorgensen and Tirado-Rives, 1988)

$$\Phi = \overset{\#\,\text{bonds}}{\sum_{i}} k_{bi}(b_i - b_{i0})^2 + \overset{\#\,\text{angles}}{\sum_{i}} k_{\theta i}(\theta_i - \theta_{i0})^2$$

$$+ \overset{\#\,\text{improper}}{\sum_{i}} k_{\tau i}(\tau_i - \tau_{i0})^2 + \overset{\#\,\text{torsions}}{\sum_{i}} k_{\omega i}\big[1 + \cos(n_i \omega_i + \delta_i)\big] \tag{2}$$

$$+ \sum_{i}^{N} \sum_{j}^{N} \big\{ Cq_i q_j / r_{ij} + \varepsilon_{ij}\big[(R_{ij}/r_{ij})^{12} - 2(R_{ij}/r_{ij})^6\big] \big\}.$$

We note that the objective function is a replaceable algorithmic component in constrained optimization; constraints defined for protein force fields with a reasonable degree of qualitative correctness allow the antlion algorithm to advance while we wait for further quantitative improvements in the protein/solvent objective function (see Section 5). Furthermore, we have used penalty functions instead of the more rigorous constraints, the distinction being that the former need not be precisely satisfied at the optimal solution.

Ultimately we hope to restore this level of rigor in a robust application of the antlion method to protein structure prediction. In order to create a generally useful constrained optimization strategy, at least part of the protein hypersurface modification algorithm must contain mathematical operations that are transferable between different polypeptides, and in particular from oligomers to higher molecular-weight polypeptides.

In this spirit of increasing model complexity, we have demonstrated the usefulness of the antlion approach on blocked alanine oligomers (Head-Gordon et al., 1990) and then on a small 26-residue polypeptide, melittin (Terwilliger and Eisenberg, 1982; Head-Gordon and Stillinger, 1993b).

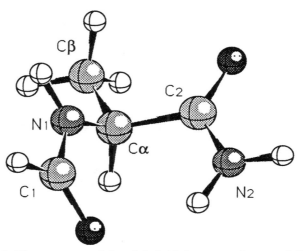

Figure 15-1. The structure of the global minimum conformer of (S)-α-(formyl-amino) propanamide.

The quantitative accuracy of structure prediction using a penalty function or constrained optimization approach can best be illustrated by the biological prototype molecule, the blocked alanine dipeptide, (S)-2-(acetylamino)-N-methylpropanamide. (Figure 15-1 displays a closely related species, (S)-a-(formylamino)propanamide, where the terminal methyls have been replaced by hydrogens.) Monte Carlo searches were performed using an empirical protein potential energy function (Momany et al., 1990). Simulated annealing and subsequent minimization on the alanine dipeptide hypersurface determined the presence of $\sim$ 40 minima. A large majority of the minima correspond to structures where a *cis–trans* isomerization of one or both peptide groups has occurred, and each such structure includes both an L- and D-isomer minimum. The remaining eight minima (four

L-isomers and four D-isomers) not represented in the preceding class are those that are traditionally defined by the two-dimensional internal coordinate torsions ϕ (C_1-N_1-C_α-C_2) and ψ (N_1-C_α-C_2-N_2) (Ramachandran et al., 1973). Figure 15-2 displays an energy contour map, showing the four L-isomer energy minima, which is generated by constraining ϕ and ψ, and allowing relaxation of the remaining degrees of freedom. The alanine dipeptide example illustrates one of our criteria for success: The empirical gas phase potential (Momany et al., 1990) used to generate Figure 15-1 is in good quantitative agreement, both energetically and structurally, with the (currently) best *ab initio* calculations (Head-Gordon et al., 1989, 1991) of the related species, (S)-α-(formylamino)propanamide (Figure 15-1). We choose to isolate the lowest-energy structure, $C7_{eq}$, which may be described as a seven-membered ring closed by an intramolecular hydrogen bond, with the side-chain methyl group equatorial to the plane of the ring.

We have considered a number of modifications to the potential energy function in order to retain only the global energy minimum (Head-Gordon et al., 1990; Head-Gordon and Stillinger, 1993b). Our criterion for a successful modification are (1) that the penalty function (constraint) explicitly or implicitly incorporate information about the tertiary structure of any peptide, (2) that the functional form of the modification is transferable across any polypeptide sequence, and (3) that a variety of conformations can be distinguished in a given segment of polypeptide ranging from the random coil to secondary structure conformers such as the α-helix and β-sheet.

The first type of modification of the gas phase potential function is to eliminate all minima where one or both peptide groups are in the *cis* conformation. The peptide torsion potential

$$V' = k_p[1 + \cos(2\omega + \pi)] \qquad (3)$$

favors minima at both $\omega = 0$ (*cis*) and π (*trans*). The obvious modification of equation (3) to favor the *trans* form is to change the multiplicity factor of 2 to 1, and to change the phase from π to 0. In order to maintain the original curvature at the minimum, we use a force constant of $4k_p$ in the modified version of equation (3). In addition, we usually desire the L-configuration of a polypeptide sequence. In order to maintain the desired chirality, we incorporate an improper dihedral function

$$V'' = k_\tau(\tau - \tau_0)^2 \qquad (4)$$

for the torsions C_α-N_1-C_2-C_β and C_α-N_1-C_2-H_α. While the V' and V'' modifications are trivial, and in some sense physically unimportant in relevant areas of configuration space for biological molecules of interest, this serves as an illustrative example for what is to follow. For the case of alanine

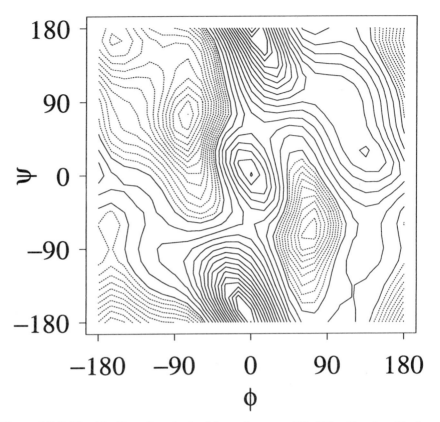

Figure 15-2. The (ϕ, ψ) surface derived from the unmodified Hamiltonian, V_0, for (S)-2-(acetylamino)-N-methylpropanamide. The ϕ and ψ variables are held fixed at each grid point (15° spacing), and all other degrees of freedom are relaxed. The dashed lines denote contours of 0.5 kcal/mole, and extend from the zero of energy (the $C7_{eq}$ conformer) to 7.0 kcal/mole. Solid contours are drawn every 1.0 kcal/mole thereafter.

dipeptide, the modifications (3) and (4) permit us to visualize transforming the energy surface in Figure 15-2 to retain only the $C7_{eq}(\phi \simeq -70°$, $\psi \simeq 70°)$ conformer. This is accomplished with the following generic penalty function:

$$V''' = k_\phi[1 - \cos(\phi - \phi_0)] + k_\psi[1 - \cos(\psi - \psi_0)], \tag{5}$$

where the modified surface (3)–(5) is exhibited in Figure 15-3. The surface modification displayed in Figure 15-3 illustrates another important feature of a successful structure prediction algorithm: the constraint parameters and functional form are well chosen in order to isolate the desired min-

imum from any starting structure. In the case of alanine dipeptide, the
$C7_{eq}$ structure is the only stationary point on the modified surface that is
a minimum: the remaining features on the map are saddle points and a
maximum. In the broader case of proteins, the successful retention of only
the native minimum should permit the use of an objective starting structure
such as an extended conformer, without necessarily appealing to sequence
or structure homologies to start off the optimization. The antlion method
is completed with unconstrained optimization on the original surface using
the constrained solution of the modified surface as the initial guess.

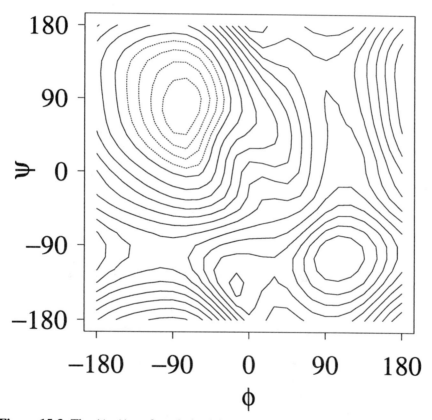

Figure 15-3. The (ϕ, ψ) surface derived from the modified Hamiltonian, V''', for
(S)-2-(acetylamino)-N-methylpropanamide. The ϕ and ψ variables are held fixed
at each grid point (15° spacing), and all other degrees of freedom are relaxed. The
dashed lines denote contours of 2.0 kcal/mole, and extend from the zero of energy
(the $C7_{eq}$ conformer) to 8.0 kcal/mole. Solid contours are drawn every 3.0 kcal/mole
thereafter.

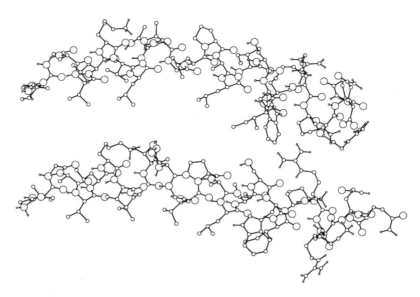

Figure 15-4. A comparison of the minimized crystal structure (*top*) and the antlion folded structure (*bottom*) of melittin. An overall rms difference of 2.45 Å between these structures is observed, while the backbone degrees of freedom show an rms difference of 1.2 Å.

The alanine dipeptide example may seem to imply that prior knowledge of the secondary and tertiary structure of globular proteins is required in order to implement the antlion approach for larger biopolymers; this need not be the case. In order to address the predictive capacity problem, we therefore adapt our antlion strategy to use neural networks (Müller and Reinhardt, 1990; Hertz et al., 1991; Churchland and Sejnowski, 1992) as a guide for designing penalty function parameters (or constraints) that retain only the native minimum. However, it should be emphasized that our use of neural networks is quite different from that conventionally required of neural networks in protein structure prediction. Traditional network approaches use the outputs of the network as the direct structure prediction (Qian and Sejnowski, 1988; Rooman and Wodak, 1988; Holley and Karplus, 1989; McGregor et al., 1989; Bengio and Pouliot, 1990; Bohr et al., 1990; Kneller et al., 1990; Friedrichs et al., 1991; Hirst and Sternberg, 1991; Vieth and Kolinski, 1991; Ferran and Ferrara, 1992; Goldstein et al., 1992; Hayward and Collins, 1992; Muskal and Kim, 1992; Stolorz et al., 1992). In our antlion optimization approach, neural networks are intended to be used as a predictor for the constraints only. Minimization first on the modified potential hypersurface and then on the unmodified hypersurface serves as the tertiary predictor. We have applied this enhanced antlion method (Head-Gordon and Stillinger, 1993b) to predicting the structure of the predominantly α-helical polypeptide melittin (Terwilliger and Eisenberg, 1982) depicted in Figure 15-4.

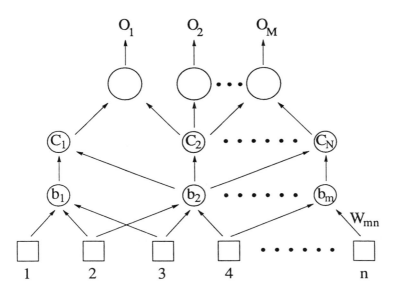

Figure 15-5. A generic feed-forward neural network architecture.

In application to the protein folding problem, neural network algorithms are required to predict patterns of local (secondary structure) and global (tertiary structure) chain folding preferences of the native protein from the amino acid sequence. The common topology of neural networks often used to predict these conformational preferences is known as feed-forward back-propagation networks with or without hidden layers (Figure 15-5) (Müller and Reinhardt, 1990; Hertz et al., 1991; Churchland and Sejnowski, 1992). In this case, each amino acid of a protein sequence is represented by a small set of input channels that are directly connected, or fed into, hidden neurons that in turn connect to output neuron(s) whose numbers represent a secondary or tertiary structure classification. The input channels generally correspond to both the amino acids whose most likely secondary or tertiary structure is being predicted, while the remainder supply a context (or window) of n amino acids preceding and succeeding this amino acid along the backbone. The learning, or training, phase of the neural network algorithm involves minimizing the function

$$E = \sum_i^N \sum_j^M \left(O_{oj}^i - O_{cj}^i \right)^2, \tag{6}$$

where M is the number of output units and N is the number of presented input patterns, O_o is the observed structure output, and O_c is the calculated output. The calculated output is usually determined as follows:

$$A_{cj}^i = \sum_k^L w_{jk} I_k^i + b_j \tag{7}$$

with an output response of

$$O_{cj}^i = \frac{1}{1 + \exp(A_{cj}^i)} \tag{8}$$

or a discontinuous response function

$$O_{cj}^i = \text{sign}(A_{cj}^i), \tag{9}$$

where L is the number of input units, I_k is the input, w_{jk} is the weight of the connection between the upstream neuron k and the downstream neuron j, and b_j is the bias associated with the output neuron j. Almost always a steepest descent algorithm is used for minimizing the function in equation (5) with respect to the free parameters w_{jk} and b_j. The parameters w_{jk} and b_j are updated, (or "back-propagated" through the network from output to input) by the following derivative expressions:

$$\Delta w_{jk} = -\gamma \partial E / \partial w_{jk},$$
$$\Delta b_j = -\gamma \partial E / \partial b_j, \tag{10}$$

where γ is a damping, or "learning," factor.

The topology of the neural network that we have used to predict the penalty function parameters for melittin is that of a simple perceptron (Müller and Reinhardt, 1990; Hertz et al., 1991; Churchland and Sejnowski, 1992), also known as feed-forward networks with no hidden layers. We have tried to exploit physically motivated ideas concerning input and output representations in order to improve the secondary structure prediction accuracy of our neural networks that are relevant to melittin. We have encoded biophysical properties into the input (amino acid sequence) representation. For the test-pilot case of melittin, the input representation for each amino acid is a five-bit binary number ordered in such a way to reflect one of the following three scales: an α-helix promotion ordering of the 20 commonly occurring amino acids deduced from substitution of these residues into a coiled coil (O'Neill and DeGrado, 1990), an α-helix promotion scale based on a statistical analysis of 60 proteins (Levitt, 1978), or a random scale generated from a normal distribution. The output is designed to be one neuron whose output value is continuous between 0 (nonhelical) and 1 (helical).

These three very simple networks were trained on a subset of an experimental database with secondary structure assignments due to Kabsch and Sander (1983). We were then able to test the three different representation networks for their predictive accuracy and robustness on the remaining proteins of the database, which included melittin. The random scale representation performed significantly more poorly than the two scales encoding biophysical properties. The promotion scale based on substitutions into an artificial coiled coil environment (O'Neill and DeGrado, 1990) was not as successful as the scale determined from a statistical analysis of globular proteins (Levitt, 1978) (which is most similar to the training and testing database). Out of the 22 possible helical residues (Kabsch and Sander, 1983) of the 26-residue peptide melittin, the random scale predicted 7 residues to be helical, the coiled-coil scale predicted 10, and the globular protein promotion scale predicted 15 correctly. These results provide direct evidence that thoughtfully designed neural networks enhance network prediction accuracy and confidence, and we return to this point in Section 4. Nonetheless, even the best neural network in this example does not adequately predict the structure of melittin in itself.

However, the deficiencies of the neural network (our best network only predicted 15 of the 22 helical residues correctly) need not preclude successful secondary structure prediction (Head-Gordon and Stillinger, 1993b). We have found two useful penalty functions forms for encouraging formation of secondary structure such as helices, turns, and sheets. The enforcement of particular ϕ_0 and ψ_0 dihedral angles, transferred from our simpler alanine dipeptide example, and the "reward" function for hydrogen bonds

$$V_{20} = \frac{q_i q_i}{r_{ij}} \qquad (11)$$

differentiates among these secondary structural classes. Based on the network prediction for melittin, the following penalty function parameters were defined. The value of ϕ_0 and ψ_0 were taken to be a perfect helix $(-57°, -47°)$, since our network predicts only helical content. The degree of helical content was indicated in the force constants k_ϕ and k_ψ, which were set equal to the output of the network (ranging from 0 to 1) described above, and scaled by a factor of 100. There are well-defined reasons for choosing the scale factor of 100 kcal/mole. We have found that the largest barrier to eliminate in the smoothing process is that due to bond-angle strain; thus penalty-function force constants must be of the same magnitude in order to compete with these barriers. We also invoked the formation of hydrogen bonds between the backbone oxygen of residue i and the backbone hydrogen of residue $i + 4$ by the use of equation (11), where $q_i = -q_{i+4}$ is the direct network output.

Using the extended structure as the initial guess (all backbone dihedral angle pairs ϕ, ψ defined as 180°, $-180°$), optimization on the modified protein potential hypersurface then proceeded with the adopted basis Newton–Raphson (ABNR) (Press et al., 1986) algorithm until a minimum was found (converged gradient of 0.005 kcal/mole Å). The minimized structure on the modified surface was then used as the starting structure on the unmodified surface, again using ABNR and a convergence criteria of 0.005 kcal/mole Å. A comparison of the predicted structure with the experimental structure provides striking confirmation that the backbone of melittin has been predicted well (Figure 15-4). It is interesting to note in this structural comparison that the side-chain degrees of freedom, where no penalty functions were defined, are remarkably similar for residues 1–10. This emphasizes one of the strengths of the antlion approach: The protein force fields themselves provide some aspect of structure prediction since the protein objective functions are parameterized for native folds. This benefit becomes more important as further progress is made in quantitative evaluation of polypeptides and proteins in their aqueous environment.

While melittin may be dismissed as a trivial protein due to a lack of rich tertiary features, it is sufficiently complex to demonstrate several important strengths of the antlion method. First, in arguments made elsewhere (Head-Gordon et al., 1990), there are approximately 10^{26} L-isomer minima on the melittin hypersurface. From an optimization point of view, the functional forms and neural network parameter predictions have provided quite robust constraints, allowing us to isolate a good approximation to the native structure using a completely extended conformer as the starting structure. In this case, the multiple minima problem is quite fierce, and global optimization techniques may be fruitless. Second, successful constrained optimization certainly relies on some threshold of performance of the predictive phase of the algorithm in order to find low-energy minima. However, the realization of 100% neural network prediction success will be unlikely (the neural networks will not allow a straight structure prediction), and the interplay of protein force fields that are parameterized for native-like folds may well overcome some deficiencies in neural network predictions.

The melittin example also provides ample evidence of certain weaknesses of the antlion algorithm. Improvement in the neural network prediction phase of the algorithm is already required since helical content is only a small component of successful prediction of even predominately α-helical proteins. Furthermore, side-chain conformational preferences must also be described (although an accurately predicted backbone may alleviate the multiple minima problem in the space of the side-chain variables [Lee and Subbiah, 1991]). Because the dimensionality of the problem rises exponentially with residue number, even optimization on the modified surface may become trapped in deep local minima starting from a highly nonoptimal initial guess such as the extended conformer. This point is well illustrated by the electrostatic side chains of melittin, which became locked in conformations in the early phase of the optimization that are in marked disagreement with experiment. Finally, the rms difference found between the antlion predicted structure of melittin and the experimental structure may be due to protein force field inadequacies, ranging from nonoptimal functional forms and parameterization to the lack of solvent representation. The following three sections provide a description of three model studies that begin to affect the solution to the above-listed failings in predicting protein structure.

3. The Poly-L-Alanine Hypothesis

The "poly-L-alanine hypothesis" asserts that there is a close correspondence, or short-distance mapping, between the backbone geometry at the native-structure free energy minimum for any polypeptide or protein of M residues, and that of a (metastable) minimum of poly-L-alanine of the same number of residues (Head-Gordon et al., 1992). Alanine is the most natural choice of residue for such a comprehensive comparison because its degree of geometric plasticity, as measured by a Ramachandran plot (Ramachandran et al., 1973), encompasses all amino acids except glycine and proline. Furthermore, it is the simplest chiral residue.

Successful validation of the poly-L-alanine hypothesis would provide the insight that the *mechanical* stability of protein native structures does not depend critically on all side-chain details, but that the energy (or free energy) stability is indeed controlled by those details. In this connection Matthews and co-workers' T4-lysozyme mutagenesis studies (Zhang et al., 1991; Eriksson et al., 1992) have a special significance, wherein replacement of several residues by L-alanine has been demonstrated to preserve native structure. Thus, the hypothesis provides a context in which to refine the protein folding problem definition as one of weak and strong sequence-structure relationships.

A wide variety of calculations has been undertaken in order to demonstrate the validity of the poly-L-alanine hypothesis. These calculations range in M, the number of residues, from very small to quite large polypeptides, and demonstrate the ability of poly-L-alanine to adopt an impressive repertoire of structural alternatives. We have demonstrated poly-L-alanine stability for several naturally occurring proteins: the 26-residue polypeptide melittin (MLT) (Terwilliger and Eisenberg, 1982), the 46-residue protein crambin (CRN) (Hendrickson and Teeter, 1981), the 58-residue protein bovine pancreatic trypsin inhibitor (BPTI) (Deisenhofer and Steigemann, 1975), the 124-residue Ribonuclease-A (RNASE) (Kartha et al., 1967), and the 158-residue superoxide dismutase (SOD) (Tainer et al., 1982).

The initial heavy-atom configurations of the above proteins were taken from the Brookhaven Protein Databank. Nonaliphatic heavy-atom centers were given hydrogens so that all hydrogen positions satisfied excluded volume and geometric constraints (Brooks et al., 1983). (Nonpolar hydrogens are represented by an extended version of the aliphatic carbon to which they are attached.) The resulting hydrogenated crystal structure was minimized using an empirical protein force field of the form in equation (2) and the following penalty function protocol. The procedure begins by placing a harmonic penalty function on all heavy atoms,

$$V_p = \sum_i k_p (r_i - r_{io})^2 \qquad (12)$$

and minimizing using the Powell algorithm (Press et al., 1986) on the hypersurface defined by Brooks et al. (1983) and equation (12). After an rms derivative convergence of 0.1 kcal/mole Å is reached (or after the completion of 200 minimization steps), the penalty function force constant, k_p, and the equilibrium value, r_{io}, are updated by reducing k_p by 5 kcal/mole Å^2 and reassigning r_{io} to be the position of i at the completion of the last minimization cycle. Once the penalty function is totally eliminated, the structure is minimized using ABNR (Press et al., 1986) until the rms derivative is less than 0.005 kcal/mole Å. Both stages of the penalty function optimization (nonbonded cutoff of 7.5 Å) update the list of nonbonded interactions every 100 steps.

To create the initial guess to the poly-L-alanine analog to the protein of interest, the resulting minimized structure was edited so that the backbone atoms and β-carbon side-chain atoms were retained; for proline and glycine, the backbone hydrogen atom and β-carbon side chain were added, respectively, with all geometric and steric constraints satisfied. This poly-L-alanine starting structure was minimized again with the penalty function protocol. Table 15-1 provides the rms differences between the crystal

Table 15-1. RMS difference between crystal struction and poly-L-alanine analog.

Sequence	Ala	Ala-Cys	Ala-Gly	Ala-Pro	AGP or AGPC
β-α-β	2.07	----	----	----	----
MLT	2.14	----	2.11	1.93	1.58
CRN	2.42	2.11	2.75	2.35	1.79
BPTI	2.29	1.80	1.73	2.14	1.53
RNAS	2.59	2.31	2.85	2.52	2.62
SOD	2.37	2.43	2.72	2.30	2.17

structure backbone and β-carbons with that of the converged, folded poly-L-alanine counterpart of the five proteins.

There is a remarkable consistency of 2.1 Å to 2.6 Å rms difference between the alanine analog of the five proteins and the proteins themselves. Much of the differences between the native structures and their poly-L-alanine images arise from small local deformations that are largely pattern-preserving, i.e., most important secondary and tertiary fingerprints are retained in the poly-L-alanine analog structures. Little collapse has resulted despite the elimination of large residues in the core. Figures 15-6 and 15-7 display ribbon structure comparisons of the RNASE and SOD proteins and their poly-L-alanine analogs. Both the rms differences and visual comparisons demonstrate the validity of the poly-L-alanine hypothesis, namely that there is a short-distance mapping of minima between the homo-sequence and the native hetero-sequence for which strong agreement is obtained along the backbone and at β-carbon side-chain positions.

Two possible lines of questions arise based on the above demonstration of the poly-L-alanine hypothesis: First, the relative importance of packing and electrostatics as one perturbs alanine into larger side chains and/or polar residues, and second, the degree to which the *specific* native sequence is required for obtaining the tertiary structure to some defined precision. As a preliminary tack we have investigated these questions by performing a penalty function minimization on hetero-sequences of the five proteins, where either (1) all cystines were retained when disulfide linkages are present in the native state, (2) all amino acids were changed to alanine except for the proline positions, (3) all residues were changed to alanine except for the glycine positions, or (4) the sequence was mutated to all alanine except for glycine, alanine, and cystine, which were retained from the native sequence. We have distinguished proline and glycine from the remaining 19 amino acids because their local conformation space should be most different from that of alanine (and the remaining residues) (Ramachandran

Figure 15-6. A backbone, ribbon structure comparison of Ribonuclease-A (RNASE) (*top*) and the RNASE native-like conformer of (ala)$_{124}$ (*bottom*). RNASE provides an example of tertiary structure with a predominance of helical secondary structure, and a well-packed hydrophobic core.

Figure 15-7. A backbone, ribbon structure comparison of Superoxide Dismutase (SOD) (*top*) and the SOD native-like conformer of (ala)$_{152}$ (*bottom*). SOD provides an example of tertiary structure with a β-barrel theme and a significant hydrophobic core.

et al., 1973). The cystines are also unusual because they can form covalent disulfide bonds nonlocal in sequence.

The minimized, hydrogenated crystal structures were edited so that one of the above subset of amino acids was retained, while the remaining amino acids were converted into L-alanine. This starting poly-L-alanine copolymer was then minimized in energy using the penalty function procedure outlined above; the resulting rms differences between the native structure and the four types of subset amino acids in their original sequence positions in the poly-L-alanine sequence are given in Table 15-1. Focusing only on the globular proteins RNASE and SOD for sake of brevity, we have found that the radius of gyration of the poly-L-alanine analogs has only decreased by 1 Å when compared to the native sequence conformation.

This seems to indicate that the loss of large side chains does not result in collapse of the poly-L-alanine structure to maximize a nonpolar core. In addition, the substitution of cystine, glycine, and proline into the poly-L-alanine matrix does not result in any significant improvement in the rms of 2.59 Å for RNASE and 2.37 Å for SOD. Further examination of the structural differences between the native sequence and poly-L-alanine structures indicate that hydrogen bonding interactions between side chains may account for the changes in rms differences for these two proteins, and that the strong structural influences of cystine and glycine found for small proteins may be only of secondary importance in globular proteins. While these are preliminary results for two isolated cases, they indicate that a systematically applied perturbative approach may uncover some general principles regarding the relative importance of packing, electrostatic interactions, and the influence of special amino acids such as proline, cystine, and glycine.

Very recent work by Matthews and co-workers (Zhang et al., 1991; Eriksson et al., 1992) has also explored the use of alanine as a "generic" residue to replace "nonessential" residues in T4 lysozyme, in order to see whether the resulting mutant is correctly folded and functional (Zhang et al., 1991). The mutant T4 lysozyme contained alanine substitutions at positions 128, 131, 132, and/or 133, so that the number of introduced alanines is a very small fraction of the total number of amino acids; in all cases except one, the mutant form was more stable than wild-type lysozyme. More recently, the same group has substituted alanine for native residues within the hydrophobic core and observed that such mutations were "cavity-creating," i.e., proteins do not collapse in order to avoid vacant space within the core (Eriksson et al., 1992). Our model poly-L-alanine studies also show such an effect in the case of the larger proteins, RNASE and SOD, where the radius of gyration was found to decrease by only 1.0 Å. The poly-L-alanine results presented here and the results of the mutagenesis experiments seem to indicate that the "nightmare scenario"—that the native state conformation relies critically on the full details of the native sequence—is not operative. Instead, the problem of protein folding could be restated as the uncovering of what are the strong sequence-structure relationships that dictate the choice of the native structure minimum. It is important to note that the hypothesis as discussed here only impacts on the relation between sequence and the native protein structure, and not about the relative importance of certain amino acids at earlier points on the kinetic pathway. In fact, those amino acids that are weakly correlated with the final folded structure may well have been critical during early folding events.

4. Neural Network Design

There are several sources of error that hamper accurate and robust neural network prediction of protein structures. First, the experimental database of known protein structures is extremely sparse compared to the entire family of possible proteins with a comparable degree of polymerization. This is a particularly insidious problem in the protein folding area since virtually all prediction methods rely to some or a significant extent on learning examples (Chou and Fasman, 1974; Lim, 1974; Levitt and Warshel, 1975; Levitt, 1976; Garnier et al., 1978; Levin et al., 1986; Gibrat et al., 1987; Qian and Sejnowski, 1988; Rooman and Wodak, 1988; Holley and Karplus, 1989; McGregor et al., 1989; Ptitsyn and Finkelstein, 1989; Bengio and Pouliot, 1990; Bohr et al., 1990; Kneller et al., 1990; Wilcox et al., 1990; Friedrichs et al., 1991; Hirst and Sternberg, 1991; Vieth and Kolinski, 1991; Eisenberg et al., 1992; Ferran and Ferrara, 1992; Godzik and Skolnick, 1992; Goldstein et al., 1992; Hayward and Collins, 1992; Hirst and Sternberg, 1992; Muskal and Kim, 1992; Scheraga, 1992; Stolorz et al., 1992; Wilmanns and Eisenberg, 1993; Head-Gordon and Stillinger, 1993b). Because dramatic expansion of the protein experimental database is not a viable option in the near future, alternative means of overcoming database degradation must be considered. Including proteins in the training set with the same representations of structural class (such as α-helix proteins, β-sheet proteins, etc.) as in the testing set has had some impact on improving secondary structure prediction (Kneller et al., 1990). While such improvement is noteworthy, it must be emphasized that robust prediction of secondary structure is necessary but not sufficient for the goal of predicting the correct protein tertiary structure. Including proteins in the training set with the same representations of sequence homology as in the testing set has been tried for tertiary structure prediction. Little success was noted here in that conventional sequence homology modeling is actually superior to neural network predictions in this case (Hirst and Steinberg, 1992).

Another source of error is that the network topologies themselves may not permit effective learning strategies, so that the network is unable to adaptively predict, or generalize to, a new data set of sequence-structure relationships. The primary criticism of the use of neural networks as a predictive tool in areas outside their original intention (how the brain functions) is the lack of rigor in defining the network topologies for the task at hand. Naive approaches often involve such errors as the use of too many network variables, so that predicted solutions are only fits in the nonlinear, least-squares sense, or that input and output representations are sufficiently ill-considered to obscure the input-output relationships. Furthermore, database sparsity is thought to be responsible for the observation that hidden layers do not improve secondary structure predictions because there is simply not enough higher order information (representative interactions between two or more amino acids) to exploit the full power of such neural network topologies (Qian and Sejnowski, 1988; Rooman and Wodak, (1988).

Finally, the search for optimal neural network solutions suffer from a serious mathematical optimization problem where network topologies with hidden layers require a global search of network variable space to determine the optimal weights and biases. The current method of choice for addressing this optimization problem in neural networks is simulated annealing; current research in this area involves the determination of an optimized cooling schedule for converging to the global minimum. However, this prescribed global optimization approach has not been demonstrated to be effective in the area of neural network prediction of protein structure as of yet.

The direct prediction of protein structure from sequence via neural networks that train on genuine protein structures is highly desirable, and the subject of active research (Qian and Sejnowski, 1988; Rooman and Wodak, 1988; Holley and Karplus, 1989; McGregor et al., 1989; Bengio and Pouliot, 1990; Bohr et al., 1990; Kneller et al., 1990; Friedrichs et al., 1991; Hirst and Sternberg, 1991; Vieth and Kolinski, 1991; Ferran and Ferrara, 1992; Goldstein et al., 1992; Hayward and Collins, 1992; Hirst and Sternberg, 1992; Muskal and Kim, 1992; Stolorz et al., 1992; Head-Gordon and Stillinger, 1993b). However, we believe that the current lack of success relative to straight statistical analysis is frustrated by the three imponderables described above. We have chosen to take a step back from recent and more ambitious attempts at protein structure prediction (ibid.) by designing networks that perform perfectly for a database complete up to a given sequence size that describes small, two-dimensional "polypeptides"

(Head-Gordon and Stillinger, 1993a; Stillinger et al., 1993). We may then speculate as to what neural network topologies may be required for longer polypeptides in three dimensions with full sequence diversity. Because the chemical interactions that give rise to the native tertiary structure are not well understood, we have chosen to strip away a great deal of this complexity in order to tackle best the problem of optimal neural networks. The two-dimensional model polypeptides that we have studied are composed of only two amino acid types, and the size regimes we have considered thus far extend from trimers up through heptamers (Head-Gordon and Stillinger, 1993a; Stillinger et al., 1993).

The potential energy function of this simple model (Head-Gordon and Stillinger, 1993a; Stillinger et al., 1993) is composed of two terms: a nonbonded interaction, and a connectivity piece that permits angle bends.

$$\Phi = K \sum_{j=2}^{n-1} \left[1 - \cos(\theta_j)\right] + 4 \sum_{i=1}^{n-2} \sum_{j=1+2}^{n} \left[r_{ij}^{-12} + f(\zeta_i \zeta_j) r_{ij}^{-6}\right] \qquad (13)$$

Here the amino acid monomers have been numbered sequentially along the bond backbone. Angle q_j measures the bend away from linearity for the two bonds impinging on monomer j. The distance between monomers i and j has been denoted by r_{ij}. Monomer species are indicated by parameters $\zeta_1 \ldots \zeta_n$, with value $+1$ for A and -1 for B. The function $f(\zeta_i, \zeta_j)$ may be written as follows:

$$f(\zeta_i, \zeta_j) = \tfrac{1}{2} - \tfrac{1}{8}(\zeta_i + \zeta_j) - \tfrac{5}{16}(\zeta_i + \zeta_j)^2; \qquad (14)$$

it equals -1 for an AA pair, $-\frac{1}{2}$ for a BB pair, and $+\frac{1}{2}$ for an AB pair. In all of the calculations reported below the bend force constant K has been assigned the value $\frac{1}{4}$. The nonbonded interaction favors most strongly AA interactions, to a lesser extent BB interactions, while AB and BA interactions are repulsive. The complex interplay of the nonbonded and connectivity potential for a given amino acid sequence resulted in native 2D structures that can be described in usual protein parlance as linear, α-helix, β-sheet, and globular conformers. This simplified model allows us the advantage of determining the global energy structure for all possible sequences in order to define a perfect and complete database, at least for 2D polypeptides up to heptamer lengths. Having surmounted the database problem, we are free to design network architectures that both accurately (100% structure prediction) and, we believe, optimally (fewest number of network variables) reproduce the observed database of sequence-structure relationships. We designed network topologies that reproduce perfectly two different tertiary structure features: the bend angles of the backbone

(Stillinger et al., 1993), and the binary distance matrix of amino acids (Head-Gordon and Stillinger, 1993a).

For sake of brevity, we only outline the general points for the distance matrix problem. In this case networks were hand-designed to reproduce whether two amino acids are in contact ($r_{ij} \leq 1.34$ bond length units, and output response of $+1$) or are not in contact (output response $= -1$) for all possible sequences for a given individual amino acid pair, σ_{ij}. This is accomplished, individually, for all possible nonbonded i, j pairs, where $i + 2 \leq j \leq n$. The benefit of this tactic is that network topologies can be designed "by hand" with specific Boolean functions, which we discuss below. Thus, given a type of Boolean function, it is easy to derive the remaining architecture to optimally reproduce the observed contacts. The multiple minimum problem in network space then reduces to a search for the best Boolean function that, together with its remaining architecture, reproduces the observations with the smallest number of network variables with 100% accuracy. Therefore we have avoided some aspect of the global optimization problem usually encountered in conventional applications of neural networks to protein structure prediction.

We carefully considered three important components of network design: input representation, neuronal output response, and a search for an optimal and robust Boolean function. We found that input representations of $+1$ and -1 for residues A and B, respectively; an output response function that is discontinuous; and the architecture depicted in Figure 15-8 discourages predictions of A and B or B and A contacts, but encourages like–like interactions (AA and BB). One important insight of this design is that it directly encodes the nonbonded interaction of our simple chemical model, so that a majority of sequences ($\sim 75\%$) for a given contact are predicted correctly with this Boolean function alone. However our interaction potential described above does not produce such a simple relation between sequence and structure as the individual Boolean function would alone indicate. The interplay of the connectivity portion of the potential with the nonbonded interactions results in an outcome for some sequences ($\sim 25\%$ in our model) where not all $A_i A_j$ and $B_i B_j$ pairs are favored, so that additional architecture is required (extra hidden neurons) to turn off those sequences where $A_i A_j$ or $B_i B_j$ are not in contact.

These additional hidden neurons that provide these corrections may also provide important insight as well. For example, long-range sequence information is required to predict even local contacts. The dedicated neuron indicated by 1, 0, 0, 0, 0, 1, 0 input weights for the σ_{13} contact for the heptamer in Table 15-2 provides such an example, where an amino acid far removed from the pair under consideration dictates quite strongly the fold outcome. Recent neural network applications that use "windows" of amino acids, i.e., local sequence information, for predicting distance matrices (Bohr et al., 1990; Friedrichs et al., 1991) of realistic polypeptides and protein structures would be deficient in two respects. First, prediction accuracy is lost to some significant extent. We have found that many corrections to the central Boolean function are possible with only one hidden neuron with input from at least this one important amino acid; this is due to common sequence information among a number of sequences that signals a folding outcome. Second, insight may be lost as well. It is quite plausible that amino acid(s) outside the window may be an especially important residue marker for signaling a particular fold, as in the σ_{13} example exhibited by our 2D polypeptides described above. In a related vein, shared hidden neurons between contact outputs may have special significance: particular amino acids (those whose weight connections to that neuron are nonzero) may be linchpin residues that determine compound structural features. For example, in the case illustrated by the heptamer contact σ_{13}, amino acid 7 determines whether contacts 13, 14, and 15 are on or off, as well as simultaneously signaling different spatial regions (contacts 13 and 46) of the native state.

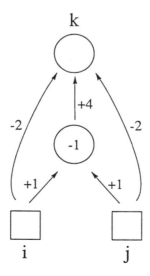

Figure 15-8. Central Boolean function for a neural network that best exploits an "opposites attract" potential energy interaction.

We have also found that a small change in sequence can result in significant changes in fold (in our model as to whether a contact is on or off),

Table 15-2. Network architectures for heptamer contact σ_{13}.

σ_{13}	weights ($i \rightarrow h$)	b_h	weights ($h \rightarrow o$)	weights ($i \rightarrow o$)	b_o
NN 3	1, 0, 1, 0, 0, 0, 0	−1	+4	−2, 0,−2, 0, 0, 0, 0	−7
	1, 1, 0, 0, 0, 0, 0	+1	+1		
	1, 0, 0, 0, 0,−1, 0	+1	+1		
	1, 0, 0,−1, 0, 0, 0	+1	+1		
	0,−1, 0, 1, 0, 0,−1	+2	+1		
	0,−1, 0, 0, 1, 0,−1	+2	+1		
	0,−1, 0, 0, 0, 1,−1	+2	+1		
	−1,−1, 0, 0,−1, 0, 1	+3	+1		
	0,−1, 0,−1, 1,−1, 0	+3	+1		

Table rows refer to the entire network architecture when using the central Boolean functions in Figure 15-8. Columns denoted by "weights" refer to the weight variables found for the optimal network for connections between the input channels and hidden layer neurons ($i \rightarrow h$), hidden neurons to output neurons ($h \rightarrow o$), and input channels connected directly to the output neurons ($i \rightarrow o$). The ordering of weights for the $i \rightarrow h$ and $i \rightarrow o$ columns imply weight subscripts w_{1j}, w_{2j}, w_{3j}, etc., where the numerical values refer to the first, second, third, etc. amino acids of a sequence as scanned from left to right, and j is the hidden or output neuron of interest. The columns denoted by b_h and b_o are the value of the bias for hidden neurons and output neurons, respectively.

and that networks do indeed have more trouble with this case. A dedicated hidden neuron with virtually full input connectivity is required to correctly predict the fold outcome. We have also found that elimination of problem sequences (the necessity of having one dedicated neuron to predict one contact of one sequence correctly) does not degrade overall performance significantly. The elimination of such dedicated hidden neurons results in a total number of incorrectly predicted sequences of 5 out of 120 for the pentamer, 13 out of 360 for the hexamer, and 31 out of 1080 for the hexamer. Thus the acceptance of some performance degradation such as this may be tolerable in a pragmatic sense, where we find a 33% reduction in the number of hidden neurons and the number of weights with only 3–4% performance loss.

Another more subtle aspect of the interplay between network design and genuine learning is the number of hidden layers that optimally (with the fewest number of network variables) and accurately predict all contact values for all sequences. Although we have explored network designs with two hidden layers, never did we find such an architecture that was more optimal than those containing only one hidden layer. This seems to be consistent with the network complexity required by our model with only two amino acid types; only with greater sequence diversity should additional hidden layers be important. By the same token, we rarely found cases where corrections to the central Boolean function involved direct connectivity from the input to the output, indicating that our 2D polypeptide model is not grossly oversimplified.

The networks devised for tertiary structure prediction of 2D polypeptides may also provide some guidance in overcoming the well-appreciated multiple minima problem in the space of the network variables (Müller and Reinhardt, 1990; Hertz et al., 1991; Churchland and Sejnowski, 1992). In protein folding applications, back-propagation algorithms are preferred in spite of the fact that only local minimum solutions are found. While researchers may train networks several times with differing initial guesses to address this deficiency of back-propagation optimization, the initial neural network topologies are always fully connected (all weight and bias variables started with nonzero values), and network variables that become zero during optimization are not "pruned out" as sometimes suggested (Müller and Reinhardt, 1990). Our optimal networks (with close to a network global solution) indicate that sparse, input-hidden neuron connections (with some weights equal to zero) are often successful at predicting the contact outcome correctly for many sequences, while dense, input-hidden neuron connectivity only predicts the contact correctly for one, or a very small number, of sequences. These network design results indicate that weights that are initially zero, or become zero during back-propagation, should remain so in order to avoid unprofitable regions of network variable space.

In summary, the construction of neural networks for our 2D polypeptides with two amino acid types indicates that genuine learning strategies are present in the networks. It is evident from the above observations that the central Boolean function described in Figure 15-8 successfully maps a basic sequence-structure relationship, namely the nonbonded interaction potential in equations (13) and (14). When the central Boolean function (or nonbonded interaction) is not sufficient for accurately describing the contact value, additional hidden neurons are then required to understand the interplay between the nonbonded interactions *and* connectivity portions of the potential. In these cases, the additional hidden neurons may provide insight into the more general aspects of the protein folding problem. The degree of network connectivity necessary between the input and these additional hidden neurons, and how many sequences are affected by these dedicated neurons, may provide a quantitative means for delineating the strong and weak sequence-structure mappings anticipated in the poly-L-alanine study (Head-Gordon et al., 1992) of Section 3. The networks presented here for our simple 2D model have demonstrated that very few contacts rely on the full sequence as input, as indicated by the small degradation in performance when dedicated neurons that correctly predict one contact value for one sequence are eliminated. Hidden neurons with sparse, input-hidden neuron connectivity, which aid in the correct contact value prediction for many sequences, imply that good network designs might be able to reveal important residues that dictate the folding outcome. Finally, thoughtful neural network design may deconvolute the problem of multiple minima in the space of the network variables by isolating the relevant region with informed initial guesses for weights and biases.

5. The Relative Roles of Sequence and Environment on Conformational Preferences of Glycine and Alanine Oligomers

Our theoretical understanding of structure and energetics of biological macromolecules has largely relied on empirical protein/solvent potential energy functions (Hagler et al., 1974; Momany et al., 1974; Brooks et al., 1983; Weiner et al., 1986; Jorgensen and Tirado-Rives, 1988; Momany et al., 1990). Because such force fields are currently a critical algorithmic component of our constrained optimization approach outlined in Section 2, it is important that we assess the qualitative and quantitative features of our chosen objective function. Empirical protein energy functions are typically of the form shown in equation (2), where the parameters are derived from a combination of theoretical and experimental data on smaller monopeptides and dipeptides (ibid.). Therefore, an understanding of the observed conformational preferences of peptides in gas, aqueous, and protein environments, is a necessary first step toward interrogating the adequacy of small model peptides as structural analogs of larger biopolymers.

Ab initio electronic structure theory refers to the first-principles determination of an optimized electronic energy at a fixed nuclear configuration (Hehre et al., 1986). Specification of a molecular structure guess, total charge, and spin multiplicity, a set of atom-centered hydrogen-like functions that describe the electronic coordinates (basis set), and an approximation to the Schrödinger equation, comprise a practical application of the theory. The molecular orbital approximation (ibid.) (Hartree-Fock [HF] theory) has been shown to provide excellent descriptions of structure when covalent bond-making (-breaking) mechanisms are not present. Correlation corrections important for obtaining reliable relative energies can be treated with a hierarchy of theories such as the Møller–Plesset series (ibid.). The quantitative description of a molecular property of interest also requires a well-calibrated choice of basis set.

We have recently completed a high-level *ab initio* gas phase study (Head-Gordon et al., 1989, 1991) of α-(formylamino)ethanamide (Figure 15-9) and (S)-α-(formylamino)propanamide (Figure 15-1). They are closely related to blocked glycine and alanine dipeptide, respectively, where the free rotor methyls of the blocked species have been replaced by hydrogens. The conformational space of these molecules, henceforth referred to as GDA and ADA, is described by the backbone dihedral angles ϕ and ψ (Ramachandran et al., 1973). For peptide chains of at least four amino acids, particular ϕ, ψ values give rise to well-defined hydrogen bond patterns of secondary structure such as the α-helix, β-sheet, and turns (Pauling et al., 1951; Ramachandran et al., 1973). Whether specific secondary structures are energetically stabilized in dipeptides is more uncertain, since hydrogen bonding at the relevant values of ϕ and ψ is not possible for these short lengths.

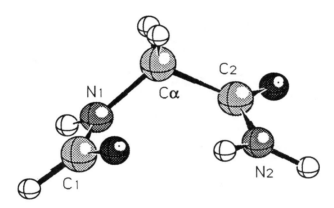

Figure 15-9. The structure of the global minimum conformer of α-(formylamino)ethanamide.

The HF method with the 6-31+G* basis (Hehre et al., 1972; Hariharan, 1974; Clark et al., 1983; Frisch et al., 1984a) is known to yield quantitatively accurate structures for chemical systems that hydrogen bond (Frisch et al., 1984b). However, given the N^4 (N = number of basis functions) computational cost dependence of the HF, such a basis is too large to evaluate a fully relaxed ϕ, ψ grid at a reasonable grid spacing. Instead, the small split valence 3-21G (Binkley et al., 1980) basis was chosen to generate (Frisch et al., 1990) the grid in 15° increments for GDA (Figure 15-10) and ADA (Figure 15-11), as a means for exploring the full conformational space. Based on the HF/3-21G grids, additional full geometry optimizations were performed (ibid.) at HF/3-21G to determine all stationary points; minima, saddle points, and maxima were characterized as such with HF/3-21G second derivative calculations (ibid.). The set of HF/3-21G stationary point structures were used in turn as starting guesses for full geometry optimization (ibid.) at the higher level of theory, HF/6-31+G*. Again, all stationary points were characterized by frequency calculations (ibid.). The best relative energies computationally feasible for systems of this size were determined with the MP2 (Hehre et al., 1986; Frisch et al., 1992) (first order) correlation method using the optimal HF/6-31+G* geometries.

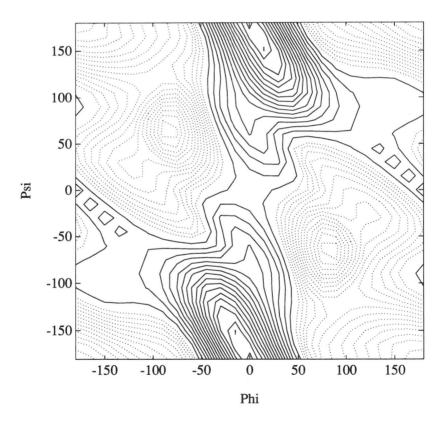

Figure 15-10. The gas phase (ϕ, ψ) surface of (S)-α-(formylamino)ethanamide generated by HF/3-21G. See Figure 15-2 for all other details.

The stationary points determined at HF/3-21G were found to be in good structural agreement (ϕ, ψ stationary point values) with the larger basis set

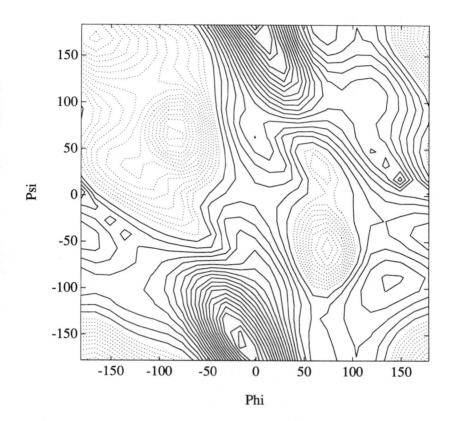

Figure 15-11. The gas phase (ϕ, ψ) surface of (S)-α-(formylamino)propanamide generated by HF/3-21G. See Figure 15-2 for all other details.

calculations, but the relative energies were deficient by comparison with the MP2 calculations. An important difference in the energetics is that shallow minima (stabilized by 1 kcal/mole or less) corresponding to secondary structure appearing on the HF/3-21G map, disappear altogether at the more reliable HF/6-31+G* level of theory for both GDA and ADA. We note that most gas phase molecular mechanics potentials show stable minima for secondary structure conformers, $\beta(\phi \simeq 120°, \psi \simeq 30°)$ and $\alpha_L(\phi \simeq 65°, \psi \simeq 35°)$ (Hagler et al., 1974; Momany et al., 1974; Brooks et al., 1983; Weiner et al., 1986; Jorgensen and Tirado-Rives, 1988). Instead, the quantitative electronic structure gas phase studies indicate that such small peptides do not intrinsically exhibit preferred stable secondary structures. Therefore, any stabilization of secondary structure conformers would strictly be a function of environment according to the highest levels of *ab initio* theory currently feasible for systems of this size.

The self-consistent reaction field (SCRF) (Onsager, 1936; Bonaccorsi et al., 1984; Tapia, 1991; Wong et al., 1991a, 1991b, 1992) method has been used to some advantage in describing the influence of a dielectric solvent on conformational equilibria of a variety of polar and ionic organic compounds. In this theory, the electric field of the solute polarizes the surrounding dielectric medium in such a way as to produce a reaction field, which in turn interacts with the charge distribution of the solute (Onsager, 1936). The primary approximations inherent in the theory are that the electrons of the solute are distinguishable from that of the (spatially) immediate surroundings of solvent, and that specific interactions (e.g., hydrogen bonding) with molecular solvent are unimportant (Bonaccorsi et al., 1984). These approximations are manifested by enclosing the solute in a low dielectric geometry such as a sphere, and replacing the surrounding molecular solvent with an appropriately chosen value of the dielectric constant.

We present here for the first time a preliminary report of HF/3-21G electronic structure calculations using the reaction-field model (Onsager, 1936) to represent the influence of aqueous environment on the full ϕ, ψ conformational space of GDA and ADA. Figures 15-12 and 15-13 provide solvent modified maps of GDA and ADA, respectively. At each ϕ, ψ grid point, a calculation of the electron density envelope (with subsequent scaling) of the optimized gas phase species was used to estimate the molecular volume (Bonaccorsi et al., 1984; Frisch et al., 1992). The spherical radius of the SCRF cavity was then defined by the cubed root of the volume, and adding 0.5 Å to account for the nearest approach of solvent molecules (Bonaccorsi et al., 1984). The maps depicted in Figures 15-12 and 15-13 were generated (Frisch et al., 1992) by partial geometry relaxation at fixed values of ϕ and ψ, in the presence of a perturbing reaction-field Hamiltonian using cavity radii values computed by the above procedure.

A visual comparison of the gas phase and solvent modified maps indicate that the influence of a solvent dielectric of 80 results in the appearance of the right-handed α-helix minimum, and a deepening of the left-handed α-helix, minimum. We anticipate that these minima should remain at higher levels of HF/SCRF because they are stabilized by at least 2 kcal/mole or more at the HF/3-21G level. An evident failure of the reaction-field representation of water is the strong maintenance of the C7 hydrogen-bonding conformers, whose influence should wane in a molecular model of solvent where alternative means for hydrogen bonding are made available. In fact, we anticipate that the β conformer should become a stable structure in a molecular aqueous solvent, since this region is quite low in energy.

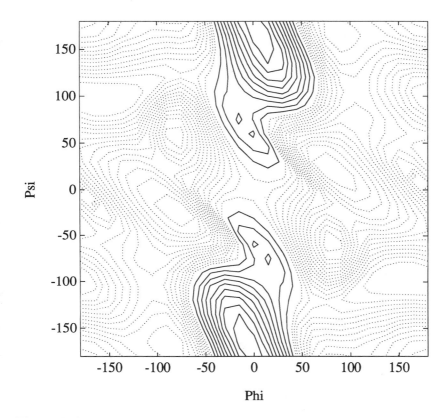

Figure 15-12. The solvent modified (ϕ, ψ) surface of (S)-α-(formylamino)ethanamide generated by HF/3-21G using a reaction-field model of aqueous environment. Each grid point involved the calculation of a cavity radius. See Figure 15-2 and text for all other details.

Nonetheless, these calculations seem to indicate that secondary structure formation in these small peptides is not intrinsic, but strictly results from the influence of environment only. Whether different side chains and/or longer peptides show a similar dependence on environment for stabilizing secondary structure minima remains an open question. Nonetheless, these results do indicate that an appropriate treatment of aqueous environment is a necessary one for obtaining quantitative descriptions of protein structure.

6. Conclusions and Future Directions

We have implemented an optimization strategy for greatly simplifying polypeptide and protein potential energy hypersurfaces in order to retain

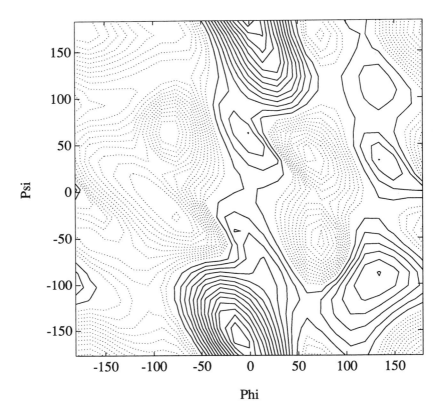

Figure 15-13. The solvent modified (ϕ, ψ) surface of (S)-α-(formylamino)propan-amide generated by HF/3-21G using a reaction-field model of aqueous environment. Each grid point involved the calculation of a cavity radius. See Figure 15-2 and text for all other details.

only one conformationally distinct minimum corresponding to a predicted native structure. The strategy involves the deformation of the objective function hypersurface, currently an empirical protein force field, in such a way that the native structure minimum forms the dominant basin on the surface. We have adapted our optimization strategy to use neural networks as a guide for designing penalty function (constraint) parameters that retain only the native globular protein minimum. The following flow diagram outlines the algorithmic components for predicting tertiary structure in any protein:

(1) amino acid sequence $\rightarrow$ neural networks;

(2) neural networks → 2° and/or 3° structure penalty parameters or constraints → define modified surface;

(3) minimization on modified surface using extended conformer as starting structure → regenerate original objective function;

(4) minimization on unmodified surface using the minimized structure found from (3) as the starting structure → converge structure to strict tolerance;

(5) predicted structure determined with atomic resolution.

Thus, the most ambitious scenario is a method that, for any polypeptide or protein, predicts atomic resolution structures using the amino acid sequence as sole input. In order to realize such an ambition, it is important that we significantly improve the three main components of the outlined algorithm: (1) neural network predictions of penalty functions (constraints) that dictate native structure, (2) incorporation of a sufficient number of quality penalties or constraints to isolate the native structure minimum, and (3) the desired use of quantitative protein and solvent force fields.

As a first step in this direction, we have completed some important simple model studies that have unveiled several interesting aspects of the underlying physics of protein folding. The conclusion that the problem of protein folding may depend only on a specific subset of amino acids at different points on the folding pathway was broadly confirmed with the poly-L-alanine hypothesis for the final native-structure minimum, and further verified by model neural networks that predict tertiary structure perfectly for model 2D chemical systems. These perfectly performing networks also strongly hinted at the fact that systematic network design can genuinely impact on improved protein structure prediction of real proteins with full sequence diversity by overcoming database inadequacies, overtraining, and the multiple minima problem in the space of the network variables. Finally, electronic structure calculations of small peptides (often used as structural models for globule proteins) have shown that secondary structure formation is a consequence of solvent environment, at least for glycine and alanine dipeptide analogs. Thus, improved quantitative algorithms of protein structure prediction must ultimately contend with the description of the protein's environment.

In keeping with our philosophy that simple model insights must converge with a structure prediction algorithm, our current research involves the transfer of learned technology from such models to real proteins. For example, we are currently assessing whether the central Boolean function architecture for our simplest 2D model can be successfully transferred to the full complexity of real proteins. We will order the twenty amino acids on a hydrophobic scale ranging from +1 (most hydrophobic) to −1 (least hydrophobic). In this case, the central Boolean architecture will at least distinguish between the unlikely hydrophobic–polar contacts from the polar–polar and hydrophobic–hydrophobic contacts, so that other hidden neurons can address the subtleties of the relative nature of the attractive non-bonded interactions such as hydrogen bonding and salt bridge formation. If a different representation is chosen where interactions are self-repelling (net electrostatic monopole assignments of the amino acids, for example) then the topology of the Boolean function would still be used, but with different weight assignments. Furthermore, we believe that our hand-designed networks will overcome some aspects of the mathematical optimization difficulties for finding optimal network solutions. We will emphasize sparse input-hidden layer connectivity as initial guesses, and prune out network variables tending toward zero, for training by back-propagation optimization algorithms. The effect of finite polypeptide length input "windows" can also be investigated with our simple model database of pentamers, hexamers, and heptamers.

Future studies will include increasing the model complexity of the neural networks and examining the problem of database degradation. We hope to judiciously incorporate more realistic chemical features into our polypeptide model in order to establish a trend of network architecture (number of hidden layers and number of hidden units within a given layer, optimal central Boolean functions, etc.) as a function of chemical complexity (the number of amino acid types and polypeptide lengths). Furthermore, because we have a database that is complete and have derived network architectures from it that are well understood, this should allow a systematic exploration of what degree of prediction accuracy is possible when particular structural classes or sequence homologies are exploited in the database, and to what degree either representation is useful. Network design features that emerge from these model studies will also be ascertained for their performance on genuine protein databases.

In the case of environmental influences on conformations of peptides, we anticipate further computational studies that allow for more complex amino acid side chains, longer polypeptide lengths, and better aqueous solvent descriptions in order to quantitate the structural preferences of real proteins. Inclusion of a molecular solvent, as in an electronic structure supermolecule approach (Madura and Jorgensen, 1986; Bash et al., 1987; Williams, 1987), would certainly overcome the primary failure of the simple reaction-field model—the lack of explicit hydrogen bonding interactions—for species such as peptides. In order to ensure that the additional studies cited above for peptide conformational preferences are feasible, alternative quantitative methods that are more inexpensive than HF need to be explored. Density functional theory, whose accuracy is comparable to MP2 and whose computational cost rises as N^2 to N^3, may well make such studies viable.

Finally, further progress in neural network design and improved force-field descriptions must funnel into a structure prediction algorithm. Successful constrained optimization certainly relies on some threshold of performance of the predictive phase of the algorithm in order to find low-energy minima. However, exhaustive numbers of constraints may not be necessary for sufficiently quantitative descriptions of the potential energy function. Where this constraint threshold exists for materials such as proteins would necessarily be a point of further exploration.

Acknowledgments. I would like to thank F.H. Stillinger, M.H. Wright, D.M. Gay, and L.W. Jelinski for collaborations and discussions, and for a terrific postdoctoral experience. I also thank M. Head-Gordon for careful reading of the manuscript.

REFERENCES

Bash PA, Field MJ, Karplus M (1987): Free energy purturbation method for chemical reactions in the condensed phase: A dynamical approach based on a combined quantum and molecular mechanics force field. *J Am Chem Soc* 109:8092

Bengio Y, Pouliot Y (1990): Efficient recognition of immunoglobulin domains from amino acid sequences using a neural network. *Computer Applications in the Biosciences* 6:319–324

Binkley JS, Pople JA, Hehre WJ (1980): Self-consistent molecular orbital methods. 21. Small split valence basis sets for first-row elements. *J Am Chem Soc* 102:939–947

Bohr H, Bohr J, Brunak S, Cotterill RMJ (1990): A novel approach to prediction of the three-dimensional structures of protein backbones by neural networks. *FEBS Lett* 261:43–46

Bonaccorsi R, Palla P, Tomasi J (1984): Conformational energy of glycine in aqueous solutions and relative stability of the zwitterionic and neutral forms. An *ab initio* study. *J Am Chem Soc* 106:1945–1950

Brooks BR, Bruccoleri RE, Olafson BD, States DJ, Swaminathan S, Karplus M (1983): CHARMM: A program for macromolecular energy, minimization, and dynamics calculations. *J Comp Chem* 4:187–217

Bryngelson JD, Wolynes PG (1987): Spin glasses and the statistical mechanics of protein folding. *Proc Natl Acad Sci USA* 84:7524–7528

Chan HS, Dill KA (1991): Polymer principles in protein structure and stability. *Annu Rev Biophys Chem* 20:447–490

Chou PY, Fasman GD (1974): Prediction of protein conformation. *Biochem* 13:222–275

Churchland PS, Sejnowski TJ (1992): *The Computational Brain*. Cambridge: MIT Press

Clark T, Chandrasekhar J, Spitznagel GW, Schleyer PVR (1983): Efficient diffuse functions augmented basis sets for anion calculations. III. The 3-21G basis set for first row elements, lithium to fluorine. *J Comp Chem* 4:294–301

Deisenhofer J, Steigemann W (1975): Crystallographic refinement of the structure of bovine pancreatic trypsin inhibitor at 1.5 Å resolution. *Acta Crystallogr, Sect B* 31:238

Eisenberg D, Bowie JU, Luthy R, Choe S (1992): Three-dimensional profiles for analysing protein sequence structure relations. *Faraday Discussions of the Chem Soc*, 25–34

Eriksson AE, Baase WA, Zhang X-J, Heinz DW, Blaber M, Baldwin EP, Matthews BW (1992): Response of a protein structure to cavity-creating mutations and its relation to the hydrophobic effect. *Science* 255:178–183

Ferran EA, Ferrara P (1992): Clustering proteins into families using artificial neural networks. *Computer Applications in the Biosciences* 8:39–44

Friedrichs MS, Goldstein RA, Wolynes PG (1991): Generalized protein tertiary structure recognition using associative memory hamiltonians. *J Mol Biol* 222: 1013–1034

Frisch MJ, Head-Gordon M, Foresman JB, Trucks GW, Raghavachari K, Schlegel HB, Robb MA, Binkley JS, Gonzalez C, Defreez DJ, Fox DJ, Whiteside RA, Seeger R, Melius CF, Baker J, Kahn LR, Stewart JJP, Fluder EM, Topiol S, Pople JA (1990): *Gaussian 90*, Gaussian Inc., Pittsburgh, PA

Frisch MJ, Pople JA, Binkley JS (1984a): Self-consistent molecular orbital methods. 25. Supplementary functions for Gaussian basis sets. *J Chem Phys* 80:3265–69

Frisch MJ, Pople JA, Del Bene JE (1984b): Molecular orbital study of the dimers $(AH_n)_2$ formed from ammonia, water, hydrogen fluoride, phosphine, hydrogen sulfide, and hydrochloric acid. *J Phys Chem* 89:3664–3669

Frisch MJ, Trucks GW, Head-Gordon M, Gill PMW, Wong MW, Foresman JB, Johnson BG, Schlegel HB, Robb MA, Replogle ES, Gomperts R, Andres JL, Raghavachari K, Binkley JS, Gonzalez C, Martin RL, Fox DJ, Defreez DJ, Baker J, Stewart JJP, Pople JA (1992): *Gaussian 92*, Revision A. Gaussian Inc., Pittsburgh, PA

Garnier J, Osguthorpe DJ, Robson B (1978): Analysis of accuracy and implications of simple methods for predicting secondary structure of globular proteins. *J Mol Biol* 120:97–120

Gibrat JF, Garnier J, Robson B (1987): Further developments of protein secondary structure prediction using information theory. *J Mol Biol* 198:425–443

Godzik A, Skolnick J (1992): Sequence structure matching in globular proteins: application to supersecondary structure and tertiary structure determination. *Proc Natl Acad Sci USA* 89:12098–12102

Goldstein RA, Luthey-Schulten ZA, Wolynes PG (1992): Protein tertiary structure recognition using optimized Hamiltonians with local interactions. *Proc Natl Acad Sci USA* 89:9029–9033

Hagler AT, Huler E, Lifson S (1974): Energy functions for peptides and proteins. I. Derivation of a consistent force field including the hydrogen bond from amide crystals. *J Am Chem Soc* 96:5319–5327

Hariharan PC, Pople JA (1974): Effect of d functions on molecular orbital energies for hydrocarbons. *Mol Phys* 27:209–14

Hayward S, Collins JF (1992): Limits on α-helix prediction with neural network models. *Proteins—Structure, Function and Genetics* 14:372–381

Head-Gordon T, Head-Gordon M, Frisch MJ, Brooks CL, Pople JA (1989): A theoretical study of alanine dipeptide and analogues. *Int J Quant Chem Biol Symp* 16:311–322

Head-Gordon T, Head-Gordon M, Frisch MJ, Brooks CL, Pople JA (1991): Theoretical study of blocked glycine and alanine peptide analogues. *J Am Chem Soc* 113:5989–5997

Head-Gordon T, Stillinger FH (1993a): Toward optimal neural networks for protein structure prediction. *Phys Rev E* 48. (In press.)

Head-Gordon T, Stillinger FH (1993b): Predicting polypeptide and protein structures from amino acid sequence: Antlion method applied to melittin. *Biopolymers* 33:293–303

Head-Gordon T, Stillinger FH, Arrecis J (1990): A strategy for finding classes of minima on a hypersurface implications for approaches to the protein folding problem. *Proc Natl Acad Sci USA* 88:11076–11080

Head-Gordon T, Stillinger FH, Wright MH, Gay DM (1992): Poly-L-alanine as a universal reference material for understanding protein energies and structures. *Proc Natl Acad Sci USA* 89:11513–11517

Hehre WJ, Ditchfield R, Pople JA (1972): Self-consistent molecular orbital methods. XII. Further extensions of Gaussian-type basis sets for use in molecular orbital studies of organic molecules. *J Chem Phys* 56:2257–61

Hehre WJ, Radom L, Schleyer PVR, Pople JA (1986): *Ab initio Molecular Orbital Theory*. New York: Wiley

Hendrickson WA, Teeter MM (1981): Structure of the hydrophobic protein crambin determined directly from the anomalous scattering of sulfur. *Nature* 290:107–113

Hertz J, Krogh A, Palmer RG (1991): *Introduction to the Theory of Neural Computations*. Redwood City, CA: Addison-Wesley

Hirst JD, Sternberg MJE (1991): Prediction of ATP-binding motifs a comparison of a perceptron type neural network and a consensus sequence method. *Prot Eng* 4:615–623

Hirst JD, Sternberg MJE (1992): Prediction of structural and functional features of protein and nucleic acid sequences by artificial neural networks. *Biochem* 31:7211–7218

Holley LH, Karplus M (1989): Protein secondary structure prediction with a neural network. *Proc Natl Acad Sci USA* 86:152–156

Jorgensen WL, Tirado-Rives J (1988): The OPLS potential functions for proteins. Energy minimizations for crystals of cyclic peptides and crambin. *J Am Chem Soc* 110:1657–1666

Kabsch W, Sander C (1983): Dictionary of protein secondary structure: Pattern recognition of hydrogen-bonded and geometrical features. *Biopolymers* 22: 2577–2637

Kartha G, Bello J, Harker D (1967): Tertiary structure of ribonuclease. *Nature* 213:862–865

Kneller DG, Cohen FE, Langridge R (1990): Improvements in protein secondary structure prediction by an enhanced neural network. *J Mol Biol* 214:171–182

Kolinski A, Skolnick J, Yaris R (1988): Monte Carlo simulations on an equilibrium globular protein folding model. *Proc Natl Acad Sci USA* 83:7267–7271

Lee C, Subbiah S (1991): Prediction of protein side-chain conformation by packing optimization. *J Mol Biol* 217:373–388

Levin JM, Robson B, Garnier J (1986): An algorithm for secondary structure determination in proteins based on sequence similarity. *FEBS Lett* 205:303–308

Levitt M (1976): A simplified representation of protein conformations for rapid simulation of protein folding. *J Mol Biol* 104:59–107

Levitt M (1978): Conformational preferences of amino acids in globular proteins. *Biochemistry* 17:4277–4285

Levitt M, Warshel A (1975) Computer simulation of protein folding. *Nature* 253: 694–698

Lim VI (1974): Algorithms for prediction of α-helical and β-structural regions in globular proteins. *J Mol Biol* 88:873–894

Madura JD, Jorgensen WL (1986): Ab initio and monte carlo calculations for a nucleophilic addition reaction in the gas phase and in aqueous solution. *J Am Chem Soc* 108:2517

McGregor MJ, Flores TP, Sternberg MJE (1989): Prediction of β-turns in proteins using neural networks. *Prot Eng* 2:521–526

Momany FA, Carruthers LM, McGuire RF, Scheraga HA (1974): Intermolecular potentials from crytal data. III. Determination of empirical potentials and application to the packing configurations and lattice energies in crystals of hydrocarbons, carboxylic acids, amines, and amides. *J Phys Chem* 78:1595–1620

Momany FA, Klimkowski VJ, Schafer L (1990): On the use of conformationally dependent geometry trends from *ab initio* dipeptide studies to refine potentials for the empirical force field CHARMM. *J Comp Chem* 11:654–662

Müller B, Reinhardt J (1990): *Neural Networks : An Introduction.* Berlin, Heidelberg: Springer-Verlag

Muskal SM, Kim SH (1992): Predicting protein secondary structure content a tandem neural network approach. *J Mol Biol* 225:713–727

O'Neill KT, DeGrado WF (1990): A thermodynamic scale for the helix-forming tendencies of the commonly occurring amino acids. *Science* 250:646–651

Onsager L (1936): Electric moments of molecules in water. *J Am Chem Soc* 58:1486–1493

Pauling L, Corey RB, Branson HR (1951): Structure of proteins two hydrogen-bonded helical configurations of the polypeptide chain. *Proc Natl Acad Sci USA* 37:205–211

Press WH, Flannery BP, Teukolsky SA, Vetterling VT (1986): *Numerical Recipes* Cambridge: Cambridge University Press

Ptitsyn OB, Finkelstein AV (1989): Prediction of protein secondary structure based on physical theory. *Protein Eng* 2:443–447

Qian N, Sejnowski TJ (1988): Predicting the secondary structure of globular proteins using neural network models. *J Mol Biol* 202:865–884

Ramachandran GN, Ramakrishnan C, Sasisekharan V (1973): Stereochemistry of polypeptide chain configurations. *J Mol Biol* 7:95–99

Rooman MJ, Wodak SJ (1988): Identification of predictive sequence motifs limited by protein structure database size. *Nature* 335:45–49

Scheraga HA (1992): Some approaches to the multiple-minima problem in the calculation of polypeptide and protein structures. *Int J Quant Chem* 42:1529–1536

Shakhnovich E, Farztdinov G, Gutin AM, Karplus M (1991): Protein folding bottlenecks a lattice monte-carlo simulation. *Phys Rev Lett* 67: 1665

Shang H, Head-Gordon T (1994): Stabilization of helices in glycine and alanine dipeptide in a reaction field model of solvent. *J Am Chem Soc* 116:1528–1532

Stillinger FH, Head-Gordon T, Hirschfeld CL (1993): Toy model for protein folding. *Phys Rev E* (In press.)

Stolorz P, Lapedes A, Xia Y (1992): Predicting protein secondary structure using neural networks and statistical methods. *J Mol Biol* 225:363–377

Tainer JA, Getzoff ED, Beem KM, Richardson JS, and Richardson DC (1982): Determination and analysis of the 2 Å structure of copper, zinc superoxide dismutase. *J Mol Biol* 160:181–217

Tapia O (1991): On the theory of solvent-effect representation. 1. A generalized self-consistent reaction field theory. *J Mol Struct (Theochem)* 226:59–72

Terwilliger TC, Eisenberg D (1982): The structure of melittin. *J Biol Chem* 257: 6016–6022

Vieth M, Kolinski A (1991): Prediction of protein secondary structure by an enhanced neural network. *Acta Biochimica Polonica* 38:335–351

Weiner SJ, Kollman PA, Nguyen DT, Case DA (1986): An all atom force field for simulations of proteins and nucleic acids. *J Am Chem Soc* 106:230–252

Wilcox GL, Poliac M, Liebman MN (1990): Neural network analysis of protein tertiary structure. *Tetrahedron Comput Methodol* 3:191–211

Williams IH (1987): Theoretical modeling of specific solvation effects upon carbonyl addition. *J Am Chem Soc* 109:6299

Wilmanns M, Eisenberg D (1993): Three-dimensional profiles from residue-pair preferences identification of sequences with beta/alpha-barrel fold. *Proc Natl Acad Sci USA* 90:1379–83

Wong MW, Frisch MJ, Wiberg KB (1991a): Solvent effects. 1. The mediation of electrostatic effects by solvents. *J Am Chem Soc* 113:4776–4782

Wong MW, Wiberg KB, Frisch MJ (1991b): Solvent effects. 3. Tautomeric equilibria of formamide and 2-pryidone in the gas phase and solution an *ab initio* scrf study. *J Am Chem Soc* 114:1645–1652

Wong MW, Wiberg KB, Frisch MJ (1992): Solvent effects. 2. Medium effect on the structure, energy, charge density, and vibrational frequencies of sulfamic acid. *J Am Chem Soc* 114:523–529

Zhang X-J, Baase WA, Matthews BW (1991): Toward a simplification of the protein folding problem a stabilizing polyalanine α-helix engineered in T4 lysozyme. *Biochem* 30:2012–2017

Zwanzig R, Szabo A, Bagchi B (1992): Levinthals paradox. *Proc Natl Acad Sci USA* 89:20–22

16

The Role of Interior
Side-Chain Packing in
Protein Folding and Stability

James H. Hurley

1. Introduction

A nearly universal feature of the structures of globular proteins is a buried core consisting of hydrophobic side chains (Chothia, 1976). The favorable free-energy of transfer of these side chains from a solvent-accessible to solvent-inaccessible state is considered the major driving force for the folding of globular proteins in aqueous solution (Dill, 1990). The large hydrophobic, buried side chains are key determinants of protein structure and stability (Alber et al., 1987; Bowie et al., 1990; Shortle et al., 1990). These side chains tend to be tightly packed (Richards, 1974) and more rigid than the rest of the protein. Steric packing interactions involving these side chains may provide constraints for protein folding (Richards, 1974). The purpose of this review is to summarize the implications of recent experiments for the role of packing interactions in protein folding and stability.

Site-directed mutagenesis, coupled with accurate thermodynamic measurements and high-resolution x-ray crystallography, has made possible tremendous advances in understanding the roles of particular intermolecular forces in protein stability. Recent reviews have covered many of the applications of these techniques to protein folding and stability in general (Alber, 1989; Dill, 1990; Pace, 1990; Creighton, 1991; Jaenicke, 1991; Matthews, 1991; Shortle, 1992), but the field has grown so rapidly that an in-depth review can now be devoted entirely to a single type of interaction. Several recent papers describing three-dimensional structures and thermal stabilities of sets of multiple-site mutants provide the most direct look to

The Protein Folding Problem and Tertiary Structure Prediction
K. Merz, Jr. and S. Le Grand, Editors
© Birkhäuser Boston 1994

date at the effects of altering packing arrangements in the hydrophobic core. The use of synthetic amino acids incorporated by peptide synthesis or mischarged tRNAs has done away with the limitation of twenty naturally occurring amino acids, and made possible unique strategies for probing packing arrangements. A truly quantitative theory of protein stability based on first principles remains elusive, but several computational studies have made inroads towards an accurate description of core packing.

"Packing interactions" and "steric complementarity" are not well-defined intermolecular forces, and are used to mean different things by different workers. In some cases, the terms may refer purely to shape complementarity between impenetrable objects. The terms as used here will usually refer to the intermolecular forces of steric repulsion and van der Waals attraction, commonly represented by the Lennard–Jones potential. Closely coupled degrees of freedom, such as bond and angle deviations, are also factored into some interpretations. The terms will generally be used in the same sense as in the study in question at the moment, with significant differences in meaning noted explicitly.

This review is not intended to address the role and magnitude of the hydrophobic effect as it contributes to protein stability. Nor is it intended to discuss the nature of the experimental methodologies used, or their relative merits and weaknesses. Protein stability depends on the equilibrium between folded and unfolded states, but this review will concentrate entirely on interactions in the folded state. Specific packing interactions are almost by definition a property of the folded state, and detailed structural information is obtainable only for this state.

2. Interpretation of Protein Structures: Evaluating Side-Chain Packing "Quality"

The structure and stability of a protein result from a balance of interactions. In principle, the three-dimensional structure could be calculated from first principles given the amino acid sequence, and the stability could be calculated from the structure. Calculation of the structure and energetics of the hydrophobic core is simply a subset of this problem. *De novo* calculation of structure and stability is not yet possible. Several semi-quantitative attributes of core-packing arrangements and sequences have been developed that partially circumvent the difficulties of *de novo* structure prediction and energy calculations from first principles. This section is devoted to analysis based on geometry alone.

Packing density was the earliest attribute of the protein core to be analyzed in detail (Richards, 1974, 1977, 1985). Packing density is defined as the ratio of the volume enclosed by the van der Waals envelope of a given molecule to the actual volume of space that it occupies. The maximum packing density for hard spheres of uniform size in three dimensions is 0.74 (Richards, 1974). Richards has developed procedures for assigning occupied volumes to atoms and groups within proteins. The simplest procedure is to locate planes bisecting atom-to-atom vectors. The planes are then used to construct the smallest possible polyhedron enclosing each atom. These polyhedra will fill space exactly. Richards also developed a modified procedure to improve volume calculation for complex molecules such as proteins, in which the van der Waals radii of different atoms are non-equal. The planes are placed perpendicular to the atom-to-atom vectors based on the ratio of the radii of the two atoms, and do not in general bisect the vector. All space is not filled exactly in the modified treatment, but in practice the error is not serious. Gellatly and Finney (1982) and Richmond (1984) have developed more mathematically rigorous volume calculation techniques, but the major conclusions regarding packing density in proteins are unaffected by these improvements. A simplified approach to packing has been developed for use in molecular modeling and structure prediction (Gregoret and Cohen, 1990), but the more elaborate approaches are better suited to cases where an accurate experimental structure is available.

Based on calculations for proteins with structures known at high resolution, the packing density in the interior of the protein is about 0.75, comparable to the packing density in crystals of small organic molecules (Richards, 1974). The value of the packing density depends strongly on the set of van der Waals radii used. In contrast, the occupied volume calculation depends weakly on the radii if the modified procedure is used, and even then only if the ratios of radii are changed significantly. Packing density ratios based on a consistent set of radii or mean occupied volumes for particular groups or atom types are thus equally useful measures of the relative denseness of a packing arrangement. The high packing density of protein interiors suggests that packing arrangements are tightly constrained, and may in turn tightly constrain the way in which a protein can fold.

An alternative way to look at the quality of a packing arrangement is to consider the area or volume of buried cavities, rather than at the volume occupied in regions that are filled. The molecular surface calculation of Connolly (1983) assigns van der Waals radii to each atom. A probe sphere with a defined radius, typically 1.2 to 1.4 Å is then rolled over the van der Waals surface. The molecular surface is traced by the surface of the probe. Any interior region that is wholly excluded from bulk solvent but large enough to accommodate the probe is considered a cavity, and the area and volume of the cavity can be calculated once the surface is defined. It is presumed that the occurrence of large cavities in a protein is destabilizing. Because the attractive dispersion interaction is a short-range force, it is expected that an atom lining the cavity wall will experience a much weaker favorable interaction with atoms across the cavity than with atoms in immediate van der Waals contact. An analogy can be drawn to the unfavorable free energy of formation of a macroscopic bubble in a liquid, the cost per unit area of which is defined as the surface tension of that liquid. Sharp et al. (1991) have proposed a model for relating macroscopic surface tension to microscopic cavity formation. Packing density as normally calculated by Richards' procedures (1974, 1977, 1985) is related to the existence of cavities because cavity volumes are normally divided between neighboring residues, rather than counted separately. This decreases the calculated packing density for atoms bordering a cavity. An unresolved problem in cavity analysis is the strong dependence of area, volume, and even the existence of a cavity, on the choice of probe size.

The tertiary template concept of Ponder and Richards (1987) provides a third geometric approach to packing analysis. The Ponder and Richards model consists of a rigid main chain, a group of flexible side chains placed in idealized conformations, and a hard-sphere potential. A central aspect of the method is a library of side-chain rotamers. Of all possible combinations of conformations about each single bond in a side chain, only a fraction are found with significant frequency in accurately determined protein structures. Lists of commonly occurring sets of conformations, termed "rotamers," have been compiled. The rigid main-chain backbone is used to place every possible side chain in each of the tabulated rotamers. Each possible rotamer is evaluated for steric overlap with the main chain and with nonmoving side chains. All possible sets of rotamers judged acceptable by this first screen are then tested for steric overlap with each other. Amino acid sequences for which an acceptable rotamer set can be generated are added to the list of sequences compatible with a given main-chain structure. The list can be evaluated by other criteria, such as the ability of the sequence to fill a packing volume close to the volume available in the native structure, or the absence of buried charges or unpartnered hydrogen-bonding groups.

The major assumptions of the tertiary template model are that the main chain will remain rigid in response to mutations, and that side chains will occupy conformations close to ideal rotamers. Modest relaxations of the main chain and shifts of side chains to strained conformations can move side-chain atoms by an angstrom or more. A relatively low energy 15° rotation in χ_1 can produce an atomic shift of more than 1.0 Å at the end of a long side chain. Acceptance of a sequence in the tertiary template model depends on the absence of any inter-atomic contacts closer than the sum of the van der Waals radii involved. In practice, this criterion severely restricts the number of amino acid sequences at interior positions that are compatible with a given tertiary structure. Inclusion of explicit hydrogen atoms increases the restrictiveness of the constraint still further. In this model, only a handful of sequences can fold into a functional three-dimensional structure. Just 14 to 34 acceptable sequences of uncharged residues were found that could be placed at five to seven interior positions in crambin, rubredoxin, and scorpion neurotoxin. In this model, packing would appear to be a major constraint both on protein folding and on possible pathways for the molecular evolution of proteins. A major focus of experiments on core packing has been to determine how severe packing restrictions are in practice.

The assumption of a rigid model has important implications for the restrictiveness of the packing constraint on folding. A given sequence might actually pack with no close contacts in a mutant structure that had accommodated the change with a combination of main-chain and side-chain adjustments. This same sequence would be disallowed in the calculation if close contacts occurred in the model based on a rigid main chain and idealized side-chain rotamers. It is impossible to anticipate every possible way in which the structure could relax in response to a mutation, but *ad hoc* approaches can be used to tune the restrictiveness of the algorithm. Hurley et al. (1992) found that by varying the amount by which atomic radii were reduced, it was possible to derive a set of reduced radii that lead to reasonable agreement with experimental data on the effects of mutations. It is found empirically that radii must be reduced more for side-chain atoms (0.6 Å) than for main-chain atoms (0.4 Å). This finding seems consistent with the common-sense notion that side chains will, on average, be more flexible than the main chain.

3. Dissecting Core Packing: Role of Individual Side Chains

Experiments aimed at evaluating the role of packing interactions are all based on creating variants of proteins with known structure, either by site-directed mutagenesis or chemical synthesis. Single mutations at a set of buried positions are used to evaluate the dependence of a given class of mutation on the local environment. More relevant to studies of packing arrangements, multiple mutations at a set of positions near each other in the three-dimensional structure can be used to evaluate the strength of interaction between positions. At a minimum, the biological activity of each variant is assessed. Ideally, one would like to have accurate thermodynamic parameters and a high-resolution crystal structure for each possible combination of mutations at a set of neighboring positions, although this is not always possible or practical for a given protein.

The earliest site-directed mutagenesis studies of core residues involved single replacements of large hydrophobic side chains, with the aim of evaluating the contribution of the hydrophobic effect to stability. At a given position, the change in free energy of folding, $\Delta\Delta G$, usually correlates well with the free energy of transfer of model compounds from aqueous to organic solvent (Yutani et al., 1987; Matsumura et al., 1988; Kellis et al., 1988; Sandberg and Terwilliger, 1989; Shortle et al., 1990; Eriksson et al., 1992a; Hellinga et al., 1992). There is approximately five-fold variation in the slope of a plot of $\Delta\Delta G$ vs. free energy of transfer for mutations in different proteins and at different positions in the same protein, however.

The crystal structures of several Leu $\rightarrow$ Ala and Phe $\rightarrow$ Ala mutant T4 lysozymes representing a range of effects on stability have resolved the apparent discrepancies in previous results (Eriksson et al., 1992a). The structure of the most destabilizing mutation, in which replacing a buried Leu with Ala destabilizes the protein by 5.0 kcal/mol, does not show significant relaxation. This mutation produces an increase in cavity surface area of 140 Å^2 compared to wild-type. Less destabilizing mutations at buried positions show a greater degree of relaxation and smaller increases in cavity area. McRee et al. (1990) have also reported structural relaxation in response to a cavity-creating mutant in Cu, Zn superoxide dismutase. There is no evidence for buried water molecules in any of the cavities created. The destabilizing effect of a large $\rightarrow$ small mutation at a buried position is strongly correlated with the surface area of the cavity remaining. The plot of $\Delta\Delta G$ vs. cavity surface area has a slope of about 20 cal mol$^{-1} \cdot$ Å^{-2} and an intercept of about 2.0 kcal/mol (Figure 16-1). The intercept is close to the value expected if the free energy of transfer between aqueous and organic solvent (uncorrected for molecular volume differences) were the appropriate model for the effect of hydrophobic side chains on protein stability and the contribution of other factors, such as side-chain conformational entropy, were minor. More relevant to this review, this study provides a direct measure of the cost of forming a cavity in the hydrophobic core of a protein. The observed cost is about half of the macroscopic surface tension of organic solvents such as benzene. This is roughly what would be expected for a cavity of atomic dimensions based on the curvature-dependent model of Sharp et al. (1991).

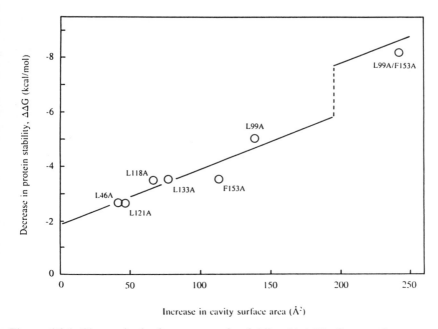

Figure 16-1. Change in the free energy of unfolding ($\Delta\Delta G$) of mutant lysozymes relative to wild-type plotted as a function of the cavity surface area created by the amino acid substitution(s). The equation of the straight line that best fits the data for the single mutants is $\Delta\Delta G = a + b\Delta A$ where $a = -1.9$ kcal/mol, $c = -0.020$ kcal/(mol Å^2), and ΔA is the increase in cavity surface area. In the vicinity of L99A/F153A an additional -1.9 kcal/mol has been added to $\Delta\Delta G$ to reflect the fact that the double mutant is roughly equivalent to two Leu-to-Ala replacements. Reprinted from Eriksson et al. (1992a) with permission from the authors and *Science*; © 1992 by the AAAS.

Since cavities destabilize proteins, presumably filling cavities might provide a means for stabilizing proteins. Karpusas et al. (1989) made two mutations, A129V and L133F, in T4 lysozyme designed to fill the two largest cavities in the wild-type structure. High resolution crystal structures were determined, and the side chains introduced by the two mutations filled the cavities essentially as intended in the design. The mutant proteins were less stable than wild-type by 0.8 and 0.3 kcal/mol, respectively, rather than more stable. Both side chains make unfavorable close contacts of about 3.0 Å to main-chain N and O atoms, and both have strained dihedral angles. There is no experimentally significant shift in the main chain to relieve close contacts. In this case, the energetic cost of moving the backbone is presumably greater than the cost of tolerating close contacts. The introduced side chains occupy standard volumes based on Richard's modified Voronoi procedure. These side chains appear to fit primarily at the expense of the pre-existing cavities they were designed to fill, and are not forced to pack at higher-than-normal density. This may illustrate a limitation of the packing density concept, in that no distinction is made between volumes of normal and distorted shape. Unfavorable close contacts may not be manifested in the packing volume if there is compensatory breathing room near some other part of the group in question. This pioneering study of the consequences of mutating smaller core residues to larger ones illustrates the complex interplay of the hydrophobic effect, cavitation, strain, and steric repulsion in determining the net effect of the mutation.

More recently, some stabilizing smaller side-chain → larger side-chain cavity-filling mutations have been isolated. A thermostable glyceraldehyde 3-phosphate dehydrogenase mutant, G316A, has been reported. Based on modeling, this mutant is predicted to stabilize the protein by filling a pre-existing cavity (Ganter and Pluckthun, 1990). Modest stabilizing effects of mutations intended to fill cavities have also been reported in *Escherichia coli* ribonuclease H (Kimura et al., 1992) and in *B. stearothermophilus* neutral protease (Eijsink et al., 1992). Other stabilizing cavity-filling mutants have been isolated from sets of multiple mutants (Wilson et al., 1992; Lim and Sauer, 1991) described later in this review. Cavity-filling mutations appear to be a general means for increasing the thermostability of proteins. The increases in melting or inactivation temperatures reported are usually 1° to 2° C or less, and the desired result is not achieved in all cases.

Perhaps the most exciting recent study in this class involved replacements at position 133 in T4 lysozyme with unnatural amino acids by the use of chemically charged suppressor tRNAs (Mendel et al., 1992). This technique allows unnatural substitutions to be made at any position in a large protein. One set of replacements was made specifically to probe packing around Leu133. Molecular dynamics simulations suggested that two possible replacements, by S-2-amino-3-cyclopentylpropanoic acid and S,S-2-amino-4-methylhexanoic acid, would not lead to unfavorable steric contacts. A third replacement, by *tert*-leucine, was expected to be unfavorable. Stability measurements confirmed these expectations. The first two replacements had essentially normal enzymatic activity and stabilized the enzyme by 1.2 and 0.6 kcal/mol, respectively, while the third had markedly reduced activity and was not further characterized. With one atom less than the phenyl group introduced by Karpusas et al. (1989), the cyclopentyl side-group in this study led to a protein more stable than the Leu → Phe mutant by 1.5 kcal/mol. This is a dramatic illustration of the potential power of this technique to open up exploration of a far greater variety of steric interactions than heretofore possible. The success of these mutations in stabilizing the protein suggests the prospects that improved packing could be a general approach to protein stabilization are better than previously thought.

Cavities in the hydrophobic core can also be filled by binding nonpolar ligands, which can stabilize a protein as do cavity-filling mutations (Eriksson et al., 1992b). Molecular modelling suggested that a benzene molecule would fit into the cavity created in T4 lysozyme by the mutant L99A. Benzene actually binds in this cavity in the manner expected, and stabilizes the enzyme by about 1.9 kcal/mol at saturating benzene concentration. The mutant F153A does not bind benzene, although one might have expected the cavity left by removal of a phenyl group to bind benzene more readily than one formed by removal of the smaller aliphatic side chain of Leu. The L99A cavity is actually larger than the F153A cavity, due to substantial relaxation of the surrounding protein structure in response to the F153A mutation. The energetics of ligand-cavity binding thus depend more on the precise shapes of the groups involved and the flexibility of the surrounding protein than on gross considerations of the volumes of groups being replaced.

A small → large mutation in T4 lysozyme, A98V, is among the most destabilizing mutations of any type isolated to date. Originally isolated from a random genetic screen for a temperature sensitive phenotype (Alber et al., 1987), A98V is 4.9 kcal/mol less stable than wild-type (Daopin et al., 1991). In contrast to the cavity-filling mutations described above, the introduced Val points directly at the main chain of an α-helix. Helices 93-106 and 143-155 are forced apart by up to 0.7 Å (Figure 16-2). This large structural change appears to be required to avoid close contacts between the introduced Val and main chain and C_β atoms on the neighboring helix. The destabilization and accompanying structural changes are thought to be so large because the contacts that produce it are between a short side chain and the main chain, hence there are no low-energy degrees of freedom by which the protein can relax (Daopin et al., 1991). The A98V mutant illustrates that, at least in some cases, changes in packing interactions can have effects on stability as large as those involving the largest effects of any other class of molecular interaction.

Richards and co-workers have taken an approach in which binding of a truncated protein, (subtilisin-treated ribonuclease A: "ribonuclease S"), to a peptide analogue of the missing fragment (S-peptide), is used to analyze the forces involved in protein stability. This approach offers two advantages, the possibility of conveniently incorporating unnatural groups into the chemically-synthesized peptide, and more importantly, the ability to measure each binding event under identical conditions. Accurate thermodynamic parameters for seven S-peptides substituted at Met-13 have been obtained by titration calorimetry (Connelly et al., 1992), and the crystal structures of the RNAase S-S-peptide complexes have been determined (Varadarajan and Richards, 1992). Free energy differences in binding are only weakly correlated with the free energies of transfer of the corresponding model compounds. Deviations from a linear $\Delta \Delta G$ vs. $\Delta \Delta G_{transfer}$ plot can probably be accounted for by differences in the packing environment. The M13F mutant introduces a bulky group that is larger than and very different in shape from the wild-type Met, and structural changes are required to accommodate the Phe. A cavity bordering Met 13 in the wild-type complex actually expands in M13F, despite the larger size of the Phe. The Phe appears to force the cavity walls apart, partially explaining the destabilization of this complex by 2.7 kcal/mol. Similar effects are seen to a lesser extent in the M13L complex, which is also less stable than would have been expected purely on the basis of hydrophobicity.

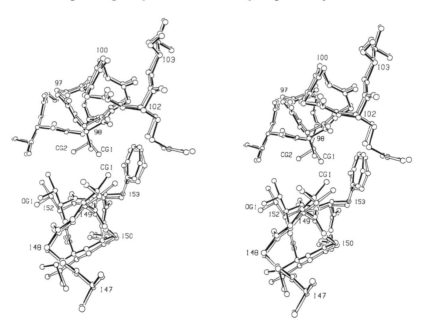

Figure 16-2. Refined structure of the A98V mutant of T4 lysozyme (open bonds) superimposed on the wild-type structure (filled bonds). Reprinted from Daopin et al. (1991) with permission from the authors and *J. Mol. Biol.*; © 1991, Academic Press.

Binding free energy changes for the substituted S-peptide complexes can rationalized in terms of structure, as for studies of free energy changes for folding of mutant proteins. Changes in other thermodynamic functions, namely ΔH, ΔS, and ΔC_p, are much harder to rationalize. It has been proposed that the ΔC_p for protein folding is closely related to the change in buried hydrophobic surface area on folding (Livingstone et al., 1991). The set of S-peptides do not obey this pattern. As Richards and co-workers point out (Varadarajan, 1992a) the binding of the S-peptide to RNAase S consists of the folding of the peptide into a helical conformation followed by binding to RNAase S, a process that is distinct from the transition of an entire protein from the completely unfolded state to the folded state. It is not clear whether compensating changes in ΔC_p in the overall folding reaction would lead to a result more consistent with the prevailing model. Difficulties were also encountered in the closely related interpretation of ΔS values, which were positive for all substitutions relative to the wild-type Met. Negative values would have been expected for mutations to less hydrophobic residues if the hydrophobic effect were the only major contributor to ΔS. The results could be partially due to the potentially large positive conformational entropy of the Met side chain in the unbound peptide, or as the authors suggest, to differences in vibrational dynamics in bound and unbound RNAase S. The M13F variant has one of the smallest ΔS values, which suggests that a significant part of the unfavorable effect of small $\rightarrow$ large mutations in sterically restricted environments could be a decrease in the entropy of the folded state due to increased rigidity.

Similar difficulties occur in interpreting calorimetric data for mutant T4 lysozymes substituted at the partially buried Ile 3 (Ladbury et al., 1992). Free-energy differences produced by mutations at this site are modest (< 3 kcal/mol), but the differences in the enthalpy of folding $\Delta\Delta H$ are as large as -13 kcal/mol. No explanation for such a large change in terms of intermolecular forces has been suggested. Unlike the peptide binding studies described above, interpreting heats of (un)folding of mutant proteins is complicated by the strong temperature dependence of this parameter and the difference in transition temperatures for each mutant. It does not appear that the calorimetric results can be explained solely based on shifted transition temperatures, since widely different $\Delta\Delta H$ values are seen for mutants with similar ΔT_m values. A major challenge remaining for mutational analysis of protein stability is to account for the effects of mutations on all thermodynamic functions in terms of intermolecular forces.

4. Cooperative Energetics in Packing

Two mutations can be shown to interact by demonstrating that their contributions to stability are not additive in a double mutant. This concept is often treated as self-evident by practitioners, but it is so fundamental to analysis of packing interactions that the assumptions involved will be stated here. Fersht and co-workers have provided exhaustive discussions of this topic, including extensions to interactions of arbitrary numbers of mutations (Horovitz and Fersht, 1992).

Consider two side chains, a and b, which interact with the rest of the protein, p, and may or may not interact with each other. If side-chain a is deleted, i.e., by mutation from a large side chain to alanine, the free energy of folding of the mutant protein will change by

$$\Delta\Delta G_a = \Delta G_{mutant} - \Delta G_{wild-type} = -\Delta G_{ab} - \Delta G_{ap},$$

where $-\Delta G_{ab}$ is the contribution to the free energy of folding of the wild-type protein by the interaction between a and b, and ΔG_{ap} includes all other contributions of a to the free energy of folding. Similarly,

$$\Delta\Delta G_b = -\Delta G_{ba} - \Delta G_{bp} = -\Delta G_{ab} - \Delta G_{bp}.$$

It is generally assumed that a and b do not interact in the unfolded state, hence ΔG_{ab} is the interaction free energy of a and b in the folded state. For the moment, it is also assumed that no new interactions are introduced as a result of the deletion at a or b.

To address the effects of multiple mutations on stability, hypothetical pathways can be constructed in which the multiple mutation is constructed in a stepwise manner by means of single mutations. The stability of the multiple mutant is uniquely defined, and independent of any hypothetical pathway by which it might be constructed. A thermodynamic cycle can thus be constructed by analogy with thermodynamic cycles involving equilibrium states connected by reversible processes. Therefore,

$$\Delta\Delta G_{ab} = -\Delta G_{ab} - \Delta G_{ap} - \Delta G_{bp'} = -\Delta G_{ab} - \Delta G_{ap''} - \Delta G_{bp}.$$

Here, p' and p'' are the single mutants in which a and b have been deleted, respectively. The degree of interaction, sometimes referred to as the cooperative free energy, between the mutations is measured by

$$\begin{aligned}
\Delta\Delta\Delta G_{ab} &= \Delta\Delta G_{ab} - \Delta\Delta G_a - \Delta\Delta G_b \\
&= \Delta G_{ab} - \Delta G_{ap} - \Delta G_{bp'} \\
&\quad - (-\Delta G_{ab} - \Delta G_{ap} - \Delta G_{ab} - \Delta G_{bp}) \\
&= (\Delta G_{bp} - \Delta G_{bp'}) + \Delta G_{ab} \\
&= (\Delta G_{ap} - \Delta G_{ap''}) + \Delta G_{ab}.
\end{aligned}$$

The final equality is due to the path-independence of the stability of the double mutant.

If the mutation does not materially effect the structure and dynamics of the protein, the interaction of a or b with the rest of the protein is not expected to change. In this case, $\Delta G_{bp} - \Delta G_{bp'} = \Delta G_{ap} - \Delta G_{ap''} = 0$, and the relation $\Delta\Delta\Delta G_{ab} = \Delta G_{ab}$ provides a measurement of the direct interaction energy between a and b. If there is no direct interaction and no coupling of the mutations via structural changes, $\Delta\Delta\Delta G_{ab} = 0$. Although non-interacting mutations must have $\Delta\Delta\Delta G_{ab} = 0$, it is possible in principle for interacting mutations to have $\Delta\Delta\Delta G_{ab} = 0$, if a variety of direct and indirect interactions cancel each other.

It is impossible to be certain that there are no structural changes without actually determining the structures of the mutants involved. Even structure determination does not allow one to entirely discount the possibility that subtle changes in protein dynamics may affect the free energy. If structures cannot be determined, or significant structural change is found, $\Delta\Delta\Delta G_{ab}$ provides information on the overall coupling between the two mutations, combining direct pairwise interaction and indirect, multi-body interaction terms.

The analysis can be generalized to increasingly subtle multi-body interactions involving mutations at three or more sites, using the $\Delta^n G$ described by Fersht and co-workers (Horovitz and Fersht, 1992). A less sophisticated, but intuitively simpler, description involves simply taking the difference between the stability change of the multiple mutant and the sum of all the component single mutations. Following the treatment for the double mutant, this $\Delta\Delta\Delta G_{ab...n} = \Delta\Delta G_{ab...n} - \Delta\Delta G_a - \Delta\Delta G_b - \cdots - \Delta\Delta G_n$ provides an overall measure of the combined direct and indirect interaction energy between a set of n side chains. A less awkward notation for $\Delta\Delta\Delta G_{ab...n}$ is $\Delta\Delta G_{coop}$, so named because it describes the contribution of cooperative interactions to the observed $\Delta\Delta G$.

In general, the non-additive free energy changes cannot be relied on to provide a direct measure of pairwise interactions in packing studies. Structures of mutant proteins altered at internal sites generally show significant structural changes, violating the condition for $\Delta\Delta\Delta G_{ab} = \Delta G_{ab}$ to be valid. Mutations of large buried residues to Ala are particularly prone to destabilize a protein and induce structural changes, so there is little point in constructing a reference protein in which multiple interior residues are mutated to Ala. Experience suggests that even an intrinsically stable protein may not fold if three or more large, buried side chains are truncated. This means that for the normal case in which a perfectly rigid non-interacting reference state is not available, the assumption that no new interactions are introduced by the mutations must be dropped. Non-additivity in a general case measures the interaction between mutations, rather than between side chains. $\Delta\Delta G_{coop}$ will be negative if favorable interactions are gained in the mutant, or if favorable interactions are lost in wild-type, or both. Unfavorable interactions in either mutant or wild-type have the opposite effect. These problems limit the interpretability of $\Delta\Delta G_{coop}$, but it is still a valuable parameter, especially taken in the context of overall stability and structure.

5. Dissecting Core Packing: Interactions between Side Chains

A complete analysis of packing interactions between side chains clearly requires construction of variants substituted at multiple interacting sites. The strength of the interaction between the sites and their dependence on structural context is evaluated by the strategy described above. If possible, the structures of single and multiple mutants are also determined and compared.

The landmark study of Lim and Sauer (1989) on alternative packing arrangements in λ repressor provides the most comprehensive data available on the constraints governing which sequences can fold into a given tertiary structure (Table 16-1). Up to seven buried hydrophobic residues were replaced simultaneously by random combinations of all 20 naturally occurring amino acids using a cassette mutagenesis technique. Functional variants were selected by a genetic screen and sequenced. Approximate conservation of hydrophobicity and conservation of side-chain volume within a range of $+2$ to -3 methylene groups were found to be the clearest requirements for an acceptable sequence. All acceptable sequences met these requirements, but only a fraction of sequences meeting these requirements were acceptable. For example, in a thorough screen of mutations at Val 36, Met 40, and Val 47 (Lim and Sauer, 1989), 5% of the $20^3 = 8000$ sequences possible meet the composition and volume constraints. The volume constraint is weak, since hydrophobicity alone reduces the allowable population to 6% of the total. Only 1.4% of all sequences are active, and only 0.3% of these have essentially wild-type activity. Many nonfunctional variants have volume and hydrophobicity changes that are smaller the for many functional variants. The authors suggest that only steric packing considerations can reasonably account for the additional constraint.

Calculations based on the modified tertiary template algorithm (Ponder and Richards, 1987; Hurley et al., 1992) found that all of the fully active variants at the three positions in λ repressor (Lim and Sauer, 1989, 1991) are sterically acceptable, none of inactive variants were acceptable, and a

Table 16-1. Reduction in allowed core sequences imposed by constraints.

Applied constraints	I (V36 M40 V47)		II (L18 V36 V47 F51)		III (L18 L57 L65)	
None	8,000	(100%)	160,000	(100%)	8,000	(100%)
Volume	5,848	(73%)	100,763	(63%)	5,052	(63%)
Composition	512	(6%)	4,096	(2.5%)	512	(6%)
Volume and composition	425	(5%)	2,846	(2%)	335	(4%)
Volume, composition, and steric:						
active sequences	110	(1.4%)	690	(0.4%)	25	(0.3%)
fully functional sequences	20	(0.3%)	270	(0.2%)	6	(0.1%)

For an experiment in which n positions are randomly altered, the total number of possible sequences is 20^n. The total number of compositionally acceptable sequences in 8^n, as only Ala, Cys, Thr, Val, Ile, Leu, Met, and Phe were found to be acceptable (Pro and Ser which were each found at only one position in one mutant are excluded). The number of sequneces allowed by the volume constraint is the number of possible sequences that have core voumes within -3 to $+2$ methylene units of the wild-type volume. The number of sequences passing all three constraints is estimated from the fraction of mutants surviving the selection. A definite lower limit for these estimates is given by the total number of active sequences actually isolated and sequenced. These values are: experiment I, 42 active, 8 fully functional; experiment II, 54 active, 22 fully functional; experiment III, 24 active, 6 fully functional. Reprinted with permission from *Nature* (Lim and Sauer, 1989); © 1989, Macmillan Magazines, Ltd.

fraction of the reduced-activity variants were acceptable. Although limited to providing a "yes" or "no" answer, the tertiary template calculation is consistent with the notion that packing is the major factor determining whether a particular sequence of hydrophobic residues will be acceptable. Hydrophobicity provides about a factor of 17 reduction in the number of possible sequences at three positions, and can be regarded as the primary determinant of sequence acceptability. In this example, packing reduces the number of allowed sequences by an additional factor of 3.5 to 17, depending on the strictness of one's definition of "acceptable." Packing constraints could be considered less important than hydrophobicity, but are clearly still an essential part of the picture.

A set of severely destabilized multiple mutations in λ repressor have recently been characterized (Lim et al., 1992). Proteins with sterically unfavorable sequences involving volume changes as large as $+6$ methylene groups can be isolated, although most are inactive. For example, V36F/M40F/V47F is completely inactive and less stable than wild-type by 3 kcal/mol. The ^{1}H NMR dispersion for this mutant was typical for a native protein (Lim et al., 1992), but high resolution structures have not yet been reported for any of the λ repressor mutants.

While Sauer and co-workers have provided the most comprehensive view of the possibilities in packing interactions, other groups have concentrated on understanding the structural and energetic aspects of packing interactions in finer detail.

A study of the buried triplet Thr 40, Ile 55, and Ser 91 in hen egg white lysozyme (Malcolm et al., 1990; Wilson et al., 1992) was initially motivated by interest in possible evolutionary intermediates between different chicken-type lysozymes. All of the eight possible combinations of T40S, I55V, and S91T were constructed, and thermal stabilities and crystal structures were determined. One variant, S91T, is markedly more stable than wild-type by 0.9 kcal/mol, which is probably due to increased hydrophobic stabilization from the filling of a pre-existing cavity in the wild-type structure. Only modest changes in structure and stability are seen in this set, and the stability changes are only weakly non-additive. This study illustrates the rather subtle effects that can be expected from a set of conservative mutations based on a family of closely related, naturally occurring protein sequences.

Essentially no cooperativity was found in a thorough study of the interaction between two nearby buried positions of bacteriophage f1 gene V protein (Sandberg and Terwilliger, 1991). The reasons for this will presumably become clear when the mutant structures are determined.

In the first serious attempt to improve protein stability by engineering favorable packing interactions in a multiple mutant, Daopin et al. (1991) partially restored the stability of the T4 lysozyme mutant A98V by mutating two neighboring sites. Mutant proteins were designed to eliminate close contacts produced when the first mutation pushed apart two interacting a-helices. The chosen mutations V149C and T152S were strongly destabilizing (by 2.2 and 2.6 kcal/mol, respectively) in the wild-type background but slightly stabilizing in the A98V background. The dramatic differences in the effect of the same mutations in two different contexts again illustrate the energetic importance of packing interactions. Some sets of mutations showed a very high degree of non-additivity, with multiple mutants in some cases more than 5 kcal/mol more stable than would have been expected for independent effects. The final triple mutant A98V/V149C/T152S was still only 0.5 kcal/mol more stable than the single mutant A98V (4.4 kcal/mol less stable than wild-type). The authors point out that main chain/side chain collisions may be an especially disruptive class of packing defect, which may be impossible to compensate by mutating surrounding side chains.

In another attempt to design favorable packing interactions in T4 lysozyme, Hurley et al. (1992) constructed a set of single to quadruple mutations chosen with the tertiary template algorithm and energy minimization. The mutations consisted of combinations of the replacements L99F, M102L, V111I, and F153L, which are at neighboring fully buried positions. With the exception of F153L, the single mutations are moderately destabilizing, decreasing stability by 0.6 to 0.7 kcal/mol. F153L is slightly stabilizing ($\Delta\Delta G = -0.3$ kcal/mol), probably because Phe153 is in a strained conformation in wild-type (Eriksson et al., 1993). The stability of the multiple mutants is highly non-additive, with the quadruple mutant and most of the triple mutants more stable than the single mutants. The triple mutant L99F/M102L/F153L is only 0.2 kcal/mol less stable than wild-type, and the quadruple mutant is 0.5 kcal/mol less stable.

The crystal structures of the single and multiple mutants at these four sites revealed several unexpected structural changes underlying the observed cooperative energetics. Most dramatic was the effect of the apparently conservative single mutation V111I (Figure 16-3). In the structure of this mutant, the central hydrogen bond of the short helix 107-114 is completely broken. This allows the introduced δ-methyl group to be pushed back to approximately the position occupied by a γ-methyl of the wild-type Val. The two halves of the helix swing out, with the ends remaining fixed. Most of the overall structure of the protein is unperturbed, but the helix 107-114 shifts by 1.5 Å rms (main chain; Figure 16-4). A change of this magnitude would have been impossible to predict by the methods used. M102L also produced an unexpected result. The introduced Leu side chain occupied a rotamer that had not been predicted to be acceptable, based on the set of residues allowed to move in the packing calculation. The structure showed this rotamer was accommodated by a rotation of the δ-methyl of Met106 from a buried position to a solvent-exposed position. Met106 was not included in the original group of flexible residues, but when the packing calculation were redone with Met106 included, the observed structure was found among the acceptable set. Two major changes were found in the structures the multiple mutants relative to the singles. In the quadruple mutant, the three replacements at positions 99, 102, and 153, allow repacking around the δ-methyl of V111I in a manner that allows the helix 107-114 to regain complete hydrogen bonding and form essentially the same structure as in wild-type. In the L99F/M102L/F153L and the quadruple mutant, the introduced Leu102 becomes less strained and moves toward the protein interior, which in turn allows the δ-methyl of Met106 to rotate to a largely buried position close to that found in wild-type.

Taken by themselves, the favorable $\Delta\Delta G_{coop}$ values (up to 1.1 kcal/mol) for these mutants do not prove that favorable interactions are introduced between side chains in the new packing arrangements in these multiple mutants. As described above, such $\Delta\Delta G_{coop}$ values can originate in the

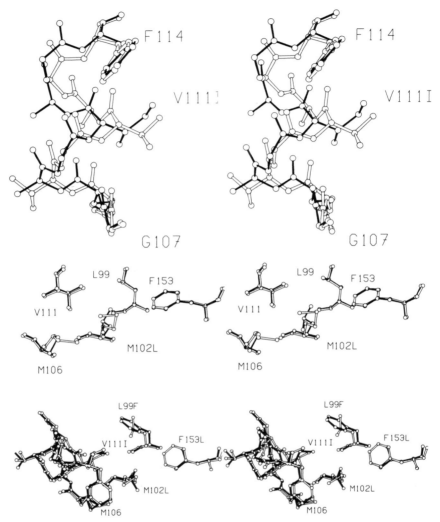

Figure 16-3. Comparison of the refined crystal structures of the V111I (*top*), M102L (*middle*), and L99F/M102L/V111I/F153L (*bottom*) mutants of T4 lysozyme (filled bonds) superimposed on the structure of (cysteine-free) wild-type lysozyme (open bonds). Reprinted from Hurley et al. (1992) with permission from the authors and *J. Mol. Biol.*, © 1992, Academic Press.

Figure 16-4. The rms shifts for the main-chain atoms of each residue for the C-terminal domains of the (A) V111I, (B) M102L, and (C) L99F/M102L/V111I/ F153L mutant of T4 lysozyme relative to the (cysteine-free) wild-type. Reprinted from Hurley et al. (1992) with permission from the authors and *J. Mol. Biol.*; © 1992, Academic Press. *See figure on following page.*

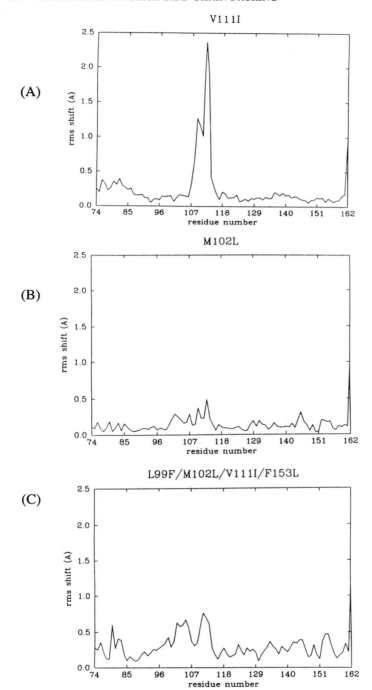

removal of favorable interactions in wild-type, as well as in their creation in a mutant. Taken together with the structural evidence for cooperative interactions and the observation that the multiple mutants are more stable than most of the singles, the observed cooperativity suggests the designs were successful in creating new, favorable packing interactions between side chains.

6. General Conclusions from Mutational Studies of Packing Interactions

The experiments described above show a rich diversity of structural and energetic effects of core packing mutations. Crystallographic studies have illustrated how a given type of mutation may result in very different consequences at different interior positions. The effects of large $\rightarrow$ small hydrophobic mutations are determined by the ability of the protein to relax to fill in a cavity. In attempts to fill pre-existing cavities by mutations or exogenous ligand binding, subtle details of the shapes of the cavity and the group introduced, and the flexibility of the surroundings, determine the outcome. Mutations that introduce steric overlaps may produce very large structural changes, or they may produce none, depending on the energetic costs of different modes of relaxation.

The meanings of the geometric attributes of packing are becoming more clear. Crystallographic measurements of a set of cavity-creating T4 lysozyme mutants show that the cost of forming a cavity in a protein interior is about 20 cal/Å (Eriksson et al., 1992a). No energetic cost has been explicitly assigned to deviations from optimal packing density, but it is apparent that structural changes tend to occur in a direction that minimizes the change in packing density (Varadarajan and Richards, 1992; Anderson et al., 1993). This is probably one mechanism whereby larger volume changes can be tolerated (Lim and Sauer, 1989, 1991) than might have been expected from a rigid main chain model (Ponder and Richards, 1987).

The energetic meaning of an "acceptable" sequence as defined by the packing criteria of Ponder and Richards (1987) and modifications thereof (Hurley et al., 1992) is becoming clear. Mutants that are very close to wild-type activity and stability are almost always "acceptable," while inactive, nonfolding, or severely destabilizing mutants are almost always not "acceptable." Mutations that have reduced activity or destabilize a protein by about 1 to 3 kcal/mol are in an intermediate class, and may be either "acceptable" or not in an unpredictable way. It may be surprising that the tertiary template approach works at all, since crystallography shows that many packing mutants exhibit large structural changes that cannot be factored into a rigid main-chain model. The Ponder and Richards (1987) model errs on the side of conservatism in rejecting sequences that can only be accommodated by substantial main-chain movements. Because these structural changes are often associated with destabilizing or activity-reducing mutations, the Ponder and Richards (1987) predictions for allowed sequences are more often right than wrong, although sometimes for the wrong reason.

The full implications of the non-additive free energies for some combinations of packing mutations are still not completely clear. Little non-additivity or structural change was found for a set of conservative mutations based on homologies occurring in nature (Malcolm et al., 1990; Wilson et al., 1992). Substantial non-additivity and structural changes were found in sets of mutations designed to increase the stability of the starting protein by introducing favorable packing interactions (Daopin et al., 1991; Hurley et al., 1992). The best-documented cases of strongly cooperative interactions are both for sets of designed mutants. No studies have been reported in which randomly generated sets of mutations have been characterized in enough detail to discuss cooperativity. It is not possible to say at present whether strong non-additivity is a widespread property of sets of multiple mutations, or something that can only be achieved by careful design or natural selection.

7. Computational Studies

The long-term goal of the mutational studies described above is a quantitative understanding of the relationship between the amino acid sequence of a protein and its structure and stability. The range of effects on stability and structure produced by various types of mutations are known, but a complete theory in terms of intermolecular forces is missing. Development of a predictive theory of protein folding and stability is perhaps the "Holy Grail" of structural biology, so there have naturally been many studies with this objective. Here, only the handful of studies directly concerned with interior packing will be reviewed.

Molecular dynamics (MD) and free-energy perturbation (FEP) techniques can be used, in principle, to calculate all of the observable effects of mutations, namely $\Delta\Delta G$, $\Delta\Delta H$, $\Delta\Delta S$, $\Delta\Delta C_p$, and changes in structure and enzymatic activity. The application of these methods in practice is limited by the accuracy of molecular mechanics potential functions and sufficient computer time for adequate sampling of conformational space. Two simulations of the effects of single mutations of buried hydrophobic residues on protein stability have recently been reported (Prevost et al., 1991; Sneddon and Tobias, 1992).

The FEP study (Prevost et al., 1991) of the I96A mutant of barnase reports satisfactory agreement between calculated and observed (Kellis et al., 1988, 1989) $\Delta\Delta G$ values. As in all FEP studies of protein stability, calculations were carried out separately for both the folded and unfolded states, with the side chain under study "alchemically" changed from one chemical group to another. The calculated free energy included several terms in this study, covalent and nonbonded interactions within the Ile and Ala side chains, and covalent and nonbonded interactions between the substituted side chains and their surroundings. Nonbonded interactions within the side chain were unimportant in this case. Perhaps surprisingly, the term involving the covalent link between the side chain and main chain contributes significantly to the free-energy difference. The authors suggest this may represent an unexpected manifestation of the hydrophobic effect, in which rearrangement of water hydrogen bonding around the Ile in the unfolded state forces the Ile into a covalently strained structure. Self-energy terms for the rest of the protein and for the solvent were apparently excluded from the total reported $\Delta\Delta G$. Because the hydrophobic effect is believed to originate in the energetics of water structure rearrangements, the exclusion of the water self-energy precludes conclusions about the hydrophobic contribution to stability. The exclusion of the self-energy of the surrounding protein also precludes conclusions about the energetics of the relaxation of the protein in response to cavity formation. Because no structures are available for the barnase cavity mutants, no comparison is possible between the actual and simulated structures. Sneddon and Tobias (1992) report an FEP study of mutations in ribonuclease T_1, but do not state whether all energy terms were included in the calculated $\Delta\Delta G$. No experimental data have been reported for the simulated mutants of ribonuclease T_1. A wealth of experimental structures and calorimetric data for packing mutants of several proteins are now available. Given the tremendous potential of the MD/FEP techniques, it is to be hoped that future simulations will make possible more challenging comparisons between experiment and theory.

Limitations on computer time have so far deterred FEP studies of large sets of mutants or of multiple mutants. Several *ad hoc* energy-based schemes have been developed in an attempt to fill the gap between descriptive attributes such as packing density, cavity area, and hard-sphere packing on one hand, and the rigorous but time-consuming MD/FEP approach on the other.

Lee and Levitt (1991) used a simplified potential consisting only of van der Waals and torsional energy terms and an elaborate conformational search procedure to calculate stabilities for the set of 78 λ repressor mutants reviewed above (Lim and Sauer, 1991). A simulated annealing algorithm was used to locate low-energy conformations of a set of six adjacent side chains by moving torsional degrees of freedom for those side chains only. The energies of the 1000 lowest-energy structures are averaged for each sequence. Only the folded state is considered in the calculations. A significant correlation was obtained between the observed ΔTm and the calculated energy for a subset of nine mutants that had been characterized in detail. Calculated energies were also reasonably consistent with the "activity grades" assigned by Lim and Sauer (1991) to all 78 mutants. Mutants in the highest grade, essentially identical to wild-type in stability and activity, all had calculated energies within ±4 kcal/mol of wild-type. Inactive mutants had calculated energies from 17 to 76 kcal/mol greater than wild-type. Energy differences for intermediate classes ranged from −3 kcal/mol to 35 kcal/mol.

As the first demonstration of a statistically significant correlation between observed and calculated properties of protein stability, the Lee and Levitt (1991) results appear to be a major advance. It is disconcerting that an algorithm that completely ignores the hydrophobic effect should do so well in predicting stability effects of hydrophobic core mutations. This may come about for several reasons. Almost all of the λ repressor mutants are destabilizing smaller → larger mutations, since the wild-type residues were the relatively small set Val, Met, Val. The replacements constructed were all equal to or larger than Val, with Ala excluded by the mutagenesis procedure. Steric repulsion is thus expected to dominate the energetics. A single fully active, stabilizing mutant was correctly predicted by the Lee and Levitt procedure. This may be possible because the simultaneous omissions of the hydrophobic effect and van der Waals interactions in the unfolded state have opposite and partially compensating effects on the calculation. One would expect the procedure to break down if tested against a large set of large → small and small → large mutations with diverse effects of stability. Certainly, the procedure would fail if mutations to polar residues were considered. Although the ranking is largely correct, the energy differences calculated are generally larger than experimental differences, and in many cases larger than the total ΔG of folding of any known protein. This is probably a consequence of the rigid model. Small violations of minimum van der Waals contact distances are penalized by the steep r^{-12} repulsive component of the Lennard–Jones potential, and cannot relax via covalent or main-chain degrees of freedom.

The factors underlying the apparent success of the Lee and Levitt (1991) predictions were recently analyzed in great detail by van Gunsteren and

Mark (1992). Two simple scales of mutational disruptiveness were invented more-or-less arbitrarily by the authors. It was found that these scales produced stronger statistical correlations with experiment than the elaborate conformational search (Lee and Levitt, 1991). Van Gunsteren and Mark (1992) plotted calculated "energies" (total disruptiveness scores) against activity class and found linear correlation coefficients of -0.796 and -0.815 for their two arbitrary disruptiveness scales, compared to -0.664 for the Lee and Levitt energies. If calculated energies are plotted against ΔT_m instead, the Lee and Levitt calculation is marginally superior. Van Gunsteren and Mark (1992) argue that there is no information in the Lee and Levitt calculation that cannot be obtained from much simpler models without the need for a computer.

Explicit energy calculations have been used in the design of experimental studies, as well as in primarily theoretical studies. Hurley et al. (1992) used energy minimization as part of an energy-based screen for promising alternative core-packing arrangements in T4 lysozyme. The minimized energy for the lowest-energy rotamer sequence for a given amino acid sequence was used as a measure of packing quality. When all degrees of freedom are allowed to relax, as in this study, the energy differences between sequences are much closer to experimentally observed magnitudes than for the rigid models such as that of Lee and Levitt (1991). The procedure also incorporated terms for side-chain conformational entropy and favorable van der Waals interactions in the unfolded state, and for the hydrophobic effect. Despite the effort to account for both the folded and unfolded states, the predicted energies were incorrect. In most cases, small $\rightarrow$ large mutations with favorable predicted $\Delta\Delta G$ values were actually destabilized.

In contrast, the signs and approximate magnitudes of predicted $\Delta\Delta G_{coop}$ values were almost all correct, although the statistical correlation between predicted and measured values is marginal (Hurley et al., 1992). The key difference between prediction of $\Delta\Delta G$ and $\Delta\Delta G_{coop}$ is that, under the assumption there are no packing interactions in the unfolded state, energy terms related to the unfolded state have no effect on $\Delta\Delta G_{coop}$. The calculated $\Delta\Delta G_{coop}$ depends only on a differences in a single term, in this case the minimized energy of the folded state. The calculated $\Delta\Delta G$ is obtained from the small difference between several terms, each of which is subject to substantial systematic error. If any one term is systematically too large or too small, the result can be that $\Delta\Delta G$ is biased towards certain residues. Correct prediction of mutational effects on stability appears to require not only accounting for all major energy terms, but properly scaling these terms together. Wilson et al. (1991) dealt with this problem in a computational procedure they developed for designing proteases with altered substrate specificity. Multivariable linear least-squares fitting against a subset of experimental data is used to derive scale factors for energy components in this technique. These scale factors have been used successfully to predict energies for a second subset of data excluded from least-squares fitting, and to design a mutant enzyme with the desired specificity change (Wilson et al. 1991). The minimized energies and other terms used in the T4 lysozyme packing study of Hurley et al. (1992) can be scaled together by a similar procedure, yielding excellent agreement ($r > 0.8$ for data not used in fitting) with experiment. It is not yet clear whether this type of parameter set is robust enough to make predictions for proteins other than the one on which it was initially developed.

Studies reported in the past two years signal that efforts have begun in earnest to understand the energetics of packing interactions based on first principles. The results of the first round of computational studies and the greatly increased pool of experimental data on packing mutations should pave the way for progress in future work.

8. Role of Packing in Protein Folding

The role of packing in protein folding remains a controversial issue. While it is not the purpose of this review to consider the driving forces for protein folding and for the uniqueness of the native fold, mutational studies have some implications for these questions. Notably, studies of λ repressor showed that only 1 in 17 sequences with acceptable hydrophobicity and volume were fully active, and less than 1 in 3 were even partially active (Lim and Sauer, 1989). This implicates packing as a major constraint on the nature of allowed sequences.

Examples such as the A98V mutant of T4 lysozyme (Daopin et al., 1991) illustrate the structural factors underlying these packing constraints. This mutant induces large-scale structural changes and destabilizes the protein by nearly 5 kcal/mol by introducing unfavorable close contacts with main-chain atoms. This is comparable to the total ΔG of folding for many proteins at ordinary temperatures. A98V is the most dramatic steric overlap mutant characterized in structural detail. There may be few examples only because no studies have been specifically aimed at determining the structural effects of steric overlaps. The high number of nonfunctional hydrophobic sequences found in the λ repressor studies suggest that destabilizations of this magnitude by packing disruptions could be relatively common (Lim and Sauer, 1989, 1991; Lim et al.,1992).

One of the earliest conclusions of mutational studies of stability was that proteins are remarkably tolerant of mutations. While many solvent-exposed sites can be mutated at will with no detrimental effects, this is clearly not the case for buried hydrophobic positions. Behe et al. (1991) have taken an extreme interpretation of the concept of tolerance to mutation, suggesting that any hydrophobic sequence can pack in the context of any tertiary structure, hence packing interactions are irrelevant in determining the fold of a protein. This proposal would provide a convenient simplification of the folding problem if correct, but it also contradicts most previous thinking about the role of packing in protein folding (Richards, 1974, 1977, 1985; Ponder and Richards, 1987). The model of Behe et al. (1991) considers preferences in surface area buried in pairwise interactions between different residue types as the hallmark of specific packing. Unlike the model of Ponder and Richards (1987), Behe et al. (1991) ignore details of side-chain conformations and atomic contacts, and do not consider non-native sequences. The model of Behe et al. (1991) cannot explain the large fraction of nonfunctional hydrophobic sequences in the λ repressor studies (Lim and Sauer, 1989, 1991) or the well-characterized individual cases of severely disruptive packing mutants (Daopin et al., 1991; Lim et al., 1992).

Experience shows the original prediction of Ponder and Richards (1987), that only a very small number of sequences would have acceptable packing, may have been too strict. The range of tolerated sequences is significantly broadened by main-chain relaxations (Daopin et al., 1991; Hurley et al., 1992; Varadarajan and Richards, 1992). Results from genetic screens and structural studies are still generally consistent with the predictions of Richards and co-workers (Richards, 1974, 1977, 1985; Ponder and Richards, 1987), particularly if the packing criteria in the tertiary template algorithm are relaxed (Hurley et al., 1992).

In the future, *de novo* protein design experiments are likely to have a major impact on thinking about the role in packing interactions in protein folding. Most *de novo* designed proteins reported to date appear to fold into molten-globule-like structures without well-defined packing interactions. The recent report of a *de novo*-designed four-helix bundle with native-like NMR dispersion at low temperature (Raleigh and DeGrado, 1992) is cause for optimism that something will soon be known about the role of packing in designed proteins.

REFERENCES

Alber T, Daopin S, Nye JA, Muchmore DC, Matthews BW (1987): Temperature-sensitive mutations of bacteriophage T4 lysozyme occur at sites with low mobility and low solvent accessibility in the folded protein. *Biochemistry* 26: 3754–3758

Alber T (1989): Mutational effects on protein stability. *Annu Rev Biochem* 58: 765–798

Anderson DE, Hurley JH, Nicholson H, Baase WA, Matthews BW (1993): Hydrophobic core repacking and aromatic–aromatic interaction in the thermostatic mutant of T4 lysozyme Ser 117 → Phe. *Protein Science* 2:1285–1290

Behe MJ, Lattman EE, Rose GD (1991): The protein-folding problem: The native fold determines packing, but does packing determine the native fold? *Proc Natl Acad Sci USA* 88:4195–4199

Bowie JU, Reidhaar-Olson JF, Lim WA, Sauer RT (1990): Deciphering the message in protein sequences: Tolerance to amino acid substitutions. *Science* 247:1306–1310

Chothia C (1976): The nature of the accessible and buried surfaces in proteins. *J Mol Biol* 105:1–14

Connelly PR, Varadarajan R, Sturtevant JM, Richards FM (1992): Thermodynamics of protein-peptide interactions in the ribonuclease s system studied by titration calorimetry. *Biochemistry* 29:6108–6114

Connolly ML (1983): Solvent-accessible surfaces of proteins and nucleic acids. *Science* 221:709–713

Creighton T (1991): Stability of folded conformations. *Current Opinion in Structural Biology* 1:5–16

Daopin S, Alber T, Baase WA, Wozniak JA, Matthews BW (1991): Structural and thermodynamic analysis of the packing of two a-helices in bacteriophage T4 lysozyme. *J Mol Biol* 221:647–667

Dill KA (1990): Dominant forces in protein folding. *Biochemistry* 29:7133–7155

Eijsink VGH, Dijkstra BW, Vriend G, Rob van der Zee J, Veltmann OR, van der Vinne B, van den Burg B, Kempe S, Venema G (1992): The effect of cavity-filling mutations on the thermostability of *Bacillus stearothermophilus* neutral protease. *Protein Engineering* 5:421–426

Eriksson AE, Baase WA, Matthews BW (1993): Similar hydrophobic replacements of leu 99 and phe 153 within the core of T4 lysozyme have different structural and thermodynamic consequences. *J Mol Biol* 229:747–769

Eriksson AE, Baase WA, Wozniak JA, Matthews BW (1992b): A cavity-containing mutant of T4 lysozyme is stabilized by buried benzene. *Nature* 355:371–373

Eriksson AE, Baase WA, Zhang X-J, Heinz DW, Blaber M, Baldwin EP, Matthews BW (1992a): Response of a protein structure to cavity-creating mutations and its relation to the hydrophobic effect. *Science* 255:178–183

Ganter C, Pluckthun A (1990): Glycine to alanine substitutions in helices of glyceraldehyde-3-phosphate dehydrogenase: Effects on stability. *Biochemistry* 29: 9395–9402

Gellatly BJ, Finney JL (1982): Calculations of protein volumes: An alternative to the Voronoi procedure. *J Mol Biol* 161:305–322

Gregoret LM, Cohen FE (1990): Novel method for the rapid evaluation of packing in protein structures. *J Mol Biol* 211:959–974

Hellinga HW, Wynn R, Richards FM (1992): The hydrophobic core of *Escherichia coli* thioredoxin shows a high tolerance to nonconservative single amino acid substitutions. *Biochemistry* 31:11203–11209

Horovitz A, Fersht AR (1992): Co-operative interactions during protein folding. *J Mol Biol* 224:733–740

Hurley JH, Baase WA, Matthews BW (1992): Design and structural analysis of alternative hydrophobic core packing arrangements in bacteriophage T4 lysozyme. *J Mol Biol* 224:1143–1159

Jaenicke R (1991): Protein stability and molecular adaptation to extreme conditions. *Eur J Biochem* 202:715–728

Karpusas M, Baase WA, Matsumura M, Matthews BW (1989): Hydrophobic packing in T4 lysozyme probed by cavity-filling mutants. *Proc Natl Acad Sci USA* 86:8237–8241

Kellis JT Jr, Nyberg K, Fersht AR (1989): Energetics of complementary side-chain packing in a protein hydrophobic core. *Biochemistry* 28:4914–4922

Kellis JT Jr, Nyberg K, Sali D, Fersht AR (1988): Contribution of hydrophobic interactions to protein stability. *Nature* 333:784–786

Kimura S, Oda Y, Nakai T, Katayanagi K, Kitakuni E, Nakai C, Nakamura H, Ikehara M, Shigenori K (1992): Effect of cavity-modulating mutations on the stability of *Escherichia coli* ribonuclease H1. *Eur J Biochem* 206:337–343

Ladbury JE, Hu Cui-Qing, Sturtevant JM (1992): A differential scanning calorimetric study of the thermal unfolding of mutant forms of phage T4 lysozyme. *Biochemistry* 31:10669–10702

Lee C, Levitt M (1991): Accurate prediction of the stability and activity effects of site-directed mutagenesis on a protein core. *Nature* 352:448-451

Lim WA, Sauer RT (1989): Alternative packing arrangements in the hydrophobic core of λ repressor. *Nature* 339:31–36

Lim WA, Sauer RT (1991): The role of internal packing interactions in determining the structure and stability of a protein. *J Mol Biol* 219:359–376

Lim WA, Farruggio DC, Sauer RT (1992): Structural and energetic consequences of disruptive mutations in a protein core. *Biochemistry* 31:4323–4333

Livingstone JR, Spolar RS, Record TR Jr (1991): Contribution to the thermodynamics of protein folding from the reduction in water-accessible nonpolar surface area. *Biochemistry* 30:4237–4244

Malcolm BA, Wilson KP, Matthews BW, Kirsch JF, Wilson AC (1990): Ancestral lysozymes reconstructed, neutrality tested, and thermostability linked to hydrocarbon packing. *Nature* 344:86–89

Matsumura M, Becktel WJ, Matthews BW (1988): Hydrophobic stabilization in T4 lysozyme determined directly by multiple substitutions at Ile3. *Nature* 344:406–410

Matthews BW (1991): Mutational analysis of protein stability. *Current Opinion in Structural Biology* 1:17–21

McRee DE, Redford SM, Getzoff ED, Lepock JR, Hallewell RA, Tainer JA (1990): Changes in crystallographic structure and thermostability of a Cu, Zn superoxide dismutase mutant resulting from the removal of a buried cysteine. *J Biol Chem* 265:14234–14241

Mendel D, Ellman JA, Chang Z, Veenstra DL, Kollman PA, Schultz PG (1992): Probing protein stability with unnatural amino acids. *Science* 256:1798–1802

Pace CN (1990): Conformational stability of globular proteins. *Trends Biol Sci* 15:14–17

Ponder JW, Richards FM (1987): Tertiary templates for proteins use of packing criteria in the enumeration of allowed sequences for different structural classes. *J Mol Biol* 193:775–791

Prevost M, Wodak S, Tidor B, Karplus M (1991): Contribution of the hydrophobic effect to protein stability: analysis based on simulations of the Ile-96 → Ala mutation in barnase. *Proc Natl Acad Sci* USA 88:10880–10884

Raleigh DP, DeGrado WF (1992): A *de novo* designed protein shows a thermally induced transition from a native to a molten globule-like state. *J Am Chem Soc* 114:10079–10081

Richards FM (1974): The interpretation of protein structures: total volume, group volume distributions and packing density. *J Mol Biol* 82:1–14

Richards FM (1977): Areas, volumes, packing, and protein structure. *Ann Rev Biophys Bioeng* 6:151–176

Richards FM (1985): Calculation of molecular volumes and areas for structures of known geometry. *Methods Enzymol* 115:440–464

Richmond TJ (1984): Solvent accessible surface area and excluded volume in proteins: Analytical equations for overlapping spheres and implications for the hydrophobic effect. *J Mol Biol* 178:63–89

Sandberg WS, Terwilliger TC (1989): Influence of interior packing and hydrophobicity on the stability of a protein. *Science* 245:54–57

Sandberg WS, Terwilliger TC (1991): Energetics of repacking a protein interior. *Proc Natl Acad Sci* USA 88:1706–1710

Sharp KA, Nicholls A, Fine RF, Honig B (1991): Reconciling the magnitude of the microscopic and macroscopic hydrophobic effects. *Science* 252:106–109

Shortle D (1992): Mutational studies of protein structures and their stabilities. *Q Rev Biophys* 25:205–250

Shortle D, Stites WE, Meeker AK (1990): Contributions of the large hydrophobic amino acids to the stability of staphylococcal nuclease. *Biochemistry* 29:8033–8041

Sneddon SF, Tobias DJ (1992): The role of packing interactions in stabilizing folded proteins. *Biochemistry* 31:2842–2846

van Gunsteren WF, Mark AE (1992): Prediction of the activity and stability effects of site-directed mutagenesis on a protein core. *J Mol Biol* 227:389–395

Varadarajan R, Connelly PR, Sturtevant JM, Richards FM (1992): Heat capacity changes for protein-peptide interactions in the ribonuclease S system. *Biochemistry* 31:1421–1426

Varadarajan R, Richards FM (1992): Crystallographic structures of ribonuclease S variants with nonpolar substitution at position 13: Packing and cavities. *Biochemistry* 31:12315–12327

Wilson C, Mace JE, Agard DA (1991): A computational method for the design of enzymes with altered substrate specificity. *J Mol Biol* 220:495–506

Wilson KP, Malcolm BA, Matthews BW (1992): Structural and thermodynamic analysis of compensating mutations within the core of chicken egg white lysozyme. *J Biol Chem* 267:10842–10849

Yutani K, Ogasahara K, Tsujita T, Sugino Y (1987): Dependence of conformational stability on hydrophobicity of the amino acid residue in a series of variant proteins substituted at a unique position of tryptophan synthase α subunit. *Proc Natl Acad Sci USA* 84:4441–4444

Keyword Index

This index was established according to the keywords supplied by the authors.
Page numbers refer to the beginning of the chapter.